THE BELLE W. BARUCH LIBRARY IN MARINE SCIENCE NUMBER 21

Organism-Sediment Interactions

Josephine Y. Aller
Sarah A. Woodin
Robert C. Aller
Editors

Published for the Belle W. Baruch Institute for
Marine Biology and Coastal Research
by the UNIVERSITY OF SOUTH CAROLINA PRESS

Copyright © 2001 University of South Carolina

Published in Columbia, South Carolina,
by the University of South Carolina Press

Manufactured in the United States of America

Library of Congress Cataloging-in-Publication Data

Organism-sediment interactions / Josephine Y. Aller, Sarah A. Woodin, Robert C. Aller, editors.
    p. cm. — (The Belle W. Baruch library in marine science ; no. 21)
    Proceedings of a symposium/workshop held in October 1998.
    Includes bibliographical references.
    ISBN 1-57003-431-1 (alk. paper)
    1. Benthic animals—Effect of sediments on—Congresses. 2. Marine sediments—Environmental aspects—Congresses. I. Aller, Josephine Y. II. Woodin, Sarah Ann, 1945–III. Aller, Robert C. IV. Belle W. Baruch Institute for Marine Biology and Coastal Research. V. Series.

QL121 .O64 2001
591.77'7—dc21   2001053060

Organism-Sediment Interactions

THE BELLE W. BARUCH LIBRARY IN MARINE SCIENCE

Dedication

This symposium volume is dedicated to Donald Cave Rhoads whose early conceptual work and pivotal review of organism-sediment interactions in 1974 set the standard for the interdisciplinary approach of future studies in the field. His intuitive understanding of organisms and benthic processes has been coupled with innovations in field studies, experimental design, and instrumentation. From the early sediment profile photographic equipment of Rhoads and Cande to the digital sediment profile technology and *in situ* sediment chemical analyzers, Don Rhoads has worked to develop efficient techniques for *in situ* documentation of organism-sediment relations as well as combined laboratory and field tests of fundamental theories. His ideas and technological advances have significantly altered the path of basic research and been widely embraced by U.S. Federal Regulatory agencies, in particular the DAMOS monitoring program of the Army Corps of Engineers to monitor effects of dredging and disposal operations on estuarine systems, and by European and Far East countries interested in monitoring and protecting coastal embayments.

Many of us have been fortunate enough to have studied and collaborated with Don during his tenure as Professor in the Department of Geology and Geophysics at Yale University and subsequently as a Senior Scientist for Science Applications International Corporation, Newport, Rhode Island. For his energy, vision, good humor, and love, we will always be appreciative.

In Memory of Howard L. Sanders
1921-2001

Our good friend and colleague passed away on February 8, 2001, at the Royal Nursing and Alzheimer's Center in Falmouth, MA. He was 79. Howard received his PhD from Yale University in 1955–yet another distinguished student who studied under the late G. Evelyn Hutchinson. Most of his professional career was spent at the Woods Hole Oceanographic Institution, from which he retired in 1986.

Howard is considered by many of us as the New World "father" of marine benthic ecology and organism-sediment relationships. He applied the concepts of J. Petersen and G. Thorson, developed from studies of European waters, to local New England waters in papers that are now considered classics (e.g., Sanders, H. L. 1956. Oceanography of Long Island Sound. X. The biology of marine bottom communities. *Bulletin of the Bingham Oceanographic Collection Yale University* 15:345–414; Sanders, H. L. 1958. Benthic studies in Buzzards Bay. I. Animal-sediment relationships. *Limnology and Oceanography* 7:63–79; Sanders H. L. 1968. Marine benthic diversity; a comparative study. *American Naturalist* 102:243–282. He and his colleagues also did classic pioneering work on deep-sea faunas (especially protobranch bivalves and primitive crustaceans) (Sanders, H. L., R. R. Hessler, & G. Hampson. 1965. An introduction to the study of deep-sea benthic faunal assemblages along the Gay Head-Bermuda transect. *Deep-Sea Research* 12:845–867). He is perhaps best known for his time-stability hypothesis for explaining global patterns in biodiversity.

To our knowledge, Howard never routinely taught a formal course in an academic setting. Nevertheless, through his mentoring and informal contact with students, he had a profound impact on all of us who had the good fortune to learn from him, either on a cruise or over coffee in his lab.

It was Howard's wish that his ashes be spread onto local waters where he did his early studies and where he spent many hours at his favorite pastime: fishing for blues, strippers, and "doormat" flounder. The very best times were those when we were fishing and not catching much but had the time to talk about sea creatures and evolution. Those were the times that I will miss most.

Goodbye old friend!

Donald C. Rhoads
(one who rarely caught fish but was well mentored)

Contents

Technological Advances in the Study of Benthic Ecology

Biogenic Modification of Physical Properties

Response of Benthos to Sedimentary Disturbances, Biological and Paleoecological Indicators of Environmental Properties and Processes

Biogeochemical Processes, Elemental and Contaminant Cycling in the Bioturbated Zone

Sedimentary Food Resources and Digestive Strategies

* Corresponding Author

Preface

Benthic organisms extensively modify the physical and biogeochemical properties of sedimentary deposits. This volume presents contributions from a symposium/workshop held in October 1998 to address the persistent and important research questions pertaining to such interactions and to identify opportunities for future investigations in the field. An additional goal was to promote active partnerships between federal regulatory agencies and the academic community in order to preserve and enhance our natural aquatic resources. The major themes of symposium contributions included examination of 1) physical and sedimentologic effects of animal-sediment interactions and their relationship to biological, geochemical, and biogeochemical processes; 2) impacts of disturbance and succession on faunal interactions and community structure using paleoecological evidence as well as living examples; 3) influences of organism-sediment interactions on microbial activities, organic matter remineralization, and nutrient cycling; 4) evidence for alternative feeding strategies based on life position, feeding habits, morphology, physiology, and manipulation of habitat biogeochemistry; and 5) application of innovative technologies for general studies and for solving specific environmental problems. Abstracts for oral presentations for which papers have not been contributed, and for poster presentations are available upon request from the senior editor.

Financial sponsorship for this symposium/workshop was provided by The National Science Foundation-Chemical Oceanography Division; The Department of the Army, New England Regulatory Division, Corps of Engineers; Science Applications International Corporation; Professor Brendan Keegan of the Martin Ryan Marine Science Institute, National University of Ireland, Galway; Department of Geology and Geophysics, Yale University; The Marine Sciences Research Center, State University of New York at Stony Brook; and the Belle Baruch Institute for Marine Biology and Coastal Research, University of South Carolina. Much appreciation goes to Joe Germano, Bob Diaz, and Dave Young for helping to make the symposium a success. The Trustees of the Belle W. Baruch Foundation granted permission for use of the Hobcaw House on the Hobcaw Barony. We wish to thank Dr. Dennis Allen and his staff at the Baruch Institute, Forest Manager Mr. George Chastain, and especially, Kathy Caufield for their cooperation and assistance with the logistics for the symposium. Finally, the efforts of many colleagues in reviewing the papers in this volume are gratefully acknowledged.

Contributors

S. Acevedo
Zoology Department
The Martin Ryan Marine Science Institute
National University of Ireland
Galway, Galway City
Ireland

Michael J. Ahrens
Marine Sciences Research Center
State University of New York at Stony Brook
Stony Brook, NY 11794-5000

Josephine Y. Aller
Marine Science Research Center (MSRC)
University of New York at Stony Brook
155 Challenger Hall
Stony Brook, NY 11794-5000

Robert C. Aller
Marine Science Research Center (MSRC)
State University of New York at Stony Brook
155 Challenger Hall
Stony Brook, NY 11794-5000

Daniel M. Alongi
Australian Institute of Marine Science
PMB No. 3
Townsville M.C.
Qld. 4810
Australia

Shirley M. Baker
Department of Ecology and Evolution
State University of New York
Stony Brook NY 11794

Present Address:
Department of Fisheries and Aquatic Sciences
University of Florida
7922 NW 23rd Street
Gainesville, FL 32653

Neal Blair
Department of Marine, Earth
  and Atmospheric Sciences
125 Jordan Hall
Faucette Drive
North Carolina State University
Raleigh, NC 27695-8208

Michael J. Bock
Darling Marine Center
University of Maine
Walpole, ME 04573

Carlton E. Brett
Department of Geology
University of Cincinnati
Cincinnati, OH 45221

Kevin B. Briggs
Naval Research Laboratory
Stennis Space Center
Stennis Space Center, MS 39529-5004

W. Russell Callender
Virginia Graduate Marine Science
  Consortium
University of Virginia
Charlottesville, VA 22903

J. Costelloe
Aqua Fact International Services Ltd.
Galway
Ireland

V. J. Cummings
National Institute of Water
  and Atmospheric Research
PO Box 11-115
Hamilton
New Zealand

G. Randy Cutter, Jr.
School of Marine Science
Virginia Institute of Marine Science
College of William and Mary
PO Box 1346
Gloucester Point, VA 23062

Timothy M. Dellapenna
School of Marine Science
The College of William and Mary
Virginia Institute of Marine Science
Gloucester Point, VA 23062

David J. DeMaster
Department of Marine, Earth
  and Atmospheric Sciences
North Carolina State University
Raleigh, NC 27695-8208

Robert J. Diaz
School of Marine Science
Virginia Institute of Marine Science
College of William and Mary
PO Box 1346
Gloucester Point, VA 23062

P. Dinneen
Galway/Mayo Institute of Technology
Galway
Ireland

Jean-Pierre Durbec
Laboratoire d'Océanographie
  et de Biogéochimie
UMR 6535 du CNRS
Centre d'Océanologie de Marseille (OSU)
Université de la Méditerranée
Campus de Luminy
case 901, 13288 Marseille cedex 9
France

William Fornes
Department of Marine, Earth
  and Atmospheric Sciences
North Carolina State University
Raleigh, NC 27695-8208
Present address:
Department of Geological Sciences
Case Western Reserve University
10900 Euclid Avenue
Cleveland, OH 44106-7216

Frédérique François
Laboratoire d'Océanographie
  et de Biogéochimie
UMR 6535 du CNRS
Centre d'Océanologie de Marseille (OSU)
Université de la Méditerranée
Campus de Luminy
case 901, 13288 Marseille cedex 9
France
Present address:
Observatoire Oceanologique de Banyuls
Universite Pierre et Marie Curie (ParisVI) -
CNRS
Laboratoire Arago
BP 44, 66651 Banyuls sur mer Cedex
France

Carl T. Friedrichs
School of Marine Science
The College of William and Mary
Virginia Institute of Marine Science
Gloucester Point, VA 23062

G. A. Funnell
National Institute of Water
  and Atmospheric Research
PO Box 11-115
Hamilton
New Zealand

Joseph D. Germano
EVS Environment Consultants, Inc.
200 W. Mercer Street, Suite 403
Seattle, WA 98119
Present Address:
Germano & Associates
12100 SE 46th Place
Bellevue, WA 98006

Ronnie N. Glud
Marine Biological Laboratory
University of Copenhagen
Strandpromenaden 5
3000 Helsingør
Denmark

M. O. Green
National Institute of Water
  and Atmospheric Research
PO Box 11-115
Hamilton
New Zealand

A. Grehan
Zoology Department
The Martin Ryan Marine Science Institute
National University of Ireland
Galway, Galway City
Ireland

J. E. Hewitt
National Institute of Water
  and Atmospheric Research
PO Box 11-115
Hamilton
New Zealand

Elizabeth K. Hinchey
School of Marine Science
The College of William and Mary
Virginia Institute of Marine Science
Gloucester Point, VA 23062

Gerhard Holst
Max-Planck-Institute for Marine Microbiology
Celsiusstr.1
28359 Bremen
Germany

Peter A. Jumars
School of Oceanography
University of Washington
Seattle, WA 98195
Present Address:
Darling Marine Center
University of Maine
193 Clark's Cove Road
Walpole, ME 04573

B. F. Keegan
Zoology Department
The Martin Ryan Marine Science Institute
National University of Ireland
Galway
Ireland

P. F. Kemp
Marine Sciences Research Center
State University of New York at Stony Brook
Stony Brook, NY 11794-5000

R. Kennedy
Zoology Department
The Martin Ryan Marine Science Institute
National University of Ireland
Galway, Galway City
Ireland

Susan M. Kidwell
Department of Geophysical Sciences
University of Chicago
5734 S. Ellis Avenue
Chicago, IL 60637

xviii Contributors

Bettina Koenig
Max-Planck-Institute for Marine Microbiology
Celsiusstr.1
28359 Bremen
Germany
Present Address:
OHB-System GmbH
Universitaetsalle 27-29
28359 Bremen
Germany

Erik Kristensen
Institute of Biology
Odense University, SDU
DK-5230 Odense M
Denmark

Michael Kuehl
Marine Biological Laboratory
University of Copenhagen
Strandpromenaden 5
3000 Helsingør
Denmark

Steven A. Kuehl
School of Marine Science
The College of William and Mary
Virginia Institute of Marine Science
Gloucester Point, VA 23062

Lisa Levin
Marine Life Research Group
Scripps Institution of Oceanography
La Jolla, CA 92093-0218

Jeffrey S. Levinton
Department of Ecology and Evolution
State University of New York
Stony Brook NY 11794

Sara M. Lindsay
Scripps Institution of Oceanography
La Jolla, CA 92093
Present Address:
School of Marine Sciences
University of Maine
Orono, ME 04469

Andrew M. Lohrer
Department of Marine Sciences
University of Connecticut
Groton, CT 06340
Present Address:
Belle W. Baruch Institute for
* Marine Biology & Coastal Research*
Marine Field Laboratory
University of South Carolina
PO Box 1630
Georgetown, SC 29442

Glenn R. Lopez
Marine Sciences Research Center
State University of New York at Stony Brook
Stony Brook, NY 11794-5000

Roberta L. Marinelli
Skidaway Institute of Oceanography
Savannah, GA 31411
Present Address:
Office of Polar Programs
National Science Foundation
Arlington, VA 22230

Chris Martin
Marine Life Research Group
Scripps Institution of Oceanography
La Jolla, CA 92093-0218

Gerald Matisoff
Department of Geological Sciences
Case Western Reserve University
Cleveland, OH 44106

Lawrence M. Mayer
Darling Marine Center
University of Maine
Walpole, ME 04573

Peter L. McCall
Department of Geological Sciences
Case Western Reserve University
Cleveland, OH 44106

D. McGrath
Galway/Mayo Institute of Technology
Galway
Ireland

Michelle Thompson Neubauer
School of Marine Science
The College of William and Mary
Virginia Institute of Marine Science
Gloucester Point, VA 23062

B. O'Connor
Aqua Fact International Services Ltd.
Galway
Ireland

I. O'Connor
Zoology Department
The Martin Ryan Marine Science Institute
National University of Ireland
Galway, Galway City
Ireland

Karla M. Parsons-Hubbard
Department of Geology
Oberlin College
Oberlin, OH 44074

Randi Pilgaard
Institute of Biology
Odense University
DK-5230 Odense M
Denmark

Gayle Plaia
Department of Marine, Earth
  and Atmospheric Sciences
North Carolina State University
Raleigh, NC 27695-8208

Jean-Christophe Poggiale
Laboratoire d'Océanographie
  et de Biogéochimie
UMR 6535 du CNRS
Centre d'Océanologie de Marseille (OSU)
Université de la Méditerranée
Campus de Luminy
case 901, 13288 Marseille cedex 9
France

Robin Pope
Department of Marine, Earth
  and Atmospheric Sciences
North Carolina State University
Raleigh, NC 27695-8208

Eric N. Powell
Haskin Shellfish Research Laboratory
Rutgers University,
Port Norris, NJ 08349

Donald C. Rhoads
Marine Technology Group
Science Applications International Corp
22 Widgeon Rd.
Falmouth, MA 02540

Michael D. Richardson
Naval Research Laboratory
Stennis Space Center
Stennis Space Center, MS 39529-5004

John A. Robbins
NOAA/Great Lakes Environmental
  Research Laboratory
Ann Arbor, MI 48104-4590

Linda C. Schaffner
School of Marine Science
The College of William and Mary
Virginia Institute of Marine Science
Gloucester Point, VA 23062

Jill L. Schmidt
School of Oceanography
University of Washington
Seattle, WA 98195

Sandra E. Shumway
Natural Sciences Division
Southampton College
Long Island University
Southampton, NY 11968

Craig R. Smith
Department of Oceanography
University of Hawaii
1000 Pope Road
Honolulu, HI 96822

Mary E. Smith
School of Marine Science
The College of William and Mary
Virginia Institute of Marine Science
Gloucester Point, VA 23062

M. Solan
Zoology Department
The Martin Ryan Marine Science Institute
National University of Ireland
Galway, Galway City
Ireland
Present Address:
Cultery Field Station, Newburgh
Ellon, Aberdeenshire AB41 0AA
Scotland

Frederick M. Soster
Department of Geology and Geography
DePauw University
Greencastle, IN 46135

George M. Staff
Department of Geology
Austin Community College
Austin, TX 78664

Georges Stora
Laboratoire d'Océanographie
  et de Biogéochimie
UMR 6535 du CNRS
Centre d'Océanologie de Marseille (OSU)
Université de la Méditerranée
Campus de Luminy
case 901, 13288 Marseille cedex 9
France

Carrie Thomas
Department of Marine, Earth
  and Atmospheric Sciences
North Carolina State University
Raleigh, NC 27695-8208

Simon F. Thrush
National Institute of Water
  and Atmospheric Research
PO Box 11-115
Hamilton
New Zealand

Yves-Alain Vetter
School of Oceanography
University of Washington
Seattle, WA 98195

Sally E. Walker
Department of Geology
University of Georgia
Athens, GA 30602

J. Evan Ward
Department of Marine Sciences
University of Connecticut
1084 Shennecossett Rd.
Groton, CT 06340

Roger Ward
Science Applications International Corp
221 Third St.
Newport, RI 02840

David S. Wethey
Department of Biological Sciences and the
Marine Science Program
University of South Carolina
Columbia, SC 29208

Robert B. Whitlatch
Department of Marine Sciences
University of Connecticut
Groton, CT 06340

Sarah A. Woodin
Department of Biological Sciences and the
Marine Science Program
University of South Carolina
Columbia, SC 29208

David K. Young
Naval Research Laboratory
Stennis Space Center
Stennis Space Center, MS 39529-5004
Present Address:
P.O. Box 399
Long Beach, MS 39560

S. Young
Zoology Department
The Martin Ryan Marine Science Institute
National University of Ireland
Galway, Galway City
Ireland

Roman N. Zajac
Department of Biology
  and Environmental Science
University of New Haven
300 Orange Avenue
West Haven, CT 06516

Organism-Sediment Interactions

The Importance of Technology in Benthic Research and Monitoring: Looking Back to See Ahead

Donald C. Rhoads, Roger Ward, Josephine Aller, and Robert Aller

Abstract: *Benthic studies in the 1800s addressed zoological and natural history investigations by means of direct observation of intertidal zonation ("marine sociology") or sampling of subtidal environments, including the deep sea, with dredges, trawls, and grab samplers. Direct observation by diving was also initiated but rarely used until development of SCUBA in the early 1940s and underwater film cameras. These methods and techniques were sufficient to address academic and fisheries questions of the day for the identification of marine organisms and their geographic and/or bathymetric distributions. Traditional sampling techniques can be broadly classified as mechanical samplers, which are still used today albeit in more sophisticated forms such as compartmentalized box cores, Van Veen and McIntyre quantitative grabs, and epibenthic sleds. Classical mechanical samplers continue to be used and will always have a role to play in "ground truth" sampling. However, a technological revolution in seafloor investigations began after 1972 with passage of the Marine Protection, Research, and Sanctuaries Act (MPRSA, Public Law 92-532), also known as the Ocean Dumping Act. Up to that time, benthic studies were focused on basic research or applied fisheries surveys. The Ocean Dumping Act widened the scope of investigations to include benthic monitoring. Agencies such as the US Environmental Protection Agency, US Army Corps of Engineers, and the National Oceanographic and Atmospheric Administration faced significant budgetary challenges in attempting to monitor numerous benthic impacts using solely classical mechanical samplers and techniques. Sampling and data work-up took too long and were too expensive. Subsequently, many remote sensing and sampling technologies have been developed, improved, or proposed to facilitate efficient habitat mapping and data collection. In large part, these technologies have been borrowed from the medical field (e.g., x-radiography and ultrasound imaging, immunosensors), military R&D developments (e.g., downward-looking sonar, side-scan sonar, laser line-scan imaging, multibeam acoustics, remotely operated vehicles (ROVs), autonomous underwater vehicles (AUVs), manned submersibles, towed sled photography and videography, satellite and areal photography, and spectroscopy), photonics research (e.g., fiber optics, planar optodes, laser imaging, sediment profile imaging, and digital imaging), and environmental magnetics (biomagnetometry). Benthic ecology is becoming increasingly interdisciplinary, encompassing the complex interplay between organisms, solid phase and porewater chemistry, nutrient fluxes, sediment transport, geotechnical properties, and the fate and effects of contaminants. The historical development of technology as it applies to the study of benthic ecosystems, the status of current technologies, and future trends are addressed. Our search for state-of-the-art sensors and sampling*

1

devices focuses on efficient monitoring of benthic biological impacts of dredging and disposal in both shallow-water and in deep-sea environments. We have reviewed the universe of existing and potential sensors, samplers, and techniques (and their deployment platforms) for the US Army Corps of Engineers and the Deep Ocean Relocation Program, Naval Research Laboratory, Vicksburg, Mississippi. Fifty-one (51) candidates were identified and placed into one of three categories of technological maturity: Category I sensors and samplers are those that exist today and can fulfill future needs with no, or little, modification (n = 12). Category II sensors exist today but require major modifications to address future needs (n = 16). Category III sensors and samplers now exist only as prototypes or concepts (n = 23).

Introduction

The purpose of this paper is to trace the historical development of technology as it applies to the study of benthic ecosystems, assess the status of current technologies, and anticipate future trends. Data for our review comes from three recent technology reviews (Rhoads et al. 1996, 1997; Valent et al. in press).

In retrospect, the role of technology in benthic studies began to change in 1972 with the passage of the Marine Protection, Research, and Sanctuaries Act (MPRSA or Ocean Dumping Act). This act required those federal management agencies responsible for protection of the environment to address systematically the potential and/or real impacts of human activities. In aquatic systems, these impacts include dredging and disposal of dredged material, sewage sludge dumping, thermal loading, marine mining, eutrophication, and impacts related to toxic effluents and runoff. Federal management agencies responsible for assessing these impacts include the US Army Corps of Engineers (USACOE), the Environmental Protection Agency (USEPA), and the National Oceanographic and Atmospheric Administration (NOAA). The potential list of monitoring sites was large and formidable in scope and cost. Measurement and observational tools were relatively primitive and inefficient, impact paradigms were wanting (as were testable hypotheses), and formal monitoring protocols were not established. Environmental impact assessment within the decade of the 1970s resulted in expensive and inefficient "shotgun" approaches to monitoring. All parameters of potential interest were measured in hope that data crunching would reveal insight into benthic responses to impacts. These early attempts largely failed. This brief background sets the stage for a technological revolution that is still taking place.

Prior to 1972, benthic investigations were largely focused on qualitative and quantitative geographic and bathymetric distributions. Applied research and monitoring were focused on fisheries issues. Sampling tools were mainly limited to traditional mechanical samplers such as dredges, grabs, nets, traps, epibenthic sleds, box cores, and gravity corers. The original designs of these samplers can be traced back to the 19th and early 20th centuries. While significant improvements have been made (e.g., Sanders & Hessler 1969), all of these devices require sampling a small area of the bottom, followed by sample recovery with shipboard preliminary processing, final laboratory processing, and data analysis. High personnel costs for sampling, processing, and analysis, and long data turnaround times were not viable alternatives for routine management decisions involving rapid and efficient assessment of the status of the environment and spatial, temporal trends at monitoring sites.

In the 30 years from the 1970s to the present time, monitoring protocols have benefitted from the development of hypothesis-based tiered monitoring (National Research Council 1990), ecologically based impact paradigms (e.g., Pearson & Rosenberg 1978; Rhoads et al. 1978), and efficient data collection techniques, along with rapid data synthesis and reporting (e.g., Rhoads & Germano 1982).

Our search for state-of-the-art sensors and sampling devices focuses on efficient monitoring of benthic biological impacts of dredging and disposal in both shallow water (inner shelf, especially estuaries and embayments; Rhoads et al. 1996, 1997) and in a deep-sea environment (Valent et al. in press). Most of our results come from Valent et al. (in press) as deep-sea monitoring represents the greatest challenge for efficient time-series monitoring. The Deep Ocean Relocation (DOR) program, otherwise known as the Abyssal Seafloor Waste Isolation Study, was undertaken by the Navy Research Laboratory, and funded by the Strategic Environmental Research and Development Program (SERDP) as part of the Compliance Trust Area.

Methods

Our DOR sensor review addresses sensor and sampling requirements for efficient monitoring of the spatial and temporal impacts of dredged material (natural sediments and their associated contaminants) that may be transported from shallow water harbors to a deep-sea disposal site by specially designed open ocean barges. At the disposal point, bagged dredged material is dropped from the barges to the abyssal seafloor. Contaminated sediments, in this hypothetical problem, are to be contained within geotextile bags to prevent dispersion in the upper layers of the ocean. Each of 20 compartments within the ocean barge is to contain 800 yd^3 of bagged sediment. Bags are released at a point above a 5,000-m deep, topographically flat, mud bottom located within a low kinetic energy abyssal plain. The interested reader is directed to Valent and Young (1995) for a more detailed description of proposed operational methods. Monitoring requirements at the deep-sea disposal site include the affected water column, bag impacts, bag rupture and spillage, dispersion, dredged material footprint, faunal colonization, chemical diagenesis, and risk of food chain contamination. Sampling includes both near-field and far-field stations.

Three coupled models were used to predict environmental impacts of bagged sediment deposited on the bottom: 1) SimDOR, a physics-based model of bag descent and short-term plume behavior, 2) geochemical models of diffusion, advection, and diagenesis, and 3) an ecological model of colonization including risk of contaminant exposure, bioaccumulation, and trophic transfer (Valent & Young 1995; Valent et al. in press).

A tiered monitoring plan was developed for all predicted physical (P), chemical (C), and biological (B) impacts. Figure 1 shows a generic example of a tiered plan beginning with a testable hypothesis. The hypothesis defines sensor engineering and performance requirements (area of coverage, resolution, accuracy, precision, threshold of detection, etc.) for candidate sensors. Tier I sensors have the potential for covering large search areas at moderate to low resolution. Evaluation of Tier I data result in well-defined management actions (ranging from no action to a requirement for higher resolution measurements). Tier II sensors have lower spatial coverage efficiencies but higher spatial, temporal resolution and more sensitive detection thresholds. Data products from Tier II monitoring are evaluated and result in one or more monitoring action(s) such as no action, use of other sensors, or revision of the measurement program and/or null hypothesis. A specific example of a tiered biological monitoring plan from the DOR program is shown in Fig. 2 with specified sensors and management options. Several interlinked tiered monitoring plans were constructed to address physical, chemical, and biological hypotheses in the DOR plan (Valent et al. in press).

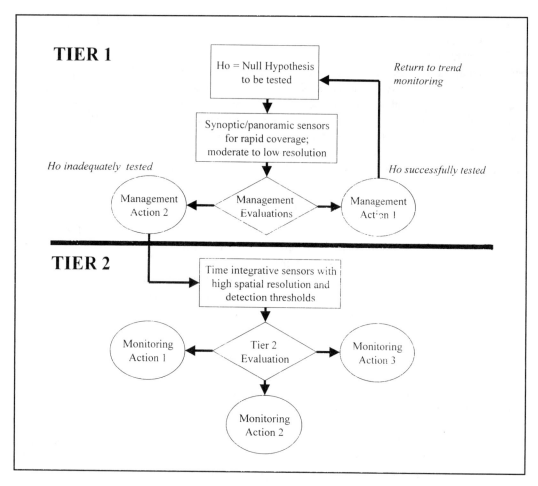

Fig. 1. A generic example of information flow in a tiered monitoring protocol. Note that Tier I sensors and techniques are reconnaissance tools with high rates of coverage while Tier II systems have higher measurement resolution and thresholds of detection at the expense of high coverage rates. (modified from Valent et al. in press)

As candidate sensor and sampling technologies were identified they were placed into one of three classes or categories based on technological maturity: 1) Category I sensors are those sensors that are currently used routinely in shallow water monitoring and could be deployed in the deep sea with minimal modification. 2) Category II sensors are sensors that are currently used in shallow water monitoring but would require extensive repackaging and/or improvement in signal detection in a deep-sea environment, and 3) Category III sensors are prototype or concept sensors that could, with robust funding, be operational within a 3-yr period. Candidate sensors and techniques were identified from professional journals, computer database searches, and telephone calls to individuals and institutions known to be active in sensor development. Each sensor, sampler, and/or technique was summarized on a data sheet with the data source(s) and technical specifications. These data sheets were then reorganized into a summary table (Table 1) arranged according to the three maturity

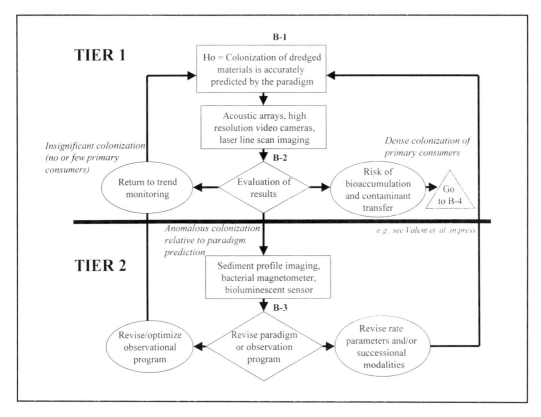

Fig. 2. An example of Tier I and Tier II candidate sensors for testing biological hypotheses associated with the DOR study. This particular DOR module addresses a deep-sea colonization model. The colonization model, in turn, is linked to other modules and hypotheses (e.g., B-4, H_o = zero or low risk of biological transport of contaminants from the disposal site to the far-field). (modified from Valent et al. in press)

categories described above and each entry was assigned to the appropriate monitoring module (Fig. 3A). In most cases, several candidate sensors, samplers, and techniques were competing for the same measurement tasks. We therefore ranked these competing sensors and techniques to highlight the most efficient candidates. Category III (prototype) sensors ranked relatively low because of their primitive state of development. The candidate sensors were ranked again using a technological "risk" adjusting factor that placed major emphasis on sensor applicability (or desirability) rather than on degree of technological development. Risk-adjusted rankings are not given here but are available in Valent et al. (in press).

Table 1 is a summary of candidate sensors arranged by technological maturity (categories I, II, and III). Each sensor, sampler, or technique is listed with a reference number (1 through 51). Reference documentation is provided in Valent et al. (in press). No ranking is implied by the numerical sequence. A brief descriptive name for each sensor is followed by its technological maturity category, range of measurements and analytes, sensitivity for detection, and the method of measurement (e.g., optical spectral, acoustic, electrochemical, membrane, magnetic, etc). Methods of deployment and recovery follow (platform architecture). The last column identifies the tiered

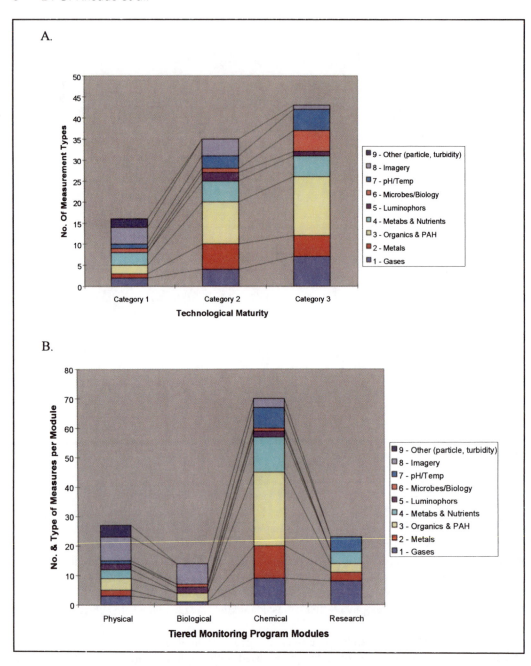

Fig. 3. A) Distribution of sensors, samplers, and techniques among the three categories of technological maturity. Category I is the most mature and Category III is the least mature. Each bar is divided into types of measurements. Note that prototype and concept sensors form the largest population (modified from Valent et al. in press). B) Distribution of sensors, samplers, and techniques among each of the tiered monitoring modules (physical, biological, chemical, and research). Note that the largest number of sensors apply to testing chemical hypotheses. Biological monitoring is the least technologically advanced.

monitoring modules (P = physical, C = chemical, B = biological) addressed by each technology. All of the sensors identified in Table 1 are capable of providing data on benthic processes. Water column sensors have been deleted from the original table as presented in Valent et al. (in press).

Twelve (no. 1–12, Table 1) Category I sensors were identified as having great potential for efficient monitoring of the impact of deep-sea dredged material disposal. Sixteen (no. 13–28) Category II sensors follow. Category III (prototype or concept) sensors form the longest list of sensors. Twenty-three (no. 29–51) sensors, samplers, and techniques, although unproven in field tests, show promise for the intended applications. Figure 3A shows the distribution of sensors, samplers, and techniques among the three categories of technological maturity along with the types of measures. Note the inverse relationship between technological maturity and numbers of candidates. Figure 3B shows the distribution of sensors, samplers, and techniques among the three tiered monitoring modules. A research module is also included to provide basic research data not gathered in each of the tiered monitoring modules. Note that the chemical module has the greatest number of options for measurement while the biological module is relatively technologically poor.

Discussion

Although this review is focused on dredged material monitoring in the deep sea, most sensors also have application to shallow-water monitoring (e.g., Rhoads et al. 1996, 1997). However, deep-sea deployment is the most technologically challenging. Therefore, if we were to refocus the review for shallow-water monitoring, all Category I and II and some Category III sensors could be immediately used without extensive repackaging in high pressure housings. Deployment architecture also could be simplified.

Tier I reconnaissance sensors (rapid search at moderate to low resolution) are well represented in Table 1. These include traditional side-scan sonar, laser line-scan (single mode), dual mode (fluorescence) laser line-scan, multibeam sonar, and traditional photography and videography plan view or sediment profile imagery. Candidates also include chemical "quick look" systems such as the Continuous Sediment Sampling System (CS^3), and REMOTS® hyperspectral UV fluorescence sediment profile imaging.

Tier II (and lower) sensors (low to moderate search rates with high resolution) dominate the candidate sensor population. One of the most interesting outcomes of the DOR sensor review is the large number of candidate prototype or concept (Category III) sensors. Most of these sensors are for characterizing solid phase and/or porewater chemistry and include optical-spectral, electrochemical, electromechanical, or magnetic measurements. Because these sensors are deployed to obtain local measurements, they are best applied to document quantitatively sediment properties at discrete stations.

For time-integrative measurements, sensors are passively exposed over a period of time until an equilibrium is established between the membrane/film/coating and ambient sediment/porewater (e.g., planar optodes, Chelex membranes, dialysis and diffusion in thin films, and acoustic wave guides or quartz microbalances). This passive time-integrating concept builds on earlier successes of porewater "peeper" technology (e.g., Hesslein 1976; Aller et al. 1998). Other sensors are "active," that is, time-series sampling is done at specific intervals separated by data gaps (burst sampling). The majority of burst samplers are based on fiber-optic spectrometry (e.g., Mach Zender interferometry wave guides, surface enhanced Raman scattering, fluorosensors). Biomagnetometry involves both passive and active measurement modes. The distributions of mobile magnetotactic bacteria

Table 1. Candidate sensors, samplers, and techniques arranged by technological maturity (categories I, II, and III) and appropriate monitoring modules (P = physical, B = biological, C = chemical). Deployment architecture abbreviations are Rover = mobile benthic sampling device, AUV = Autonomous Underwater Vehicle, ROTV = Remotely Operated Towed Vehicle, and POD = stationary tripod or quadrapod for Eularian measurements. Reference citations for each candidate sensor can be found in Valent et al. (in press). Number of sensor, technique (1–51) does not indicate any ranking.

No.	Sensor/Technique	Maturity	Measures	Sensitivity	Type	Remarks	Monitoring Module
1	Biomarker Procedure	I	Cytochrome P-450, Mono-oxygenase (pollutants/carcinogen)	High sensitivity (presence/absence)	Electro-chemical	Ship/Rover/AUV/ROTV - trap/recover	C, B
2	Molecular Marker Procedure	I	LABs, surfactants, coprostanol, PAHs (sediments)	21.9‰	Electro-chemical	Ship/Wire - trap/recover	C
3	Grab Respirometer (Rover/AUV/ROTV)	I	O_2	15–20%	Electro-chemical	Rover/AUV/ROTV/Stationary pod B real-time	Research
4	Photo/Video Imagery	I	Photo imagery (20 m²)	3 x 5 mm resolution	Optical/spectral	Ship/Rover/AUV/ROTV	P, B
5	Acoustic Array (kHz): physical (~20) biological (~150) tracking (~12)	I	Backscatter/imagery (100 m range); sensitivity to particulates TBD	2 cm resolution	Acoustic	a) ROTV/AUV/Rover/ Vertical array - real-time b) Vertical array B - real-time c) Ship B real-time	P, B
6	REMOTS	I	Biology, grain size (sediment)	0.062 mm	Optical/spectral	Ship/Wire/Rover B real-time	P, B, C
7	Acoustic Side Scan	I	Bags, thick deposits, sediment classification	10 cm resolution	Acoustic	Rover/AUV/ROTV - real-time	P, B

Table 1. Continued.

No.	Sensor/Technique	Maturity	Measures	Sensitivity	Type	Remarks	Monitoring Module
8	Smith/Benthic Chamber	I	Organics, PAHs, metals, PCBs, temperature, salinity, gases, pH, nutrients (sediment, water)	to ppm	Electro-chemical, Optical/spectral	Pod - real-time	C, Research
9	Acoustic Positioning Network (Long Baseline)	I	Position	unknown	Acoustic	Pod - real-time	P
10	Sediment Trap	I	Sediment flux	to ppm	Membrane	Ship/Wire B sample/recover	P
11	Water Transfer System (filter)	I	Larvae/eggs suspended in water	N/A	Sampler	Ship/Wire B sample & return for lab analysis	B
12	Submersible Chemical Analyzer (SCANNER)	I	Silicates & sulfides	0.1 micromolar	Optical/spectral	Ship/Rover/AUV/ROTV B real-time & pod for time-series	C
13	Diffusion Gradients in Thin Films (DGT)	II	Metals (sediment, water)	mm gradients & concentrations	Membrane	Ship/Wire B sample/recover	C
14	Fiber Optic Chemical Sensors (FOCS)	II	Chloroform, hydrocarbons, pH, CO_2, O_2, ammonia, Al, cyanide, SO_2 (sediment, water)	to regulatory limits	Optical/spectral	Ship/Wire B real-time	C
15	Membrane, Semi-permeable	II	Dissolved organic, bioavailable (water)	Unknown	Membrane	Ship/Wire B sample/recover	C

Table 1. Continued.

No.	Sensor/Technique	Maturity	Measures	Sensitivity	Type	Remarks	Monitoring Module
16	Bioluminescent Biosensor	II	Hg^{+2} (bioavailable)	nanoB micro molar	Optical/ spectral	Rover/AUV/ROTV B real-time	C
17	Continuous Sediment Sampling System (CS³)	II	Metals (sediment)	ppm	Gamma	ROTV B real-time	P, C
18	Benthic Flux Sampling Device (BFSD)	II	Organics, PAHs, metals, PCBs, temperature, salinity, gases, pH, nutrients (sediment, water)	ppm	Electro-chemical, Optical/ spectral	Pod B real-time	C, Research
19	Portable Synchronous UV Scanning Spectrofluorometer	II	PAHs (water, sediment)	ppm	Optical/ spectral	Rover B real-time	P, C
20	Time-resolved Fluorescence Spectroscopy	II	PAHs, nutrients, metabolites (sediment, water)	ppt–ppm	Optical/ spectral	Rover/AUV/ROTV real-time	P, C
21	REMOTS UV Imaging Spectrometer	II	Organics, PAHs, biology, grain size (sediment, water), luminophores	ppm	Optical/ spectral	Ship/Wire - real-time	P, B, C
22	Semipermeable Membrane Device (SPMD)	II	Organics, PAHs (water)	< 30 fg l⁻¹	Membrane	Ship/Wire - Deploy/Recover	C
23	Laser Line Scan	II	Bags, sediment classification	cm resolution	Optical/ spectral	AUV/ROTV - real-time	P, B

Table 1. Continued.

No.	Sensor/Technique	Maturity	Measures	Sensitivity	Type	Remarks	Monitoring Module
24	Dual Mode Laser Line Scan	II	Bags, sediment classification, PAHs, organics, luminophores	cm resolution, ppm	Optical/spectral	AUV/ROTV - real-time	P, B
25	Multibeam Sonar	II	Bags, thick deposits	50 cm resolution	Acoustic	ROTV - real-time	P
26	Gill Chamber	II	Organics, PAHs, metals, PCBs, temperature, salinity, gases, pH, nutrients (sediment, water)	to ppm	Electro-chemical	Pod - real-time	C, Research
27	Flow-through Infrared Sensor	II	VOCs & PCBs	unknown	Optical/spectral		P, C
28	Potentiometric Stripping Analysis (PSA)	II	Trace metals (Hg, Cu, Cd, Zn, Pb) in water	ppb	Electo-chemical	Rover/AUV/ROTV - real-time	C
29	Opto-Chemical Sensor (OCS)	III	O_2, pH, CO_2 (water)	0.2% gas concentration, > 0.1 pH unit	Optical/spectral	Ship/Rover/ AUV/ ROTV B real-time	C
30	Remote Fiber Spectroscopy	III	Colormetric organochlorides, inorganic (nitrates, sulfates, cyanides, ferri/ferro cyanide, chloroform (sediment, water)	Depends on optode	Optical/spectral	Rover/AUV/ROTV - real-time	C
31	Plate Optodes B fluorosensor	III	Calcium, CO_2, O_2, pH (sediments); fluorescent organics	Varies	Optical/spectral	Ship/Wire/Rover - real-time	P, C

Table 1. Continued.

No.	Sensor/Technique	Maturity	Measures	Sensitivity	Type	Remarks	Monitoring Module
32	DGT B Chelex Resin	III	Metals (water, sediment); zinc, Ni, Fe, Cu	unknown	Membrane	Rover/AUV/ROTV - deploy/recover	C
33	Film Sensor B Plastic, Colorimetric	III	CO_2 (water)	unknown	Optical/ spectral	Rover/AUV/ROTV - deploy/recover; lifetime is hours	C, Research
34	Optomembrane, Ion-sensitive	III	O_2, pH (water, sediment)	unknown	Optical/ spectral	Rover/AUV/ROTV - deploy/recover	C
35	Optical Fluorsensor (phase modulated)	III	pH, CO_2	unknown	Optical/ spectral	Rover/AUV/ROTV - real-time	Research
36	Integrated Optical Sensor (FOCS); Mach Zehnder	III	PAHs, organics (CL, phenols, phosphates, solvents, alcohol, nutrients) (sediment, water)	to 20 ppm	Optical/ spectral	Rover/AUV/ROTV - real-time	C
37	Fluorosensor - Opto-Chemical Sensor/Film	III	pH, CO_2, O_2 (water)	0.2% gas concentration, > 0.1 pH unit	Optical/ spectral	Rover/AUV/ROTV - real-time (minutes)	Research
38	Dialysis & Diffusion Equilibrium in Thin Film (DDEET)	III	Trace metals peeper (water)	ppm?	Membrane	Rover/AUV/ROTV - deploy/recover	C

Table 1. Continued.

No.	Sensor/Technique	Maturity	Measures	Sensitivity	Type	Remarks	Monitoring Module
39	Surface Enhanced RAMAN Scattering (SERS)	III	PAHs (water, vapor)	ppm	Optical/ spectral	Rover - water/sediment profiling	C
40	Immuno Electro-Chemical Graphite Sensor	III	PAHs (water)	ng ml^{-1}	Electro-chemical	Rover/AUV/ ROTV - real-time	C
41	Wave Guide: Mach-Zehnder Interferometer (MZI)	III	PAHs (water)	very sensitive	Optical/ spectral	Ship/Wire/Rover/ROTV - Sample & flow-through analysis	C
42	Reflectometric Interference Spectroscopy (RIS)	III	PAHs (water)	very sensitive	Optical/ spectral	Rover/AUV/ROTV - real-time	C
43	MEM: Microcantalever Vapor Sensor (MVS)	III	PAHs (water)	picograms	Electro-mechanical	Ship/Wire - Sample & return for lab analysis	C
44	Wave Guide: Quartz Crystal Microbalance	III	Organics (water)	ppm	Optical/ spectral	Rover/ROTV - real-time	C
45	Wave Guide: Pizoelectric Quartz Crystal Immunosensor	III	Pesticides, PAHs (water)	ng ml^{-1}	Optical/ spectral	Rover/ROTV - real-time	C

Table 1. Continued.

No.	Sensor/Technique	Maturity	Measures	Sensitivity	Type	Remarks	Monitoring Module
46	Fluorosensor: Remote Spectral Imaging System (RSIS)	III	Organics, PAHs (water, sediments)	ppm	Optical/ spectral	Rover/ROTV - real-time	C
47	Surface Acoustic Wave Sensor (SAW)	III	Organics, PAHs (vapor)	ppm	Acoustic	Rover - sampling	C
48	Fiberoptic Immunosensor	III	PAHs (water, sediment)	ppm	Optical/ spectral	Rover/AUV/ ROTV B real-time	C
49	Chemical Nose	III	Organics, PAHs (vapors)	Unknown (mg ml^{-1}?)	Optical/ spectral	Rover/AUV/ROTV - real-time	C
50	Biomagnetometer	III	Microbes, O_2 (sediment)	3-5% O_2 saturation; 0.01 Gauss	Magnetic	Rover - deploy/expendable	P, C, Research
51	Liquid Phase Chemi-luminescence (CL)	III	Metal inorganic ions, biomolecules, carcinogens (water)	up to femtomolar (10^{-9})	Optical/ spectral	Ship/Rover/AUV/ROTV B real-time & pod for time-series	C

populations in the sediment column time-integrate the position of the redox boundary over periods of days to weeks. Magnetic sensor scanning of the sediment column is then done in "burst" mode to document changes in the depth distribution of these biological magnets. Another microbial sensor is based on genetically engineered bioluminescent bacteria. These bacteria are contained within small benthic chemostats; subsets of these chemostats are used as controls while others are periodically exposed to ambient water. A digital video camera monitors the intensity of bioluminescence that is inversely proportional to the presence of contaminants (e.g., metals).

The somewhat arbitrary division of time-averaging measurements versus instantaneous measurement is emphasized here because many management questions are best addressed by time-averaging (integrating) approaches, for example, mapping the time-averaged physical and chemical footprint of dredged material or long-term release of contaminants into the water column or redistribution of particulate phases. Instantaneous, or burst sampling, best serves to address research and/or modeling efforts where quantitative measurements provide data for coefficients involving short-term flux rates. Instantaneous measurements taken over a long time period lend themselves to research questions or Tier II monitoring while time-integrative sensors provide critical threshold data for management decisions based on predetermined parameters such as presence/absence, effects/no effects, change/no change.

Summary and Conclusions

We have shown that a potentially large population of candidate sensors exists for efficient monitoring of both deep-sea and shallow-water environments. The list of sensors continues to grow and Table 1 is already incomplete based on new technologies that have appeared since completion of the DOR sensor review. We conclude from our review that: (1) Benthic ecological, biological sensors lag far behind candidate physical and chemical sensors in terms of both numbers of types and technological maturity. (2) The greatest potential for narrowing this technology gap is to develop biological, ecological sensor candidates in maturity categories II and III. (3) Long-term (weeks to months) time-integrative sensors are of high value to answering many management questions such as presence/absence, above or below threshold, prograding or retrograding succession, trends in bioaccumulation, etc. Instantaneous measurements that characterize conditions at one location at one instant in time tend to be relatively inefficient for testing hypotheses related to long-term trend monitoring.

Finally, with so many candidate sensors, why are so many monitoring programs still using 19th century methods? The acceptance of new technologies and methods by the USEPA, USACOE, and NOAA is slow. The reason for this conservatism is based on protecting the integrity of existing Standardized Operating Procedures (SOPs) and databases. New, relatively unproven protocols are threats to established Quality Assurance/Quality Control (QA/QC) procedures. The senior author (DCR) has had first-hand experience with this conservatism. Sediment-profile imaging first appeared in a peer-reviewed journal in 1971 (Rhoads & Cande 1971). The first applied demonstrations of this technology took place in the early 1980s (Rhoads & Germano 1982, 1984), but not until the mid to late 1980s was this methodology (also known as REMOTS® or Remote Ecological Monitoring Of The Seafloor) used routinely for monitoring the ecological and sedimentological impacts of dredged material, marine mining, eutrophication, and hypoxia. The decade of the 1990s has seen increasing acceptance of the technology worldwide. In summary, sediment-profile imaging, computer image analysis has taken about 15 yr to mature and be accepted. As shown in this paper, technology does not appear to be holding back the evolution of more efficient and responsive means for carrying out environmental monitoring. Rather, it appears to be the inertia of existing protocols. Perhaps a quote from Abraham Maslow, a developmental psychologist (1908–1970) best summarizes this point: "When the only tool you own is a hammer, every problem begins to resemble a nail." Given past experience with REMOTS® it is likely that adoption of new protocols and paradigms may be driven more by fiscal expedience than by a technological imperative.

Acknowledgments

The US Army Corps of Engineers sensor reviews were prepared by Science Applications International Corporation (SAIC) for the Coastal Ecology Branch, Environmental Laboratory, Waterways Experiment Station, Vicksburg, Mississippi. Thanks are extended to R. Ward, J. A. Muramoto, D. Carey, and E. J. Saade for their contributions. Barbara Hecker (Hecker Environmental Consulting) contributed analytical methods to this effort.

The Deep Ocean Relocation (DOR) project (Phase II), was funded by the Navy and carried out by the Navy Research Laboratory (NRL), Stennis Space Center, Stennis, Mississippi. This program involved several workshops on monitoring and sensor requirements by the whole project team. P. J. Valent and D. K. Young headed the Naval Research Laboratory project team. Other Navy participants were F. Bowles, D. Bibee, P. Fleischer, B. Green, Maria Kalcic, and W. Sawyer. T. Nelson and J. Stamates represented the NOAA. Norman Francingues, Jr. represented the USACOE, Waterways Experiment Station. Contributors from SAIC include D. Carey, P. Dirnberger, M. Fritz, D. Rhoads, and J. Scott as well as SAIC subcontractors J. Aller and R. Aller (Marine Science Research Center, University of New York at Stony Brook). James Blake of ENSR helped in developing the biological impact model. N. Clesceri, H. Ehrlich, S. Komisar, and M. Leister of the Rensselaer Polytechnic Institute and R. Jahnke of the Skidaway Institute of Oceanography provided geochemical modeling. Specific sensor reviews were contributed by Y. Furukawa of the University of Southern Mississippi, Wm. Jones of the University of Maryland Biotechnology Institute, G. Rowe of Texas A&M University, A. Palowitch of Point Conception, and A. Yayanos of the University of California, Scripps Institute of Oceanography (SIO). A. Nehorai, University of Illinois at Chicago, and K. Smith, SIO, provided system architecture designs and sensor-array deployment optimization modeling.

Literature Cited

Aller, R. C., P. O. J. Hall, P. D. Rude, & J. Y. Aller. 1998. Biogeochemical heterogeneity and suboxic diagenesis in hemipelagic sediments of the Panama Basin. *Deep-Sea Research I* 45:133–165.

Germano, J. D. & D. C. Rhoads. 1984. REMOTS® sediment profiling at the Field Verification Program (FVP) disposal site, p. 536–544. *In* Montgomery, R. L. & J. W. Leach (eds.), Dredging and Dredged Material Disposal, v. 1. American Society of Civil Engineers, New York.

Hesslein, R. H. 1976. An in-situ sampler for close interval pore water studies. *Limnology and Oceanography* 21:912–914.

National Research Council (NRC). 1990. Managing troubled waters. The role of marine environmental monitoring. National Academy Press, Washington, D.C.

Pearson, T. H. & R. Rosenberg. 1978. Macrobenthic succession in relation to organic enrichment and pollution of the marine environment. *Oceanography and Marine Biology Annual Review* 16:229–311.

Rhoads, D. C. & S. Cande. 1971. Sediment profile camera for in situ study of organism-sediment relations. *Limnology and Oceanography* 16:110–114.

Rhoads, D., C., D. Carey, E. J. Saade, &. B. Hecker. 1997. Capabilities of laser line scan technology for aquatic habitat mapping and fishery resource characterization. Technical Report EL-97-7. US Army Corps of Engineers, Environmental Impact Research Program, Waterways Experiment Station, Vicksburg, Mississippi.

Rhoads, D. C. & J. D. Germano. 1982. Characterization of benthic processes using sediment imaging; An efficient method of Remote Ecological Monitoring of the Seafloor (REMOTS® System). *Marine Ecology Progress Series* 8:115–128.

Rhoads, D. C., P. L. McCall, & J. Y. Yingst. 1978. The ecology of seafloor disturbance. *American Scientist* 66:577–586.

Rhoads, D. C., J. A. Muramoto, & R. Ward. 1996. A review of sensors appropriate for assessment of submerged coastal habitats and biological resources. Technical Report EL-96-10. US Army Corps of Engineers, Environmental Impact Research Program, Waterways Experiment Station, Vicksburg, Mississippi.

Sanders, H. & R. R. Hessler. 1969. Ecology of the deep-sea benthos. *Science* 163:1419–1424.

Valent, P. J. & D. K. Young. 1995. Abyssal seafloor isolation environmental report. Navy Research Laboratory NRL/MR/7401-95-75-76.

Valent, P. J., D. K. Young, A. W. Green, & D. C. Rhoads. In press. Sensors and system architecture for monitoring dredged material relocation to the abyssal plain. Navy Research Laboratory (NRL) Report MR//7401-98-8212. Stennis Space Center, Mississippi.

In Situ Measurement of Organism-Sediment Interaction: Rates of Burrow Formation, Abandonment and Sediment Oxidation, Reduction

Robert J. Diaz and G. Randy Cutter, Jr.

Abstract: The development of sediment profile cameras made possible the collection of point data on in situ organism-sediment processes and observations of burrows, tubes, feeding voids, and other biogenic structures. To extend our understanding of biogenic structures and their rates of formation and abandonment and the effects of organisms on sediment geochemistry, we modified a sediment profile camera to collect a time-lapse series of images at one point. From two deployments (one for 18 d in the York River, Virginia, at an 18 m, muddy, meso-polyhaline transition site, and another for 2 d in Ware River, Virginia, at a 1 m, sandy, high mesohaline site) we found the average rate of burrow production to be 7 mm h^{-1} in mud and 4 mm h^{-1} in sand. However, for short periods the rates of burrow production could exceed 100 mm h^{-1} and burrow loss 74 mm h^{-1}. The primary burrower at both sites was <u>Nereis</u> <u>succinea</u>. Burrows constructed near the sediment surface in the tidal resuspension layer were reformed during periods of deposition and had a median life span of 5 h. During periods of surface sediment erosion the surface fauna extended their burrows downward. Burrows extended into anoxic sediments but were oxidized by active or passive ventilation. Oxidation of newly constructed burrows proceeded on the order of 0.3 to 2.0 h. The maximum extent of sediment oxidation away from burrow walls was on the order of 1–2 mm. <u>Nereis</u> burrows in muddy sediment were backfilled with material from the resuspension layer. In sand, however, their burrows appeared to collapse. Burrows of capitellid-like worms did not become oxic even hours after formation. After apparent abandonment, the capitellid-like burrows lasted from 3.4 h to 15.8 h. The occurrence of a hypoxic event at the muddy site did not appear to alter rates of burrow formation but did effect macrofaunal behavior. During hypoxia, the odds of a nereid worm being at or on the sediment surface was 18.8 times higher than during normal oxygen conditions.

Introduction

The role of macrofauna in influencing sediment physical and geochemical processes has long been known (Fager 1964; Aller 1978; Yingst & Rhoads 1978; Rhoads & Boyer 1982) but is not as well understood as the role sediment properties play in structuring benthic communities (Petersen 1913; Thorson 1957; Rhoads 1974; Snelgrove & Butman 1994). In particular the burrowing life style is important in geochemical exchanges across the sediment-water interface because there are so many burrowing species that perforate sedimentary "fabric." While most burrowers concentrate their

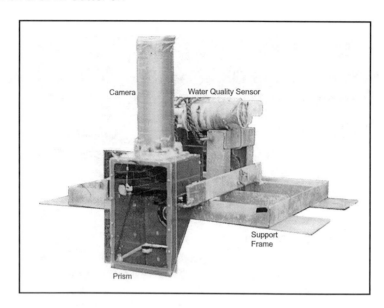

Fig. 1. Sediment profile camera configured for long-term time-lapse deployment on low-profile frame with water quality sensors.

activities in the top few centimeters of sediment, significant numbers penetrate deeper into the sediment (Rhoads 1967; Frey 1968). Aller (1978) was one of the first to quantify the importance of burrows to geochemical processes and found burrows to be key biogenic structures that influence many diagenetic reactions in sediments (Aller & Aller 1998).

Information on in situ subsurface activities of fauna (feeding, irrigation, burrow complexity, depth, construction rate, behavior) is difficult to obtain but is important to understanding processes such as sediment mixing and solute transport. Most work on form and extent of burrows in the field has been done by careful sectioning of core samples (Hertweck 1986), x-radiograph of thin-core sections (Schaffner 1990), or in situ resin casting (Meadows et al. 1990). Laboratory studies have relied primarily on observation through thin aquaria (Rosenberg et al. 1991) or x-rays (Charbonneau & Hare 1998) to reveal biogenic details. While all these techniques provide information on biogenic structures and faunal activities, there could be a problem with the methods disturbing natural sedimentary fabric or organism behavior, particularly in soft muddy sediments (Jones & Jago 1993). Also, only direct dissection or thin aquaria reveal fine structure, oxidation state, or animal behavior associated with burrowing actives.

With the development of sediment profile cameras, point data on in situ organism-sediment processes could be collected and burrows, tubes, feeding voids, and other biogenic structures observed (Rhoads & Cande 1971). To extend our understanding of biogenic activities, the rates of burrow formation and abandonment, and the effects of organisms on sediment geochemistry, we modified a sediment profile camera to collect a time-lapse series of images at a single point. We sampled a deep (18 m), soft mud bottom in York River and a shallow (1 m) sand bottom in Ware River, both tributaries of Chesapeake Bay. The site in York River had low polyhaline salinity during our study and experienced periodic hypoxia. The site in Ware River was high mesohaline.

Materials and Methods

A standard sediment profile camera (Hulcher Model Minnie) was converted into a time-lapse camera with the addition of an interval timer (Canon model TM-1). In time-lapse mode the camera could take 880 images of the same 15.5 x 25 cm sediment column. Set to take one image every 30 min and with 100 feet of film, the camera would run for about 18 d. The camera and prism were mounted to a low-profile aluminum frame to minimize flow disturbance and to prevent the camera from sinking into the bottom (Fig. 1). We used Fujichrome 100 Professional D slide film on both deployments. After development and mounting, each slide image was digitized for animation and measurement of size, shape, and position of biogenic structures and fauna. The entire time series was analyzed for the appearance of biogenic structures and oxidation state. Biogenic structures that had a lumen were considered to be burrows. Oxidation state of the sediment was determined from its color: brown sediment was considered oxidized and blue-gray sediment was considered anoxic (Fenchel 1969). Each structure or organism was indexed and followed through the time series with the assumption that a single individual was responsible for each burrow. Assumptions made when applying statistical tests for significance of odds ratios were that (1) each image was an unbiased two-dimensional representation of the burrower's habitats, (2) burrows were not connected, (3) each worm was an individual, and (4) every burrow in every image whether a worm was present or absent was a Bernoulli trial. Data were analyzed using SAS Proc Freq and Proc Logistic.

Image processing and measurement was accomplished using several techniques. Images were displayed using Adobe Photoshop and NIH Image. NIH Image and Object-Image (a spin-off of NIH Image) were used for digital measurement of the York River images and to create registered animations for the time-lapse series from the York and Ware rivers. Animating the series proved very useful for detecting and interpreting developments in biogenic features and changes in the sediments over time. Measurements for the Ware River images were made directly from projected images.

Image features characterized by visual and digital analyses included: number, name, and position of worms visible; number and identification of burrows present; maximum depth of burrow below the sediment-water interface; burrow wall condition; position of the burrow relative to anoxic sediments; burrow addition or loss; total burrow length and length of the open sections of burrows; presence or absence of the current indicator; relative concentration and size of material in suspension; number and type of epifauna; presence of detritus; and camera tilt angle after disturbance. The sediment-water interface contour and burrow paths were also traced through time to depict their relative changes.

Muddy Bottom, Deep Water Deployment

On July 3, 1996, divers placed the camera in 18 m of water on a soft mud bottom in the lower York River (76°30.20'W, 37°14.56'N). This was the location where Diaz et al. (1992) placed a telemetering water quality data buoy in 1989 to study the effects of periodic hypoxia on benthic fauna. The interval timer was set for one picture every 30 min. Between 1600 h and 1630 h on July 4 the camera and frame were disturbed by being lifted from the bottom and dropped. The problem was discovered when divers attached a water quality sensor on July 9. On July 10 the camera was redeployed. A second water quality sensor was added on July15. On July 21, 18 d later, the camera ran out of film and was retrieved. Prior to retrieval the sediment immediately in front of the prism was collected with a 10 x 15 x 15 cm core, washed through a 0.5-mm sieve, preserved, and all individuals were identified.

Synoptic water quality and depth data were logged from July 9 to the end of the deployment. Every 10 min, two Hydrolab DataSondes logged salinity, temperature, depth, and dissolved oxygen using sensors next to each other and 35 cm from the sediment surface. Current movement was to be recorded by using a lead weight hanging from a string in front of the prism window. The angle of deflection from vertical was to be used to estimate current speed and direction. However, the weight was too small, and therefore whenever there was current, it was out of the image. It was present during slack tides, and therefore used to gage flow versus no flow conditions.

SANDY BOTTOM, SHALLOW WATER DEPLOYMENT

On July 3, 1998, the camera was placed in 1 m of water on a sand bottom in the Ware River (76°26.82'W, 37°22.78'N). On July 6 the camera was retrieved. No water quality or benthic data were collected during this deployment.

Results

DEEP WATER DEPLOYMENT

The 18-d time series covered the interval July 3–21, 1996, and was divided into three parts. The first part prior to disturbance of the camera yielded 53 images at half-hour intervals. The second, following the disturbance yielded 277 images at half-hour intervals. The third, acquired after camera system redeployment, yielded 225 hourly interval images. On redeployment the system was disturbed, therefore images for the first 22 h were not analyzed.

Burrowing activity was detected within 11.5 h and 6.0 h of the initial deployment and camera disturbance, Part 1 and Part 2 of the series, respectively . Disturbance of sediment on redeployment, Part 3, did not allow determination of initial burrow development. During Part 1, the day prior to camera disturbance, four burrows were constructed (Table 1) and one worm was seen several times. The worm appeared to be a nereid. This was consistent with results of Neubauer (1993) and with the sediment core sample collected in front of the prism that yielded seven *Nereis succinea*. All except one of the burrows (A) appeared to be nereid structures. Burrow A lasted only 2 h and was constructed entirely within the oxic unconsolidated surface sediment layer, with a maximum depth below the sediment-water interface (SWI) of 16 mm. Burrow B was constructed by observed worm B, and was actively maintained for 8 h, reaching a maximum depth of 36 mm below the SWI during the period. Burrow C was a small structure extending 21 mm below the SWI. Burrow D was short-lived (1 h) and constructed within 7 mm of the SWI in the unconsolidated oxic sediment region. The maximum depths of these burrows were variable primarily because of changes in the elevation of the SWI. The surface layer was constantly in a state of flux, accreting and eroding, from a combination of tidal currents and bioturbation from fish and crabs (Fig. 2). Some changes in the SWI represent artifacts of the camera structure that altered flow and appeared to attract mobile fauna, and some changes represent actual interface dynamics.

Sections of burrows, except part of A, that were within the oxic zone had oxic walls for the life span of the burrow. The burrow segments constructed below the redox potential discontinuity (RPD) layer became oxidized at rates related to worm activity. When worm B was active (worm seen in the image), burrow walls below the RPD layer were oxic within the time interval between images (0.5 h), so oxidation may have occurred more quickly. When part of the burrow remained vacant (no worm in image) but still open, the walls below the RPD became oxic after 3 h. Sections of burrow

Table 1. Burrows formed during York River time-lapse deployment. The series ran for 440 h starting July 3 at 12:00 (0 h). Time present includes 0.5 h added to difference between first and last seen times to adjust for burrows that were only present in one image. Surface (burrow position) indicates that the burrow had connection with the sediment-water interface, subsurface indicates no connection detectable.

Name	Likely Taxon	Burrow Position	Burrow First Seen (h)	Burrow Last Seen (h)	Time Present (h)
		Part 1 of series started at 0 h			
A	??	Surface	11.5	12.5	1.5
B	Nereid	Surface	12.0	28.0	16.5
C	Nereid	Surface	17.0	28.0	11.5
D	Nereid	Surface	26.5	27.0	1.0
		Part 2 of series, started at 28.5 h			
E	Nereid	Surface	34.5	68.0	34.0
F	Nereid	Subsurface	76.5	77.0	1.0
G	Nereid	Subsurface	77.0	79.5	3.0
H	Nereid	Surface	79.0	101.0	22.5
EE	Nereid	Surface	91.5	92.5	1.5
FF	Nereid	Surface	98.5	101.0	3.0
GG	Nereid	Surface	104.0	105.5	2.0
HH	Nereid	Surface	108.0	108.5	1.0
J	Nereid	Surface	110.5	116.5	6.5
KK	Nereid	Surface	114.5	114.5	0.5
K	Nereid	Surface	118.0	165.0	47.5
LL	Nereid	Surface	124.5	129.0	5.0
MM	Nereid	Surface	133.0	161.0	28.5
NN	Nereid	Surface	154.5	155.0	1.0
L	Nereid	Subsurface	164.5	169.0	5.0
		Part 3 of series, started at 204 h			
M	Nereid	Surface	204.0	440.0	236.5
PP	Nereid	Surface	204.0	204.0	0.5
QQ	Nereid	Surface	212.0	215.0	3.5
Q	Nereid	Surface	244.0	288.0	44.5
R	Nereid	Surface	244.0	327.0	83.5
S	Nereid	Surface	306.0	315.0	9.5
T	Nereid	Surface	335.0	380.0	45.5
N	Nereid	Surface	402.0	417.0	15.5
P	Nereid	Surface	440.0	440.0	0.5

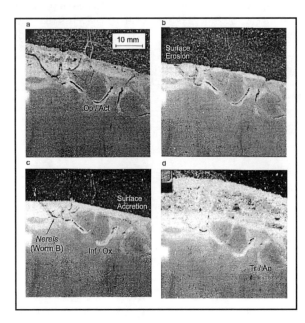

Fig. 2. Images from York River, Virginia, deployment. (a) Open and active (Op/Act) burrow segments of worm B, time 0400; (b) Surface sediments eroded by tidal currents, time 0630; (c) Worm B present near sediment surface, burrow segment to the right is infilling but still oxic (Inf/Ox), surface sediments are accreting, time 1000; (d) Infilled burrow segment is in transition to anaerobic (Tr/An), accreted sediment has been burrowed by worm B, time 1300. Animation of this sequence can be found at http://www.vims.edu/~cutter/benthicecol.html

B infilled and eventually disappeared as walls and infilled sediments gradually became anoxic and reverted to the color of background sediments, but other parts of B remained oxic until the camera was disturbed (Fig. 2).

In the 16-h interval following the appearance of the first burrow, the four burrows observed averaged a total length of 128 mm per image (SD = 39) with a maximum of 210 mm (Table 2). The rate of burrow production, defined as the change in total burrow length per time, during this period had a low overall average of 7 mm h^{-1} (SD = 43). Partitioned into net gains and losses, burrow production averaged 36 mm h^{-1} (SD = 41) and 22 mm h^{-1} (SD = 22) respectively. For short periods the maximum burrowing gains occurred at rates of about 100 mm h^{-1} and losses about 70 mm h^{-1} (Table 2). The ratio of total length of open burrow segments to total burrow lengths ranged from 0.5 to 1.0 (Table 2). Burrow life spans (time from their appearance to when they were undetectable) ranged from 0.5 h to 236 h, with a median life span of 5.0 h. The average burrow life span of 22.6 h (SD = 46.5) was skewed by two burrows (M and R) that were present for much of the third part of the time series (Table 1).

Burrows went through two phases during their life spans, an active phase and an abandonment phase, both related to worm activity or lack of activity. For example, for 10 h after initial observation, worm K was seen occupying its burrow. K was last observed 10.5 h after initial observation in a secondary branch. One hour later infilling of the apparently abandoned burrow with surficial sediments began. Complete infilling occurred 19.5 h later, 30 h into the life of the burrow. The burrow was still evident after infilling because the hues belonging to the recently oxidized sediments contrasted the hues of the surrounding anoxic sediments. The hues of the infilled burrow sediment virtually matched those of the anoxic subsurface sediments by 25.5 h after abandonment. However, part of the burrow feature extended deeper into darker blue-gray sediments, and was still visually distinct from its surroundings 41.5 h after initial observation. After 47.5 h the life of burrow K was considered ended even though traces of the deepest parts of the burrow were still visible against surrounding sediments when the camera was retrieved for maintenance and repositioning.

Bottom temperature ranged from 23°C to 26°C and salinity from 14 psu to 25 psu during the deployment. From July 9 to July 12 the study site was periodically subjected to hypoxia (DO < 2 mg l^{-1}), but on July 12 Hurricane Bertha arrived in the study area which resulted in the reoxygenation of

Table 2. Summary of burrowing activity for Part I of York River deployment. First burrow appeared 11.5 h after the camera was deployed. Burrow production is the increase or decrease in total burrow length per hour. Proportion of burrow open is the ratio of open burrow length, not infilled, to total length.

Time (h)	Number of burrows	Total length of burrows (mm)	Burrow production (mm h^{-1})	Proportion of burrow open
11	0	0	.	.
12	2	95	95	1.00
13	2	198	104	0.91
14	2	194	-4	0.87
15	2	210	16	0.63
16	2	136	-74	0.87
17	3	123	-13	0.79
18	3	113	-10	0.82
19	3	123	9	0.81
20	3	96	-27	0.76
21	3	101	5	0.52
22	3	107	6	0.82
23	3	83	-24	0.74
24	3	123	39	0.80
25	3	113	-9	0.74
27	4	123	5	0.66
28	4	111	-12	0.67

bottom waters. Dissolved oxygen remained high until late on July 19 when another hypoxic event started; this event continued to the end of the deployment (Fig. 3). Oxidation state of the burrows did not seem to be affected by hypoxia; however, nereid worm burrowing activity and behavior was affected. During hypoxia, burrowing activity was concentrated in the top 1–2 cm of surface sediments; prior to hypoxia, burrowing activity extended down to at least 6 cm. The occurrence of worms extending their bodies into the water column or laying on top of the sediment significantly increased, with the odds ratio of a worm being present at or on the surface during hypoxia being 18.8 (95% confidence interval of 8.2 to 42.6).

SHALLOW WATER DEPLOYMENT

A total of 171 images were taken at a 20-min interval from 17:37 EST on July 3 to 05:57 on July 6, 1998. Over this 57-h period subsurface macrofauna constructed 25 burrows (Table 3). While no faunal sample was collected, the two dominant burrowing species in size and number in the Ware River are *Nereis succinea* and *Heteromastus filiformis* (personal observations). Sediment grain size was uniform fine sand as determined from the sediment profile images.

A small anaerobic burrow formed within the first 20-min of the deployment and after an hour four burrow structures were present. Burrowing activity peaked 8–18 h after deployment, with the majority of burrows formed during this interval (Table 3). Nereid worms remained close to the sediment surface, maximum burrowing depth was 2.5 cm. Capitellid worms ranged from 1.4 cm below the sediment surface down to 8 cm (Table 4). The presence of worms in images was significantly associated with daylight hours but declined with each day. The odds ratio of a worm being present during daylight, controlling for day, was 3.6 (2.4–5.3, 95% confidence interval) over night-time presence. The odds declined significantly by 0.5 (0.4–0.6, 95% CI) for each consecutive day.

The sediment was disturbed by a blue crab (*Callinectes sapidus*) that started digging a pit 5 h after deployment (Fig. 4). The crab eventually dug a 9-cm-deep pit, which was still present at the end of the deployment, on the right side of the prism window. Burrowing organisms were not able to keep up with the crab disturbance and were lost from view during pit excavation. The crab dug up the burrows of six worms, but it could not be determined if the worms were eaten.

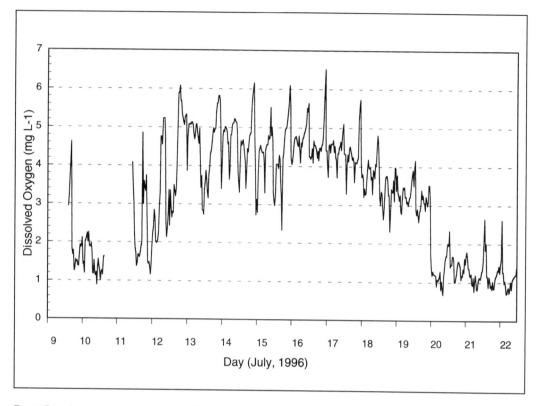

Fig. 3. Dissolved oxygen concentrations from the 18-m muddy site in the York River, Virginia.

Wind-generated wave turbulence alternately eroded and deposited sediment in front of the prism window. Over the first 24 h the sediment surface was eroded 7–10 mm across the 10-cm width of the prism window not affected by the crab pit and 6–20 mm during the second 24 h. Burrowers were able to keep up with surface erosion from waves by extending burrow networks downward. Nereid D constructed a series of short looping burrows as surface sediments eroded downward (Fig. 4). Over the 57 h, worm D constructed a total of 270 mm of burrow with the average burrow length per image being 32 mm (SD = 15 mm) and average depth of burrow being 11 mm (SD = 5). The deepest and fastest burrowers appeared to be capitellids. Worm K burrowed 54 mm in two successive 20-min intervals to a sediment depth of 81 mm. The range of burrow depths for other capitellids and nereids was 18–50 mm and 3–25 mm, respectively (Table 4).

Burrows were of two basic types, those that became oxidized and those that remained anoxic. The oxic burrows were larger in diameter, likely created by nereids, and once formed tended to last through the time series until eroded away. Rate of oxidation varied from < 20 min, with burrows appearing oxic on first image occurrence, to 1.3 h for first traces of oxic sediment to appear (Table 3). Oxidation of worm L's burrow was particularly interesting because it proceeded from the deep end of the burrow to the surface. After contacting the prism faceplate at about 4-cm sediment depth (Fig 4), worm L burrowed to the sediment surface. Within 0.3 h of reaching the surface, oxidation of the burrow proceeded from the bottom of the burrow toward the surface. After 1.3 h the entire burrow was oxic. Active ventilation by worm L likely drew oxygenated water from other sections of

Table 3. Burrows formed during Ware River time-lapse deployment. The series ran for 57 h beginning July 3 at 17:37 EST (0 h). For Burrow Position, surface indicates that the burrow had connection with the sediment-water interface, subsurface indicates no connection detectable. Time Present includes 0.3 h added to the difference between the first and last seen times to adjust for burrows that were only present in one image. Time to Oxic is the interval from burrow first seen to the formation of brown oxidized sediment particles in the burrow. Worm Last Seen is the hour the worm was last observed in the images: always = worm always present, never = worm was never observed in the burrow. Time to Vanishing is the time interval from worm last seen to disappearance of the burrow (burrow last seen). Cause of vanishing was either crab burrowing activity, surface sediment erosion, or collapse of the burrow.

ID	Likely Taxon	Burrow Position	Burrow First Seen (h)	Burrow Last Seen (h)	Time Present* (h)	Time to Oxic (h)	Worm Last Seen (h)	Time to Vanishing (h)	Cause of Vanishing
J	Nereid	Surface	0.3	56.3	> 56.3	1.0	always		present at end
D	Nereid	Surface	0.3	56.3	> 56.3	1.0	always		present at end
G	Nereid	Surface	7.7	14.7	7.3	0.3	8.4	6.7	crab
X	Nereid	Surface	7.7	17.0	9.7	0.3	17.0	0.2	crab
B	Nereid	Surface	9.7	19.7	10.3	3.0	always		crab
T	Nereid	Surface	12.0	17.7	6.0	0.3	always		crab
Y	Nereid	Surface	16.0	22.3	6.7	0.3	never		erosion
R	Nereid	Surface	23.7	56.3	> 33.0	1.3	always		present at end
A	Nereid	Surface	54.3	56.3	> 2.3	0.7	always		present at end
S	Nereid	Surface	26.3	56.3	> 30.3	NO	always		present at end
L	Nereid	Surface	50.3	56.3	> 6.3	0.3	always		present at end
E	??	Surface	0.3	2.3	2.7	NO	never		erosion
Z	??	Surface	1.3	1.3	0.3	NO	never		crab
H	??	Surface	3.0	4.3	1.7	NO	never		erosion
N	??	Surface	3.0	3.0	0.3	NO	always		erosion
V	Nereid	Subsurface	2.3	4.3	2.3	1.0	always		crab
C	Capitellid	Subsurface	0.0	4.3	4.7	NO	3.4	1.4	collapse
K	Capitellid	Subsurface	2.3	24.7	22.7	NO	12.0	13.0	collapse
F	Capitellid	Subsurface	2.7	20.0	17.7	NO	12.7	7.7	collapse
I	Capitellid	Subsurface	7.0	19.6	13.0	NO	13.9	6.0	collapse
M	Capitellid	Subsurface	7.7	25.0	17.7	NO	9.6	15.8	collapse
P	Capitellid	Subsurface	14.0	17.7	4.0	NO	14.6	3.4	collapse
O	Capitellid	Subsurface	53.7	56.3	> 3.0	NO	54.0		present
Q	??	Subsurface	3.0	4.0	1.3	NO	never		erosion
U	??	Subsurface	9.7	10.0	0.6	NO	always		collapse

* time > value indicates that the burrow was still present when the series ended
NO = never oxic

Table 4. Length and depth statistics for burrows from Ware River time-lapse deployment. Taxa are nereid-like (N), capitellid-like (C), or unknown (?). N is the number of images (taken every 20 min) in which the burrow was seen. Min is minimum and Max is maximum length or depth measured. Burrow depths were measured from the sediment surface of each image and include changes due to erosion and redeposition. SD is the standard deviation.

ID	Taxa	N	Burrow Length				Burrow Depth			
			Min (mm)	Max (mm)	Mean (mm)	SD (mm)	Min (mm)	Max (mm)	Mean (mm)	SD (mm)
S	N	42	14	53	34.3	8.7	-10	-19	-15.1	1.9
D	N	111	4	66	32.0	14.7	-3	-21	-11.2	4.7
J	N	112	12	61	26.5	9.6	-6	-21	-11.1	2.9
A	N	3	23	26	24.7	1.5	-11	-11	-11.0	0.0
R	N	20	10	25	22.5	4.0	-6	-10	-8.0	1.5
V	N	7	15	25	19.1	3.7	-13	-24	-21.6	4.0
G	N	22	7	28	14.0	4.4	-12	-25	-17.6	3.4
B	N	9	7	17	10.0	2.9	-19	-23	-20.9	1.2
X	N	29	4	11	7.4	1.7	-4	-11	-7.9	1.9
T	N	18	3	7	4.7	1.1	-3	-6	-4.7	0.9
K	C	61	22	71	66.2	8.6	-55	-81	-76.0	3.3
F	C	46	4	58	51.8	11.6	-23	-46	-39.9	6.1
M	C	53	45	45	45.0	0.0	-35	-40	-36.0	0.7
O	C	4	7	19	13.5	5.0	-14	-14	-14.0	0.0
I	C	39	9	20	12.6	2.4	-39	-50	-41.2	3.1
C	C	14	5	8	6.1	0.8	-18	-19	-18.4	0.5
P	C	12	4	6	4.8	0.6	-26	-31	-28.3	1.4
N	?	7	12	42	36.6	10.9	-38	-41	-39.4	1.3
H	?	5	6	28	22.5	9.3	-8	-11	-9.2	1.3
E	?	8	5	6	5.9	0.4	-7	-8	-7.7	0.5
Q	?	4	6	7	6.5	0.6	-15	-20	-16.5	2.4

its burrow gallery to cause this pattern of oxidation. All other burrows in both sand and mud appeared to oxidize from the surface down. Anoxic burrows were formed by smaller free-burrowing worms, probably capitellids, and tended to disappear within hours of the worm leaving the faceplate.

Discussion

BURROWING

The overall rate of burrow production (gains minus losses) was low in both habitats, 7 mm h^{-1} in deep muddy sediments and 4 mm h^{-1} in shallow sandy sediments for the first 24-h period. However, these low rates do not reflect the actual activity of the worms. Rather, they depict a time-averaged, near-equilibrium condition where burrow excavation is balanced by infilling, collapse, and loss to erosion. Thus the calculated rates suggest that worms were active elsewhere, not just in the

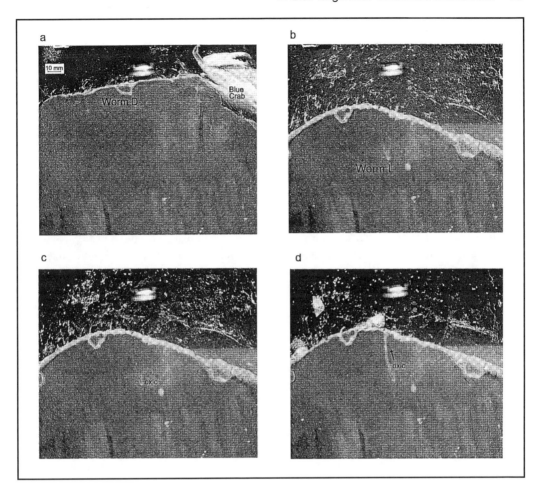

Fig. 4. Images from Ware River, Virginia, deployment. (a) After 9.3 h, blue crab digs a pit, looping the burrow of worm D near the center of the image; (b) after 49.3 h, worm L first appears, the burrow of worm D has shifted to the left about 20 mm; (c) After 49.7 h, worm L has burrowed to the sediment surface, but the burrow is only oxic at the bottom; (d) After 53.0 h, worm L has oxygenated its burrow with a ventilating current of water from somewhere within its burrow gallery. The burrow was not oxygenated from the opening above. Animation of this sequence can be found at http://www.vims.edu/~cutter/benthicecol.html

segments of the burrows observed, and that burrows were continually subject to destructive forces. Slightly lower excavation rates would have led to an equilibrium state where gain equaled loss and burrow development would appear to be zero; however, such a case would be an artifact of time-averaging episodic events. The standard deviation of burrow production provided a measure of how dynamic burrowing was, where gains and losses did not negate one another. In both deep mud and shallow sand, the standard deviation of burrow production was about 42–43 mm h^{-1}. For the deep, muddy site, two periods of large increases in burrow length and one period of large losses heavily influenced the SD during the first day. At the shallow sandy site, gain and loss of burrow length was about uniform with time.

The ratio of open burrow segment length to total burrow length depicted the degree of recent infaunal activity. When worms were active, the ratio was high (close to 1), and burrow segments were kept open through movement and clearance of material. As worm activity declined, burrows infilled and collapsed, and the ratio was low. Burrowers did keep pace with the eroding sediment surface in both sandy and muddy areas. As the surface was eroded, burrows were built deeper relative to older structures but at about the same depth below the SWI. Burrows constructed near the sediment surface in the tidal resuspension layer typically collapsed or were lost to erosion when tidal currents were high; however, they were reformed during periods of slack current. The life span of shallow surface burrows was short: a median of 5 h in mud and 7 h in sand, and dependent upon burrow position relative to the sediment-water interface and worm activity.

Oxidation of newly constructed burrows could proceed quickly, from less than the time interval of the images in both habitats, to 1.6 h in sandy sediment and 3.0 h in muddy sediment. Rate of burrow oxidation was dependent upon a combination of worm activity and physical irrigation. The maximum extent of sediment oxidation away from burrow walls was on the order of 1–2 mm in both mud and sand. The nereid burrows in mud appeared to remain oxidized on the order of days after abandonment. The sand time series was not long enough to follow the oxidation state of abandoned nereid burrows. After abandonment of burrows by capitellid worms, burrows in sand lasted 1.4–15.8 h. In mud, the average life span of a burrow was 24 h; however, burrow life spans ranged from 0.5 h to > 264 h.

The importance of burrows to sediment diagenetic processes is determined primarily by two factors: (1) how long the burrow remains open, which is a function of the occupant's activities (Aller & Aller 1998), and (2) how long it takes anaerobic processes to reduce oxidized sediments that infill abandoned burrow structures. The oscillation of redox conditions within burrows, which Aller and Aller (1998) found to stimulate diagenetic reactions, did occur in both muddy and sandy sediments. However, different burrowing behavior may also contribute to different reaction rates. We found that in addition to maintaining burrow galleries, worms (1) developed new burrows at an average rate of 4–7 mm h^{-1} and for short periods exceeded 100 mm h^{-1}, and (2) varied activity level within the three-dimensional geometry of their galleries, based on the assumption that when the worm was not visible in an image it was occupying another part of its burrow gallery.

HYPOXIA

The onset of hypoxia during the last 2 d of York River deployment did not appear to alter rates of burrow formation. Seasonal hypoxia, typically lasting on the order of months, has been found to reduce biogenic mixing of sediments (D'Andrea et al. 1996), but the effects of shorter hypoxic events on sediment mixing is unknown.

Infaunal behavior was affected by hypoxia. A nereid worm was about 18 times more likely to be at or on the sediment surface during hypoxia versus normoxia. This is typical macroinvertebrate behavior in response to hypoxia (Diaz & Rosenberg 1995), but mobile predators (toad fish, gobie, blue crab, grass shrimp), which typically avoid hypoxic bottom water (Pavela et al. 1983; Pihl et al. 1991; Breitburg 1992), were also observed. When predators are in hypoxic water their feeding behavior ceases or occurs at rates lower than in normoxia (Baden et al. 1990). However, predators are quick to recover from hypoxia and are known to prey on hypoxia-stressed infauna (Pihl et al. 1992; Nestlerode & Diaz 1998). A longer time series that encompassed the beginning and end of a hypoxic event would be needed to provide further insight into the animal-animal and animal-sediment interactions during hypoxia.

Acknowledgments

Contribution 2333 of the Virginia Institute of Marine Science.

Literature Cited

Aller, R. C. 1978. Experimental studies of changes produced by deposit feeders on pore water, sediment and overlying water chemistry. *American Journal of Science* 278:1185–1234.

Aller, R. C. & J. Y. Aller. 1998. The effect of biogenic irrigation intensity and solute exchange on diagenetic reaction rates in marine sediments. *Journal of Marine Research* 56:905–936.

Baden, S. P., L. O. Loo, L. Pihl, & R. Rosenberg. 1990. Effects of eutrophication on benthic communities including fish – Swedish west coast. *Ambio* 19:113–122.

Breitburg, D. L. 1992. Episodic hypoxia in Chesapeake Bay: Interacting effects of recruitment, behavior, and physical disturbance. *Ecological Monographs* 62:525–546.

Charbonneau, P. & L. Hare. 1998. Burrowing behavior and biogenic structures of mud-dwelling insects. *Journal of the North American Benthological Society* 17:239–249.

D'Andrea, A. F., N. I. Craig, & G. R. Lopez. 1996. Benthic macrofauna and depth of bioturbation in Eckernfoerde Bay, southwestern Baltic Sea. *Geo-Marine Letters* 16:155–159.

Diaz, R. J. & R. Rosenberg. 1995. Marine benthic hypoxia: A review of its ecological effects and the behavioural responses of benthic macrofauna. *Oceanography and Marine Biology Annual Review* 33:245–303.

Diaz, R. J., R. J. Neubauer, L. C. Schaffner, L. Pihl, & S. P. Baden. 1992. Continuous monitoring of dissolved oxygen in an estuary experiencing periodic hypoxia and the effect of hypoxia on macrobenthos and fish. *Science of the Total Environment*, Supplement 1992, p. 1055–1068.

Fager, E. W. 1964. Marine sediments: Effects of a tube-building polychaete. *Science* 143:356–359.

Fenchel, T. 1969. The ecology of marine microbenthos. IV. Structure and function of the benthic ecosystem, its chemical and physical factors and microfauna communities with special reference to the ciliated Protozoa. *Ophelia* 6:1–182.

Frey, R. W. 1968. The lebensspuren of some common marine invertebrates near Beaufort, North Carolina. I. Pelecypod burrows. *Journal of Paleontology* 42:570–574.

Hertweck, G. 1986. Burrows of the polychaete *Nereis virens* Sars. *Senckenbergia Maritima* 17:319–331.

Jones, S. E. & C. F. Jago. 1993. In situ assessment of modification of sediment properties by burrowing invertebrates. *Marine Biology* 115:133–142.

Meadows, P. S., J. Tait, & S. A. Hussain. 1990. Effects of estuarine infauna on sediment stability and particle sedimentation. *Hydrobiologia* 190:263–266.

Nestlerode, J. A. & R. J. Diaz. 1998. Effects of periodic environmental hypoxia on predation of a tethered polychaete, *Glycera americana*: Implication for trophic dynamics. *Marine Ecology Progress Series* 172:185–195.

Neubauer, R. J. 1993. The relationship between dominant macrobenthos and cyclical hypoxia in the lower York River. Masters Thesis, College of William and Mary, Williamsburg, Virginia.

Pavela, J. S., J. L. Ross, & M. E. Chittenden, Jr. 1983. Sharp reductions in abundance of fishes and benthic macroinvertebrates in the Gulf of Mexico off Texas associated with hypoxia. *Northeast Gulf Science* 6:167–173.

Petersen, C. G. J. 1913. Valuation of the sea. II. The animal communities of the sea-bottom and their importance for marine zoogeography. *Report of the Danish Biological Station to the Board of Agriculture* 21:1–44.

Pihl, L., S. P. Baden, & R. J. Diaz. 1991. Effects of periodic hypoxia on distribution of demersal fish and crustaceans. *Marine Biology* 108:349–360.

Pihl, L., S. P. Baden, R. J. Diaz, & L. C. Schaffner. 1992. Hypoxia-induced structural changes in the diet of bottom-feeding fish and crustacea. *Marine Biology* 112:349–361.

Rhoads, D. C. 1967. Biogenic reworking of intertidal and subtidal sediments in Barnstable Harbor and Buzzards Bay, Massachusetts. *Journal of Geology* 75:461–476.

Rhoads, D. C. 1974. Organism sediment relations on the muddy sea floor. *Oceanography and Marine Biology Annual Review* 12:263–300.

Rhoads, D. C. & L. F. Boyer. 1982. Effects of marine benthos on physical properties of sediments. A successional perspective, p. 3–51. *In* McCall, P. L. & M. J. S. Tevesz (eds.), Animal-Sediment Relations. Plenum Press, New York.

Rhoads, D. C. & S. Cande. 1971. Sediment profile camera for in situ study of organism-sediment relations. *Limnology and Oceanography* 16:110–114.

Rosenberg, R., B. Hellman, & B. Johansson. 1991. Hypoxic tolerance of marine benthic fauna. *Marine Ecology Progress Series* 79:127–131.

Schaffner, L. C. 1990. Small-scale organism distributions and patterns of species diversity: Evidence for positive interactions in an estuarine benthic community. *Marine Ecology Progress Series* 61:107–117.

Snelgrove, P. V. R. & C. A. Butman. 1994. Animal-sediment relationships revisited: Cause versus effect. *Oceanography and Marine Biology Annual Review* 32:111–177.

Thorsen, G. 1957. Bottom communities (sublittoral or shallow shelf). *Geological Society of America Memoir* 67, p. 461–534.

Yingst, J. Y. & D. C. Rhoads. 1978. Sea floor stability in central Long Island Sound. Part II. Biological interactions and their potential importance for seafloor erodibility, p. 245–260. *In* Wily, M. A. (ed.), Estuarine Interactions. Academic Press, New York.

Reflections on Statistics, Ecology, and Risk Assessment

Joseph D. Germano

Abstract: *Despite all the advances in theoretical ecology over the past four decades and the huge amounts of data that have been collected in various regional marine monitoring programs, we still do not know enough about how marine ecosystems function to be able to make valid predictions of impacts before they occur, accurately assess ecosystem "health," or perform valid risk assessments. The current struggle to establish sediment quality criteria and the crises in fisheries management in the northwest Atlantic on both Georges Banks and the Grand Banks are excellent examples of our lack of knowledge and predictive power concerning ecosytem dynamics. One of the biggest impediments to the interpretation of ecological data and the advancement of our understanding about ecosystem function is the desire of marine scientists and policy regulators to cling to the ritual of null hypothesis significance testing (NHST) with mechanical dichotomous decisions around a sacred 0.05 criterion. Ecologists would benefit from abandoning statistical significance testing and take heed of lessons learned in the fields of decision theory, clinical psychology, and medical epidemiology to structure a new approach for the emerging field of ecological risk assessment.*

Introduction

With the emergence of ecological risk assessment as a new field of study in applied ecology in the 1990s, a wide variety of researchers, consultants, and regulators hopped on the bandwagon in the hopes that this new discipline would provide some quantitative, objective insights to the diagnosis of ecosystem health. While a variety of textbooks (e.g., Calabrese & Baldwin 1993; Suter 1993), regulatory guidance (USEPA 1998), and periodicals (e.g., *Environmental Toxicology and Chemistry; Human and Ecological Risk Assessment*) are available for both instruction and information exchange, some investigators in this field are starting to voice discomfort with some of the current practices in ecological risk assessment (Power & McCarty 1997).

My own evaluation of ecological risk assessments, based on actual studies as well as published books and articles, has elicited the same vague discomfort I have experienced in the field of marine ecology from performing benthic community monitoring and contaminated sediment assessments. This discomfort, simply stated, is that despite the huge amounts of data that have been collected in various regional marine monitoring programs, we still do not know enough about how marine ecosystems function to be able to make valid predictions of impacts before they occur, accurately assess ecosystem "health," or perform valid risk assessments. The current situation with the decimation of commercial fishing stocks in the northwest Atlantic Ocean on both Georges Banks and the Grand Banks is a good example of our lack of knowledge or predictive power concerning ecosystem

33

dynamics. As an environmental consultant, I can derive some comfort from knowing that our inabilities in these areas will provide long-term job security. However, repeated trips down well-worn paths, which have provided little insight in the past (e.g., collection and analysis of more benthic samples, repeated bioassay tests with unknown relevance to actual field impacts), have made me question the wisdom of continuing to support the status quo.

During the past eight years, most of the scientific papers I have read have been from clinical psychology journals, and they have served as a springboard for further readings in the fields of statistics, decision theory, and expert judgment. I have been overwhelmed by the similarity among the fields of psychology, social science, and ecology in terms of the applications of decision theory or approach to problem diagnosis. In all three disciplines, researchers are dealing with phenomena whose mechanisms are poorly understood. Just as there are no known universal hard and fast "laws of nature" in psychology or social science, we are operating in a similar void as far as our understanding of how marine ecosystems function. It is extremely difficult for a psychologist or psychiatrist to predict the occurrence of mental disease or correctly diagnose character and personality disorders and then prescribe the most effective course of treatment (particular type of therapy? specific type of medication? both?). It is equally difficult for a marine scientist to predict when the next algal bloom or red tide will occur or diagnose ecosystem health and prescribe the most effective remediation treatment. One of the main impediments to progress in environmental resource management is that the classical statistical models that most of us were taught in graduate school have proved to be of little use in providing insight into ecosystem process or function. Ecologists would do well to take heed of lessons learned in the fields of decision theory, clinical psychology, and medical epidemiology to structure a new approach to ecological risk assessment. While the details of this new approach are outlined elsewhere (Germano 1999), I will frame the problem and provide a brief overview of some alternative approaches for ecologists to improve their diagnostic accuracy.

Initial Premise

Because we do not have a valid model for how marine ecosystems truly function, scientists have been gathering data for the past century trying to gain such an insight. I think one of the biggest impediments to the interpretation of ecological data and the advancement of our understanding about ecosystem function has been the desire of marine scientists (biologists, ecologists, chemists, toxicologists, etc.) and policy regulators to cling to the ritual of null hypothesis significance testing (NHST) with "mechanical" dichotomous decisions around a sacred 0.05 criterion. The continued blind application and misinterpretation of the "Fisherian" school of statistics (which all of us were taught in graduate school) appears to have stifled or limited our understanding of complex systems; psychologists and social scientists recognized these limitations years ago (e.g., Berkson 1938; Bakan 1966; Carver 1978), but few marine ecologists are aware of them (see Germano 1999 for more details).

The standard statistical tests that most ecologists and risk assessors employ are a small subset of what is available; Fig. 1 (from Raiffa 1968) presents a taxonomic overview of the field of statistics. There are three main branches to Fig. 1—Data Analysis, Decision Approaches, and Inference Approaches—and the key proponents for each approach are identified at the terminal ends of the branches.

The first step with which the ecologist or risk assessor is typically faced is that of Data Analysis. The task at hand for those venturing down this branch is not to draw inferences from the subsample to the population at-large, but instead to glean insights and explore relationships that help structure

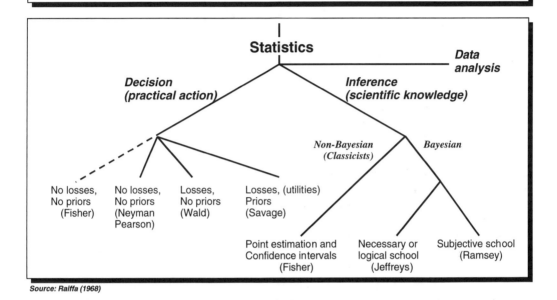

An Overview of Statistics

Source: Raiffa (1968)

Figure 1. A taxonomic overview of the field of statistics.

the data. Bringing order to chaos or finding a pattern or relationship in what first appears as disconnected parts is a necessary first step to lead one from the world of nightmare to the world of knowledge (Hutchinson 1953). The best guide I can recommend for this voyage is Tukey (1977).

The remaining statistical problems can usually be assigned to either of the remaining branches, depending on whether or not they are problems of *decision*, where the alternatives can be listed and the consequences can be evaluated, or problems of *inference*, where questions like "What do these data say? What have we learned? What should we now think?" are addressed (see Raiffa [1968] for a discussion of all branches in Fig. 1). Ecologists and risk assessors need to focus on the Data Analysis and the Inference branches; where most of us have gone astray is to proceed directly to the Inference branch and take a blind turn down the Classicists or Fisherian terminus without even being aware of the Bayesian branch. This has stifled rather than promoted insights in the field.

The Problem with Statistical Significance Testing

The main problem, simply stated, is that statistical significance does not tell us what we want to know. Because we are desperately trying to find out or prove what we want to know, we either ignore or misunderstand what NHST does and instead think it is telling us exactly what we want. What we are constantly trying to find out either through our research or monitoring studies is "Given these data, is my research hypothesis (H_1) true?" Most of us will recall that the p-value from a statistical test (such as the t test or F test) is a probability or the proportion of the time we can expect to find mean differences as large as or larger than the difference we get when sampling from the same population assumed under the null hypothesis. So, to re-phrase our initial underlying desire, what we would like to know from our statistical significance test is "Given these data, what is the probability that my research hypothesis (H_1) is true?" In our more sophisticated moments, we may realize that the statistical tests we are performing are testing the null hypothesis (H_o), so we may unconsciously think the p-value is telling us, "Given these data, what is the probability that the null hypothesis (H_o) is true?" However, what the p-value is telling us is "Given that H_o is true, what is the probability of these (or more extreme) data?" The following four questions are *not* equivalent, and *all* that the p-value indicates from NHST is a response to the fourth question:

1. Given these data, is my research hypothesis (H_1) true?
2. Given these data, what is the probability that my research hypothesis (H_1) is true?
3. Given these data, what is the probability that the null hypothesis (H_o) is true?
4. Given that the null hypothesis (H_o) is true, what is the probability of these (or more extreme) data?

The implications of an investigator designing experiments or interpreting data while confusing these four questions are not trivial. This is only one of the four major fantasies about statistical significance testing (Germano 1999). Statistical significance (our convention of $p \leq 0.05$) simply means statistical rareness (Carver 1978). Results are considered "significant" because they would occur rarely in random sampling from a population under the conditions of the null hypothesis. In other words, a statistically significant result means the probability is low we would get the result obtained *given that the null hypothesis is true* (Question 4).

Given this stark reality, statistical significance, by itself, means little or nothing; for the majority of applied environmental monitoring projects where statistical tests are employed, deep down we do not believe for a minute that the null hypothesis is really true, for example, an H_o that states "There is no difference in sediment PAH concentration at the disposal site and the reference site." As Bakan (1966) pointed out, a quick glance at any set of statistics on total populations would confirm the rarity of the null hypothesis in nature. Even if we are operating under conditions where H_o is true, statistically significant but meaningless results can always be obtained if one takes enough samples; a recent monitoring report submitted to our company for review cited "significant differences" in seawater pH between sampling surveys as a major finding in the conclusions (pH of 8.1 versus 8.3). Carver (1978) stated in his excellent review article that real problems occur when statistical significance testing is used to make inferences, a trend all too common these days in the world of environmental data analysis, such as is done in the interpretation of bioassay results or in the continuing search for sediment quality criteria.

It is not at all surprising that ecologists and risk assessors have had a difficult task in attempting to diagnose ecosystem health accurately. Environmental problems are "sticky" (Bardwell 1991), and there are a variety of impediments that ecologists face in their search for valid predictors of environmental health (Germano 1999). A good example of one such impediment is the bias of preconceived notions: when an investigator sets out to test a hypothesis or search for environmental impacts, more often than not they seek confirmatory evidence exclusively (Chapman & Chapman 1967; Nisbett & Ross 1980). Any nonconfirming evidence, when obtained, is either ignored, under weighted, or dismissed (Faust 1986b), giving both a biased research design that will heavily favor the investigator's desired outcome and a distorted view of reality.

Because most researchers and regulators are unaware of the impacts these impediments have on the strength of any conclusions that have been reached, it is difficult to envision any reason to do things differently. Fortunately, there are a variety of approaches that have been discovered by investigators in the fields of psychology and decision theory that provide some hope for a new way of assessing environmental risk and understanding ecosystem dynamics.

A Potential Solution

While there are six suggested methods (Germano 1999) that will help ecologists improve their diagnostic accuracy of environmental health and facilitate greater insights about how ecosystems really do function so that predictive assessments can be made, one of the most promising methods for avoiding the pitfalls of NHST is the application of Bayesian statistics. I would refer the interested reader to the host of journal articles or textbooks for more details (e.g., Schlaifer 1961; Edwards et al. 1963; Galen & Gambino 1975; Iversen 1984; Berger 1985). While Bayesian statistics are more popular and better known in the realm of business statistics for performing cost-benefit analyses, they also provide the appropriate framework for dealing with medical, psychological, and ecological data for decision analysis and validating predictive models.

Bayes theorem is a simple and fundamental fact about probability applied to a field of statistics: probability is orderly opinion, and inference from data is nothing other than the revision of such opinion in the light of relevant new information (Edwards et al. 1963). The prior odds are expressed as a ratio of the likelihood of the hypothesis being true divided by the likelihood of the hypothesis being false; once some new information or data are obtained, the prior odds must be modified in light of this new information (Arkes 1981). To do this, the prior odds are multiplied by the likelihood ratio, which is the probability of obtaining that piece of datum if the hypothesis were true divided by the probability of obtaining the datum if the hypothesis were false. When the prior odds are multiplied by the likelihood ratio, we get the posterior odds. The posterior odds are the probability that the hypothesis is true given this piece of information divided by the probability that the hypothesis is not true given this piece of information. So, the whole formula is

$$\text{Posterior Odds} = \text{Likelihood Ratio} \times \text{Prior Odds}$$

The crucial component of a Bayesian analysis compared to a classical Fisherian approach is the incorporation of the prior odds, or base rates. How common or rare is the phenomenon we are trying to detect? I am convinced that abandoning NHST and switching to Bayesian models would have the biggest single impact on improving our ability to assess ecosystem health, interpret bioaccumulation results, assess sediment quality criteria, or perform valid risk assessments. By employing Bayesian models and incorporating base rates, we would gain a much more realistic assessment of the validity of the indicators we are measuring. What this means is that if the condition we are trying to measure is rare (low prior odds), then our assessment technique, unless it is 100% accurate, will primarily generate false positives (conclude that condition is present when it is not).

As an example, suppose the base rate of flounder liver lesions from sediment contamination in a population is 1 per 100,000. Suppose toxicologist Jones just plays the base rates and always concludes the condition is never present, without spending the money to do any histopathological analyses. Toxicologist Smith, however, has jumped on the biomarker bandwagon and is willing to diagnose the condition. For argument's sake, let us say Smith never misses a true case of sediment-induced flounder liver lesions (i.e., he never gets a false negative); however, he does make a false positive identification for 1 in 1,000 cases (i.e., he says the condition is present when it is not).

We should recognize before proceeding further that we have endowed Toxicologist Smith with remarkable diagnostic powers that exceed most analytical chemistry labs and without a doubt all toxicology labs (i.e., a 0% false negative and a 0.1% false positive rate). An extensive regional sampling survey is done (similar to the USEPA's EMAP program) and the results are applied to the 100,000 flounders that were caught. Toxicologist Jones misses the 1 case when the condition is actually present (a false negative error), and Smith misses the 0.1% x 99,999 cases in which the condition is not present, or makes about 100 false positive errors. Therefore, Smith is wrong 100 times more often than Jones. While this example is extreme, one of the points that Faust (1984, 1986a,b, 1989) has made repeatedly in his various publications cannot be ignored: *unless a test can surpass the diagnostic hit rate achieved by the base rates alone, it will decrease instead of increase diagnostic accuracy.* Should we use sediment quality criteria that correctly predict toxicity 90% of the time? The answer may seem obvious, but it cannot be determined unless one knows the base rates in the population of interest. It is ludicrous to think we will improve our diagnostic accuracy in ecological investigations or interpret results in a meaningful fashion if we continue to ignore base rates; it is important to always keep in mind that posterior odds are essentially a contest between prior odds and the likelihood ratio (therefore, base rates cannot be ignored).

The attractiveness of Bayesian methods resides in its parallel construct with research methodology. Research is cumulative, and while t tests or F tests are based on the "onceness" of the experiment, Bayesian methods permit the use of knowledge or results from earlier research in the formulation of results from new research. However, critics continually counter that subjective prior distributions have no place in an objective scientific analysis. Very few statistical analyses even approximate objectivity. Both Iverson (1984) and Berger (1985) point out that classical statistics are just as subjective as Bayesian analysis, and that subjectivity is expressed in the choice of the significance level and the choice of the statistical model used for the final analysis (which can have a much greater impact on the outcome than the choice of a prior distribution). The use of priors in Bayesian methods brings the subjective aspects of the analysis out in the open; the analyst is forced to express personal opinions and biases, and each reader can make their own assessment about the reasonableness of the priors. Anyone is free to apply their own set of priors on the data if they so desire to see how it affects the posterior distribution, but for large samples, the prior distributions typically have little or no effect on the posterior distribution (principle of stable estimation; see Iverson [1984] for more details).

There do exist well-developed techniques for calculating prior odds, but they are not for the statistically unsophisticated; a quick glance at the formulas in Chapter 3 of Berger (1985) will confirm this. Instead of shunning a Bayesian approach because these calculations may be more advanced than what most ecological investigators are used to dealing with (they are not one of the statistical function choices in an Excel® spreadsheet or SPSS routines), it would be much more fruitful for ecologists to enlist the help and collaboration of statisticians trained in Bayesian methods. Admitting that we need help with statistics will probably be one of the most difficult tasks for ecologists and risk assessors, because this is something that most of us have always managed (properly or improperly) to do on our own. The irony of this type of attitude was best reflected in the following interchange that took place in a statistics discussion group on the Internet:

> *Request to Internet statistics discussion group:*
>
> *A physician friend, who is doing research, is looking for a statistics book that is easy to use (by nonstatisticians) and which will cover the appropriate methods (medicine, biostat, ?). He is frustrated by the usual offerings which tend to be written for degreed statisticians. Please email me with details on suggested texts for his purposes. Any advice is appreciated.*

Reply:

Would you please ask your physician friend if he can find me a good textbook on neurosurgery? My wife has been having headaches lately, and I want to explore in there, but every time I find a textbook on neurosurgery it seems to assume that the reader is a degreed M.D., and knows about things such as histology, anatomy, neuroanatomy, neuroendocrine systems, and a bunch of stuff that I'm just sure isn't really necessary to do competent brain surgery. I don't want any fancy stuff, just the stuff that I can use to remove the part of the brain that causes headaches....

Our task as ecologists, risk assessors, or environmental scientists would be to acquire a rudimentary understanding of Bayesian methods in order to provide the professional statistician with the information they need to perform their analyses, and then we can interpret the results in a relevant context for the particular ecosystem under investigation. However, I am not advocating a mere substitution of Bayesian methods for NHST as a basis for automatic inference; it's only one of six suggested techniques needed to improve diagnostic accuracy (Germano 1999).

Conclusions

Our current ability to diagnose ecosystem health accurately or to make valid predictions of recovery and/or the effectiveness of alternative remediation treatments ranges from being extremely limited in many cases to nonexistent in others. Every one will freely admit that ecological systems are extremely complex, and this simple statement has important implications for the development of ecological indicators (Bernstein unpublished); worthwhile indicators cannot be developed without a valid model or thorough understanding of how ecosystems function. To admit that the misinterpretation and misapplication of NHST has been a large part of our problems is, as Bakan (1966) stated so aptly in his wonderful review of the crisis of statistical testing in psychological research, "to assume the role of the child who pointed out that the emperor was really outfitted only in his underwear."

With recent advances in electronic image processing, the routine use of personal computers, the ability for rapid data collection with in-situ long-term monitoring arrays, and the development of powerful graphical databases such as geographic information systems, marine ecologists (in fact, scientists in all disciplines) will witness a radical change during the next decade in the way we work and our understanding of the world around us. I am convinced that we will be able to achieve new heights in our knowledge and predictive capabilities of ecosystem structure and function, largely because of the advances in computer technology. Investigators will have access to a virtual wealth of information in the large environmental databases that will be integrated at a higher systems level than ever before.

With the recent advances in hardware and software for personal computers and electronic imaging, desktop machines are for the first time becoming capable of information synthesis at a level of integration comparable to humans, but with the distinct added advantage of being capable of handling, manipulating, and synthesizing millions of bits of data more than we are capable of processing. Computer workstations based on visualization operating systems will remove most, if not all, of the intimidating factors for the uninitiated using the machines for the first time. These image-based operating systems will make these machines accessible to countless people who otherwise would not start to explore the potential these systems hold. The framework illustrated in Fig. 2 outlines why this

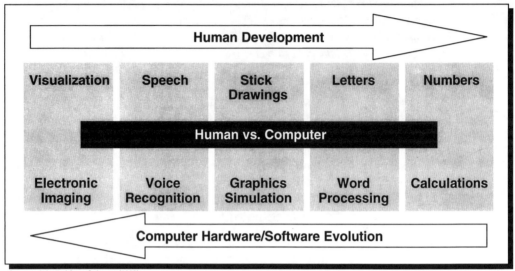

Source: Adapted from Schwarz (1990)

Figure 2. Ontogeny of information synthesis.

is so. If you think about human development, when a baby is born, no one has to teach the child to see; their perception and understanding of the world are learned through the processing of visual images. As children get older, they will learn to speak and then start making crude stick drawings; eventually they will master the alphabet and letters, and then finally the manipulation of numbers and complex mathematical functions. With computers, the ontogeny has been just the opposite: at their "birth," they could manipulate numbers and perform numerical calculations faster than any human being. After about two decades, they finally "developed" the ability to do word processing, followed by graphics-simulation hardware and software development and speech synthesizers. And just recently, electronic imaging and visualization software have been developed to a level where it is economical and accessible to the millions of personal computer workstations. Computers are finally able to integrate information at a systems level comparable to the way we understand and perceive our surroundings, making the synergistic interchange between human and machine more powerful than it has ever been.

These machines can be the key to achieving a new level of understanding about ecosystem structure and function, allowing us for the first time to be able to truly assess ecosystem "health" or make valid risk assessments and realistic impact predictions. However, this new level of understanding will not happen overnight, and I also am certain it will not happen if we continue to do research and data interpretation as we have in the past.

The lure of being able to make more new types of measurements through electronic image analysis or gather reams of data and display them in geographic information systems is a powerfully seductive one. As one who has used image analysis routinely in my work over the past 15 years with sediment profile photography (Rhoads & Germano 1982, 1986), I can attest to the ease with which one can fall prey to this. David Ehrenfeld at Rutgers University has made the observation that just because computers can do things very fast, there is no evidence that we can make the leap from doing things fast to really understanding what is going to happen next on the basis of the information we supply (Jensen 1995). Unless we think about why we are collecting these data and how they are going to be used, we are going to be no better off in 5–10 years than we are now in terms of being able to assess ecosystem health or protect and manage endangered resources, no matter how sophisticated our workstations are.

Abandoning NHST does not mean that all the data we've collected to date are of no use. Alternative hypotheses could be formulated and tested with a wealth of on-line data; there exist enormous environmental databases that have been compiled by the USEPA, NOAA, or the US Army Corps of Engineers (ODES, STORET, NS&T, ERED, etc.). The advances in electronic imaging workstations combined with the wealth of data already available would allow investigators to have access to and sort variables of interest rapidly and efficiently so that they could accurately examine covariation for proposed predictors or criteria. The usefulness of suggested predictors could then be assessed with Bayesian methods. Finally, the derivation of any predictive variable or measurement technique could then be followed by cross-validation studies. This type of follow-up is essential, because a predictive assessment technique should be shown to work where it is needed, that is, in cases where the outcome is unknown. If a method is intended only for local use or in the region in which it was developed, then the investigator can partition a representative sample from that setting into derivation and cross-validation groups. If broader application is intended, then new cases should be representative of the potential setting and populations of interest.

Combining the advances in computer technology with different assessment methodologies would allow us to explore questions about ecosystem function in a manner never before accessible. We would be wise to take heed of the lessons learned in the fields of clinical psychology, decision theory, and medical diagnostics (Germano 1999) to help us reach new plateaus of understanding in ecological assessment. If we have the strength to let go of the familiar, abandon NHST, and apply alternative assessment methods that have led to advances in other fields, we will finally have the potential to actually manage our environmental resources instead of being able merely to describe or confirm after the fact that an area is indeed polluted.

Literature Cited

Arkes, H. E. 1981. Impediments to accurate clinical judgment and possible ways to minimize their impact. *Journal of Consulting and Clinical Psychology* 49:323–330.

Bakan, D. 1966. The test of significance in psychological research. *Psychological Bulletin* 66:423–437.

Bardwell, L. V. 1991. Problem-framing: A perspective on environmental problem-solving. *Environmental Management* 15:603–612.

Berger, J. O. 1985. Statistical Decision Theory and Bayesian Analysis. Second Edition. Springer-Verlag. New York, New York.

Berkson, J. 1938. Some difficulties of interpretation encountered in the application of the chi-square test. *Journal of the American Statistical Association* 33:526–542.

Calabrese, E. J. & L. A. Baldwin. 1993. Performing Ecological Risk Assessments. Lewis Publishers, Chelsea, Michigan.

Carver, R. P. 1978. The case against statistical significance testing. *Harvard Educational Review* 48:378–399.

Chapman, L. J. & J. P. Chapman. 1967. Genesis of popular but erroneous psychodiagnositc observations. *Journal of Abnormal Psychology* 72:193–204.

Edwards, W., H. Lindman, & L. J. Savage. 1963. Bayesian statistical inference for psychological research. *Psychological Review* 70:193–242.

Faust, D. 1984. The limits of scientific judgment. University of Minnesota Press. Minneapolis, Minnesota.

Faust, D. 1986a. Learning and maintaining rules for decreasing judgment accuracy. *Journal of Personality Assessment* 50:585–600.

Faust, D. 1986b. Research on human judgment and its application to clinical practice. *Professional Psychology: Research and Practice* 17:420–430.

Faust, D. 1989. Data integration in legal evaluations: Can clinicians deliver on their premises? *Behavioral Sciences & the Law* 7:469–483.

Galen, R. S. & S. R. Gambino. 1975. Beyond Normality: The Predictive Value and Efficiency of Medical Diagnoses. Wiley, New York, New York.

Germano, J. D. 1999. Ecology, statistics, and the art of misdiagnosis: The need for a paradigm shift. *Environmental Reviews* 7:167–190.

Hutchinson, G. E. 1953. The concept of pattern in ecology. *Proceedings of the Academy of Natural Sciences of Philadelphia* 105:1–12

Iversen, G. R. 1984. Bayesian Statistical Inference. Sage University Paper series on Quantitative Applications in the Social Sciences, 07-043. Sage Publications, Beverly Hills and London.

Jensen, D. 1995. Limiting the future: An interview with David Ehrenfeld. *The Sun* 240:3–6.

Nisbett, R. E. & L. Ross. 1980. Human Inferences: Strategies and Shortcomings of Social Judgment. Prentice-Hall, Englewood Cliffs, New Jersey.

Power, M. & L. S. McCarty. 1997. Fallacies in ecological risk assessment practices. *Environmental Science and Technology* 31:370A–375A.

Raiffa, H. 1968. Decision Analysis. Random House, New York, New York.

Rhoads, D. C. & J. D. Germano. 1982. Characterization of organism-sediment relations using sediment profile imaging: An efficient method of Remote Ecological Monitoring Of The Seafloor [REMOTS System]. *Marine Ecology Progress Series* 8:115-128.

Rhoads, D. C. & J. D. Germano. 1986. Interpreting long-term changes in benthic community structure: A new protocol. *Hydrobiologia* 142:291–308.

Schlaifer, R. 1961. Introduction to Statistics for Business Decisions. McGraw-Hill, New York, New York.

Schwarz, R. D. 1990. The electronic imaging industry: Where we stand now. *Advanced Imaging* 5:17–19.

Suter, G. W. II. 1993. Ecological Risk Assessment. Lewis Publishers, Chelsea, Michigan.

Tukey, J. W. 1977. Exploratory Data Analysis. Addison-Wesley Publishing Company, Reading, Massachusetts.

United States Environmental Protection Agency (USEPA). 1998. Guidelines for Ecological Risk Assessment. EPA/630/R-95/002F. April 1998. Final. Risk Assessment Forum, Washington, D.C.

Sources of Unpublished Materials

Bernstein, B. A framework for trend detection: Coupling ecological and managerial perspectives. Presented at the International Symposium on Ecological Indicators. Fort Lauderdale, Florida. 16–19 October 1990.

Sediment Profile Imagery as a Benthic Monitoring Tool: Introduction to a 'Long-Term' Case History Evaluation (Galway Bay, West Coast of Ireland)

B. F. Keegan, D. C. Rhoads, J. D. Germano, M. Solan, R. Kennedy,
I. O'Connor, B. O'Connor, D. McGrath, P. Dinneen, S. Acevedo,
S. Young, A. Grehan, and J. Costelloe

Abstract: While there has been significant growth in marine sediment profile imagery (REMOTS® Sediment Profile Imagery) since the 1980s, reports of back-to-back Sediment Profile Imagery (SPI) with more traditionally generated benthic data are still quite rare. In May 1987, a SPI camera array was deployed at coastal sites in Ireland and France as part of a demonstration project funded by the European Union. At each location, its effectiveness was evaluated against a background of ground truth data provided by grab and core sampling. This paper introduces data obtained in Galway Bay (west coast of Ireland) from sediment profile images collected in 1987 and in subsequent benthic recharacterizations (up to 1997) of some of the same biotopes, and compares selected SPI returns with those provided by the more traditional methods. The findings corroborate the view that SPI is a powerful sampling tool that is at its best when used in concert with a detailed, calibrated base derived from quality traditional sampling and sample processing.

Introduction

Since the 1970s, extraordinary growth has been seen in the application of digital image processing concepts, not least in the area of oceanography. Where the soft seafloor is concerned, the sediment profile camera of Rhoads and Cande (1971) was the 'turnkey' in opening up a whole new era in interrogating marine benthic processes and events. Indeed, REMOTS, or SPI (Sediment Profile Imagery) as it is more usually termed in Europe, has become virtually synonymous with D. C. Rhoads, his colleagues and associates (see, *inter al.*, Rhoads et al. 1981; Rhoads & Germano 1982, 1986; Diaz et al. 1994). In spite of its widespread use in variously focused benthic investigations in estuaries, coastal waters, the deep sea, and even fresh water (e.g., Revelas et al. 1987; Boyer & Shen 1988; Diaz & Schaffner 1988; Boyer & Hedrick 1989; O'Connor et al. 1989; Nichols et al. 1990; Diaz & Gapcynski 1991; Krieger et al. 1991; Grehan et al. 1992, 1998; Keegan et al. 1992; Rumohr & Schoman 1992; Rumohr et al. 1992, 1996; Valente et al. 1992; Schaffner et al. 1992; Diaz et al. 1993; Bonsdorff et al. 1996; Nilsson & Rosenberg 1997; Diaz 1999), SPI is only now getting the widespread recognition it deserves. While this has something to do with acknowledged limitations in image interpretation, there remain certain impediments linked to the size and weight of the device, as well as to its restriction to use in muds and muddy sands. The relatively high cost of the most basic SPI assembly is perhaps most telling of all. This last point is borne out by the fact that the authors are aware of much more SPI reportage than resides in the public domain and scientific literature. SPI has tended to be used in activities promoted more by government and the wealthier commercial environmental consultancies than by the more traditional research sector.

Correspondingly, many findings are either found in the "grey" literature or can only be accessed with great difficulty, if at all. Linked in part to this situation is the fact that, with some notable exceptions (e.g., Grizzle & Penniman 1991; Diaz et al. 1994; Rosenberg 1995; Rumohr et al. 1996), there are very few published accounts of back-to-back imagery-derived and traditionally generated benthic (faunal) data.

In May 1987, a REMOTS® camera array was deployed at coastal sites in Ireland and France as part of a European Union funded "demonstration" project. At each location, its effectiveness was evaluated against a background of ground truth data provided by grab and core sampling. This paper introduces data obtained in Galway Bay (west coast of Ireland) from sediment profile images, taken then and in subsequent benthic recharacterizations (up to 1997) of some of the same biotopes, and compares selected SPI returns with those deriving from more traditional sampling methods.

The recharacterizations are part of an intensive effort to validate and improve SPI image interpretation from an ecological standpoint, via a number of in situ and laboratory techniques. The main aims of this program are (1) to develop a model of the net community porewater and particle bioturbation at two well-documented locations (the Mutton Island and Margaretta sites), and (2) to investigate the nature of temporal change in diagnostic SPI parameters. Both short-term (time lapse) and long-term (monthly occupancy) SPI studies are carried out in concert with regular traditional recharacterizations (i.e., grab and/or core sampling of fauna and sediments) and water column profiling. Modelling is being done through numerical simulation of porewater and particle tracer experiments performed on selected numerically prominent and in situ stands of fauna. The targeted species are assumed to be the key species (sensu Lewis 1978) or ecological dominants in their respective communities.

Study Area

Galway Bay has its axis on the 53°11' N parallel and is approximately delimited by the 9°0' W and 9°55' W meridians (Fig. 1a). The Aran Islands slant northwest-southeast across the mouth of the bay and afford some shelter from the prevailing southwesterly and west winds. Eastward from the islands the water depth declines steadily from 65 m. For the most part, the bottom contours are regular with a few localized shoal patches. Deposit substrates range from the finest muds to coarse terrigenic and biogenic gravels. The area to the north of the bay is drained by a number of small rivers and by the extensive Corrib system. Drainage from the eastern and southern shores is more difficult to define as the watercourses are largely underground. The Corrib outflow is considerable, if variable, and is typically dispersed over the northern sector of the bay. This exaggerates the thermoclines and haloclines that develop in the general water-body during times of prolonged calm. Circulation in the bay is counterclockwise with water entering through the South Sound and leaving through the North Sound. The freshwater outflow is directed along the northern perimeter, a feature that is manifested in a pronounced silt-clay gradient in the sediment, ranging from > 10% in the northeastern sector to < 2% in the southwest. Recorded sediment organic carbon content in the inner reaches of the bay ranges from 0.12% to 4.3% and tends to be correlated positively with the silt-clay component.

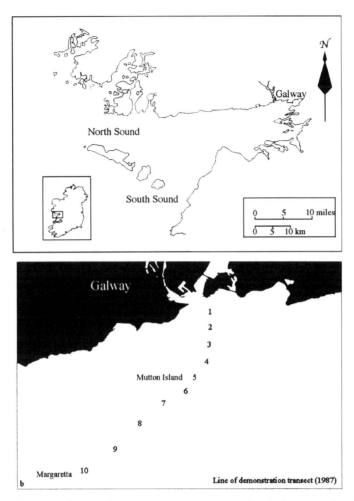

Fig. 1. (a) Map of study location; (b) line of 1987 demonstration transect showing long-term sites Mutton Island (5) and Margaretta (10).

Concentrations of copper, zinc, and iron in the sediment are at or below the average for unpolluted, inshore deposits. Higher values for lead are attributed to air and dredged spoil (harbor approach channel) dispersal of this metal's powdered oxide, which was bulk shipped from Galway City docks throughout the 1970s. At Galway City, predicted mean spring and mean tidal ranges are 4.3 m and 1.9 m, respectively.

The fauna of the Greater Galway Bay area is perhaps better documented than any other Irish inshore area of similar areal extent (see, *inter al.,* Keegan & Könnecker 1973; Keegan 1974a,b; Keary & Keegan 1975; Keegan et al. 1976, 1985; Shin et al. 1982; Heffernan et al. 1983; O'Connor et al. 1983, 1986a,b, 1992; O'Foighil et al. 1984; Souprayen et al. 1991; Acevedo & Keegan 1995; Grehan & Keegan 1995). For the purposes of this paper, the reportage is confined to a transect line in the inner bay, approximating the Corrib's main river outflow pathway (Fig. 1b). While predominantly estuarine and neritic in character, this area is subject to periodic intrusions of oceanic water.

Post-1987 benthic recharacterizations in Galway Bay have been largely confined to two sites from the original line of transect. The inner site (the Mutton Island site; site 5 on the 1987 transect; Fig. 1b) lies near the newly designated outfall location for Galway City's new domestic sewerage scheme. The outer site (the Margaretta site; site 10 on the 1987 transect; Fig. 1b) has been monitored more or less continually since 1974 (except for the period October 1976 to October 1978).

Methods

A Benthos Sediment Profile Camera (model 3731, Benthos Inc., North Falmouth, Massachusetts, USA) was used in the 1987 demonstration and in some monitoring work immediately thereafter. An Irish-built system (SPI I), closely modelled on the Benthos Inc. device and using a standard 35 mm SLR camera, has been used since 1990. Initially SPI measurements of all physical and some biological parameters were measured directly from the film negatives using a video digitizer and computer image analysis system, laterally, color diapositives have been scanned directly into the computer. More recently, and underpinning the recharacterization, imaging has also been effected with (a) a diver-operated profiler fitted with a time-lapse facility and (b) a digital profiling camera (SPI II) rated to depths > 5,000 m (see Grehan et al. 1998). At its most basic, the diver-operated camera is a lightweight (aluminum) version of the central column (prism and camera with electronics housing) from the original framed device. SPI II is built around a Kodak DCS 410 (1524 x 1012 CCD array) digital camera with a Nikon N-090s camera-back. Here, the basic SPI I design has been modified so that routine servicing and maintenance can be performed without removing the main pressure housing containing the camera and batteries. As presently configured, up to 100 images can be taken per single deployment. Downloaded on deck, retrieved images are immediately available for computer interrogation. Downward facing cameras, delivering pre-impact images of the sediment surface, are regularly mounted on the remotely operated systems.

A customized version of the software package, Image Analyst, is used for image analysis, with the derivative data archived in a commercially available relational database (Filemaker Pro). Images are prepared for analysis using Adobe Photoshop. The actual analytical protocol still closely follows that of Rhoads and Germano (1982), to the inclusion of the Organism-Sediment Index (OSI). The latter, in turn, incorporates discrimination of the stage component to the well-documented successional paradigm for soft marine sediments associated with Pearson and Rosenberg (1978) and Rhoads and Germano (1986).

For the 1987 demonstration, faunal and sediment samples were taken with a Reineck corer. For every station sampled , the faunal returns retained on a 0.1-mm mesh were identified to species level (where possible) and enumerated. An arbitrary 'successional stage' designation was given to each station that reflected the known biological and/or ecological attributes of the more common species. Normal classificatory analysis was used to determine faunal groupings along the demonstration transect, and, the dissimilarity form of the Bray-Curtis coefficient was used to measure interstation resemblance. Stations within the resulting matrix were grouped by hierarchical agglomerative clustering using flexible sorting with a cluster-intensity coefficient of fl = -0.25.

For the recharacterizations, faunal samples were taken with a modified 0.1-m^2 van Veen grab and sieved on a 0.5-mm mesh. Two stainless steel mesh trap doors have been incorporated into the upper surfaces of the grab 'scoops' to reduce the bow wave effect and allow for ready measurement of digging depth. Sieve catch was stored in 10% neutral-buffered formaldehyde for 3 mo to facilitate standardization of biomass. A 1% Eosin-scarlet red solution was added to the samples to aid location of macrofauna. Throughout, the faunal data have been used in the computation of certain community parameters with respect to each station: species diversity, H', species evenness, J (Pielou 1975), and species richness (Margalef 1958).

Abundance Biomass Comparison (ABC) plots were generated on the numeric and biomass data from both sites (Warwick 1986; Lambshead et al. 1993), where the species were ranked in order of numeric and gravimetric importance on the X-axis (logarithmic scale), with percentage dominance on the Y-axis (cumulative scale). For biomass (wet weight) determination, biological material was processed as recommended by Rumohr (1999).

Based on traditional sieve analysis, cumulative size frequency curves have been drawn for grab-taken sediment samples and size distribution parameters (i.e., graphic mean, sorting, and skewness) calculated according to Folk (1974). More recently, a Malvern Mastersizer X has been used for laser diffraction particle sizing. Sediment organic carbon content is typically determined using the chromic acid oxidation method (CAOV) described in Holme and McIntyre (1984). Porosity is based on sediment water content.

The data analysis package Primer (Clarke & Warwick 1994) was used to investigate the association between faunal abundance patterns and SPI data. Here, the top 10 numerical dominants were used to produce a Bray-Curtis similarity table (Bray & Curtis 1957) of the 4th-root-transformed faunal returns. This matrix was represented two-dimensionally using an MDS ordination. SPI parameters used in an attempt to explain faunal patterns are listed in Table 1. Log10-transformed SPI data were two-dimensionally represented by an MDS ordination based on Euclidean distance. The weighted Spearman harmonic rank correlation (ρ_w) procedure (BioEnv; Clarke & Ainsworth 1993) was used to quantify the match between the faunal and SPI data, based on the rank similarity matrices. Combinations of SPI parameters were considered at steadily increasing levels of complexity, that is, from 1 to n parameters at a time (n = 6). The value of ρ_w ranges from +1 to -1, corresponding to cases where the two sets of ranks are in complete agreement or disagreement, respe typically, ρ_w will be positive. Because the SPI and faunal similarity ranks are based on a large number of interdependant similarity calculations, they are not mutually independent variables. This precludes ctively. Values close to 0 correspond to an absence of any match between the two patterns, but referring ρ_w to standard statistical tables of Spearman's rank correlations to assess whether two patterns are matched significantly.

Table 1. Combinations of the six SPI parameters, taken n at a time, yielding the best matches of faunal abundance and SPI similarity matrices for each n as measured by weighted Spearman harmonic rank correlation (r_w). Succ = successional stage (Rhoads and Germano 1986), aRPD = mean apparent RPD depth, Mz = Sediment particle size Graphic Mean, OSI = Mean Organism Sediment Index score, Penet.= Mean prism penetration depth, SBR= surface boundary roughness.

	Variable Combinations	r_w for n combinations
1	Succ.	0.17
2	Succ., aRPD	0.04
3	Mz, Succ., aRPD	-0.05
4	Mz, OSI, Succ., aRPD	-0.09
5	Penet., Mz, OSI, Succ., aRPD	-0.17
6	Penet., SBR, Mz, OSI, Succ., aRPD	-0.34

Results

On May 13 and 14, 1987, as part of the original demonstration project, five replicate images were taken at each of the 10 stations in Fig. 1b. SPI findings are summarized in Fig. 2a-f. Most stations had sediments consisting of very fine (4-3 phi) to fine (3-2 phi) sands. Station 9 (2-1 phi) had the coarsest sediment, and the finest material was at stations 3, 4, and 10 ($\geq$ 4 phi). Bioturbation is an important process in determining sedimentary fabrics, except at stations 2 and 9. Station 2 was internally stratified and may represent an area of periodic high sedimentation. Strong bottom currents affect station 9 and bioturbation is apparently not an important process here. The lowest kinetic energy stations were 3, 4, 5, 6, and 10; these stations have fine-grained organic sediments that are thoroughly bioturbated. A major kinetic gradient exists from station 6 (low) to 9 (high), and from station 9 (high) to station 10 (low). There is a notable increase in the apparent redox potential discontinuity (RPD) values between station 1 (0.04 cm) and station 6 (3.15 cm). This likely reflects an organic enrichment gradient extending seaward from Galway Harbour (i.e., along the path of the River Corrib outflow). Station 5 appears to be located at the marginal edge of this enrichment gradient. Stations 2 through 4 are dominated by dense aggregations of stage I taxa. This is biological evidence for the enrichment gradient inferred from the apparent RPD depths. It could not be ascertained if the elongate deep-dwelling polychaetes observed at station 1 were stage I or stage III taxa; these were tentatively mapped as stage III. Stations 5 and 6 consisted of mixtures of stage I and II seres. These two stations apparently lie along the outer edge of the enriched area. Stations 7 and 8 had a tubicolous infauna, suggesting that the bottom is more stable than at station 9; all three stations were rippled. Station 10 is a deeply bioturbated sediment containing a dense assemblage of infaunal ophiuroids.

Organism-Sediment Index (OSI) values could not be calculated for station 9. Stations 1 through 4 had OSI values $\leq$ 6. It is suggested that these stations experience high rates of organic loading, resulting in shallow RPD depths, with most replicates from these stations dominated by stage I seres. Stations 5 and 6 had intermediate OSI values, suggesting, again, that they lie near the margin of the enriched area. Stations 7, 8, and 10 had high indices, indicating that they are located in a relatively undisturbed benthic habitat.

Sediment mean particle size, percentage silt-clay, and organic carbon content for stations 1 to 10, which were derived from the ground truth exercise supporting the 1987 demonstration, are shown in Fig. 3. One hundred and forty-five species of macrobenthic animals were identified. This figure also presents the computed community indices for the full suite of stations, and displays the results of the classification procedure that delineates two main groups of stations: 1, 2, 3, and 4 and 5, 6, 7, 8, and 10. Station 9 has an anomalous character with respect to the both groups.

The classification procedure, in addition to identifying two main groups of stations and the relative discreteness of station 9, shows something of a sequential progression from the innermost station (1) in the mouth of the River Corrib to station 10 at the western extremity of the sampling transect. This 'gradient' is taken to reflect the diminishing influence of the Corrib inflow. Previous workers (e.g., Shin et al. 1982) have commented on the controlling effects of the river on faunal distribution patterns in Galway Bay. Station group 1 (stations 1, 2, 3, and 4) reflects the more obvious enriching effects of the inflow and has a benthic fauna typical of such situations. Station group 2 (stations 5, 6, 7, 8, and 10) is broadly referable to the boreo-Mediterranean *Amphiura filiformis* community (see Thorson 1957; Buchanan 1963). The successional staging shows stations 5 and 6 to have something of a transitional character. Whilst the underlying causality cannot be readily identified,

Table 2. Margaretta Station. A) Summary data on the water column for the period October1966 to November 1997; B) Representative (December 1996) biological data returned in 5 × 0.1 m² van Veen grabs; C) Representative (December 1996) SPI data (based on 10 images).

A. Physical Environment	Mean	Min	Max
Water			
Bottom Current Speed (cm s⁻¹)	15.63	0.67	60.17
Salinity (off bottom)	34.18	33.59	35.24
Sediment			
Graphic mean		2.32	4.68
Sorting		1.43	2.58
Skewness		-0.37	0.40
% Org.C (CAOV)		0.56	2.28
Bulk Porosity	32.39	28.72	40.03

B. Community parameters	n = 5	$\bar{x}$± 95% CI	
No. of Species		44.8 ±	4.3
No. of Individuals		826 ±	348
Diversity (H')		3.89 ±	0.16
Evenness (J')		0.71 ±	0.04
Species Richness (D)		4.56 ±	0.24
Traditional successional stage		3	

C. SPI parameters	n = 10	$\bar{x}$ ± 95% CI	
Apparent RPD (cm)		5.44 ±	0.53
Prism Penetration (cm)		15.24 ±	3.19
Successional stage		III ±	0
Surface Boundary Roughness		1.62 ±	1.39
Grain size Maj. mode (ø)		2-3 ±	0
OSI score		11 ±	0

the separate faunal status of station 9 was to be expected from the granulometric returns: the highest gravel (28.66%) and coarse sand (52.51%), and the lowest silt-clay (2.60%) and organic carbon (0.56%) of all the stations sampled. Reflecting the known biological and/or ecological attributes of its more common species, an arbitrary "successional stage" designation is given to each station. This designation is shown above that based on image analysis in Fig. 3.

While the main reportage on the re-occupations of the Mutton Island and Margaretta stations lies outside the scope of this paper, selected imagery and other findings deriving from them are presented in Tables 1 to 3 and Figs. 4 to 7. Station profiles in terms of nonbiological data (over the

Following Pages: Fig. 2. Summary REMOTS® data maps from 1987 demonstration: (a) sediment major mode; (b) mean depths of prism penetration (cm); (c) benthic "process" map; (d) mean apparent RPD depths (cm); (e) infaunal successional stages; (f) mean Organism Sediment Index (OSI) values.

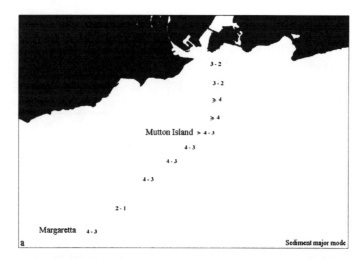

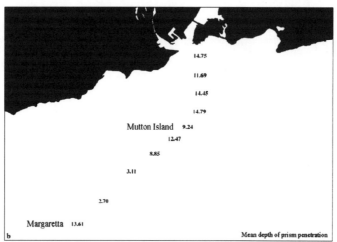

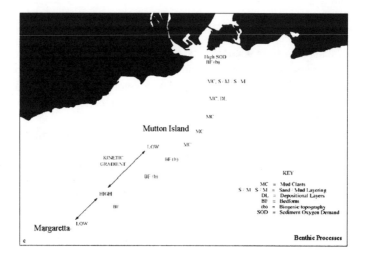

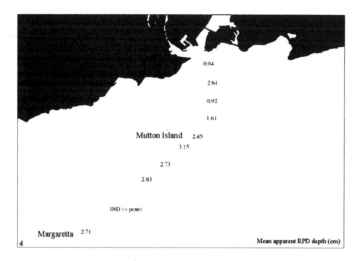

0.04

2.61

0.92

1.61

Mutton Island 2.45

3.15

2.73

2.83

IND (> pene)

Margaretta 2.71

d   Mean apparent RPD depth (cm)

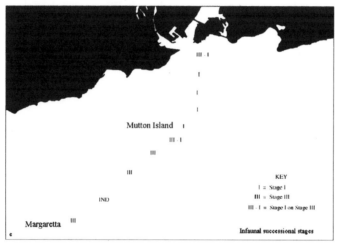

III - I

I

I

I

Mutton Island I

III - I

III

III   KEY

I = Stage I

III = Stage III

IND   III - I = Stage I on Stage III

Margaretta III

e   Infaunal successional stages

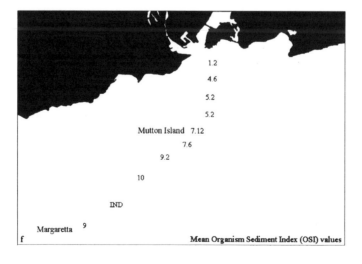

1.2

4.6

5.2

5.2

Mutton Island 7.12

7.6

9.2

10

IND

Margaretta 9

f   Mean Organism Sediment Index (OSI) values

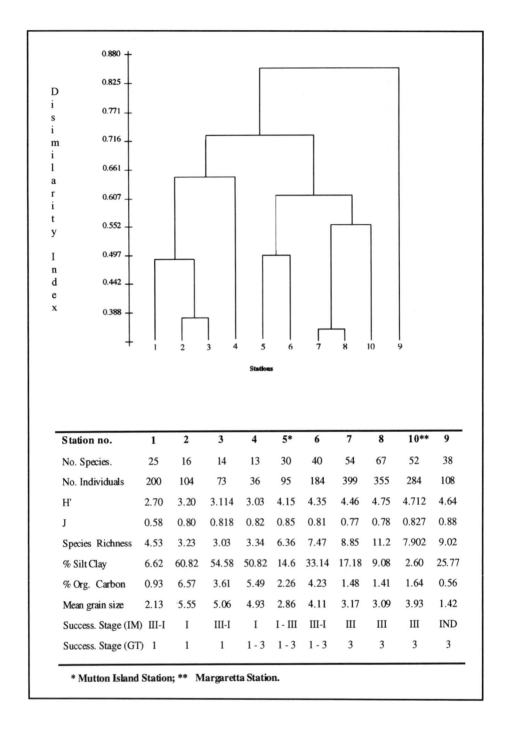

Station no.	1	2	3	4	5*	6	7	8	10**	9
No. Species.	25	16	14	13	30	40	54	67	52	38
No. Individuals	200	104	73	36	95	184	399	355	284	108
H'	2.70	3.20	3.114	3.03	4.15	4.35	4.46	4.75	4.712	4.64
J	0.58	0.80	0.818	0.82	0.85	0.81	0.77	0.78	0.827	0.88
Species Richness	4.53	3.23	3.03	3.34	6.36	7.47	8.85	11.2	7.902	9.02
% Silt Clay	6.62	60.82	54.58	50.82	14.6	33.14	17.18	9.08	2.60	25.77
% Org. Carbon	0.93	6.57	3.61	5.49	2.26	4.23	1.48	1.41	1.64	0.56
Mean grain size	2.13	5.55	5.06	4.93	2.86	4.11	3.17	3.09	3.93	1.42
Success. Stage (IM)	III-I	I	III-I	I	I - III	III-I	III	III	III	IND
Success. Stage (GT)	1	1	1	1 - 3	1 - 3	1 - 3	3	3	3	3

* Mutton Island Station; ** Margaretta Station.

Fig. 3. Station dendrogram and data deriving from traditional sampling in the 1987 demonstration. Faunal data based on pooled (0.1 mm²) Reineck corer returns, sieved at 1.0 mm. IM (for imagery) based on SPI images; GT (for ground truth) based on the known biological and/or ecological attributes of the more common species returned in the cores.

Table 3. Mutton Island station A) Summary data on the water column for the period October1966 to November 1997; B) Representative (December 1996) biological data returned in 5 × 0.1 m² van Veen grabs; C) Representative (December 1996) SPI data (based on 10 images).

A. Physical Environment	Mean	Min	Max
Water			
Bottom Current Speed (cm s⁻¹)	11.18	0.325	6.62
Salinity (off bottom)	31.68	20.71	34.56
Sediment			
Graphic mean		1.43	4.22
Sorting		0.9	2.24
Skewness		0.13	0.47
% Org.C (CAOV)		0.32	1.78
Bulk Porosity	29.67	25.74	41.59

B. Community parameters	$\bar{x} \pm$ 95% CI, n = 5		
No. of Species	22	±	2.3
No. of Individuals	261.4	±	122.4
Diversity (H')	3.08	±	0.44
Evenness (J')	0.69	±	0.08
Species Richness (D)	2.56	±	0.36
Traditional successional stage	2		

C. SPI parameters	$\bar{x} \pm$ 95% CI, n = 10		
Apparent RPD (cm)	4.13	±	0.49
Prism Penetration (cm)	12.89	±	1.39
Successional stage	III	±	0
Surface Boundary Roughness	0.64	±	0.27
Grain size Maj. mode (ø)	3.1	±	0.37
OSI score	8.8	±	0.3

period October 1996 to November 1997) and representative faunal and SPI data (December 1996) are given in Tables 1 and 2. Interestingly, on this occasion, the only images from Mutton Island were stage III. Table 3 shows the outcome of the BioEnv procedure.

In the abundance-biomass comparisons presented in Figs. 4 and 5 (together with their respective databases), the crossing of the K-dominance curves at Mutton Island point to moderate pollution, and the separate curves at the Margaretta site (the biomass above the numbers' curve) attest to unpolluted conditions (Warwick 1986). These plots do not incorporate deep burrowing species (e.g., the decapods *Upogebia deltaura*, *U. stellata*, and *Calianassa subterranea,* and the bivalve *Thracia pubescens*), which were not taken by grab but whose presence was revealed by suction sampling (particularly at the Margaretta site). Their inclusion would consolidate the ABC exposition for both sites.

Figure 6 presents selected sediment profile images from Mutton Island and the Margaretta station. Figure 7 shows the two-dimensional MDS ordinations representing (a) faunal abundance patterns, (b) combined SPI survey parameters, (c) successional staging (Rhoads & Germano 1986),

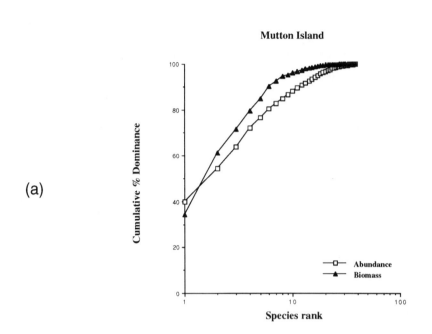

(b) December 1996, Mutton Island: Numeric Dominance

Species Rank	Species	Abundance	% Total Abundance	Cumulative % Total Abundance
1	Longipedia coronata	521	39.86	39.86
2	Nephtys hombergii	190	14.54	54.40
3	Prionospio spp.	123	9.41	63.81
4	Mysella bidentata	108	8.26	72.07
5	Oligochaeta spp.	59	4.51	76.59
6	Thyasira flexuosa	50	3.83	80.41
7	Perioculodes longimanus	30	2.30	82.71
8	Melinna palmata	28	2.14	84.85
9	Lumbrineris fragilis	25	1.91	86.76
10	Capitellidae spp.	19	1.45	88.22

(c) December 1996, Mutton Island: Gravimetric Dominance

Species Rank	Species	Biomass(g)	% Total Biomass	Cumulative % Total Biomass
1	Philine aperta	5.91128	34.24	34.24
2	Nephtys hombergii	4.64863	26.93	61.16
3	Pagurus bernhardus	1.80718	10.47	71.63
4	Thyasira flexuosa	1.38434	8.02	79.65
5	Amphiura filiformis	0.92562	5.36	85.01
6	Melinna palmata	0.89699	5.20	90.21
7	Lanice conchilega	0.41742	2.42	92.62
8	Moerella pygmaea	0.33399	1.94	94.56
9	Lumbrineris fragilis	0.14750	0.85	95.41
10	Euclymene lumbricoides	0.13865	0.80	96.22

Fig. 4. Mutton Island station. (a) Abundance-biomass curves based on the numeric and biomass data in (b) and (c).

(d) the apparent RPD, and (e) successional staging and apparent RPD combined, for the two stations. The representations are based on data taken from re-occupancies in 1990, 1996, and 1997.

Discussion

In the broadest terms, the findings in Galway Bay confirm that SPI is a 'powerful benthic sampling tool.' Within the limitations of its current design features, its performance well measures up to the claims that had been made on its behalf. Without making any allowance for the amenability of the substrate to SPI mapping or such constraints as small-scale environmental heterogeneity, there is a very high level of direct intrastation conformity with respect to the returns on successional staging (i.e., as between imagery and grab and core sampling) in the 1987 demonstration. Essentially both SPI and core-sampling support the same pattern of station separation along the line of transect. For all that, traditional faunal samples regularly return animals, sometimes in abundance (and not necessarily small), that do not appear in the images and could give a greater information return if they had been visually registered. For now, and in this same regard, it must be said that the Galway-based authors have greater confidence in diagnosing the more hard and fast seres (i.e., stages I and III of Rhoads & Germano 1986), than in apparent transitions between, or admixtures of, seres (I on III, III on I, etc.). Because of a refinement of the traditional sampling approach (i.e., sieving levels, change of sampling device), the demonstration data cannot be compared directly with those produced in the course of the station re-occupations. That said, it is clear that the essential character of the Mutton Island and Margaretta stations has endured over the intervening decade. This is especially, and immediately, evident in the representative faunal data together with the coincident imagery (Tables 1 & 2) and the abundance-biomass comparisons shown in Figs. 4 and 5.

From the limited exposition of the site re-occupations, faunal MDS shows that the Margaretta station is relatively constant over time, and Mutton Island is much more variable. SPI MDS also groups the Margaretta returns closely together, unlike those from Mutton Island. Interstation comparisons reveal that Mutton Island and Margaretta can be very similar or very different.

BioEnv picks successional staging as the SPI parameter that best matches the faunal data. The weighted harmonic mean (ρ_w) is low (0.17) but produces a good visual match to the faunal data; the Mutton Island sample that is most similar to those from the Margaretta is actually incorporated into the Margaretta group. This illustrates the well-documented admixture of stage I and stage III that typifies the Mutton Island station. The ordination based on the aRPD (apparent redox potential discontinuity) produces a pattern similar to that for the successional stage, but groups the Margaretta returns more tightly, even though ρ_w is lower (0.04). This attests to one of the underlying assumptions of SPI, that a higher successional stage and deeper mean aRPD are associated. Overall, however, the best match of the two patterns is poor, possibly reflecting a lack of adequate data to properly interrogate SPI. More extensive faunal data than the top 10 dominants would have allowed more subtle variation in the communities to be explored. Further, more occupancies would have allowed for more thorough assessment of correspondence between SPI parameters and faunal distributions. These shortfalls will be addressed in future publications, which will incorporate the returns from nine occupancies of the targeted stations. Clearly, this current exercise does not include extreme sedimentary habitats or faunal communities. Rather, it is concerned with different facies of the same faunal community, one more stressed than the other, but with similar sedimentary habitats.

In these circumstances, it is not surprising that it is the operator-assigned successional stage rather than the empirically measured aRPD that best reflects the faunal pattern. Other workers have

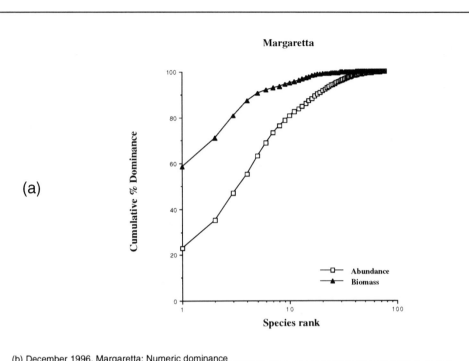

(a)

(b) December 1996, Margaretta: Numeric dominance

Species Rank	Species	Abundance	% Total Abundance	Cumulative % Total Abundance
1	*Philomedes brenda*	952	23.06	23.06
2	*Pholoe inornata*	506	12.26	35.32
3	*Amphiura filiformis*	478	11.58	46.90
4	*Mysella bidentata*	350	8.48	55.38
5	*Prionospio* spp.	329	7.97	63.35
6	*Ampelisca brevicornis*	230	5.57	68.92
7	*Capitellidae* spp.	185	4.48	73.40
8	*Eudorella truncatula*	121	2.93	76.33
9	*Lumbrineris fragilis*	98	2.37	78.71
10	*Thyasira flexuosa*	83	2.01	80.72

(c) December 1996, Margaretta: Gravimetric Dominance

Species Rank	Species	Biomass(g)	% Total Biomass	Cumulative % Total Biomass
1	*Amphiura filiformis*	84.26523	58.57	58.57
2	*Echinocardium cordatum*	17.74521	12.34	70.91
3	*Turritella communis*	14.10019	9.80	80.71
4	*Leptopentacta elongata*	9.48764	6.60	87.30
5	*Glycera tridactyla*	4.82173	3.35	90.66
6	*Melinna palmata*	2.05990	1.43	92.09
7	*Ophiura ophiura*	1.20561	0.84	92.93
8	*Lumbrineris fragilis*	1.01300	0.70	93.63
9	*Scalibregma inflatum*	1.00644	0.70	94.33
10	*Astropecten irregularis*	0.96884	0.67	95.00

Fig. 5. Margaretta station. (a) Abundance-biomass curves based on the numeric and biomass data in (b) and (c).

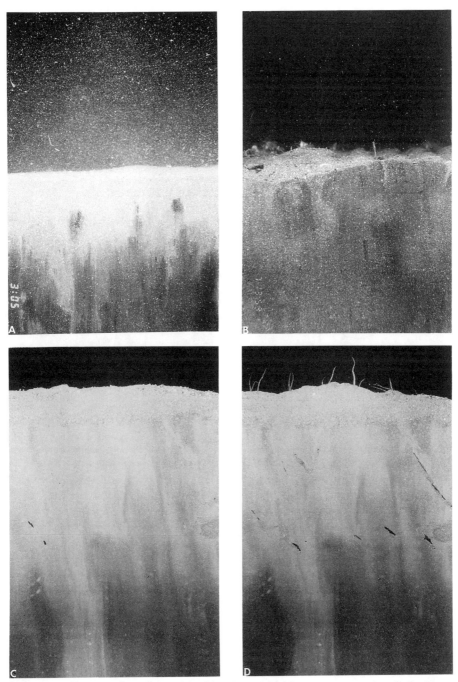

Fig. 6. Sediment profile images from Mutton Island (a & b) and the Margaretta station (c & d). The Mutton Island images show variations between a stage I and a stage III assemblage, whilst the Margaretta consistently returns stage III images. Frames c & d are taken from a time-lapse series, where frame c was taken 10 h before frame d, with the latter showing burrows and arms of the ophiuroid *Amphiura filiformis*.

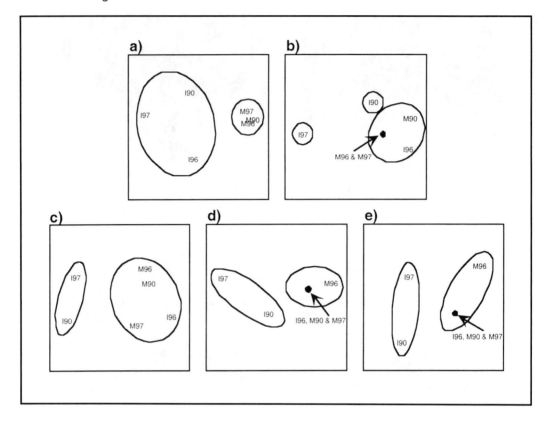

Fig. 7. Two-dimensional MDS ordinations representing (a) faunal abundance patterns, (b) combined SPI survey parameters, (c) successional staging (d) apparent RPD, and (e) successional staging and apparent RPD combined, for the Mutton Island (I) and Margaretta (M) stations. Data taken from single re-occupancies in 1990, 1996, and 1997. Ordination stress values are 0.01, 0.01, 0.00, 0.00, and 0.00 respectively.

re-addressed the successional staging paradigm with respect to their own particular study locations. Rumohr et al. (1996) divided the successional continuum into five distinct stages, but these were proffered as specific to the Baltic Sea. The present authors are particularly interested in the Benthic Habitat Quality (BHQ) index of Nilsson and Rosenberg (1997) that, they postulate may be suitable for many boreal and temperate areas. Taking into account the variation experienced within time-lapse deployments, this index was applied to the Margaretta station returns. The Margaretta assemblage is numerically dominated by the ophiuroid *Amphiura filiformis*, which tends to be active (feeding) when stronger currents associated with the ebb and flow of the tide are experienced. BHQ Index and OSI values (Rhoads & Germano 1986) were calculated for each SPI shot taken at 30-min intervals over a 24-h occupancy. The OSI score of 11 remained constant over the sampling period (suggesting stage III, equilibrium conditions). On the other hand, the BHQ score ranged from 9 to 15, crossing the boundary between stage 2 and stage 3 in the scheme of Nilsson and Rosenberg (1997). The BHQ takes account of the presence or absence of fauna and any associated sedimentary modification to assign each image a score that, in turn, determines successional stage. The OSI uses a successional stage determined by the operator, with the apparent characteristics of the sedimentary matrix being used to compute a score, without direct reference to the presence or absence of fauna. So, in this case, it would appear the BHQ is more sensitive and of greater utility in describing small-scale temporal change than the OSI. On the other hand, the OSI is a more robust index and possibly is more appropriate to the commercial and/or large-scale monitoring activities in which SPI has been more commonly used up to now. Certainly, the BHQ takes some of the subjective element out of staging an image, and would appear to be a more useful index for interrogating animal-sediment relations per se.

The original developers (see, *inter alia*, Revelas et al. 1987) stressed that REMOTS technology was "not developed as a replacement for conventional benthic monitoring tools, but as a survey and reconnaissance technique to optimize the efficiency of benthic monitoring programmes." Broad areal coverage and rapid data return allowed stations for other sampling devices (both biological and physical) to be located on the basis of 'almost real time' or at least very recent technical information. Clearly, the advent of the digital camera has further enhanced this attribute. The authors caution against being over parsimonious in scaling down traditional sampling programs. SPI is at its most useful when used in concert with a detailed calibrated database derived from quality traditional sampling and sample processing.

In terms of its current design, the size of the prism in the SPI array impedes penetration in all but the softer, less compact sediments. When, for the sake of enhanced penetration, it becomes necessary to use the full complement of lead weights (66 kg), the system becomes difficult to handle on smaller craft with restricted lifting equipment. Size and, correspondingly, weight could be reduced if the prism could be replaced to act more as a slim 'digging blade,' the whole exposed face of which could be digitally scanned in situ. Such a blade would not only facilitate easier and deeper penetration but also extend the use of SPI to more compact, fine to medium sands. The authors have already experimented with a suitable casing that penetrated these more resistant deposits to depths exceeding 55 cm. However, a physically robust scanner that will tolerate the shock of impact and have a level of resolution adequate to the purpose remains to be identified.

Acknowledgments

Albert Lawless, John Galvin, and David Burke are gratefully acknowledged for their technical input. This work was partly funded within the provisions of European Union contracts B-04-08 and MAS3-CT96-5035.

Literature Cited

Acevedo, S. & B. F. Keegan. 1995. Reproductive cycle of *Thyasira flexuosa* (Montagu, 1830) (Bivalvia: Thyasiridae) in Inner Galway Bay, West Coast of Ireland, p. 143–158. *In* Keegan, B. F. & R. O'Connor (eds.), Irish Marine Science 1995. Galway University Press Ltd., Galway, Ireland.

Bonsdorff, E., R. J. Diaz, R. Rosenberg, A. Norkko, & G. R. Cutter, Jr. 1996. Characterisation of soft-bottom benthic habitats of the Åland Islands, northern Baltic Sea. *Marine Ecology Progress Series* 142:235–245.

Boyer, L. F. & J. Hedrick.1989. Submersible-deployed video sediment-profile camera system for benthic studies. *Journal of Great Lakes Research* 15:34–45.

Boyer, L. F. & E. J. Shen. 1988. Sediment-profile camera study of Milwaukee Harbour sediments. *Journal of Great Lakes Research* 14:444–465.

Bray, J. R. & J. T. Curtis. 1957. An ordination of the upland forest communities of southern Wisconsin. *Ecological Monographs* 27:325–349.

Buchanan, J. B. 1963. The bottom fauna communities and their sediment relationships off the coast of Northumberland. *Oikos* 14:154–175.

Clarke, K. R. & M. Ainsworth. 1993. A method of linking multivariate community structure to environmental variables. *Marine Ecology Progress Series* 92:205–219.

Clarke, K. R. & R. M. Warwick. 1994. Change in Marine Communities: An Approach to Statistical Analysis and interpretation. Natural Environmental Research Council, United Kingdom.

Diaz, R. J. 1992. Ecosystem assessment using estuarine and marine benthic community structure, p. 76–85. *In* Burton, A. (ed.), Contaminated Sediment Toxicity Assessment. Lewis Publishers, Boca Raton, Florida.

Diaz, R. J., G. R. Cutter, & D. C. Rhoads. 1994. The importance of bioturbation to continental slope sediment structure and benthic processes off Cape Hatteras, North Carolina. *Deep-Sea Research II* 41:719–734

Diaz, R. J. & P. C. Gapcynski. 1991. Sediment surface and profile image (SPI) analysis applied to the rapid assessment of benthic habitats, p. 319–322. *In* Keegan, B. K. (comp.), Activity Report 1988–1991. COST 647. Coastal Benthic Ecology. Commission of the European Communities, Brussels.

Diaz, R. J., L. J. Hansson, R. Rosenberg, P. C. Gapcynski, & M. A. Umger. 1993. Rapid sedimentological and biological assessment of hydrocarbon contaminated sediments. *Water, Air and Soil Pollution* 66:251–266.

Diaz, R. J. & L. C. Schaffner. 1988. Comparison of sediment landscapes in Chesapeake Bay as seen by surface and profile imaging, p. 222–240. *In* Lynch, M. P. & E. C. Krome (eds.), Understanding the Estuary: Advances in Chesapeake Bay Research. Pub. 129, CBT/TRS 24/88.

Folk, R. L. 1974. Petrology of Sedimentary Rocks. Hemphill Publishing Company, Austin, Texas.

Grehan, A. & B. F. Keegan. 1995. Temporal variability of a *Melinna palmata* (Ampharetidae) dominated polychaete assemblage in Inner Galway Bay, west coast of Ireland, p. 175–190. *In* Keegan, B. F. & R. O'Connor (eds.), Irish Marine Science 1995. Galway University Press Ltd. Galway.

Grehan, A. J., B. F. Keegan, M. Bhaud, & A. Guille. 1992. Sediment Profile Imaging of soft substrates in the western Mediterranean: The extent and importance of faunal reworking. *Comptes redus de Académie des Sciences, Paris* 315:309–315.

Grehan, A. J., M. McKillen, & B. F. Keegan. 1998. Customisation of rapid visual reconnaissance technology for use within the deep sea sediment profile imagery, p. 1,115–1,119. *In* Oceans '98. Conference Proceedings. Oceanic Engineering Society of the Institute of Electrical and Electronics Engineers. Nice, France.

Grizzle, R. E. & C. A. Penniman. 1991. Effects of organic enrichment on estuarine macrofaunal benthos: A comparison of sediment profile imaging and traditional methods. *Marine Ecology Progress Series* 74:249–262.

Heffernan, P., B. O' Connor, & B. F. Keegan. 1983. Population dynamics and reproductive cycle of *Pholoe minuta* (Fabricus) (Polychaeta: Sigalionidae) in Galway Bay, west coast of Ireland. *Marine Biology* 73:285–291.

Holme, N. A. & A. D. McIntyre. 1984. Methods for the Study of the Marine Benthos. 2nd. Edition. Blackwell Scientific Publications, Oxford, United Kingdom.

Keary, R. & B. F. Keegan. 1975. Stratification by infaunal debris: A structure, a mechanism and a comment. *Journal of Sedimentary Petrology* 45:128–31.

Keegan, B. F. 1974a. The macrofauna of maerl substrates on the west coast of Ireland. *Cahiers de Biologie Marina* 15:513–530.

Keegan, B. F. 1974b. Littoral and benthic investigations on the west coast of Ireland - III. The bivalves of Galway Bay and Kilkieran Bay. *Proceedings of the Royal Irish Academy* 74B:85–123.

Keegan, B. F. & G. Könnecker. 1973. In situ quantitative sampling of benthic organisms. *Helgoländer wissenschaftliche Meeresuntersuchungen* 24:256–263.

Keegan, B.F., B. O'Connor, D. McGrath, & G. Könnecker. 1976. The *Amphiura filiformis-Amphiura chiajei* community in Galway Bay (west coast of Ireland) - A preliminary account. *Thalassia Yugoslavia* 12:189–198.

Keegan, B. F., B. D. S. O'Connor, & G. F. Könnecker. 1985. Littoral and benthic investigations on the west coast of Ireland - XX. Echinoderm Aggregations. *Proceedings of the Royal Irish Academy* 85B:91–99.

Keegan, B. F., B. D. O'Connor, D. McGrath, P. Dinneen, E. O'Céidigh, & P. Leighton. 1992. Comparison and interpretation of infaunal distribution patterns in Killary Harbour (west coast of Ireland) as revealed by traditional sampling and by sediment profile imagery (REMOT™), p. 21–24. *In* Gillooly, M. & G. O'Sullivan (eds.), Lough Beltra 1988. Proceedings of the 5th Annual Lough Beltra Workshop. Department of the Marine, Dublin, Ireland.

Krieger, Y., S. Mulsow, & D. C. Rhoads. 1991. Organic enrichment of the seafloor: Impact assessment using a geographic information system, p. 363–372. *In* Piollman, W. & A. Jaescheke (eds.), Informatik fur den Umweltschutz. Proceedings of a symposium at Wien, Austria, September 1990. Springer, Berlin.

Lambshead, P. J. D., H. M. Platt, & K. M. Shaw. 1983. The detection of differences among assemblages of marine benthic species based on an assessment of dominance and diversity. *Journal of Natural History* 17:859–874.

Lewis, J. R. 1978. The impact of community structure on benthic monitoring studies. *Marine Pollution Bulletin* 9:64–67.

Margalef, D. R. 1958. Informational theory in ecology. *General Systems* 3:36–71

Nichols, M., R. J. Diaz, & L. C. Schaffner. 1990. Effects of hopper dredging and sediment dispersion, Chesapeake Bay. *Environmental Geology and Water Sciences* 15:1–43.

Nilsson, H. C. & R. Rosenberg. 1997. Benthic habitat quality assessment of an oxygen-stressed fjord by surface and sediment profile images. *Journal of Marine Systems* 11:249–264.

O'Connor, B., T. Bowmer, & A. Grehan. 1983. Long-term assessment of the population dynamics of *Amphiura filiformis* (Echinodermata: Ophiuroidea) in Galway Bay, west coast of Ireland. *Marine Biology* 75:279–286.

O'Connor, B., T. Bowmer, D. McGrath, & R. Raine. 1986a. Energy flow through an *Amphiura filiformis* (Echinodermata: Ophiuroidea) population in Galway Bay, west coast of Ireland: A preliminary investigation. *Ophelia* 6:351–357.

O'Connor, B. D. S., J. Costelloe, B. F. Keegan, & D. C. Rhoads. 1989. The use of REMOTS® technology in monitoring coastal enrichment from mariculture. *Marine Pollution Bulletin* 20:384–390.

O'Connor, B. D. S., D. McGrath, & B. F. Keegan. 1986b. Demographic equilibrium: The case of an *Amphiura filiformis* assemblage on the west coast of Ireland. *Hydrobiologia* 142:151–158.

O'Connor, B, D. McGrath, G. Könnecker, & B. F. Keegan. 1993. Benthic macrofaunal assemblages of greater Galway Bay. Biology and environment. *Proceedings of the Royal Irish Academy* 93:127–136.

O' Foighil, D., D. McGrath, M. E. Connelly, B. F. Keegan, & M. Costelloe. 1984. Population dynamics and reproduction of *Mysella bidentata* (Bivalvia: Galeommatacea) Galway Bay, Irish west coast. *Marine Biology* 81:283–291.

Pearson, T. H. & R. Rosenberg. 1978. Macrobenthic succession in relation to organic enrichment and pollution of the marine environment. *Oceanography and Marine Biology Annual Review* 16:229–311.

Pielou, E. C. 1977. Mathematical Ecology. John Wiley & Sons, New York.

Revelas, E. C., J. D. Germano, & D. C. Rhoads. 1987. REMOTS : Reconnaissance of benthic environments, p. 2,069B2,083. *In* Coastal Zone '87, American Society of Civil Engineers, Seattle, Washington.

Rhoads, D. C. & S. Cande. 1971. Sediment profile camera for in situ study of organism-sediment relations. *Limnology and Oceanography* 16:110–114.

Rhoads, D. C. & J. D. Germano. 1982. Characterisation of benthic processes using sediment profile imaging: An efficient method of Remote Ecological Monitoring of the Seafloor (REMOTS™ System). *Marine Ecology Progress Series* 8:115–128.

Rhoads, D. C. & J. D. Germano. 1986. Interpreting long-term changes in benthic community structure: A new protocol. *Hydrobiologia* 142:291–308.

Rhoads, D. C., J. D. Germano, & L. F. Boyer. 1981. Sediment profile imaging: An efficient method of remote ecological monitoring of the seafloor (REMOTS™ System). *Oceans* September 1981:561–566.

Rosenberg, R. 1995. Benthic marine fauna structured by hydrodynamic processes and food availability. *Netherlands Journal of Sea Research* 34:303–317.

Rumohr, H. 1999. Soft bottom macrofauna: Collection, treatment, and quality assurance of samples. ICES Techniques in Marine Environmental Science no. 27 (revision of no. 8). International Council for the Exploration of the Sea, Copenhagen, Denmark.

Rumohr, H., E. Bonsdorff, & T. H. Pearson. 1996. Zoobenthic succession in Baltic sedimentary habitats. *Archive of Fishery and Marine Research* 44:179–214.

Rumohr, H. & H. Schomann. 1992. REMOTS sediment profiles around an exploratory drilling rig in the southern North Sea. *Marine Ecology Progress Series* 91:303–311.

Rumohr, H., H. Schomann, & T. Kujawski. 1992. Sedimentological effects of the Great Belt crossing as revealed by REMOTS photography, p. 135–139. *In* Bjornestad, E., L. Hagerman, & K. Jensen (eds.), Proceedings of the 12th Baltic Marine Biologists Symposium. Olsen and Olsen. Fredensburg, Denmark.

Schaffner, L. C., P. Jonsson, R. J. Diaz. R. Rosenberg, & P. Gapcynski. 1992. Benthic communities and bioturbation history of estuarine and coastal systems; Effects of hypoxia and anoxia. *The Science of the Total Environment* Suppl.1992:1,001–1,016.

Shin, P. K, M. E. Connelly, & B. F. Keegan. 1982. Littoral and benthic investigations on the west coast of Ireland. XV. The macrobenthic communities of North Bay (Galway Bay). *Proceedings of the Royal Irish Academy* 82 B:133–152.

Souprayen, J., J. C. Dauvin, F. Ibanez, E. Lopez-Jamar, B. O'Connor, & T. H. Pearson. 1991. Long-term trends of subtidal macrobenthic communities: Numerical analysis of four north-western European sites, p.265–437. *In* Keegan, B. F. (ed.), Space and Time Series Data Analyses in Coastal Benthic Ecology. An analytical exercise organised within the framework of the COST 647 Project on Coastal Benthic Ecology. Commission of the European Communities, Brussels.

Thorson, G. 1957. Bottom communities (sublittoral or shallow shelf). *The Geological Society of America Memoir* 67:461–534

Valente, R. M., D. C. Rhoads, J. D. Germano, & V. J. Cabelli. 1992. Mapping of benthic enrichment patterns in Narragansett Bay, Rhode Island. *Estuaries* 15:1–17.

Warwick, R. M. 1986. A new method for detecting pollution effects on marine macrobenthic communities. *Marine Biology* 92:557–562.

Imaging of Oxygen Distributions at Benthic Interfaces: A Brief Review

Bettina Koenig*, Gerhard Holst, Ronnie N. Glud, and Michael Kuehl

Abstract: Recently, new experimental techniques to map two-dimensional oxygen distributions in benthic communities have been developed. The oxygen distributions are visualized by the use of planar sensor foils with an oxygen-sensitive fluorophore layer, containing a photo-stable ruthenium complex that is reversibly quenched by oxygen. Recording the fluorescence intensity from the sensor foil using a light-sensitive digital camera system allows the two-dimensional oxygen distribution to be quantified over an area of several square centimeters at a spatial resolution better than 50 μm. Larger areas can be covered by changing the optics of the system; however, the spatial resolution decreases. Since the first primitive approaches, more sophisticated systems have been developed, in which both fluorescence intensity and fluorescence lifetime imaging is possible in one as well as two dimensions (planar optodes). We have applied planar optodes for mapping the oxygen dynamics in benthic systems exhibiting various degrees of heterogeneity (e.g., biofilms, microbial mats, and sediments with and without significant faunal activity). The technical developments of sensors, camera systems, and applications of the imaging techniques are reviewed here.

Introduction

The benthic sediment-water interface is an important and active horizon of the marine environment. Intense production and degradation of organic matter takes place within a narrow zone, leading to a dynamic exchange of solutes between the benthic community and the overlying water as well as with the sediment below this dynamic layer. The interface is not a flat horizon but is typically characterized by an extensive heterogeneity, both at micro (Jørgensen & Des Marais 1990) and macro scales (Aller et al. 1998; Glud et al. 1998a). For example, the heterogeneity is related to variability in settling rates of organic carbon, temperature, and microbial and macrofaunal activity. In order to study the millimeter-to-centimeter-thick oxic zone of benthic communities, measuring techniques with a high spatial and temporal resolution are required. Oxygen microelectrodes were introduced to aquatic biology by Revsbech et al. (1980), and they for the first time enabled detailed studies at a sub-millimeter resolution of surface sediments.

Recently, a new optical microsensor (microoptode) was introduced to marine ecology (Klimant et al. 1995). Compared to microelectrodes, microoptodes have the main advantages of cheap and easy construction, and superior long-term stability. They now represent a realistic alternative to microelectrodes even on in situ measuring platforms (Glud et al. in press a). The measuring principle of O_2 microoptodes is based on dynamic fluorescence quenching by oxygen. The fluorescence dye tris (4,7-diphenyl-1,10-phenantrolin)-ruthenium(II) perchlorate (Ru-DPP) is immobilized in a polymer matrix and coated on a thin fiber-optic tip. However, even though both microelectrode and microoptode techniques offer high spatial and temporal resolutions, microprofiles only represent a one-dimensional approach, measuring oxygen concentration at a single point in a three-dimensional heterogeneous sample. The difficult task of describing or overcoming heterogeneity of benthic communities has been performed by multiprofiling (Jørgensen & Des Marais 1990) or by using sensor arrays (Holst et al. 1997). Often, this is time consuming, requires large and sophisticated set-ups, complicates many in situ applications, and is in many instances almost impossible. For this reason, and to obtain solute distributions in two dimensions, oxygen planar optodes have been developed.

The technique of the oxygen planar sensor is based on the same principle as for oxygen microoptodes. However, instead of fixing the sensor matrix (incorporating the luminescence dye) on an optical fiber tip, the sensor matrix is immobilized on transparent support foils (Glud et al. 1996). The foils can, in combination with CCD (charged coupled device) cameras and imaging techniques, resolve O_2 gradients in two dimensions. Here, we present a brief review of planar optodes and their applications to marine science, as well as results from the first measurements obtained by a newly developed camera measuring system that enables both luminescence intensity and luminescence lifetime (rate of luminescence decay) based measurements.

Theory

Both the new and the old imaging systems for high resolution O_2 distributions in one and two dimensions, discussed below, are based on the commonly known effect of dynamic quenching of luminescent indicators by oxygen (Stern & Volmer 1919; Kautsky 1939). Oxygen molecules diffuse into a sensor matrix where they react reversibly with a fluorescent dye (Ru-DPP) described as dynamic fluorescence quenching. The effect of this reaction is a measurable decrease in fluorescence intensity with rising oxygen concentrations. The CCD camera monitors the emitted fluorescence intensity signal. According to the timing scheme (Fig. 1) and the imaging processing, the resulting image either corresponds to the fluorescence intensity or the fluorescence lifetime (rate of fluorescence decay) (Holst et al. 1998). Lifetime imaging is based on the fact that the oxygen concentration is related to the rate of the fluorescence decay. The relation between oxygen concentration (c) and the fluorescence intensity (I) and the time decay rate (τ), respectively is given by the modified Stern-Volmer equation (Bacon & Demas 1987)

$$\frac{I}{I_0} = \frac{\tau}{\tau_0} = \left[\frac{a}{(1 + Ksv * c)} + (1 - a) \right]$$

where, c = oxygen concentration
   I = the luminescence intensity, I_0 in absence of O_2
   τ = the time decay rate, τ_0 in absence of O_2
   Ksv = the coefficient expressing the quenching efficiency of the fluorophore
   a = the nonquenchable fraction of the luminescence

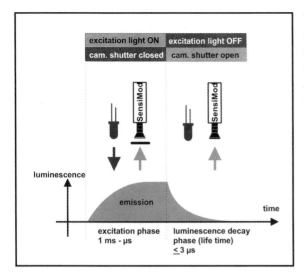

Fig. 1. Principle for image acquisition of fluorescence lifetime imaging. The timing (range of ms to μs) starts with excitation light ON & the camera shutter closed. The excitation light is then switched OFF and the camera shutter is opened (with a possible delay).

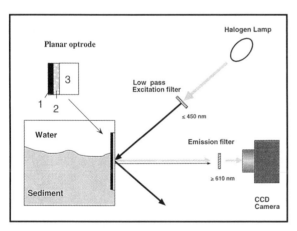

Fig. 2. Schematic illustration of the experimental set-up (Glud et al.1996). An enlargement of the three-layer planar optode is shown in the upper left corner: supporter foil (3), sensing layer (2), optical insulation (1).

The parameter a is related to the type of matrix used and has to be determined before sensor application. To evaluate oxygen concentrations from fluorescence the value of Ksv has to be determined first, via a simple two-point calibration (Glud et al. 1996).

Applications with the Old Planar Optode System

MATERIALS AND METHODS

The imaging system described by Glud et al. (1996) for oxygen measurements was based on intensity measurements. As an oxygen-quenchable fluorophore, the dye tris (4,7-diphenyl-1,10-phenantrolin)-ruthenium(II) perchlorate (Ru-DPP) was applied (Bacon & Demas 1987; Klimant & Wolfbeis 1995; Hartmann & Ziegler 1996). The fluorescence dye was dissolved together with scattering particles in plasticized PVC (Preininger et al. 1994). This solution was spread on a 175-μm thick polyester foil (©Mylar, DuPont, Germany). The thickness of the sensing layer was approximately 10 μm, and was covered by a second black silicon layer (20 μm) for optical insulation. This was necessary in order to avoid scattering effects of the biological sample and a potential stimulation of photosynthesis. Silicon was used because it is highly permeable to oxygen.

The planar optode was glued to the inside of a small Plexiglas frame (Fig. 2). To excite the fluorophore, light from a halogen lamp filtered with a blue glass-filter (450 nm) was used. After passing through an emission filter, the emitted fluorescence light was collected by a CCD camera. Camera timing and data acquisition were controlled via a personal computer. A spatial resolution of approximately 26 μm was achieved with the system. The initial calibration measurements were performed using seawater flushed with

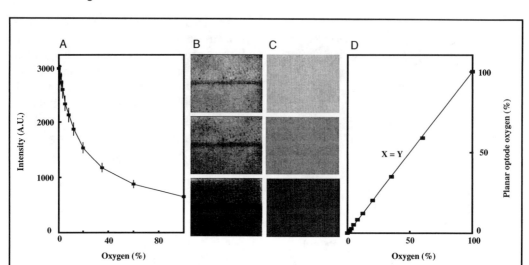

Fig. 3. (A) Average fluorescence intensity as a function of O_2 saturation in water; standard deviation shown as error bars. (B) Three original images obtained at different O_2 saturations. The darker line is due to sensor inhomogeneity. (C) Three examples of pixel by pixel calculated images at different O_2 saturations expressed on a linear gray-scale from 0 to 255. (D) The average O_2 concentration measured by the planar optode as a function of the actual O_2 concentration.

defined gas mixtures of oxygen and nitrogen. Correlation between averaged fluorescence intensity and O_2 saturation of the water is shown in Fig. 3A. The intensity decline with rising oxygen saturation resulted in a hyperbolic curve, which could be described by the modified Stern-Volmer equation. The standard deviation of the fluorescence intensity varied between 5% and 8% of the average signal of all pixels. This signal to noise ratio is mainly influenced by heterogeneity in the sensor layer, which is due to uneven distribution of the fluorophore, agglutination of scattering particles, and an uneven field of excitation light. A further source of measured heterogeneities could be unevenly distributed clear silicon used to glue the optode to the aquarium wall (Fig. 3B). Every pixel of the image can be treated as a single one-dimensional optical sensor. As a consequence, pixel by pixel calibration was performed to reduce the influence of the aspects mentioned above. The result of a pixel to pixel calibration is demonstrated in Figs. 3C,D.

APPLICATION TO MARINE SEDIMENTS

The first planar optode measurements in aquatic sediments were performed in intertidal sediment (Glud et al. 1996). The overlying water was flushed with either 12% or 24% oxygen-saturated water and, after equilibration was established, a series of images were taken (Fig. 4). To determine the position of the sediment surface, sodium dithionite grains were deposited on the sediment surface. Due to reduction of any O_2 in their immediate vicinity, the grains appeared as single blue dots on the image. The sediment-water interface was estimated by the dots on subsequent images (thick black line in Fig. 4A,B). Oxygen penetration varied between 0.54 mm and 2.62 mm at 12% O_2 saturation and between 1.8 mm and 3.5 mm at 24% O_2 saturation.

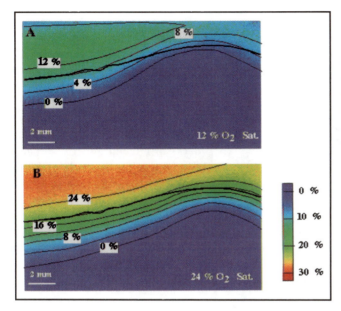

Fig. 4. Calibrated images of the O_2 distribution at the sediment-water interface with (A) 12% and (B) 24% O_2 saturation in the overlying water. The images represent an area of 17.0 x 10.2 mm. The surface is denoted with a thick black line. The O_2 concentration is expressed on a linear color-scale of 256 colors.

MEASUREMENTS ON A BIOFILM

As an example of a complex benthic community, a biofilm developing from an inoculated waste-water treatment plant was examined. Planar optodes were fixed in a flume and as biofilms gradually developed, the two-dimensional O_2 distribution at the base of the biofilms was investigated (Glud et al. 1998b). The O_2 distribution of a 400-μm thick, 13-d-old biofilm is shown in Fig. 5. As flow increased from 6.2 cm s^{-1} to 35.1 cm s^{-1}, the average O_2 saturation increased from 0% to 23.1%. Some areas of the biofilm were more sensitive to changes in flow velocities than others (Fig. 5). Flow-sensitive sites coincided with voids of anoxic 'islands' cell clusters as observed through a stereomicroscope. The complex structure ensured efficient nutrition to the base of the film and the community as such (Costerton et al. 1994; de Beer et al. 1994). Due to the high spatial heterogeneity of biofilms it is normally not appropriate to study them with microsensors. One alternative is to use planar optodes.

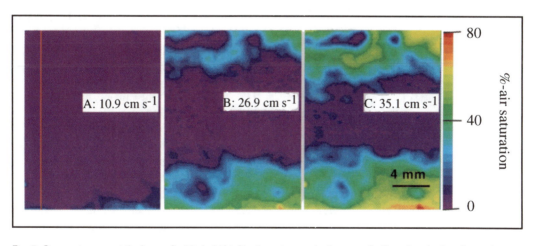

Fig. 5. Oxygen images at the base of a 13-d-old biofilm (growing on the bottom of a flow chamber) at flow velocities of (A) 10.9 cm s^{-1}, (B) 26.9 cm s^{-1}, and (C) 35.1 cm s^{-1}. The oxygen concentration is expressed on a linear color-scale of 256 colors.

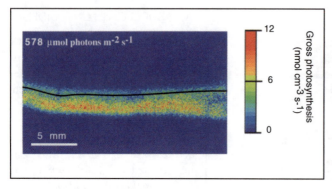

Fig. 6. Gross photosynthetic rates determined by the light-dark shift technique (Revsbech & Jørgensen 1983) at irradiance of 578 μmol photon $m^{-2} s^{-1}$. The thick horizontal line indicates the position of the mat surface.

Very recently Glud et al. (in press b) applied planar oxygen optodes in a comparable experimental set-up to study gross and net photosynthesis in a heterogeneous cyanobacterial mat. The gross photosynthetic rates of the mat were determined by applying the light-dark-shift technique to the mat (Revbech & Jørgensen 1983; Glud et al. 1992). The gross rates showed a surface layer with low activity and a deeper horizon dominated by micro colors with a relatively high activity (Fig. 6).

The New System

MATERIALS AND METHODS

To overcome heterogeneity in sensors and light field, a pixel by pixel calibration was applied in this set-up. In the case of lifetime imaging, intensity variations due to photobleaching or variable indicator concentrations are negligible since the decay rate of fluorescence is independent of, for example, fluorochrome concentrations (Holst et al. 1998). This new system, still under development, consists of a fast-gated CCD camera (SensiMod PCO) with an electronic-operated on-chip shutter and a blue excitation light source (Fig. 7). Two different light sources have been used, a combination of eight blue-light emitting diodes (LED, BP280CWPB1K, DCL Components Limited) or a Xenon flashlamp (A0021F, Oxygen Enterprise) combined with a blue bandpass filter. A close to

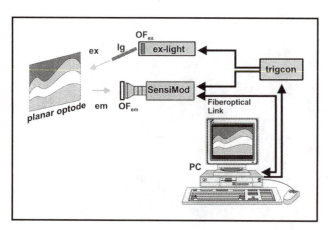

homogeneous light field was created by a fiberoptic ring light (light spot of 50 mm diameter at a distance of 50 mm from the ring light, Schölly Fiberoptic GmbH). The ring was mounted in a light-tight housing in front of the camera and connected to the light source (Holst et al. 1998). The area covered by one image is approximately 32 mm x 24 mm and corresponds to a theoretical spatial resolution of 50 μm^2 per pixel. As for the old system, applying other optics can change both the covered area and the spatial resolution.

Fig. 7. Schematic drawing of the imaging system combined with a planar optode. The imaging system consists of a CCD camera, a PC, and a trigger control unit working together with an excitation light source, optical filters (OF), and a ring light as light guide (lg).

The camera is connected to a PC, which is the controlling unit for image acquisition, storage, display, and timing. To ensure precise timing of excitation light source switching and image acquisition, essential for accurate lifetime measurements, the PC adjusts the necessary timing signals via a connected pulse-delay generator (DG535, SRS Stanford Research Systems) as described in detail by Holst et al. (1998). A schematic drawing of the set-up is shown in Fig. 7.

Concurrent with the improvement of the technical part of the system, the matrix of the planar oxygen optode has been improved. An organically modified sol-gel matrix is in part characterized by the pore size and its water insolubility (Hoshino & Mackenzie 1995; McEvoy et al.1996). During the sol-gel process, hydrolysis and polycondensation reactions of silica alkoxides take place, allowing the preparation of a unique, noncrystalline material with tunable properties. Matrices with high oxygen permeability and a hydrophobic character can be obtained by varying the sol-gel process. Adding substituted organic precursors induces flexibility and lower brittleness of the matrix. The preparation of the organically modified sol-gel was further improved to achieve a long-term stability and foil homogeneity, overcoming the main problems of plasticized PVC-based sensors.

For preparation of the planar optode, the fluorescence dye Ru-DPP (1% wt) was dissolved with scattering particles (33% wt) in the sol-gel matrix (66% wt) and spread as a thin layer (~10 μm after solvent evaporation) on a polyester foil (125 μm) as described by Glud et al. (1996). In contrast to Glud et al. this oxygen sensor has been applied without optical insulation. Because of the milky appearance of the sensor, caused by the scattering particles, the measurements are not affected by scattering from sediment particles (data not shown). While taking an intensity image, the excitation light is switched on for < 1 ms. During such short exposure, it is unlikely for photosynthetic oxygen stimulation within the sample to occur (Host et al. 1995).

MEASUREMENTS IN BIOIRRIGATED SEDIMENTS

The newly developed imaging system was applied in the intensity mode and calibrated as described above. The very first data of the new measuring device were obtained from a microbial mat sampled at shallow water depth at Gullmars Fjord (Sweden). At the sampling location the bottom water was practically stagnant. The sample, taken by a core with a flat glass window, was kept at in situ conditions. Two-dimensional oxygen distributions were measured during a 24-h cycle. During recording, the core and the measuring set-up were maintained outside the laboratory so that the sample was exposed to natural irradiance.

Oxygen distributions within the mat at 1:00 P.M. and at 1:00 A.M., respectively are shown in Fig. 8. The original size of the observed area is 12.5 mm x 18.5 mm, and the surface is approximately at the level of 14 mm (Fig. 8). The different oxygen penetration depths of the two mats reflected the effect of the two light extremes. In daylight, high levels of oxygen supersaturation were observed in the mat at a depth of approximately 4 mm. The overlying water was supersaturated. At night, high O_2 values were only observed close to the mat surface, while the overlying water was still supersaturated. In contrast to Glud et al. (1996), a relief picture (in the way of a contrast picture of the surface) was taken to detect the surface of the mat.

Nonsensitive spots in the planar optode or sand grains between the sensor and the core wall most likely caused the small dots (marked with arrows) that were visible in the images. Nonsensitive sites could result from evaporation of the solvent out of the sensing matrix. In the future, applying other solvent mixtures and a more homogeneous spreading of the matrix-dye mixture will minimize this effect.

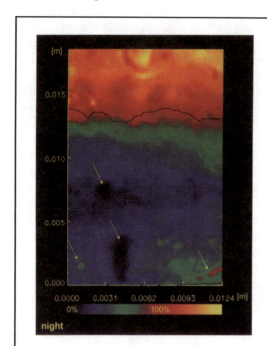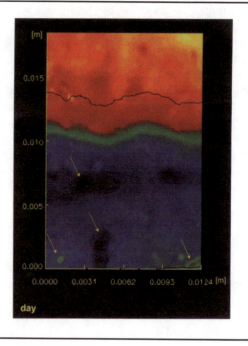

Fig. 8. Images of the oxygen distribution in the fjord mat during a 24-h cycle at 1:00 A.M. (day) and 1:00 P.M. (night). The image represents an area of 12.5 x 18.5 mm with a spatial resolution of 50 μm. The surface is denoted as a thick black line and the O_2 concentration is expressed in a linear color-scale of 256 colors. The dots marked with arrows are insensitive spots or sand grains between sensor foil and glass window.

Future Plans

In the near future, we will optimize measurements in a luminescence decay-rate-based mode. Further plans are to develop planar sensors for other chemical variables, for example, pH and CO_2, and to combine several fluorochromes in one sensor for fast detection of multiple chemical parameters. A pH planar optode could work based on a fluorescence indicator with a pH-sensitive adsorption maximum in the visible part of the spectrum and a nearby pH-independent wavelength. The second wavelength can be used as referencing system (Kohls et al. 1997). Developments at modular and flexible in situ use (for example, benthic landers) will be initiated. There are also plans to detect concentration changes continuously over time as has been done by Glud et al. (1998b).

Acknowledgments

We acknowledge the financial support from the European Commission (Microflow CT 970078) and the Max-Planck Society.

Literature Cited

Aller, R. C., P. O. J. Hall, P. D. Rude, & J. Y. Aller. 1998. Biogeochemical heterogeneity and suboxic diagenesis in hemipelagic sediments of the Panama Basin. *Deep Sea Research* 45:133–165.

Bacon, J. R. & J. N. Demas. 1987. Determination of oxygen concentrations by luminescence quenching of polymer immobilised transition-metal complex. *Analytical Chemistry* 59:2780–2785.

Costerton, J. W., Z. Lewandowski, D. de Beer, D. Caldwell, D. Korber, & G. James. 1994. Biofilm—The customised microniche. *Journal of Bacteriology* 176:2137–2142.

De Beer, D., P. Stoodley, F. Roe, & Z. Lewandowski. 1994. Effects of biofilm structures on oxygen distribution and mass transport. *Biotechnology and Bioengineering* 43:1131–1138.

Glud, R. N., N. B. Ramsing, & N. P. Revsbech. 1992. Photosynthesis and photosynthesis coupled respiration in natural biofilms quantified with oxygen microsensors. *Journal of Phycology* 28:51–60.

Glud, R. N., N. B. Ramsing, J. K. Gundersen, & I. Klimant. 1996. Planar optodes: A new tool for fine scale measurements of two-dimensional O_2 distribution in benthic communities. *Marine Ecology Progress Series* 140:217–226.

Glud, R. N., O. Holby, F. Hofmann, & D. E. Canfield. 1998a. Benthic mineralisation in Arctic sediments (Svalbard). *Marine Ecology Progress Series* 173:237–251.

Glud, R. N., I. Klimant, G. Holst, O. Kohls, V. Meyer, M. Kuehl, & J. K. Gundersen. In press a. Adaptation, test and in situ measurements with O_2 microoptodes on benthic landers. *Deep Sea Research*

Glud, R. N., M. Kuehl, O. Kohls, & N. B. Ramsing. In press b. Heterogeneity of oxygen production and consumption in a photosynthetic microbial mat as studied by planar optodes. *Journal of Phycology*

Glud, R. N., C. M. Santegoeds, D. de Beer, O. Kohls, & N. P. Ramsing. 1998b. Oxygen dynamics at the base of a biofilm studied with planar optodes. *Aquatic Microbial Ecology* 14:223–233.

Hartmann, P. & W. Ziegler. 1996. Lifetime imaging of luminescent oxygen sensors based an all-solid-state technology. *Analytical Chemistry* 68:4512–4514.

Holst, G., M. Kühl, & I. Klimant. 1995. A novel measuring system for oxygen microoptodes based on a phase modulation technique. *SPIE Proceedings* 2508:387–398.

Holst, G., R.N. Glud, M. Kuehl, & I. Klimant. 1997. A microoptode array for fine-scale measurements of oxygen distribution. *Sensors and Actuators B* 38-39:122–129.

Holst, G., O. Kohls, I. Klimant, B. Koenig, M. Kuehl, & T. Richter. 1998. A modular luminescence lifetime imaging system for mapping oxygen distribution in biological samples. *Sensors & Actuators B* 51/1-3:163–170.

Hoshino, Y. & J. D. Mackenzie. 1995. Viscosity and structure of ormosil solutions. *Journal of Sol-Gel Science & Technology* 5:83–92

Jørgensen, B. B. & D. Des Marais. 1990. The diffusive boundary layer of sediments: Oxygen microgradients over a microbial mat. *Limnology and Oceanography* 35:1353–1355.

Kautsky, H. 1939. Quenching of luminescent by oxygen. *Faraday Society* 35:216–219.

Klimant, I., V. Meyer, & M. Kuehl. 1995. Fiber-optic microsensors, a new tool in aquatic biology. *Limnology and Oceanography* 40:1159–1165.

Kohls, O., I. Klimant, G. Holst, & M. Kühl. 1997. Development and comparison of pH microoptodes for use in marine systems. *SPIE Conference on Progress in Biomedical Optics* 2978:82–91.

Kuehl, M., Y. Cohen, T. Dalsgraad, B. B. Jørgensen, & N. P. Revsbech. 1995. Microenvironment and photosynthesis of zooxanthellae in scleractinian corals studied with microsensors for O_2, pH, and light. *Marine Ecology Progress Series* 117:159–172.

McEvoy, A. K., C. M. McDonagh, & B. D. MacCraith. 1996. Dissolved oxygen sensor based on fluorescence quenching of oxygen-sensitive ruthenium complexes immobilised in sol-gel-derived porous silica coatings. *Analyst* 121:785–788.

Preininger, C., I. Klimant, & O. S. Wolfbeis. 1994. An optical fiber sensor for biological oxygen demand. *Analytical Chemistry* 66:1841–1846.

Revsbech, N. P., J. Sørensen, T. H. Blackburn, & J. P. Lomholt. 1980. Distribution of oxygen in marine sediments measured with microelectrodes. *Limnology and Oceanography* 25:403–411.

Revsbech, N. P. & B. B. Jørgensen. 1983. Photosynthesis of benthic microflora measured by the oxygen microprofile method: Capabilities and limitations of the method. *Limnology and Oceanography* 28:749–756.

Stern, O. & M. Volmer. 1919. Ueber die Abklingzeit der Fluoreszenz. *Physikalische Zeitschrift* 20:183–188.

A New Model of Bioturbation for a Functional Approach to Sediment Reworking Resulting from Macrobenthic Communities

Frédérique François*, Jean-Christophe Poggiale, Jean-Pierre Durbec, and Georges Stora

Abstract: We developed a mechanistic model to characterize, quantify, and predict sediment reworking resulting from macrobenthic communities. It is a time- and space-dependent mixing model using ordinary differential equations and based on functional grouping of individuals that make up a community. It uses five elementary models, each formalizing the sediment reworking resulting from one of the five functional groups defined in the literature: the biodiffusers, the upward-conveyors, the downward-conveyors, the regenerators, and the gallery-diffusers. This model describes the temporal changes of tracer distribution in a 2-D sediment section inhabited by a defined set of organisms. It is a model well suited for (1) characterizing, checking, and predicting the sediment reworking resulting from a well-defined macrobenthic community and studying its variations according to different environmental conditions, (2) testing the validity of the sediment mixing coefficients estimated with classical advection-diffusion models by adjustment with 1-D experimental profiles, (3) proposing for some macrobenthic communities a more appropriate model than the classical advection-diffusion models, (4) studying the roles of species diversity, functional composition, and functional richness of macrobenthic communities in bioturbation, and (5) studying the effect of the specific variability within functional groups on the bioturbation process.

Introduction

In marine and lacustrine sediments, bioturbation is one of the major processes affecting ecosystem functions. Through activities such as gallery diging, tube irrigation, movement, and fecal pellet egestion, macrobenthic communities alter the primary structure of sedimentary deposits. They induce particle and porewater displacements and so affect the physical, chemical, and biological properties of the substratum (e.g., Rhoads 1974; Rhoads et al. 1978; Aller & DeMaster 1984; Eckman 1985; Gérino 1990; Meadows & Meadows 1991; Gilbert et al. 1995, 1996). In particular, their activity determines whether a particle (e.g., organic-matter particle, sediment-associated contaminant) remains buried in the substratum, degrades, enters food webs, or is released to the water column. Thus, bioturbation influences fluxes of organic matter, nutrients, and contaminants at the sediment-water interface and within the sediment column (e.g., Aller 1982; Kristensen & Blackburn 1987; Gérino 1990; Aller 1994; Kure & Forbes 1997; Madsen et al. 1997).

73

Characterizing and quantifying bioturbation is essential to understanding the fate of the matter that settles on the sediment surface. To this purpose, several models have been developed. Most of them adopt an advection-diffusion formulation where the sediment reworking resulting from an entire macrobenthic community is integrated in an overall term of diffusion and advection (Goldberg & Koide 1962; Guinasso & Schink 1975; Robbins et al. 1979; Berner 1980; Fisher et al. 1980; Aller 1982; Boudreau 1986). Some of these models also take into account nonlocal transport when sediment motions are not random but directed and/or when average distance traveled during mixing events is essentially of the same order as the full mixed-layer depth (Boudreau 1986).

While these models are good at fitting experimental data and thus at quantifying the bioturbation processes that have occurred at a site, their efficacy for making predictions is limited. They do not explicitly link the actual sediment mixing events to organism activities. Wheatcroft et al.'s (1990) proposed step lengths and rest period decomposition of the biodiffusion coefficient (D_b) are a first attempt to link specific mechanisms of biogenous sediment mixing with a commonly used bioturbation coefficient. To progress in this way, we have developed a mechanistic model based on a functional approach to the different individuals that make up a community. This approach makes it possible to study the community-bioturbation relation by varying, for instance, the species diversity, the functional composition, or the functional richness of a set of species populating a given area. Five functional groups, each including species with analog mixing modes, have been defined according to the literature (e.g., Boudreau 1986; Smith et al. 1986; Gardner et al. 1987; François et al. submitted). And, a mechanistic model of sediment reworking has been developed for each one (François et al. 1997, submitted): the biodiffusers, the upward-conveyors, the downward-conveyors, the regenerators, and the gallery-diffusers. The bioturbation model we present here makes use of different elementary mechanistic models according to the functional composition and richness of the macrobenthic community studied. It is a mechanistic time- and space-dependent mixing model using ordinary differential equations and describing the changes of tracer concentration in a 2-D sediment section inhabited by a defined macrobenthic community.

The Bioturbation Model

DEFINITION OF THE DIFFERENT FUNCTIONAL GROUPS

Five functional groups can be distinguished; each of these is described below and illustrated in Fig. 1.

Biodiffusers include species whose activities result in diffusive transport of sediment: they move sediment particles in a random manner over short distances. Movements under the sediment surface of the bivalves *Ruditapes decussatus* (Linné) and *Venerupis aurea* (Gmelin), for instance, lead to this type of mixing (Gérino 1992; François et al. 1999), as do those of the amphipod *Pontoporeia hoyi* (Smith) described by Robbins et al. (1979).

Upward-conveyors include head-down, vertically oriented species that remove sediment at depth in the substratum and expel it at the sediment-water interface. They cause both "active" and "passive" transport of sediment. In active transport, sediment is moved from the bottom up through their gut. In passive transport, sediment moves all around them from the interface to the bottom of their feeding zone due to sediment discharge at the sediment-water interface and to the subsidence of the sediment column in the feeding cavity. This behavior has been described accurately for tubificid oligochaetes inhabiting lake sediments (e.g., Fisher et al. 1980).

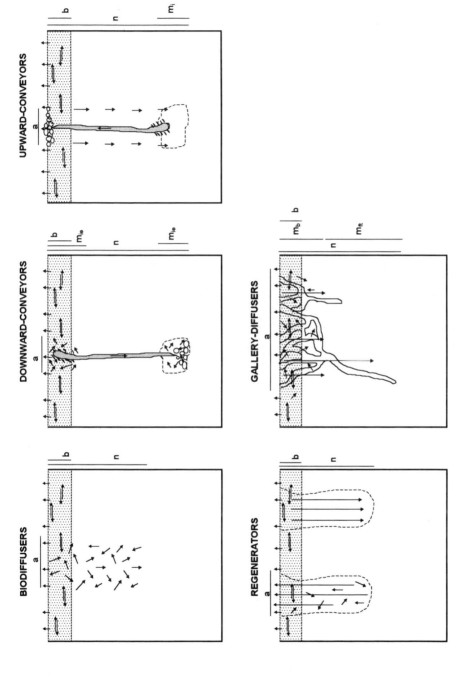

Fig. 1. Schematic description of the physical and biological reworking mechanisms of (1) the biodiffusers, (2) the upward-conveyors, (3) the downward-conveyors, (4) the regenerators, and (5) the gallery-diffusers. The parameters: a is the width of the organism mixing zone, n is the depth of the organism mixing zone, m_{ie} is the height of the ingestion-egestion zone of the downward-conveyor organism, m_i is the height of the ingestion zone of the upward-conveyor organism, m_{ft} is the height of the tube bottom zone of the gallery-diffuser organism, m_b is the height of the diffusion zone of the gallery-diffuser organism, and b is the height of the first row of the matrix. Arrows represent movement of sediment particles.

Downward-conveyors include head-up, vertically oriented species causing "active" transport of sediment through their gut from the sediment-water interface to their egestion depth. Smith et al. (1986) have described this behavior pattern for worms of the phylum Sipunculida.

Regenerators include digging species that "transfer sediment at depth to the surface where it is washed away and replaced by sediment of surficial signature" (Gardner et al. 1987). This behavior has two effects: an output of sediment to the water column during digging and a net movement of surficial sediment to the bottom of the burrow after it has been deserted. Gardner et al. (1987) cite, for example, the regeneration process of the fiddler crab *Uca pugilator* (Smith).

Gallery-diffusers include species that dig gallery systems in sediment, leading to biodiffusive tracer distribution in the upper layers of the sediment, intensively drilled by organisms, and leading to an advective transport of matter from the surface to the deep part of the tubes due to egestion of feces and to the dragging of particles down to the tube bottom by animal movements. The polychaete *Nereis diversicolor* (O.F. Müller) reworks the sediment by this process (François et al. submitted).

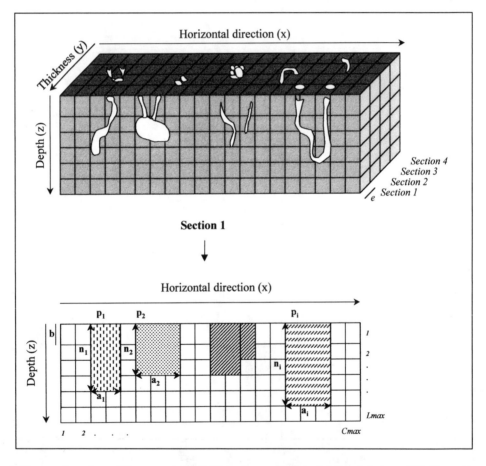

Fig. 2. Passage from the sedimentary column (3 D) to the matrix of the functional diagram (2D). The position of organism i in the matrix is designated p_i, a_i is the width of the mixing zone of organism i , n_i is the depth of the mixing zone of organism i , b is the height of the first row of the matrix, and $Lmax$ and $Cmax$ are the total number of rows and columns of the matrix. The different pattern designs indicate different functional groups.

Functional Diagram

As for the elementary models (François et al. 1997), the sediment column is divided into sections of e units thick (Fig. 2). The reworking processes are considered invariable in this direction. Each section is divided into cells in two directions, vertically and horizontally. Cells are the same size except for the height of the first-row cells. Their height, b, varies with the scale of local water currents, which induce lateral mixing processes in the upper layer of sediment. The cells of the first column are linked to those of the last to ensure horizontal sediment continuity. Each organism that reworks the sediment occupies a matrix a_i cells wide and n_i cells deep, which is initialized at cell p_i located at the sediment-water interface. Activity zones of different individuals can be joined side by side but never overlap. Each organism reworks its area of sediment according to its functional group mixing mechanisms and with its specific characteristics (e.g., depth of activity, mixing intensity; Table 1).

Mixing is ensured by fluxes between cells according to the mode of transport that proceeds: for example, biodiffusion, advection, or ingestion of sediment. Matrix cells without organisms are also subject to fluxes: physical mixing and output to the water column for first-row cells and tracer degradation for all the matrix cells. In our approach, the different organisms of the community are randomly distributed across the matrix. The part of the matrix without individuals is divided into vacant zones with randomly defined width and depth equal to the matrix height. Organism zones and vacant zones are then randomly distributed along the matrix in such way that the different zones do not overlap and the whole matrix is occupied. This process can be repeated many times during calculations to simulate the movements of organisms in the sediment.

Mathematical Formulation

As a first approximation, mixing mechanisms of separate individuals are considered additive. Biological interactions such as between species or functional groups are not taken into account. Only physical mixing due to local water currents ensures material fluxes in the spaces between individuals. The mathematical formulation of the bioturbation model is thus the set of the functional group's equations plus the equations for the vacant zones. For each cell of the matrix, there is an ordinary differential equation such as

$$\frac{dQ(l,c,t)}{dt} = \sum_{k=1}^{K_{l-1,c}} V_{l-1,c}^{k} \, Q(l-1,c,t) + \sum_{k=1}^{K_{l+1,c}} V_{l+1,c}^{k} Q(l+1,c,t)$$

$$+ \sum_{k=1}^{K_{l,c-1}} V_{l,c-1}^{k} Q(l,c-1,t) + \sum_{k=1}^{K_{l,c+1}} V_{l,c+1}^{k} Q(l,c+1,t)$$

$$- \sum_{k=1}^{K_{l,c}} V_{l,c}^{k} \, Q(l,c,t) + \sum S$$

where:

$Q(l,c,t)$ is the quantity of tracer contained in the cell in row l and column c, at time t (unit of mass, M)

$K_{i,j}$ is the number of fluxes affecting the cell (l,c) and coming from the cell (i,j)

$V_{i,j}^{k}$ is the mixing coefficient characterizing the flux number k getting out of the cell (i,j) and affecting the cell (l,c)

$\sum S$ is the sum of the variations that affect the cell (l,c) but that do not depend on the instantaneous quantity of tracer contained in cells (MT^{-1}), such as variations due to the advection in bulk of a portion of sediment.

Considering, for instance, a cell of the first row of a section of sediment inhabited by an upward-conveyor (Fig. 3), the variation equation is

$$\frac{dQ(l,c,t)}{dt} = Dh\ (Q(l,c+1,t) +\ Q(l,c-1,t))$$
$$-\ (2Dh + f + dg)\ Q(l,c,t)$$
$$+\ \frac{R_{ie}}{a^2 m}\ \sum_{\substack{i=n-b\\-m+2}}^{n}\ \sum_{j=p}^{p+a-1} Q(i,j,t) - \frac{R_{ie}}{ab}\ QT(l,c,t)$$

The output fluxes affecting the cell (l,c) are degradation of the tracer (dg), output to the water column (f), exchanges due to the physical mixing occurring between cells of the first layer of b cells of height (Dh), and output due to the bulk fall of a quantity (QT) of sediment into the ingestion cavity at depth $(R_{ie}/ab\ QT(l,c,t))$. The inputs come from the joined cells $(l-1,c)$ and $(l+1,c)$ because of physical mixing (Dh), and from the cells of the ingestion cavity due to egestion of fecal pellets at the sediment surface $(R_{ie}/a^2m\ \Sigma_i\ \Sigma_j\ Q(i,j,t))$. R_{ie} is the rate of sediment ingestion and egestion of the upward-conveyor organism (T^{-1}), and a and m are respectively the width of its mixing zone and the height of its ingestion cavity, both expressed in number of cells. More details are given in François et al. (1997).

SIMULATIONS

Numerical solutions for $Q(l,c,t)$ were computed using the Runge Kutta method (RK4) with a fixed step determined according to the conditions imposed by the upward-conveyors inhabiting the sediment. As described by François et al. (1997), the modeling of the advection process due to the subsidence of the sediment column in the cavity dug at depth by the animal imposes a time step for each organism, depending on their mixing characteristics. At the start of each simulation, the smallest time step of all upward-conveyors is determined and used for the whole community. At the end of the simulation, the quantities of tracer in each cell and at each time step are available. This makes it possible to represent the tracer distribution in two dimensions (vertically and horizontally) and the tracer profile (quantity versus depth) in the sediment section at each time.

In the present paper, parameter values are not specified. They have been chosen either (i) in the range of those estimated by François et al. (1999, submitted) using elementary models (François et al. 1997) and the results of in vitro experimentation

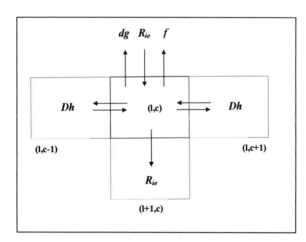

Fig. 3. Fluxes affecting the first-row cells of the reworking zone of an upward-conveyor organism. (l,c) designates the row and the column of the cell considered, Dh is the physical mixing rate, f is the output to the water column rate, dg is the degradation rate of the tracer, and R_{ie} is the ingestion-egestion rate of the upward-conveyor organism.

Table. 1. Notation, definition, and units of the different parameters used in the bioturbation model.

	Abbreviations	Definitions	Units
		Variable	
	$Q(l,c,t)$	Quantity of tracer contained in the cell in row l and column c at time t.	Mass^{-1}
		Parameters	
Matrix	C_{max}	Total number of columns in the matrix	Number of cells
	L_{max}	Total number of rows in the matrix	Number of cells
	c	Number of the column considered	(-)
	l	Number of the row considered	(-)
	b	Height of the first row	Number of cells
Physical	Dh	Physical diffusion rate	Time^{-1}
	f	Output to the water column rate	Time^{-1}
Biological			
– Position	a^i	Width of the mixing zone of organism i	Number of cells
	n^i	Depth of the mixing zone of organism i	Number of cells
	p^i	Organism i position in the matrix	(-)
	m_b^i	Height of the diffusion zone of the gallery-diffuser organism i	Number of cells
	m_{ft}^i	Height of the tube bottom zone of the gallery-diffuser organism i	Number of cells
	m_{ie}^i	Height of the ingestion-egestion zone of the downward-conveyor organism i	Number of cells
	m_i^i	Height of the ingestion zone of the upward-conveyor organism i	Number of cells
– Mixing	R_d^i	Biodiffusion rate of organism i	Time^{-1}
	R_t^i	Biotransport rate of the gallery-diffuser organism i	Time^{-1}
	R_{ie}^i	Ingestion and egestion rate of conveyor organism i	Time^{-1}
	ff	Expulsion rate of sediment to the water column of the regenerator i during its active stage	Time^{-1}
Biogeochemical	dg	Degradation rate of the tracer	Time^{-1}

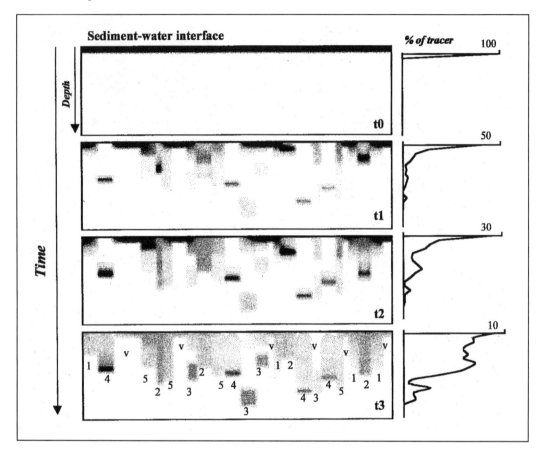

Fig. 4. Changes over time of the 2-D distribution and 1-D profile of a tracer initially deposited at the sediment-water interface of a sediment section reworked by a community of 20 theoretical individuals shared among the five functional groups. (1) biodiffuser, (2) upward-conveyor, (3) downward-conveyor, (4) regenerator, (5) gallery-diffuser, and (v) a vacant zone. Notation and units are listed in Table 1.

with polychaetes and bivalves, (ii) according to particulate mixing parameters obtained from the literature (Karickhoff & Morris 1985; Gérino 1992), or (iii) arbitrarily when no information was available.

APPLICATIONS

Application: *Characterizing, checking, and predicting sediment reworking due to a well-defined macrobenthic community, and studying its variations according to different environmental conditions.*

Knowledge of the different species of an experimental community and of their individual mixing characteristics allows simulation of the sediment reworking due to this set of individuals under different tracer input conditions (transient or constant). Figure 4 shows the 2-D distributions and the

1-D profiles of a tracer initially shared equally among the cells of the first row of the matrix subjected to the sediment reworking of a theoretical community of 20 individuals apportioned among the five functional groups previously defined. The 2-D distributions present the changes in tracer patchiness over time induced by this community, and the 1-D profiles show the mean percentages of tracer present at each level for each time. The 2-D distributions can be compared to results of experimentation carried out with flat transparent cores and [137]cesium-labeled clay layers periodically scanned (as perfected by Matisoff 1995); 1-D profiles are the common results of core analysis.

This approach will allow investigation of the interactions between different mixing modes experimenting with combinations of different macrobenthic species and comparing them to simulation results.

This model is also a good tool to interpret the effect of environmental perturbations (e.g., phytoplanktonic bloom, hydrocarbon or other contaminant deposition) on the bioturbation process. It provides a basis for testing hypotheses such as an alteration in the ethological characteristics of the organisms (e.g., depth or intensity of reworking), a change in the dominant species, or a change in the functional composition and/or richness of the community.

Application: Testing the validity of the sediment mixing coefficients estimated with classical advection-diffusion models by adjustment with 1-D experimental profiles.

Most field measurements or laboratory experiments with cores lead to 1-D tracer profiles. They are instantaneous records of the tracer mean concentration at each level of the cores. Horizontal patchiness due to the different organism activities is ignored, as are the high-frequency events that lead to the observed tracer distribution. The nondetermination of the present macrobenthic species or lack of knowledge of their reworking mode and intensity can lead to false interpretation and so to false quantification of the bioturbation process. Figure 5 shows tracer profiles of sets of organisms having strongly differing organism associations. Although the mechanisms brought into play, and sometimes the 2-D tracer distributions, are quite different, the profiles look very similar. Consideration of the 1-D profiles alone leads to the conclusion of similar sediment reworking modes and calculation of similar values of biodiffusion mixing rates in spite of the widely differing dynamics of the tracer in the sediment column. Different dynamics can have very different implications for biochemical processes insofar as they determine change over time of the environmental conditions of tracer particles (such as oxidation-reduction potential or microbial communities).

The bioturbation model does not allow inferring of the specific mechanisms of reworking of the different individuals of the community, but it does give the opportunity to simulate the 2-D tracer distribution due to these organisms. It thus provides a basis for judging the fit of the 1-D tracer profiles calculated with classical advection-diffusion models to the observed distributions.

Application: Proposing for some macrobenthic communities a more appropriate model than the classical advection-diffusion models.

Diagenetic models need bioturbation modeling to integrate the influence of macrobenthic organisms on porewater and particle distributions in sediment. The classical advection-diffusion models are limited for making predictions insofar as they do not explicitly link the actual sediment mixing events to the organism's activities. The different rapid mixing events are not taken into account, so a lot of information concerning the instantaneous environmental conditions of the tracer

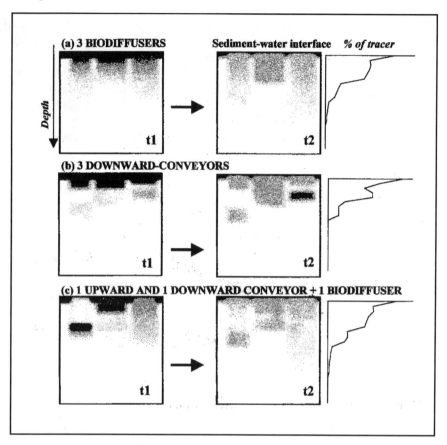

Fig. 5. The 2-D distributions and 1-D profiles of a tracer initially deposited at the sediment-water interface of a sediment section reworked by three different sets of three individuals. Despite the different reworking characteristics of the three sets, profiles look similar. Notation and units are listed in Table 1.

(e.g., O_2, microbial community) is ignored. For that, our mechanistic approach would fill some gaps left by advection–diffusion models. Moreover, our model offers a means to study the community-bioturbation relation, which is necessary for a better understanding and modeling of the bioturbation process. In particular, the model would give answers to some questions: Does species diversity, functional composition, or functional richness control bioturbation? Are there keystone species and/or keystone functional types? Does it depend on the community?

We still lack knowledge about the diversity-ecosystem functioning relation. The classic view that biodiversity begets superior ecosystem function is being called into question increasingly often. Recently, Tilman et al. (1997) and Hooper and Vitousek (1997) have shown with herbaceous communities that, more than species diversity, functional composition (defined as the kind of the different functional groups) and functional richness (defined as number of functional groups) are the principal factors explaining ecosystem functioning. They showed that productivity and nutrient cycling are controlled to an overwhelming extent by the functional characteristics of the dominant plants and that the immediate benefits of species-richness within functional groups remain weak. To provide tests of such relations, experiments are required. In complement to field and laboratory

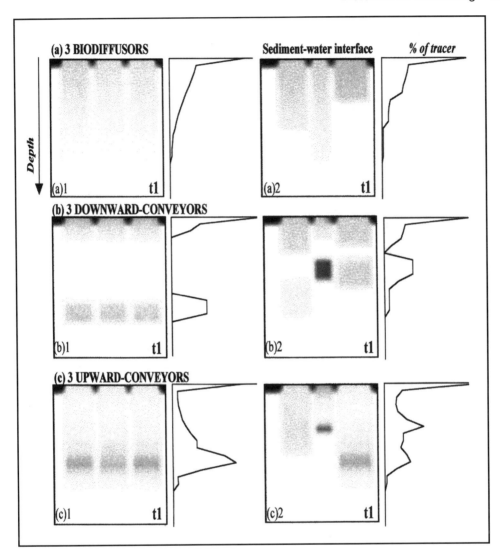

Fig. 6. The 2-D distributions and 1-D profiles of a tracer initially deposited at the sediment-water interface of a sediment section reworked by three different sets of (1) three identical functional analogs and (2) three different functional analogs. According to the specific characteristics of the individuals, the tracer profiles due to functional analogs differ. Notation and units are listed in Table 1.

experiments, community models constitute well-suited tools to study the relations between ecosystem function and diversity. An example is the mechanistic model presented by Loreau (1998) of a spatially structured ecosystem in which plants compete for a limiting soil nutrient. This model shows that plant species richness does not necessarily enhance ecosystem processes but that complementarity among species in the space they occupy belowground and positive correlation between mean resource-use intensity and diversity could generate such an effect. Our bioturbation model constitutes a well-suited tool to study the diversity-ecosystem functioning relation for benthic ecosystems. It enables computer experiments to be carried out in which the species diversity, the functional diversity, and the functional composition are directly controlled. This type of experiment may also lead to determination of the keystone species and/or keystone functional types, which have a major impact on the bioturbation process of the macrobenthic community.

Is specific variability within functional groups an important factor in the community bioturbation process? Results presented by François et al. (1999) show that two similar species of bivalve mollusc, *Ruditapes decussatus* and *Venerupis aurea*, cohabiting in muddy sand, both rework the sediment in a biodiffuser mode but with different intensities. *R. decussatus* mixes the sediment at a biodiffusion rate twice as high as *V. aurea* but expels sediment to the water column at a rate only one-half as large. These differences lead to a bigger capacity and a faster rate of burial of the material delivered to the sediment surface.

R. decussatus buries 41.8% of the material arriving at the sediment surface versus 20.5% for *V. aurea*. Of the matter that arrives at the sediment surface, 20.5% is buried in 4.9 d by *R. decussatus* and in 28.6 d for *V. aurea*. This variability can strongly affect tracer patchiness and profiles in sediment when cohabiting species of the same functional group have different mixing characteristics. Differences in depth and intensity of mixing induce tracer patchiness in the sediment column compared with the horizontal uniformity observed for identical species (Fig. 6). Profiles are also affected by specific variability insofar as the characteristic profiles of each functional group (François et al. 1997) are strongly perturbed and so more difficult to interpret. So, for some sets of species, specific variability within functional groups seems to have a significant effect on sediment reworking.

The answers to these questions should improve understanding of the community-bioturbation relation and thus help to provide a more appropriate model for bioturbation than the advection-diffusion models.

Conclusion

We have developed a mechanistic model that calculates the 2-D distribution of a tracer in a sediment section mixed by a set of organisms whose modes and biological characteristics of reworking are known. It allows prediction of the tracer distribution in sediment mixed by a macrobenthic community. This model is a suitable tool to investigate tracer patchiness due to organisms inhabiting sediment whose 1-D tracer profiles are known and so to test the validity of the diffusion analogy for bioturbation. Moreover, this approach opens up new perspectives for the study of the community-bioturbation relation. It enables computer tests of the relation between bioturbation process and, for example, species diversity, functional composition, functional richness, or specific variability. This approach could lead to determination of keystone species and/or keystone functional groups whose activities have a major impact on the sediment mixing caused by the whole community. Understanding of the community-bioturbation relation is essential for further progress in bioturbation and diagenesis modeling. Laboratory experiments to characterize and quantify sediment reworking due to the different species of macrobenthic communities and to test results of the various simulations are needed.

Acknowledgments

This work was supported by the G. D. R. HYCAR (CNRS, Universite, and ELF-Aquitaine).

Literature Cited

Aller, R. C. 1982. The effects of macrobenthos on chemical properties of marine sediment and overlying water, p. 53–102. *In* McCall, P. L. & M. J. S. Tevest (eds.), Animal-Sediment Relations. Plenum Press, New York.

Aller, R. C. & D. J. DeMaster. 1984. Estimates of particle flux and reworking at the deep-sea floor using $^{234}Th/^{238}U$ disequilibrium. *Earth and Planetary Science Letters* 67:308–318.

Aller, R. C. 1994. Bioturbation and remineralization of sedentary organic matter: Effects of redox oscillation. *Chemical Geology* 114:331–345.

Berner, R. A. 1980. Early Diagenesis: A Theoretical Approach. Princeton University Press, Princeton, New Jersey.

Boudreau, B. P. 1986. Mathematics of tracer mixing in sediment. I-Spatially-dependent, diffusive mixing. II: Non local mixing and biological conveyor-belt phenomena. *American Journal of Science* 286:161–238.

Eckman, J. E. 1985. Flow distribution by an animal tube mimic affects sediments bacterial colonization. *Journal of Marine Research* 43(2):419–435.

Fisher, J. B., W. L. Lick, P. L. McCall, & J. A. Robbins. 1980. Vertical mixing of lake sediments by tubificid oligochaetes. *Journal of Geophysical Research* 85:3997–4006.

François, F., K. Dalegre, F. Gilbert, & G. Stora. 1999. Variabilité spécifique à l'intérieur des groupes fonctionnels: Étude du remaniement sédimentaire de deux bivalves Veneridae, *Ruditapes decussatus* et *Venerupis aurea*. *Comptes Rendus de l'Académie des Sciences, Paris, Ser. 3 (Sciences de la vie)* 322:339–345.

François, F., J. C. Poggiale, J.P. Durbec, & G. Stora.1997. A new approach for the modeling of sediment reworking induced by a macrobenthic community. *Acta Biotheoretica* 45:295–319.

Gardner, L. R., P. Sharma, & W. S. Moore. 1987. A regeneration model for the effect of bioturbation by fiddler crabs on ^{210}Pb profiles in salt marsh sediments. *Journal of Environmental Radioactioactivity* 5:25–36.

Gérino, M. 1990. The effects of bioturbation on particle distribution in Mediterranean coastal sediment. Preliminary result. *Hydrobiologia* 207:251–258.

Gérino, M. 1992. Etude expérimentale de la bioturbation en milieu littoral et profond. Quantification des structures de bioturbation et modélisation du remaniement biologique du sédiment. Thése de Doctorat de l' Université Aix-Marseille II, France.

Gilbert, F., P. Bonin, & G. Stora. 1995. Effect of bioturbation on denitrification in a marine sediment from the west Mediterranean littoral. *Hydrobiologia* 304:49–58.

Gilbert, F., G. Stora, & J. C. Bertrand. 1996. In situ bioturbation and hydrocarbon fate in an experimental contaminated Mediterranean coastal ecosystem. *Chemosphere* 33(8): 1449–1458.

Golberg, E. D. & M. Koide. 1962. Geochronological studies of deep-sea sediments by the ionium/thorium method. *Geochimica et Cosmochimica Acta* 26:417–450.

Guinasso, N. L. & D. R Schink. 1975. Quantitative estimates of biological mixing rates in abyssal sediments. *Journal of Geophysical Research* 80(21):3032–3043.

Hooper, D. U. & P. Vitousek. 1997. The effects of plant composition and diversity on ecosystem processes. *Science* 277:1,302–1,305.

Karickhoff, S. W. & K. R. Morris. 1985. Impact of tubificid oligochaetes on pollutant transport in bottom sediments. *Environmental Science & Technology* 19(1):51–56.

Kristensen, E. & T. Blackburn. 1987. The fate of organic carbon and nitrogen in experimental marine sediment systems. Influence of bioturbation and anoxia. *Journal of Marine Research* 45:231–237.

Kure, L. V. & T. L. Forbes. 1997. Impact of bioturbation by *Arenicola marina* on the fate of particle-bound fluoranthene. *Marine Ecology Progress Series* 156:157–166.

Loreau, M. 1998. Biodiversity and ecosystem functioning: A mechanistic model. *Proceeding of the Naturalist Academy of Science USA* 95:5632–5636.

Madsen, S. D., T. L. Forbes, & V. E. Forbes. 1997. Particles mixing by the polychaete *Capitella* species 1: Coupling fate and effect of the particle-bound organic contaminant (fluoranthene) in a marine sediment. *Marine Ecology Progress Series* 147:129–142

Matisoff, G. 1995. Effects of bioturbation on solute and particle transport in sediments, p. 201–272. *In* Allen, H. E. (ed.), Metal Contaminated Aquatic Sediments. Ann Arbor Press, Chelsea, Michigan.

Meadows, P. S. & A. Meadows. 1991. The geotechnical and geochemical implications of bioturbation in marine sedimentary ecosystems. *Symposium of the Zoological Society of London* 63:157–181.

Rhoads, D. C. 1974. Organism-sediment relations on the muddy sea floor. *Oceanography and Marine Biology an Annual Revue* 12:263–300.

Rhoads, D. C., P. L. McCall, & J. Y. Yingst. 1978. Disturbance and production on the estuarine seafloor. *American Journal of Science* 66:577–586.

Robbins, J.A., P. L. McCall, J. B. Fisher, & J. R. Krezoski. 1979. Effects of deposit feeders on migration of ^{137}Cs in lake sediments. *Earth and Planetary Science Letters* 42:277–287.

Smith, J. N., B. P. Boudreau, & V. Noshkin. 1986. Plutonium and ^{210}Pb distributions in Northeast Atlantic sediments, subsurface anomalies caused by nonlocal mixing. *Earth & Planetary Science Letters* 28(1):15–28.

Tilman, D., J. Knops, D. Wedin, P. Reich, M. Ritchie, & E. Siemann. 1997. The influence of functional diversity and composition on ecosystem process. *Science* 277:1300–1302.

Wheatcroft, R. A., P. A. Jumars, C. R. Smith, & A. R. M. Nowell. 1990. A mechanistic view of the particulate biodiffusion coefficient. Step lengths, rest periods and transport direction. *Journal of Marine Research* 48(1):177–207.

Unpublished Materials

François, F., M. Gérino, G. Stora, J. P. Durbec, & J. C. Poggiale. A functional approach of the sediment reworking due to gallery digging macrobenthic organisms: Modeling and application with the polychaete *Nereis diversicolor*.

The Role of Suspension-Feeding Bivalves in Influencing Macrofauna: Variations in Response

S. F. Thrush*, V. J. Cummings, J. E. Hewitt, G. A. Funnell, and M. O. Green

Abstract: *Suspension-feeding bivalves are known to affect near-bed hydrodynamics and rates of sedimentation, erosion, and biodeposition. The pinnid bivalve* Atrina zelandica *(Gray) is a large suspension-feeding bivalve that can grow up to 30 cm long and 12 cm wide and form dense biogenic reefs in a variety of near-shore, soft-sediment habitats. This animal should provide a good model to identify the influence of benthic suspension feeders on macrofaunal communities via modification of physical properties. However, a multi-site* Atrina *density manipulation experiment revealed highly variable effects. Variability in the relationship between the density of* Atrina *and macrofauna was also apparent in a broad-scale survey carried out in conjunction with the experiment. Analysis of relationships among aspects of the* Atrina *landscape (density, size, and distance between individuals) observed at different sample grains, and macrofauna (number of individuals, taxa, deposit feeders, and suspension feeders) found by the survey indicated that stronger effects would have been apparent in larger experimental plot sizes. Also, the size and spatial arrangement of* Atrina *was more important than density in influencing macrofauna. Meta-analysis of the influence of environmental characteristics on the strength of the relationship between* Atrina *and macrofauna demonstrated the importance of sediment particle size and near bed flow characteristics for both survey and experiment. These results emphasise the importance of understanding how environmental processes operating over different spatial and temporal scales influence the strength and direction of interactions between* Atrina *and the macrofaunal community.*

Introduction

Field experiments are a potentially powerful technique that enables relationships between different processes influencing the structure and function of ecological systems to be tested. However, it is now recognized that the importance of many ecological processes is scale-dependent (Schneider & Piatt 1986; Wiens 1989; Legendre 1993; Dayton 1994; Hall et al. 1994; Thrush et al. 1997). This implies that relationships observed as a result of experimental manipulations may vary depending on the location of the study in space and time, as well as with the spatial and temporal scales of manipulation. Given the effort required to conduct manipulative field experiments in marine soft-sediment ecosystems, it is unlikely that manipulative experiments at multiple sites and times will become commonplace. Thus, it is imperative that techniques are developed to enable experimental results to be placed into a broader context. In this paper we illustrate the problems with interpreting the results of a multisite experiment designed to assess the role of large suspension-feeding bivalves in influencing macrofaunal composition and present a technique to enable some assessment of the generality of results.

Suspension-feeding bivalves can modify their local environment (Dame 1993). Theoretical and flume studies have demonstrated the modification of boundary roughness due to the physical structure of the animals or water jets associated with inhalant or exhalant feeding currents (Ertman & Jumars 1988; Monismith et al. 1990; O'Riordan et al. 1993; O'Riordan et al. 1995; Pilditch et al. 1997). Field studies have also shown that all these physical effects can modify boundary flows over beds of suspension-feeding bivalves (Frechette et al. 1989). The shells of bivalves also provide hard surface habitats in soft-sediment environments and provide further potential for modification to flows and benthic boundary conditions associated with the growth of encrusting organisms. Shells can also provide a refuge from large predators. Further modification to the sedimentary environment around suspension-feeding bivalves can result from the production of faeces and pseudofaeces that modify sediment deposition and organic loading (Bernard 1974; Kautsky & Evans 1987). Thus, local habitat modifications due to the presence of beds of suspension-feeding bivalves have the potential to influence sedimentation and erosion rates, biogeochemical fluxes, and ecological processes in soft-sediment systems.

Despite this variety of mechanisms whereby suspension-feeding bivalves could directly or indirectly influence the density of macrofauna, specific experimental field manipulations of the density of suspension-feeding bivalves have failed to produce consistent results (Olafsson et al. 1994). Woodin (1976) postulated that at high densities suspension feeders should inhibit recruitment by filtering potential colonizing larvae. But experimental tests of this hypothesis are often equivocal, either because experimental densities were not sufficient to induce this effect, the effect was offset when larval densities were enhanced by boundary layer hydrodynamics (Ertman & Jumars 1988; Andre et al. 1993), or planktonic larvae were not important colonists (Commito 1987; Commito & Boncavage 1989). However, there may also be a mismatch between the scales at which experiments are feasible and the scale at which larval depletion is induced.

Recently, we conducted a series of studies to assess the role of *Atrina zelandica,* a large suspension-feeding pinnid bivalve, in influencing macrobenthic community structure. Observations by divers and high-resolution side-scan sonar have revealed large patches (100s of meters) of *Atrina* in both sandy and muddy sediments. Typically, the anterior ½ – ⅔ of the *Atrina*'s shell protrudes out of the seafloor (approximately 7 cm). We have demonstrated that, depending on the density and spatial structure of the *Atrina* beds, on scales of 100s of meters, they can modify boundary flows (Green et al. 1998). In fact, *Atrina* densities can be sufficiently high to generate skimming flow (i.e., when *Atrina* are sufficiently dense, turbulent eddies between individuals cause flow to skim over the surface of the *Atrina* bed, producing a very low drag coefficient). Potentially, such flows have deleterious effects on food and larval supply and on gas exchange. Preliminary surveys conducted in both sandy and muddy habitats have demonstrated significant differences in the composition of nematode and macrobenthic assemblages inside and outside of *Atrina* patches (Warwick et al. 1997; Cummings et al. 1998). Differences in the results of these surveys between the mud and sand habitats indicated that the influence of *Atrina* on macrofauna was being modified by factors related to location, such as hydrodynamics and sediment particle size. Thus, to further investigate the role of *Atrina* in influencing macrobenthic community composition, we conducted a density manipulation experiment and extensively surveyed a series of natural *Atrina* beds.

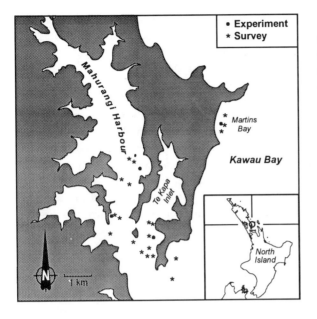

 markers on the map: • Experiment * Survey; Mahurangi Harbour; Martins Bay; Kawau Bay; Te Kapa Inlet; N; 1 km; North Island

Fig. 1 Study sites in Mahurangi Harbour and Kawau Bay, North Island, New Zealand.

In this paper we set out to (1) address whether the experimental plot size was appropriate to identify the local effects of *Atrina* on macrofauna; (2) determine if differences in experimental treatment effects were similar to differences in the effect of *Atrina* on macrofauna observed in broad-scale surveys; and (3) determine whether differences in the environmental characteristics of sites could help to explain the variability in the strength of treatment effects. These are relevant questions to ask when attempting to interpret multisite experimental results given that broader scale environmental variables can confound the simple interpretation of treatment effects. Moreover, we anticipated that modification of boundary flows will be scale-dependent and thus hydrodynamic changes induced by *Atrina* in small experimental plots may not mimic flows over larger patches. Thus, combining the survey and experimental data could identify factors that are contributing to the variability in macrofaunal response to *Atrina*.

Methods

To encompass habitats with variations in hydrodynamic conditions and sediment particle size, field studies were conducted in Mahurangi Harbour and along the adjacent coastline of Kawau Bay (Fig. 1). This area is on the east coast of the North Island of New Zealand, about 50 km north of Auckland.

Details of the experiment and survey design and sampling will be presented elsewhere (Cummings et al. in press; Hewitt et al. in prep.). Briefly, the experiment consisted of three treatments of transplanted *Atrina* (7.4 ± 0.5 cm shell width) at densities of 0, 7.5, and 75 individuals m^{-2}. The plot size was 4 m^2. Four sites, three in Mahurangi Harbour (Te Kapa, Mid-Harbour, and Upper Harbour) and one in Kawau Bay (Martins Bay) were selected (Fig. 1). At each site, treatments were replicated four times in a randomized block design. Initially, six experimental sites were established, although two were destroyed by storms before the first sampling (i.e., January 1996). Sampling was performed January 18–20, 1996, September 17–20, 1996, and March 17–19, 1997. On each sampling, three cores (10 cm diam. x 10 cm deep) were collected from random locations within each plot to assess the abundance of macrofauna.

Samples were also collected from the top 2 cm of sediment in each plot from two of the four blocks to assess surficial sediment particle size and organic matter content. Samples of sediment for particle size analysis were digested, in 6% hydrogen peroxide for 48 h to remove organic matter, and dispersed using Calgon (0.5% [mass:volume] sodium hexametaphosphate), before being analyzed by a Mastersizer laser system at 0.5-phi intervals from 15 phi to -1 phi. Organic matter was measured from samples collected by drying the sediment at 90°C for 48 h, and then combusting for 3 h at 550°C. Plaster blocks were used to provide a comparative assessment of water flux 2 cm above the bed in two experimental blocks at each site. Plaster blocks consisted of 2 cm thick slices of cast plaster (7 cm diameter) that were oven dried (60°C) to a constant weight and then varnished on the bottom and sides (see Commito et al. 1995 for details of method).

A survey of the relationship between *Atrina* and macrofaunal density was also conducted in January 1996. Twenty randomly chosen sites with *Atrina* populations were sampled, 15 in Mahurangi Harbour and five on the adjacent coast (Fig. 1). Each site was surveyed by sampling along two 20-m transects, one perpendicular and one parallel to the main hydrodynamic conditions (tidal flows or waves). Along each transect, 10 cm diam. x 10 cm deep cores were placed by divers so as to encompass the maximum variability in *Atrina* density. Before the cores were collected, the transects were videoed so that relationships between *Atrina* and macrofaunal density could later be assessed. At each site, one pooled sample from the top 2 cm of sediment was collected to assess sediment particle size and organic matter content; it was analyzed as described above.

Additional environmental information included water depth, seabed drag coefficient (C_{100}), and flow. The seabed drag coefficient reflects the proportion of mean-flow kinetic energy at 1 m above the seabed that is dissipated in the benthic boundary layer by turbulence. The C_{100} was calculated based on equations provided by Green et al. (1998) for different densities of *Atrina*. Depth-averaged and tidally averaged flow speed was derived from a layered three-dimensional hydrodynamic model of Mahurangi Harbour and the surrounding area (Oldman & Black 1997).

Here we present some results of univariate generalized linear modelling from both studies to assess how useful the survey was in generalizing and interpreting the experimental results; the specific results of the experiment and survey will be presented in other papers. However, some explanation is needed of how the video and core data collected during the survey were used to assess the influence of sampling grain on the evaluation of *Atrina* relationships with macrofauna. Grain is the area of an individual sample (Isaaks & Srivastava 1989; Wiens 1989; He et al. 1995). The minimum grain was based on one 0.5 m x 0.4 m video image quadrat with a core in the middle of the image. The grain was increased by successively adding video quadrats adjacent to the minimum grain quadrat in each direction along the transect line. Thus, the analysis provided samples based on grains of 1, 3, 5, 7, and 9 adjacent video quadrats and resulted in a maximum grain of 0.5 x 4.5 m. At each grain, *Atrina* density, size (shell width), and minimum and maximum distance between *Atrina* were regressed against the four macrofaunal variables, according to the model selection criteria described in Hewitt et al. (in prep.). We present the most parsimonious model describing the relationship between the *Atrina* landscape and the four macrofaunal aggregate variables (number of taxa, number of individuals, number of suspension feeders, and number of deposit feeders).

To assess the role of the environmental variables in affecting the variability (strength) of relationships between *Atrina* and macrofaunal variables in both the experiment and the survey we used meta-analysis (Rosenthal 1991; Gurevitch & Hedges 1993). First, the relationship between *Atrina* density and the number of individuals, number of taxa, number of suspension feeders and number of deposit feeders for each survey site were determined using generalized linear modelling. In all cases the best models had normal error structures, with either identity or log link functions. Pearson's r from each of these analyses was used as a measure of variability (strength) of the relationship, and used as the dependent variable in a multiple regression with the environmental data

Table 1. Ranges of variables found in the experiment and the survey. Macrofaunal densities are based on a mean of three cores for the experiment and individual cores for the survey.

	Experiment			Survey		
	range	max	min	range	max	min
% Medium sand	53.2	60.5	7.3	75.3	78.0	2.7
% Fine sand	24.9	46.8	21.9	32.6	43.0	10.4
% Clay	3.9	6.2	2.3	11.5	11.6	0.2
% Silt	42	65.0	13	69	75	6
Flow (m s^{-1})	0.25	0.3	0.05	0.3	0.35	0.05
Water depth (m)	5.0	8.0	3.0	6.5	9.5	3.0
% organic content	1.3	2.1	0.8	3.6	4.6	1.0
Atrina size (cm)	0.25	8.0	7.75	9.0	13.0	4.0
C_{100} ($\times$ 1,000)	–	–	–	740	795	55
Plaster block erosion (g h^{-1})	0.34	0.58	0.24	–	–	–
Deposit feeders (individuals core^{-1})	78	81	3	95	95	0
Suspension feeders (individuals core^{-1})	39	39	0	38	38	0
Individuals core^{-1}	145	152	7	175	185	10
Number of taxa core^{-1}	32	38	6	25	32	7

for each site. Table 1 outlines the type and range of environmental variables available. The most parsimonious model was derived by backward selection with a $p \geq 0.15$ exit criteria. Terms were only removed from the model if doing so did not result in a significant increase in deviance (Crawley 1993). Analyses were performed using SAS/Insight (1993).

We performed the same type of analysis on the experimental data to see if the environmental variables affecting the strength of relationships between *Atrina* and macrofauna in the survey were consistent with those in the experiments. In this analysis C_{100} could not be calculated because the experimental sites were situated in the middle of 100-m patches of bare sediment and the benthic boundary layer would not have been fully developed over the experimental plots. C_{100} was replaced by weight loss of plaster blocks, this is a comparative measure of water flux over the experimental plots well within the benthic boundary layer.

Results

Generalized linear modelling of the experimental results revealed a number of significant interactions (two-way and three-way) among site, date, and treatment effects for number of individuals, number of taxa, number of suspension feeders, and number of deposit feeders (Table 2).

Table 2. Results of Generalized Linear Modelling of experimental effects (Treat) over all sites and all times. Normal error structures and identity links were used for all variables except suspension feeders, for which a Poisson error structure and a log link were used. For models using normal error structures, values given are the mean squares and the Type III probability value. For the Poisson error structure, Pearsons c^2 and Type III probability values are given. Degrees of freedom are given in brackets.

	Number of individuals		Number of taxa		Suspension feeders		Deposit feeders	
	MS	p	MS	p	Pearsons x^2	p	MS	p
Error (379)	400.7		17.0		1058.6		167.8	
Site (3)	91835	0.0001	224.7	0.0001		0.0001	12412	0.0001
Date (2)	85.1	0.8087	221.0	0.0001		0.0001	5521	0.0001
Block (3)	144.2	0.7820	24.45	0.2314		0.6605	388.4	0.0756
Treat (2)	1569	0.0208	10.9	0.5258		0.0027	392.7	0.0978
Site*Date*Treat (12)	673.7	0.0688	73.3	0.0001		0.0268	108.5	0.8017
Site*Date (6)	26000	0.0001	72.5	0.0004		0.0001	4584.6	0.0001
Site*Treat (6)	3343.5	0.0001	27.9	0.1348		0.0008	653.3	0.0009
Date*Treat (4)	425.2	0.3756	78.4	0.0012		0.4246	44.8	0.8992

Re-analysis of the data for each site, and time if appropriate, to identify main effects showed significant treatment effects (Table 3). Treatment effects on the number of individuals and number of deposit feeders were significant only at Martins Bay, with a consistent effect across all sampling occasions of lower abundances associated with high *Atrina* density. However, for the number of taxa and suspension feeders, the significant relationships with *Atrina* density changed in direction with both time and site (Table 3). The significance of treatment effects demonstrates that our experiment was sufficiently powerful to detect effects and that *Atrina* do influence macrofaunal community composition.

The survey revealed that, over all sites, different aspects of the *Atrina* landscape (size, density, and distance between individuals) tended to influence the different macrofaunal aggregate variables over different spatial scales (Table 4). Generally the grain at which the *Atrina* landscape variables were important was greater than 2 m. The only exception to this was for deposit feeders, where maximum distance calculated over a 0.5-m grain was important but only in conjunction with size and density measured over larger grains. Although the amount of variability explained by the different aspects of the *Atrina* landscape are low, this may be due to differing responses in different flow environments (Hewitt et al. in prep.). While *Atrina* density was important for the number of taxa and number of deposit feeders, other aspects of the *Atrina* landscape were also important. No relationships could be identified between the *Atrina* landscape and the number of suspension feeders. This implies that spatial arrangement of individuals is at least as important as density alone and is influencing different macrofaunal variables over different spatial scales. As density was the only factor manipulated in the experiment, subsequent analyses focus on this alone.

Meta-analysis of the survey data revealed that a small selection of the available environmental variables had a consistent effect on the strength of relationships between *Atrina* density and macrofauna. Medium sand, fine sand, and clay reduced the variability of the relationship between *Atrina* density and the number of taxa, individuals, and deposit feeders (Table 5). These particle size variables worked in combination with C_{100}, flow speed, and *Atrina* size, which increased variability in relationships, although flow speed and *Atrina* size were only important for the number of taxa and deposit feeders, respectively. For these three macrofaunal variables, the strongest role for environmental variables in influencing the variability of relationships with *Atrina* were apparent for the number of taxa and number of deposit feeders (Table 5). None of the environmental variables were useful in explaining the relationship between the density of *Atrina* and suspension feeders.

A similar set of variables influenced the strength of the relationship between *Atrina* density and the four macrofaunal variables in the experiment (Table 6). Medium and fine sands had an important influence on the strength of the relationship between *Atrina* density and the number of taxa, individuals, and deposit feeders. Plaster block weight loss was also important in explaining the strength of experimental effects for these macrofaunal variables. Flow speed was again an important variable for the number of deposit feeders; however, in the experiment it had a positive effect while in the survey it had a negative effect. Flow speed was also important for the number of taxa. *Atrina* size had a negative effect on the number of individuals and deposit feeders. Clay was not important in the experiment, probably because of the small range of the variable (Table 1). Again, no environmental variables were found to be influencing the strength of the relationship between suspension feeders and *Atrina* density.

Table 3. Summary of Generalised Linear Modelling of experimental effects for all sites and times. Values are Type III probability values for the experimental effect; when p < 0.05, the result of multiple comparisons follows. C = controls (0 *Atrina* m^{-2}), L = low density (7.5 *Atrina* m^{-2}), H = high density (75 *Atrina* m^{-2}). Letters connected by underlining are not significantly different. When a significant date*treatment interaction was found, results for each date are given. Normal error structures and identity links were used for all variables except suspension feeders where a Poisson error structure and a log link were used.

Site & Date	Number of individuals	Number of taxa	Suspension feeders	Deposit feeders
	p	p	p	p
Martins Bay				
all	0.0004 C L>H			0.0167 CL>H
January 1996		0.0033 L > H C	0.0007 L>C>H	
September 1996		0.135	0.0068 C>L>H	
March 1997		0.0334 C > H L	0.0001 C>H L	
Mid Harbour				
all	0.1154			0.309
January 1996		0.0301 H > C L	0.7103	
September 1996		0.3404	0.2003	
March 1997		0.959	0.779	
Te Kapa				
all				0.4603
January 1996	0.076	0.4969	0.1514	
September 1996	0.0917	0.2419	0.3017	
March 1997	0.3052	0.2317	0.2642	
Upper Harbour				
all	0.594		0.5684	
January 1996		0.1193		0.277
September 1996		0.0197 L C H		0.815
March 1997		0.0303 H C L		0.1167

Table 4. Optimal scales of sample grain to determine relationships between macrofauna and the density and spatial arrangement of *Atrina* as determined by the survey. Distances given are the maximum and minimum distances between individuals and size is the maximum width of *Atrina*.

Dependent variable	Coefficient sign	Variable	Grain (m)	R^2
Number of individuals	+	minimum distance	4.5	0.2
	+	*Atrina* size	2.5	
Number of taxa	+	*Atrina* size	4.5	0.23
	−	*Atrina* density	3.5	
Suspension feeders	−	maximum distance	4.5	0.1
Deposit feeders	+	maximum distance	0.5	0.15
	+	*Atrina* size	2.5	
	−	*Atrina* density	4.5	

Discussion

In this paper we set out to explain the apparent variability in the role of *Atrina* in influencing macrofaunal community structure. To do this we addressed whether the plot size used in the experiment was appropriate for identifying the local effects of *Atrina* on macrofauna; whether the variable effects detected in the experiment reflected those found in broad-scale surveys; and also whether the observed variability in the strength of experimental treatment effects could be explained by differences in the environmental characteristics among sites.

The analyses of the strengths of relationships between *Atrina* and macrofauna in relation to sample grain imply that we might expect stronger effects in larger patches of *Atrina*. Although the regression models derived from our broad-scale survey only explained between 10% and 23% of the variance, they provide some indication that we should be cautious in extrapolating relationships based on experiments with a plot size of 2 m x 2 m. Nevertheless, this analysis did not indicate the experiment was conducted at completely the wrong scale. We characterized grain using the longest linear dimension of the sampling unit. We expected this to be a better reflection of the amount of variability encompassed by the sampling unit than characterization based on area, because the video quadrats were rectangular and orientated with respect to hydrodynamic conditions. Moreover, this analysis was performed over the whole data set and there may be differences inside and outside of the harbour. Thus we should expect to see some effects of *Atrina* on macrofauna in the 4-m² experimental plots.

Interestingly, the survey investigation of relationships between macrofauna and the *Atrina* landscape (Hewitt et al in prep) also revealed that the influence of *Atrina* on macrofauna was driven by a combination of density and spatial pattern (i.e., size of gaps between individuals and between patches of *Atrina*). Spatial relationships were not manipulated in the experiment, even though other studies have indicated spatial arrangement can influence a number of ecological processes (e.g., Thrush 1991; Thrush et al. 1997). The survey revealed that the spatial arrangement of individual *Atrina* was a more powerful explanatory variable than density alone. This implies that important factors affecting the relationship between *Atrina* and macrofauna are influenced by the size of the

Table 5. Effect of environmental variables on the variability of the relationship between macrofauna and density of *Atrina*, as represented by Pearson's r, found in the survey. No adequate model could be developed for suspension feeders. C_{100} = drag coefficient. Flow = depth-averaged and tidally-averaged flow speed.

Dependent variable	Environmental variables	Parameter estimate	p	R^2
Number of taxa	% medium sand	1.71	0.0010	0.8
	% fine sand	2.08	0.0014	
	% clay	12.57	0.0015	
	Atrina size	-0.18	0.0022	
	C_{100}	-0.018	0.0045	
Number of individuals	% medium sand	1.37	0.0160	0.53
	% fine sand	1.76	0.0148	
	% clay	10.56	0.0177	
	C_{100}	-0.016	0.0312	
Deposit feeders	% medium sand	1.42	0.0033	0.76
	% fine sand	1.82	0.0029	
	% clay	11.88	0.0023	
	flow	-40.63	0.0122	
	C_{100}	-0.012	0.0428	

gaps between individuals. Small-scale turbulence around individuals influences very localized rates of erosion and deposition (Eckman & Nowell 1984; Shimeta & Jumars 1991; Pilditch et al. 1997), which for suspension-feeding bivalves will affect the deposition and resuspension of faeces and pseudofaeces.

Our experimental effects were variable in space, and this variability was not due to any experimental artefact as it was also apparent in the survey results. Analysis of the importance of environmental variables in explaining variability in the strength of relationships between each of the macrofaunal variables (i.e., the number of taxa, individuals, and deposit feeders) and *Atrina* density revealed similar environmental variables were important in both the experiment and the survey. Furthermore, the same variables were generally important for both the experiment and the survey. The environmental variables were not useful in explaining the strength of the relationship between *Atrina* and the number of suspension feeders, possibly because of the low densities of suspension feeders in the three harbour sites. Nevertheless, these results help us explain variability in experimental effects and provide confidence in the generality of experimental results. The survey meta-analysis revealed that we could expect sites with sediments of high medium-to-fine sand and clay content and low C_{100} to have the clearest treatment effects. Identifying such sites is complicated by the relationship between C_{100} and *Atrina* (viz., C_{100} increases with density until skimming flow is induced) (Green et al. 1998). Nevertheless, the experimental sites at which we could not detect any significant treatment effects (Te Kapa and Upper Harbour) were located in areas of fine muddy sediments. The experimental site with the coarsest sediment (Martins Bay) revealed the strongest treatment effects on macrofauna. This is consistent with an earlier survey that also revealed stronger effects of *Atrina* on macrofaunal community composition in sandy sediments (Cummings et al. 1998).

Table 6. Effect of environmental variables on the variability of the relationship between macrofauna and density of *Atrina*, as represented by Pearson's r, found in the experiment. No adequate model could be developed for suspension feeders. Plaster = plaster block weight loss.

Dependent variable	Environmental variables	Parameter estimate	p	R^2
Number of taxa	% medium	0.15	0.0472	0.93
	sand	0.53	0.0524	
	% fine sand	43.8	0.0104	
	plaster	22.7	0.0546	
	flow			
Number of individuals	% medium	0.20	0.0904	0.93
	sand	0.53	0.0680	
	% fine sand	37.9	0.0191	
	plaster	-39.0	0.0240	
	Atrina size			
Deposit feeders	% medium	0.17	0.0479	0.99
	sand	0.44	0.0572	
	% fine sand	38.8	0.0119	
	plaster	26.2	0.0435	
	flow	-19.1	0.1173	
	Atrina size			

Although we can use the meta-analysis to explain the variation in strength of relationships identified by the experiment, as with any regression, the environmental variables used might reflect direct causal relationships or may be surrogates for some other unmeasured factors. It is interesting that the meta-analysis revealed that effects are more variable in the finer sediments. As coarser sediments are generally low in organic matter content, sands may make relationships less variable if there were threshold relationships among macrofauna, *Atrina*, and sediment organic matter content due to the relative importance of biodeposition by *Atrina*. We would expect biodeposition to be highest at the Upper Harbour site and lowest in Martins Bay. The interaction of sediment organic content and pseudofaeces production are unlikely to influence macrofauna in a linear fashion, which would explain why silt or sediment organic content were not detected as important in the regression models. This may also account for the sediment particle size and hydrodynamic variables working together but in opposite directions in these models.

The experiment described in this paper was designed to assess the role of a large suspension-feeding bivalve on macrofauna community composition, based on preliminary surveys that revealed differences in community composition inside and outside of *Atrina* patches (Warwick et al. 1997; Cummings et al. 1998) and differences in flow conditions over *Atrina* beds of varying density (Green et al. 1998). The experiment, which was designed to assess the variability in response moving down an estuary and on to the open coast, entailed sampling over time to assess effects associated with recruitment events. The experiment revealed significant treatment effects, the strength and direction of which were variable in both space and time. Conducting experiments at more than one location has the potential to improve the generality of results. Certainly, the generality of theories based on studies conducted at one or only a few locations is questionable (Wiens 1976; Dayton 1984, 1994; Dayton & Tegner 1984; Underwood & Denley 1984; Foster 1990; Brown 1995), especially where the biological effects are driven by a number of interacting processes. Differences in habitat, particularly changes in sediment type, usually reflect changes in a variety of factors (see Snelgrove & Butman 1994 for a recent review) and this will make it difficult to isolate single processes causing variation in local biological interactions. Our experiment was conducted over a reasonably large scale, especially when the encompassed variation in environmental conditions is considered. Analysis of the significant treatment effects alone did not provide any generality because other factors operating over broader scales (e.g., variations in flow and sediment particle size) constrained the outcome of local ecological processes. Combining surveys and experiments and assessing the role of environmental variables in affecting the strength of relationships between *Atrina* and macrofauna did provide some insight. This emphasizes the importance of collecting appropriate co-variables, particularly for broad-scale studies.

Atrina may be engineers of macrobenthic community composition, but the strength of their influence is dependent on environmental characteristics. Simple generalizations on the ecological role of suspension feeders in soft-sediment ecosystems may not be possible when there are a variety of local mechanisms interacting with broader scale processes. Suspension-feeding bivalves can be highly effective at removing particles from the water (e.g., Officer et al.1982; Newell 1988; Loo & Rosenberg 1989; Hily 1991); thus influencing chemical and particle fluxes. However, the importance of different processes interacting to produce variable responses to *Atrina* described in these studies reflect the type of problems that can arise when trying to scale-up and assess the role of such organisms as ecosystem engineers. It is important to develop an understanding of local mechanisms; however, if processes are scale-dependent, it is also important to place understanding of the fine-scale detail into a broader framework that indicates how local processes may vary under the influence of other factors.

Acknowledgments

Rod Budd, Andy Hill, Terry Hume, Carly Milburn, and Stephanie Turner helped with the diving. Bob Whitlatch helped with the survey. Electra Kalaugher and Carly Milburn helped with the sample processing and identifications. John Oldman modelled water flow for our sites.

Literature Cited

Andre, C., P. R. Jonsson, & M. Lindegarth. 1993. Predation on settling bivalves by benthic suspension feeders: The role of hydrodynamics and behaviour. *Marine Ecology Progress Series* 97:183–192.

Bernard, F. R. 1974. Annual biodeposition and gross energy budget of mature Pacific oysters, *Crassostrea gigas. Journal of the Fisheries Research Board of Canada* 31:185–190.

Brown, J. H. 1995. Macroecology. University of Chicago, Chicago, Illinois.

Commito, J. A. 1987. Adult-larval interactions: Predictions, mussels and cocoons. *Estuarine, Coastal and Shelf Science* 25:599–606.

Commito, J. A. & E. M. Boncavage. 1989. Suspension-feeders and coexisting infauna: An enhancement counterexample. *Journal of Experimental Marine Biology and Ecology* 125:33–42.

Commito, J. A., S. F. Thrush, R. D. Pridmore, J. E. Hewitt, & V. J. Cummings. 1995. Dispersal dynamics in a wind-driven benthic system. *Limnology and Oceanography* 40:1513–1518.

Crawley, M. J. 1993. GLIM for Ecologists. Blackwell Scientific Publications, Oxford.

Cummings, V. J., S. F. Thrush, J. E. Hewitt, & S. J. Turner. 1998. The influence of *Atrina zelandica* (Gray) on benthic macroinvertebrate communities in soft-sediments habitats. *Journal of Experimental Marine Biology and Ecology* 228:227–240.

Cummings, V. J., S. F. Thrush, J. E. Hewitt, & G. A. Funnell. In press. Variable effects of a large suspension-feeding bivalve on infauna: Experimenting in a complex system. *Marine Ecology Progress Series*

Dame, R. F. (ed.). 1993. Bivalve Filter Feeders in Estuarine and Coastal Ecosystem Processes. NATO ASI Series, Springer-Verlag, Berlin.

Dayton, P. K. 1984. Processes structuring some marine communities: Are they general?, p. 181–200. *In* Strong, D. R., D. Simberloff, L. G. Abele, & A. B. Thistle (eds.), Ecological Communities Conceptual Issues and the Evidence. Princeton University Press, Princeton, New Jersey.

Dayton, P. K. 1994. Community landscape: Scale and stability in hard bottom marine communities, p. 289–332. *In* Giller, P. S., A. G. Hildrew, & D. Raffaelli (eds.), Aquatic Ecology: Scale, Pattern and Processes. Blackwell Scientific, Oxford.

Dayton, P. K. & M. J. Tegner. 1984. The importance of scale in community ecology: A kelp forest example with terrestrial analogs, p. 457–483. *In* Price, P. W., C. N. Slobodchikoff, & W. S. Gaud (eds.), A New Ecology: Novel Approaches to Interactive Systems. John Wiley & Sons, New York.

Eckman, J. E. & A. R. Nowell. 1984. Boundary skin friction and sediment transport about an animal-tube mimic. *Sedimentology* 31:851–862.

Ertman, S. C. & P. A. Jumars. 1988. Effects of bivalve siphonal currents on the settlement of inert particles and larvae. *Journal of Marine Research* 46:797–813.

Foster, M. S. 1990. Organisation of macroalgal assemblages in the Northeast Pacific: The assumption of homogeneity and the illusion of generality. *Hydrobiologia* 192:21–33.

Frechette, M., C. A. Butman, & W. R. Geyer. 1989. The importance of boundary-layer flows in supplying phytoplankton to the benthic suspension feeder, *Mytilus edulis* L. *Limnology and Oceanography* 34:19–36.

Green, M. O., J. E. Hewitt, & S. F. Thrush. 1998. Seabed drag coefficients over natural beds of horse mussels (*Atrina zelandica*). *Journal of Marine Research* 56:613–637.

Gurevitch, J. & L. V. Hedges. 1993. Meta-analysis: Combining the results of independent experiments, p. 378–398. *In* Scheiner, S. M. & J. Gurvitch (eds.), Design and Analysis of Ecological Experiments. Chapman and Hall, New York.

Hall, S. J., D. Raffaelli, & S. F. Thrush. 1994. Patchiness and disturbance in shallow water benthic assemblages, p. 333–375. *In* Giller, P. S., A. G. Hildrew, & D. Raffaelli (eds.), Aquatic Ecology: Scale, Pattern and Processes. Blackwell Scientific, Oxford.

He, F., P. Legendre, C. Bellehumeur, & J. V. LaFrankie. 1995. Diversity pattern and spatial scale: A study of a tropical rain forest of Malaysia. *Environmental and Ecological Statistics* 1:265–286.

Hily, C. 1991. Is the activity of benthic suspension feeders a factor controlling water quality in the Bay of Brest? *Marine Ecology Progress Series* 69:179–188.

Isaaks, E. H. & R. M. Srivastava. 1989. Applied Geostatistics. Oxford University Press, Oxford, England.

Kautsky, N. & S. Evans. 1987. Role of biodeposition by *Mytilus edulis* in the circulation of matter and nutrients in a Baltic coastal ecosystem. *Marine Ecology Progress Series* 33:201–212.

Legendre, P. 1993. Spatial autocorrelation: Trouble or new paradigm? *Ecology* 74:1659–1673.

Loo, L.-O. & R. Rosenberg. 1989. Bivalve suspension-feeding dynamics and benthic-pelagic coupling in an eutrophicated marine bay. *Journal of Experimental Marine Biology and Ecology* 130:253–276.

Monismith, S. G., J. R. Koseff, J. K. Thompson, C. A. O'Riordan, & H. M. Nepf. 1990. A study of model bivalve siphonal currents. *Limnology and Oceanography* 35:680–696.

Newell, R. I. E. 1988. Ecological changes in Chesapeake Bay: Are they the result of overharvesting the American oyster, *Crassostrea virginica*?, p. 534–546. *In* Lynch, M. P. & E. C. Krome (eds.), Understanding the Estuary: Advances in Chesapeake Bay Research. Chesapeake Research Consortium, Solomons, Maryland.

Officer, C. B., T. J. Smayda, & R. Mann. 1982. Benthic filter feeding: A natural eutrophication control. *Marine Ecology Progress Series* 9:203–210.

Olafsson, E. B., C. H. Peterson, & W. G. Ambrose, Jr. 1994. Does recruitment limitation structure populations and communities of macroinvertebrates in marine soft sediments: The relative significance of pre- and post-settlement processes. *Oceanography and Marine Biology Annual Review* 32:65–109.

Oldman, J. W. & K. P. Black 1997. Mahurangi Estuary numerical modelling. Report Number ARC 60208/1. National Institute of Water and Atmospheric Research, Hamilton, New Zealand.

O'Riordan, C. A., S. G. Monismith, & J. R. Koseff. 1993. A study of concentration boundary-layer formation over a bed of model bivalves. *Limnology and Oceanography* 38:1712–1729.

O'Riordan, C. A., S. G. Monismith, & J. R. Koseff. 1995. The effect of bivalve excurrent jet dynamics on mass transfer in a benthic boundary layer. *Limnology and Oceanography* 40:330–344.

Pilditch, C. A., C. W. Emerson, & J. Grant. 1997. Effect of scallop shells and sediment grain size on phytoplankton flux to the bed. *Continental Shelf Research* 17:1869–1885.

Rosenthal, R. 1991. Meta-Analytic Procedures for Social Research. Sage Publications, Newbury Park, California.

SAS/Insight. 1993. Users' Guide, Version 6. second ed. SAS Institute Inc., Cary, North Carolina.

Schneider, D. C. & J. F. Piatt. 1986. Scale-dependent correlation of seabirds with schooling fish in a coastal ecosystem. *Marine Ecology Progress Series* 32:237–246.

Shimeta, J. S. & P. A. Jumars. 1991. Physical mechanisms and rates of particle capture by suspension feeders. *Oceanography and Marine Biology: An Annual Review* 29:191–257.

Snelgrove, P. V. R. & C. A. Butman. 1994. Animal-sediment relationships revisited: Cause versus effect. *Oceanography and Marine Biology: An Annual Review* 32:111–177.

Thrush, S. F. 1991. Spatial patterns in soft-bottom communities. *Trends in Ecology and Evolution* 6:75–79.

Thrush, S. F., D. C. Schneider, P. Legendre, R. B. Whitlatch, P. K. Dayton, J. E. Hewitt, A. H. Hines, V. J. Cummings, S. M. Lawrie, J. Grant, R. D. Pridmore, & S. J. Turner. 1997. Scaling-up from experiments to complex ecological systems: Where to next? *Journal of Experimental Marine Biology and Ecology* 216:243–254.

Underwood, A. J. & E. J. Denley. 1984. Paradigms, explanations and generalizations in models for the structure of intertidal communities on rocky shores, p. 151–180. *In* Strong, D. R., D. Simberloff, L. G. Abele, & A. B. Thistle (eds.), Ecological Communities Conceptual Issues and the Evidence. Princeton University Press, Princeton, New Jersey.

Warwick, R. M., A. J. McEvoy, & S. F. Thrush. 1997. The influence of *Atrina zelandica* Gray on nematode diversity and community structure. *Journal of Experimental Marine Biology and Ecology* 214:231–247.

Wiens, J. A. 1976. Population responses to patchy environments. *Annual Review of Ecology and Systematics* 7:81–120.

Wiens, J. A. 1989. Spatial scaling in ecology. *Functional Ecology* 3:385–397.

Woodin, S. A. 1976. Adult-larval interactions in dense faunal assemblages: Patterns of abundance. *Journal of Marine Research* 34:25–41.

Unpublished Materials

Hewitt, J. E., S. F. Thrush, P. Legendre, V. J. Cummings, & A. Norkko. In prep. Integrating heterogeneity from different scales: A multi-resolution study along a physical gradient of the effect of a suspension-feeding bivalve on benthic macrofauna.

Turbidites and Benthic Faunal Succession in the Deep Sea: An Ecological Paradox?

David K. Young, Michael D. Richardson*, and Kevin B. Briggs

Abstract: Characteristics of benthic faunal succession following turbidity flows in the deep sea will vary according to the composition of turbidite materials, the spatial scales of deposition, the structure of initial benthic communities, and the frequency of depositional events. Despite a number of uncertainties regarding these effects, we make several generalizations in order to stimulate research on successional responses of benthic fauna to such episodic events. We find no support for a hypothesis formulated on the speculation by Heezen et al. (1955) that there should be, "... a high [positive] correlation between nutrient-rich turbidity current areas and a high standing crop of abyssal animals." There is no definitive evidence for the extensive 'mining' of deeply deposited sediments reported by Jumars and Gallagher (1982) as being a foraging strategy for deposit-feeding benthos in turbidite sediments. The time for complete recovery of the benthic fauna following an episodic deposition of material on the scale of a turbidity flow is postulated to be hundreds to thousands of years. If viewed as a matrix of dynamic events, the effects of turbidity flows on abyssal fauna are on larger spatial and longer temporal scales than most other deep-sea disturbances and therefore provide the biological framework upon which many of the smaller and shorter term effects are superimposed. In interpreting and reconstructing causes for distributions of deep-sea organisms, benthic ecologists should incorporate a similarly holistic view (i.e., the turbidity current paradigm) as sedimentologists in order to better understand how the abyssal benthic fauna has adapted to episodic disturbances over geologic time.

Introduction

Results from early deep-sea explorations (e.g., Agassiz 1892; Mortensen 1938; Bruun 1953) provided the scientific framework for the views of Bruce Heezen, Maurice Ewing, and Robert Menzies that turbidity currents play a potentially significant role in abyssal productivity. In speculating about the role of turbidity currents in the nutrition of deep-sea benthos, they state that there should be, "...a high [positive] correlation between nutrient-rich turbidity current areas and a high standing crop of abyssal animals" (Heezen et al. 1955, p. 180). This statement, considered herein as a testable hypothesis, has never been critically examined. In this paper, we test the hypothesis based on results from recent research that quantifies the relationships between organically rich turbidite deposits and biomass in deep-sea fauna.

Jumars and Wheatcroft (1989) state that turbidite sediments should be important sources of organic enrichment of deep-sea fauna but note that successional sequences and times for recovery from the initial disturbances are unknown. Jumars (1993) later questions what successional infor-mation could be resolved from such 'unsteady sequences' as turbidites. He suggests that, "A great deal more will be learned for biology when stratigraphers become braver in interpreting jumbled sedimen-tary sequences complete with erosive gaps and turbidites" (Jumars 1993, p. 301). We agree with Jumars' suggestion. In the present paper, we use the data from well-studied abyssal plains to interpret characteristics of faunal succession following episodic depositions of sediment by turbidity currents in the deep sea. We recommend areas of research requiring further attention and suggest that a new viewpoint needs to be taken by benthic ecologists who undertake such work.

Whereas this paper focuses on animal-sediment relationships in turbidite deposits, we acknowledge the research showing a strong positive correlation among pelagic sedimentation rates, particulate organic carbon flux, and bioturbation rates in parts of the abyssal ocean dominated by pelagic deposition (Jahnke & Jackson 1992). The low biomass of abyssal benthos is recognized to result from the extremely low fluxes of organic matter reaching the great depths of the deep sea (reviewed by Rowe 1983). In pelagic sediments, increases of organic carbon flux from water column transport of detritus to the abyssal seafloor correlate with higher bioturbation rates and greater depths of biogenic reworking (Trauth & Sarnthein 1997). High rates of organic flux are also positively correlated with higher abundance, greater biomass, and larger mean body size of deposit-feeding benthic fauna in the deep sea (Smith 1992). Benthic standing stock is less well correlated with percent organic matter than with organic flux rates, as most organic matter found below the upper few centimeters is refractory in nature and of little nutritional value (Richardson & Young 1987). Deposit-feeding megabenthos, though less abundant than the smaller macrobenthos, dominate spatial and temporal scales of bioturbation, thereby controlling diffusion and advection of particles and solutes, which affect rates of sediment diagenesis (Wheatcroft et al. 1990).

Background

A Short History of Deep-Sea Benthic Investigations

The understanding of deep-sea biological processes has evolved as scientists have obtained new information on biotic interactions and rates of deep-sea processes as a result of technological advances in seafloor experimentation and of sampling this remote environment. Changes in our understanding of the deep-sea fauna may be viewed in terms of the following significant events.

Despite rare collections by a few scientists showing that life existed in abyssal ocean depths, the deep sea was assumed to be a biological desert during times prior to the *Challenger* expedition (pre-1872). This "azoic theory," usually attributed to Edward Forbes, was reviewed in detail by Mills (1983). Inaccessibility of the deep sea fostered the predominant opinion of an environment in which perpetual darkness, lack of plants, low temperature, and high pressure precludes the existence of life.

Extensive sampling of diverse deep-sea fauna during the H.M.S *Challenger* expedition (1872–1876) refuted the azoic theory. From analyses of these samples emerged a new view of the deep sea being inhabited by relict fauna (i.e., ancient life forms that have changed little over geologic time) and of the abyssal ocean as a "refugium" where processes proceed at very slow rates (reviewed by Menzies & Imbrie 1958). Charles Wyville Thomson reported on qualitative collections from this first deep-sea expedition on a global scale, showing similarities of the abyssal fauna (i.e., "cosmopolitan" species) among ocean basins (Thomson 1878).

The intensive period of international, worldwide expeditions during the 1950s–1960s was characterized by the first quantitative benthic sampling of the deep sea (reviewed by Mills 1983). A picture developed of a sparse, though diverse and speciose, fauna sustained by a slow rain of detritus, where changes occur very slowly and biota are sensitive and susceptible to environmental change (i.e., "biologically accommodated") (Sanders 1968). Deep-sea species were seen as members of a fragile fauna that evolve slowly, grow slowly, and are long-lived.

Benthic colonization experiments during the 1970s–1980s, beginning with the serendipitous experiment provided by the accidental sinking of the research submersible *Alvin*, fostered a new understanding of the rates of microbial and other benthic processes in the deep sea (reviewed by Gage & Tyler 1991). The early in situ experiments revealed rates of years for the benthic recolonization of trays of azoic sediments by opportunistic species (Grassle 1977). Subsequent experimental manipulations via mechanical disturbances and organic enrichments demonstrated that recolonization may only require weeks to months and suggested that trays may cause artificial conditions that impede rates of recolonization (Grassle & Morse-Porteous 1987; Smith 1992). The short-term nature of these experiments precluded assessment of the time required for the fauna to reach equilibrium (prior to disturbance) conditions.

A more temporally and spatially dynamic picture emerged from the sediment trap measurements and observations of "falls" of organic matter in the 1970s–1980s. Naturalists have often observed that organic material such as wood, seaweeds and seagrass are rapidly colonized and often eaten by deep-sea fauna (reviewed by Gage & Tyler 1991). Sediment trap measurements pioneered by Honjo (1978) and others revealed pulsed, seasonal inputs of organic matter from algal blooms, which are rapidly fed upon by specialized surface deposit-feeders. Observations by remote video (Isaacs 1969) and research submersibles (Smith 1986) showed that "opportunistic" scavengers (bathypelagic fish and benthic invertebrates) rapidly congregated on and fed upon fish carcasses.

In 1977 there was an unexpected discovery of large, chemoautotrophic faunas associated with deep-sea hydrothermal vents (Corliss et al. 1979). The vent fauna is characterized by low diversity, rapid rates of colonization and very high growth rates (Lutz et al. 1994). Such observations led John Gage to state (Gage 1991, p.87), "...low reproductive production, slow growth and great age is by no means the general rule in the deep sea." The commonly held viewpoint today is that vent-type communities may occur wherever sufficient organic matter can create the reducing conditions required for chemoautotrophic production in the otherwise oxic, food-limited deep sea (Gooday & Turley 1990). Vent communities can be considered as one of a "parallel" grouping of marine fauna (Gage & Tyler 1991) associated with other sources of reduced compounds, such as submarine seeps, organically rich reduced sediments, whale carcasses, and sunken cargoes. The reduced compounds serve as energy sources for chemolithoautotrophic bacteria, which, in turn, provide food for this fauna (reviewed by Young & Valent 1997).

Episodic Deposition in the Deep Sea

We re-evaluated our understanding of deep-sea biological processes when recently confronted with having to assess the potential impact of benthic succession following the disposal of dredged material on the abyssal seafloor (Young & Valent 1996). There was no existing body of knowledge with which to assess this problem directly (reviewed by Valent & Young 1995). We therefore made the analogy between sudden deposition of dredged material on the abyssal seafloor and episodic deposition of sediment by turbidity flows in order to assess the ecological effects and the time for recovery of an abyssal fauna affected by such an event (Young & Richardson 1998).

Relationships of Episodic Depositional Events to Biological Processes in the Deep Sea

Episodic depositional events caused by localized slides or slumps, such as those associated with the 1929 Grand Banks earthquake (Heezen & Ewing 1952), and with large-scale turbidity flows have produced the abyssal plains of the world's oceans (Weaver et al. 1987). It is clear that deep-sea fauna have had to cope with such periods of destruction and recovery from episodic depositional events over the course of geologic time, particularly during periods of glacially lowered sea-level (Bowles et al. 1998). Young and Richardson (1998) conclude that the time for recovery of the benthic fauna from episodic depositions on the scale of a turbidity flow, including adjustment to the changed sedimentary and geochemical environment, will vary according to the sedimentary characteristics of the turbidite material itself, and also to the spatial scales (including areal extent and thickness), the composition of the initial benthic community, and the frequency of the depositional events.

Individual turbidite deposits resulting from turbidity flows may have volumes as large as 120 km^3 (Weaver et al. 1987) and depths up to 100 m, although thicknesses of centimeters to meters are more common, especially at the distal ends far from continental landmasses (Pilkey 1987). Spatial scales are generally related to basin morphologies, bathymetric gradients, and distances from continental margins, and are dependent on quantities of source material and the magnitude of the stimulus triggering the depositional event (Gorsline 1980).

Time periods between turbidity currents producing turbidite layers in the deep sea are quite variable. Various indices, such as times between cable breaks and stratigraphic markers (ash layers, faunal changes, and radioisotopic dating), have been used to determine recurrence intervals for turbidity flows. Piper and Normark (1983) report an inverse relation between layer thickness and the frequency of depositional events in ancient turbidites. This inverse relationship is noted for turbidites generated by both seismic and sedimentological processes. The sparse data on published flow periodicities are summarized in Table 1 for events from the Holocene and earlier ages.

There are relatively few observations or measurements of bioturbation in locations of the turbidity flows referenced in Table 1. For the Congo Submarine Canyon with frequent turbidity flows, Heezen and Hollister (1971, p.301) note that the time for benthic animals "...to track newly deposited mud is measured in weeks and not in years." Areas of the deep sea that are subject to frequent depositional events are apparently characterized by benthic fauna that have adapted over time to these periodic disturbances (Aller & Aller 1986). In the case of the San Pedro and Santa Monica basins, where anoxic and hypoxic conditions prevail, there are no living macrofauna and sediment sequences are undisturbed by bioturbation (Gorsline & Emery 1959). For the Carpathian Flysch Formation, Sujkowski (1957) reported fossil evidence of allochthonous, broken, shelly fossils within turbidites that were deposited at slope depths on an average of once every 4,000 yr. Graptolites and pteropods are common in the pelagic layers that are presumed deposited at time intervals between the turbidites.

Notable exceptions to the paucity of information about bioturbation in turbidite locations are the Cascadia Channel-Abyssal Plain in the Northeast Pacific (Griggs et al. 1969), the Madeira Abyssal Plain in the Northeast Atlantic (Weaver et al. 1987), and the Venezuela Abyssal Plain in the Caribbean Sea (Richardson et al. 1985). By examining published evidence from these turbidite sedimentary provinces, in which depositional and bioturbation rates have been studied, we hope to gain a better understanding of the spatial and temporal scales of change and recovery of a deep-sea fauna from the impact of episodically deposited sediment in abyssal depths.

Table 1. Recurrence intervals for turbidity currents during Holocene and earlier ages. Lithified fossil series are identified by Formation (Fm.). Locations and references are given for cited time periods.

Location	Time Period (years)	Reference
Congo Submarine Canyon	2	Heezen 1963
Magdalena River Delta	2	Menard 1964
Sea of Japan	50	Chough 1984
Cal. Borderland Basins		
Canyons	1–10	Gorsline & Emery 1959
Basin Center	400	Gorsline & Emery 1959
Cascadia Basin		
Upper Channel	410–510	Griggs & Kulm 1970
Lower Channel	580–1.5K	Griggs & Kulm 1970
Tongue of the Oceans	500–10K	Rusnak & Nesteroff 1964
Algero-Balearic Basin	1K–3K	Rupke & Stanley 1974
Columbus Basin	3K–6K	Bornhold & Pilkey 1971
Madeira Abyssal Plain	1K–65K	Weaver & Rothwell 1987
Martinsburg Fm.	500	McBride 1962
Tyee Fm.	500–1K	Lovell 1969
Melbourne Trough Fm.	2K	Moore 1971
Carpathian Flysch Fm.	4K	Sujkowski 1957

Cascadia Channel and Abyssal Plain

Griggs et al. (1969) discuss variability of depths and frequencies of burrowing within successive turbidite layers and report abundance and biomass of benthic fauna collected from surface sediments. With the exception of three of nine turbidite layers where there are distinct burrows to depths of 25 cm or more, Griggs et al. (1969) report that active feeding is mostly limited to the upper 5 cm of sediment. Griggs et al. (1969, p.167) further note that, "...no correlation exists between thickness and the frequency [of turbidite layers] and penetration depth of burrows." There is no definitive evidence for the extensive "mining" of deeply deposited sediments, which was suggested by Jumars and Gallagher (1982) as being an optimal foraging strategy for deposit-feeding benthos in turbidite sediments. Further, despite there being more abundant benthos (i.e., chiefly annelids) in the organically enriched turbidite sediments of the Cascadia Channel than in the adjacent hemipelagic sediments of the Cascadia Abyssal Plain, benthic biomass is not significantly higher (Griggs et al. 1969) in the turbidite sediments (Table 2). The annelids in the Cascadia Channel sediments are probably the small "opportunistic" types of species commonly found as early colonizers of new substrata in the deep sea (Grassle & Morse-Porteous 1987). Thus, the Heezen et al. (1955) hypothesis predicting a high correlation between nutrient-rich turbidity current areas and a high standing crop of abyssal animals is not supported by these data.

Madeira Abyssal Plain

According to Weaver et al. (1987), the central Great Meteor East (GME) area of the Madeira Abyssal Plain is the "...most extensively surveyed area of abyssal plain in the world's ocean." The reason for this extensive survey is because the GME area (located between 30°30' N and 32°30' N, and 23°30' W and 26°00' W at 5,400 m depth) in the North Atlantic Ocean was considered as a potential site for the deposition and isolation of high-level radioactive waste (Nuclear Energy Agency 1984).

The sediments of the GME area consist of meter-thick (up to 5 m) tur-bidite sequences separated by thin (centimeter to decimeter) layers of pelagic clays, marls, and oozes (deLange et al. 1987). The sediments show recurrent patterns of turbidites "capping" under-lying pelagic sediments (previously well oxygenated and bioturbated). This is evidenced by the sharp contrast of the darker color and higher organic carbon content of the overlying turbidite layers compared with that of the pelagic

Table 2. Comparisons of mean faunal biomass and percent sediment organic content among abyssal plains that are both affected by and unaffected by recent turbidite sedimentation. Data for Venezuela Basin are from Richardson and Young (1987) and Briggs et al. (1985); for Eastern North Atlantic from Thurston et al. (1994), Santos et al. (1994), Wolff et al. (1995), and Thomson et al. (1997); and for the Cascadia Abyssal Plain from Griggs et al. (1969). Differences among sampling and analyses techniques permit comparisons only within each of the three abyssal plain areal groupings. Abbreviations for abyssal sites are as follows: PAP = Porcupine Abyssal Plain, MAP = Madiera Abyssal Plain, GME = Great Meteor East. As indicated by Rice et al. (1994) and later confirmed by Rice (1997), both the MAP and GME sites were affected by a recent (approximately 1,000 year old) turbidity flow. See text for discussion of abyssal sites.

Abyssal Plain	Abyssal Sites		
Venezuela Basin	Hemipelagic	Pelagic	Turbidite
Meiofauna (mg wet weight m^{-2})	450	198	151
Macrofauna (mg wet weight m^{-2})	350	70	135
Megafauna (g wet weight/h haul)	1,108	887	806
Sediment organic content (%)			
upper 10 cm (mean)	0.54	0.30	0.70
upper 40 cm (range)	0.4–0.7	0.1–0.4	0.4–1.55
Eastern N. Atlantic	PAP	MAP	GME
Megafauna (g $\times$ $10^3 m^{-2}$)			
Otter Trawl	189		11
Epibenthic sled	169	5	4
Sediment organic content (%)	0.26–0.30	0.3–0.4	1.24–1.40
Cascadia Abyssal Plain	Abyssal Plain	Turbidite	
Station #	24	24A	
Macrofauna			
(no. m^{-2})	330	1,011	
(mg wet weight m^{-2})	1,820	2,200	
Sediment organic content (%)	1.7	2.2	

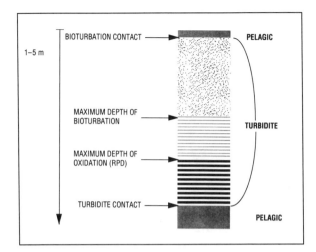

Fig. 1. Generalization of a typical sediment profile from the Great Meteor East area of the Madeira Abyssal Plain consisting of thick turbidite sequences separated by thin layers of pelagic clays, marls, and oozes showing recurrent patterns of turbidites "capping" underlying pelagic sediments (Turbidite Contact). The redox potential discontinuity (RPD) interfaces are preserved in the turbidite layers (Maximum Depth of Oxidation). Bioturbation affects only the upper portion of the turbidite layers (Maximum Depth of Bioturbation). Tops of turbidite layers, showing the interface with pelagic sediments, are defined where visible burrows occupied more than 50% of the sediment cores (Bioturbation Contact). See text for full description [modified from Thomson et al. 1987].

layers (Weaver & Rothwell 1987). Thomson et al. (1987) describe varying depths of a preserved redox potential discontinuity layer in the turbidite layers depending on the thickness of the turbidite layer, percent organic matter, and the period of time permitting an "oxidation front" to move progressively downward into the sediment (Fig. 1). Figure 1 shows that the interface of each turbidite layer with the overlying pelagic sediments is defined by visible burrows resulting from bioturbational activities (Weaver & Rothwell 1987). In only two turbidite layers does bioturbation affect sediments deeper than 20-cm in the sediment (Table 3). There is no correlation between the organic carbon content of turbidite layers and the depth of burrowing or intensity of biogenic reworking. In fact, turbidite sequence "g", showing the greatest depths of burrowing (Table 3), is an organic-poor turbidite layer (deLange et al. 1987). The depths of biogenic reworking in successive turbidite layers are variable (Jarvis & Higgs 1987), as are those in the Cascadia Channel discussed previously. The hiatus of bioturbation caused by a volcanic ash layer (turbidite sequences b_1 and d) is similar to effects of cessation of biogenic sediment reworking caused by anoxia (Savrda & Bottjer 1991).

Much of the surface sediment of the GME area is composed of a recent turbidite layer (Wilson et al. 1985) characterized by having no recognizable veneer of overlying pelagic sedimentation and minimal bioturbation (Table 3), which is limited to surficial sediments (Jarvis & Higgs 1987; Thomson et al. 1987). This turbidite layer resulted from a turbidity flow that occurred about 1,000 yr before present (Thomson & Weaver 1994). Huggett (1987) reports that the turbidite layer could be distinguished from adjacent abyssal pelagic sediment on the basis of surface lebensspuren (i.e., life traces), suggesting basic differences in faunal foraging behavior between the two sedimentary provinces. The dominant lebensspuren on the turbidite sediments are meandering traces attributed to mobile foragers and small spokes produced by sessile fauna; whereas, the pelagic sediments are characterized by irregular traces attributed to mobile foragers and medium-size spokes produced by sessile fauna. Huggett (1987) postu-lated that these different foraging strategies result from differences in near surface organic carbon (0.12–0.15% organic carbon in pelagic sediments versus 0.25% organic carbon in the turbidites) (Table 2). He also suggested that the turbidite sediments are too young (1,000 yr) to have been totally reworked by benthic fauna.

Thurston et al. (1994) report smaller size megafauna with a lesser overall standing crop and lower biomass from a site in the GME area of the Madeira Abyssal Plain (centered at 31°15' N,

Table 3. Periods of time, depth, and intensity of bioturbation between successive turbidite depositional events on the Madeira Abyssal Plain. Layered turbidite deposits are identified as lettered sequences (a–j) (from Jarvis & Higgs 1987).

Turbidite Sequence	Time Period (years)	Bioturbation, Burrowing
a	1K	minimal surficial mixing
a_1	11K	20 cm depth burrows
b & b_1	18K*	negligible mixing
d & d_1	48K*	weakly burrowed
e	12K	surficial mixing only
f	54K	burrows to 20 cm common and to 35 cm depth occasional
g & g_1	65K	prominent to 50 cm depth
h	57K	burrows to 15 cm depth
j	29K	extensive mixing; 12 cm depth burrows

* Periods of volcanic ashfall

25°25' W at 5,440 m water depth) compared with a site in the Porcupine Abyssal Plain (PAP) that was unaffected by recent turbidite flows (centered at 48°50' N, 16°30' W at 4940 m water depth) (Table 2). Thus, no increase in benthic biomass is found at the GME area that could be attributed to an increase in organic carbon in the surface turbidite layer. These findings are contrary to the hypothesis of Heezen et al. (1955). Rice (1997 personal communication) hypothesizes that a higher quality of phytodetritus at the PAP site could account for the differences of benthic biomass between sites, but reports (Rice et al. 1994) that a significantly higher flux of suspended matter has not been measured at the PAP site compared with the GME site. Accumulation rates for pelagic sediments in the GME area vary between 0.1 and 1.0 cm kyr^{-1} (Weaver & Rothwell 1987), so it is apparent that there has been insufficient time for pelagic sedimentation to cover the turbidite deposit, especially given surficial mixing of sediments by benthic fauna. In noting the shallowness of the surface mixed layer, Thomson and Weaver (1994, p.3) state that, "...a deeper mixed layer might have been expected if this [more organic matter] had attracted higher biological activity."

Thurston et al. (1994, p. 1344) state that it was 'unlikely' that residual effects of the turbidite deposit are affecting the megafaunal community at the GME area. Evidence for long-term effects of turbidity flows upon abyssal animal communities, as discussed by Young and Richardson (1998), suggest that low densities and biomass of megafauna at the GME site could be caused by such an episodic event even after 1,000 yr. In fact, the small individual sizes and overall low biomass of megafauna at the GME site likely result from residual effects of the relatively recent turbidity flow in this area of the Madeira Abyssal Plain. This conclusion, though discounted by Rice (1997, personal communication), remains as the most parsimonious explanation for the observed results.

Venezuela Abyssal Plain

The turbidite site is located in the central Venezuela Abyssal Plain (13°45' N; 67°45' W) of the Caribbean Sea at the distal end of ponded turbidity flows from the adjacent Colombia Basin, which originated at the Magdalena River Delta (see Table 1). These turbidity flows deposited layers of organic-rich (up to 1.55% organic carbon) terrestrial materials onto an organic-poor pelagic sediment (Table 2).

The sediment profiles at the turbidite site are characterized by a largely undisturbed, 14-cm-thick turbidite layer possessing different properties than the layers of highly bioturbated, pelagic deposits above and below it (Young & Richardson 1998). The turbidite layer exhibits typical graded bedding and contains sediment of coarser grain size and lower porosity than the pelagic layers (Briggs et al. 1985). The organic carbon content of the thickest turbidite layer, though variable, is very high for deep-sea sediment (Waples & Sloan 1980). The rapid deposition of sediments associated with the turbidite flow apparently produced "...a hiatus in bioturbation activity," and preserved intact gradients of sediment properties in buried pelagic layers which were "...created and maintained by past biological activity" (Richardson et al. 1985). X–radiograph profiles of the interface between the turbidite layer and the buried pelagic layer (Briggs et al. 1985) exhibit vertical failed escape burrows, suggesting that motile burrowers attempted, without success, to reach the new sediment surface.

The sedimentation rate of pelagically derived material at the Venezuela Basin turbidite site following the episodic turbidity flow event was 7.2 cm kyr^{-1}, as determined by activity profiles of uranium and thorium series isotopes (Cole et al. 1985). This rate suggests that the latest turbidity flow event occurred about 2,000 yr ago. Radionuclide profiles show highest biogenic mixing rates (900 cm^2 kyr^{-1}) for the upper 4–8 cm of the pelagic layer overlying the turbidite layer (Li et al. 1985). As with the example from the Madeira Abyssal Plain (Huggett 1987), the lebensspuren from the turbidite province of the Venezuela Abyssal Plain are markedly different from those of adjacent sedimentary provinces (Young et al. 1985), with lower diversity of types and a higher concentration of sedentary verses mobile forms compared to pelagic and hemipelagic sites.

The faunal differences between the turbidite and the other two provinces within the Venezuela Basin are striking. Trawl data presented by Briggs et al. (1996) indicate that in comparison with the pelagic and hemipelagic provinces, the turbidite province has lower species diversity, lower species richness, and fewer deposit feeders. Furthermore, sponges and holothurians dominate the trawl catches of the turbidite province, in contrast to the mollusks, decapods, and fishes that dominate at the adjacent provinces.

Biochemical analyses (Baird & White 1985; Shaw & Johns 1985) show that the layered turbidite sediment of the Venezuela Abyssal Plain contains labile organic compounds that are nutritious to deposit feeders. The biomass of meiofauna and macrofauna (Richardson & Young 1987) and of megafauna (Briggs et al. 1996) is low in relation to the organic matter potentially available to consumers at this site (Table 2). As with the examples from turbidite deposits of the Cascadia Channel and those of the Madeira Abyssal Plain, the hypothesis that nutrient enrichment by turbidity currents results in a high standing crop of abyssal animals (Heezen et al. 1955) is not supported by data from the turbidite province of the Venezuela Basin.

Benthic Faunal Succession in Deep-Sea Turbidites

If an analogy with shallow-water, benthic successional stages (sensu Rhoads et al. 1978) is applicable, early benthic 'pioneers', or opportunists, in the deep sea should be followed by a benthic fauna that feeds at greater depths within the sediment. In shallow-water marine benthic environments, equilibrium assemblages comprise long-lived, large, deposit-feeding individuals that reestablish themselves only after several years following a major disturbance (Rhoads & Germano 1982). Young and Richardson (1998) speculate that the time to reach the stage of an equilibrium benthic assemblage in deep-sea turbidite deposits is on the order of hundreds to thousands of years. Given sufficient time, the resulting equilibrium benthic assemblage should consist of more abundant and larger animals that have higher bioturbation rates and deeper mixing depths than the fauna that existed prior to the depositional event.

Sequences (A) and (B) of Fig. 2 show two hypothetical scenarios for benthic faunal succession of organically rich turbidite depositions in the deep sea. Both scenarios assume sufficiently thick depositions that destroy the benthic fauna by burial (Young & Richardson 1998), while the resulting relict burrows are preserved in situ as described by paleontologists (see overview by Seilacher 1991).

Sequence (A) of Fig. 2 presumes that the turbidite layer is completely mixed with the overlying pelagic sediments by benthic fauna. This scenario shows that the deep-burrowing benthic fauna has extensively "mined" the turbidite layer assuming that the energy expended to make deep burrows has metabolically repaid the cost of making them (sensu Jumars & Wheatcroft 1989). The stratigraphic result of this activity (final member of sequence A, Fig. 2) is that the turbidite layer is destroyed and only those relict burrows below the depth of deepest biogenic reworking are preserved. This end-member stage is a likely result predicted by optimal foraging theory, assuming that no hiatus prevents the recruitment by benthic fauna and that sufficient time is available to completely rework the turbidite layer. Such a stratigraphic result, however, has not been seen in any of the turbidite sequences reviewed in this paper (see Table 1 and Table 3).

Sequence (B) of Fig. 2 presumes that a delay in benthic faunal recolonization of the turbidite layer results in an initial hiatus of bioturbation following turbidite deposition. The preservation potential of sedimentary event layers, dependent upon competing processes of sediment reworking and rates (and depths) of deposition, therefore requires that a time-lag (long initial rest period, sensu Wheatcroft 1990) be introduced. The large areal extent of influence by turbidites and the slow modes of faunal recruitment can reduce the rates of recolonization of new substrata, as discussed by Young and Richardson (1998). A time lag can also be the result of insufficient oxygen to support aerobic respiration, particularly following episodic introductions of organically rich material (Jahnke 1998). Savrda and Bottjer (1991) show that laminated strata in fossil turbidite deposits are preserved during periods of anoxia and that higher bioturbation activities and deeper biogenic reworking of subsequently deposited sediment take place after oxygenation of bottom water is reestablished. The past presence of near-surface reducing conditions in turbidite sediments of the Venezuela Abyssal Plain is revealed by a thin, crusty iron-rich layer between the surficial pelagic and buried turbidite layers (Briggs et al. 1985). The turbidite layer with volcanic ash from the Madeira Abyssal Plain (Table 3) shows a similar hiatus of biogenic mixing by benthic fauna. The stratigraphic result of a delay in benthic faunal recolonization (see final member of sequence B, Fig. 2) is a higher probability of preservation of turbidite and pelagic layers. The resultant stratigraphy, of layered sequences of unmixed layers interspersed with bioturbated layers, is what is most commonly observed in deep-sea turbidite sediments (Piper & Stow 1991) and in turbiditic sedimentary rocks (Sepkoski et al. 1991).

Conclusions and Speculations

We do not find support for the hypothesis formulated from the speculation of Heezen et al. (1955) that there should be, "...a high [positive] correlation between nutrient-rich turbidity current areas and a high standing crop of abyssal animals." We also find little evidence for extensive deep deposit "mining" of turbidite sediments suggested by the optimal foraging theory (Jumars & Gallagher 1982). In fact, the mere existence of layered stratigraphic sequences so commonly observed in deep-sea turbidites and turbiditic sedimentary rocks depends upon incomplete sediment mixing.

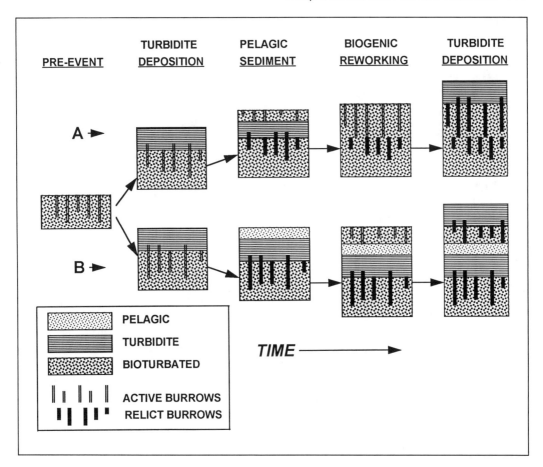

Fig. 2. Two hypothetical scenarios (A and B) for benthic faunal succession of turbidite depositions in the deep sea. Pelagic, turbidite, and bioturbated (both pelagic and turbidite) layers of sediment are shown; active and relict burrows of benthic fauna are depicted. The following sequences are shown: initial conditions prior to turbidity flow (*Pre–event*) with bioturbated sediment and active burrows; turbidite deposition burying benthic fauna (*Turbidite Deposition*) and capping off "escape" active burrows (both A and B); subsequent pelagic sedimentation (*Pelagic Sediment*) showing either bioturbation and active burrows by initial "pioneering" benthic species (A) or a delay in benthic faunal recolonization with a hiatus of bioturbation (B) [note: relict burrows are preserved in both A and B]; bioturbation by benthic fauna (*Biogenic Reworking*) showing active burrows and either complete mixing of pelagic and turbidite sediments (A) or incomplete mixing of surficial pelagic sediment (B); and subsequent deposition by a turbidity flow (*Turbidite Deposition*) burying benthic fauna, preserving relict burrows, and capping off either bioturbated sediment layer (A) or bioturbated, pelagic, and turbidite layers (B). See text for full description.

We suggest that evoking the models of "equilibrium" and successional stages for the benthic fauna of turbidite deposits in the deep sea is not heuristic and may not be appropriate. In considering the high unpredictability of turbidite currents, by definition, deep-sea benthos may not often have sufficient time to adapt to such episodic disturbances, as have shallow-water species. Perhaps viewing benthos of deep-sea turbidite deposits as "relics of former disasters" (sensu Johnson 1972) is more testable and inherently useful than considering benthic succession in these sediments as leading to equilibrium communities. As Jumars (1993, p. 301) asked, "Do geological old segments of the record of unsteady sedimentation [as turbidite sequences] suggest a fauna that was more poorly evolved to deal with unsteady sedimentation?" Our analysis of observations reported in this paper is supportive of this suggestion.

The turbidity current paradigm (Kuenen & Migliorini 1950) produced a dramatic change in the sedimentologists' view of deep-sea sedimentation patterns from one being dominated by a tranquil ocean with a slow rain of particulate matter to one with episodic but widespread and massive movements of sediment (Stow 1985). Benthic ecologists, in interpreting and reconstructing causes for distributions of deep-sea fauna should incorporate a similar view to that of sedimentologists to better understand how the abyssal benthic fauna has responded to such episodic catastrophes over geologic time. As stated by Einsele et al. (1991, p.14), "...biological responses of benthic communities are...but an integral part of dynamic stratigraphy". The long time scales of benthic faunal succession in the deep sea suggest the importance of turbidity flows as episodic disturbances in determining benthic community structure and, perhaps, in influencing evolutionary processes. Angel and Rice (1996) speculate that turbidity flows, while causing local extinction of benthic communities by burying them, also create mosaics of habitat in time and space that may contribute to the high species diversity of benthic fauna found in the deep sea.

Our analysis of data reviewed in this paper argues for a more all encompassing view of faunal life processes on the abyssal seafloor. One that not only acknowledges the importance of inputs of organic matter from detrital rain, geothermal vents, chemical seeps, organic falls, and various anthropogenic sources, but also inputs from episodic disturbances such as turbidity flows. If viewed as a matrix of dynamical events, the effects of turbidite flows on abyssal fauna occur on larger spatial scales and at longer temporal scales than most other habitat modifications. It is upon these large–scale disturbances that many of the smaller and shorter term effects are superimposed. These effects may be best interpreted from a combined paleontological and neontological approach to the stratigraphic study of deep-sea sediments. Such an approach, as pioneered by Donald Rhoads, has proven useful in comparative studies of fossil and recent shallow-water benthic communities (e.g., Rhoads 1970, 1974; Rhoads & Morse 1971; Rhoads et al. 1972; Rhoads & Boyer 1982; Larson & Rhoads 1983). More research is needed to better understand deep-sea geochemical and biological processes and their interactions in abyssal sedimentary environments as influenced by paleoceanographic events.

Acknowledgments

The authors would like to thank Donald C. Rhoads, whose body of innovative research in marine benthic ecology and paleontology inspired us to write this paper. We appreciate the thoughtful reviews by Bob Aller, Dolf Seilacher, and an anonymous reviewer. The time to write this paper was supported by the Coastal Benthic Boundary Layer Program of the Office of Naval Research and core programs of the Naval Research Laboratory (NRL). The original research was supported by the Strategic Environmental Research and Development Program of the US Department of Defense and the Deep Ocean Relocation Program of the Defense Advanced Research Projects Agency. NRL Contribution number PP/7431-98-006.

Literature Cited

Agassiz, A. 1892. Report on the dredging operations off the west coast of Central America to the Galapagos. Bulletin Museum Comparative Zoology, Harvard, Cambridge, Massachusetts. Vol. 23, No. 1.

Aller, J. Y. & R. C. Aller. 1986. Evidence for localized enhancement of biological activity associated with tube and burrow structures in deep-sea sediments at the HEBBLE site, western North Atlantic. *Deep-Sea Research* 33:755–790.

Angel, M. V. & T. L. Rice. 1996. The ecology of the deep ocean and its relevance to global waste management. *Journal of Applied Ecology* 33:915–926.

Baird, B. H. & D. C. White. 1985. Biomass and community structure of the abyssal microbiota determined from the ester–linked phospholipids recovered from Venezuela Basin and Puerto Rico Trench sediments. *In* Young, D. K. & M. D. Richardson (eds.), Benthic Ecology and Sedimentary Processes of the Venezuela Basin: Past and Present. *Marine Geology* 68:217–231.

Bornhold, B. D. &, O. H. Pilkey. 1971. Bioclastic turbidite sedimentation in Columbus Basin, Bahamas. *Bulletin of the Geological Society of America* 82:1341–1354.

Bowles, F. A., P. R. Vogt, & W. Jung. 1998. Bathymetry (Part I), sedimentary regimes (Part II), and abyssal waste disposal potential near the conterminous United States. *Journal of Marine Systems* 14:211–239.

Briggs, K. B., M. D. Richardson, & D. K. Young. 1985. Variability in geoacoustic and related properties of surface sediments from the Venezuela Basin, Caribbean Sea. *In* Young, D. K. & M. D. Richardson (eds.), Benthic Ecology and Sedimentary Processes of the Venezuela Basin: Past and Present. *Marine Geology* 68:73–106.

Briggs, K. B., M. D. Richardson, & D. K. Young. 1996. The classification and structure of megafaunal assemblages in the Venezuela Basin, Caribbean Sea. *Journal of Marine Research* 54:705–730.

Bruun, A. 1953. Problems of life in the deepest deep sea. The "Galathea" Expedition. *Geographical Magazine* 26:1–16.

Chough, S. K. 1984. Fine–grained turbidites and associated mass flow deposits in the Ulleung (Tsushima) back–arc basin, East Sea (Sea of Japan). *In* Stow, D. A. V. & D. J. W. Piper (eds.), Fine–grained Sediments: Deep-water Processes and Facies. *Geological Society London Special Publication* 15:185–196.

Cole, K. H., N. L. Guinasso, M. D. Richardson, J. W. Johnson, & D. R. Schink. 1985. Uranium and thorium series isotopes in Recent sediments of the Venezuela Basin, Caribbean Sea. *In* Young, D. K. & M. D. Richardson (eds.), Benthic Ecology and Sedimentary Processes of the Venezuela Basin: Past and Present. *Marine Geology* 68:167–185.

Corliss, J. B., J. Dymond, L. I. Gordon, J. M. Edmond, R. P. von Herzen, R. D. Ballard, K. Green, D. Williams, A. Bainbridge, K. Crane, & T. H. van Andel. 1979. Submarine thermal springs on the Galapagos Rift. *Science* 203:1073–1083.

deLange, G. J., I. Jarvis, & A. Kuijpers. 1987. Geochemical characteristics and provenance of late Quaternary sediments from the Madeira Abyssal Plain, N. Atlantic, p. 147–165. *In* Weaver, P. P. E. & J. Thomson (eds.), Geology and Geochemistry of Abyssal Plains. Geological Society Special Publication 31. Blackwell Scientific, London.

Einsele, G., W. Ricken, & A. Seilacher. 1991. Cycles and events in stratigraphy—Basic concepts and terms, p. 1–19. *In* Einsele, G., W. Ricken & A. Seilacher (eds.), Cycles and Events in Stratigraphy. Springer–Verlag, Berlin Heidelberg, Germany.

Gage, J. D. & P. A. Tyler. 1991. Deep-Sea Biology: A Natural History of Organisms at the Deep-Sea Floor. Cambridge University Press, New York.

Gage, J. D. 1991. Biological rates in the deep sea: A perspective from studies on processes in the benthic boundary layer. *Reviews in Aquatic Sciences* 5:49–100.

Gooday, A. J. & C. M. Turley. 1990. Responses by benthic organisms to inputs of organic material to the ocean floor: A review. *Philosphical Transactions of the Royal Society London A* 331:119–138.

Gorsline, D. S. 1980. Deep-water sedimentological conditions and models. *Marine Geology* 38:1–21.

Gorsline, D. S. & K. O. Emery. 1959. Turbidity-current deposits in San Pedro and Santa Monica Basins off Southern California. *Bulletin of the Geological Society of America* 70:279–290.

Grassle, J. F. & L. S. Morse–Porteous. 1987. Macrofaunal colonization of disturbed deep-sea environments and the structure of deep-sea benthic communities. *Deep–Sea Research* 34:1911–1950.

Grassle, J. F. 1977. Slow recolonization of deep-sea sediment. *Nature* 265:618–619.

Griggs, G. B., A. G. Carey, & L. D. Kulm. 1969. Deep-sea sedimentation and sediment fauna interaction in Cascadia Channel and on Cascadia Abyssal Plain. *Deep-Sea Research* 16:157–170.

Heezen, B. C. 1963. Turbidity currents, p. 742–775. *In* Hill, M. N. (ed.), The Sea, Vol. 3. Interscience, New York.

Heezen, B. C. & W. M. Ewing. 1952. Turbidity currents and submarine slumps, and the 1929 Grand Banks earthquake. *American Journal of Science* 250:849–873.

Heezen, B. C., M. Ewing, & R. J. Menzies. 1955. The influence of submarine turbidity currents on abyssal productivity. *Oikos* 6:170–182.

Heezen, B. C. & C. D. Hollister. 1971. The Face of the Deep. Oxford Universty Press, New York.

Honjo, S. 1978. Sedimentation of materials in the Sargasso Sea at a 5,367 m deep station. *Journal of Marine Research* 36:469.

Huggett, Q. J. 1987. Mapping of hemipelagic versus turbiditic muds by feeding traces observed in deep–sea photographs, p. 105–112. *In* Weaver, P. P. E. & J. Thomson (eds.), Geology and Geochemistry of Abyssal Plains. Geological Society Special Publication 31, Blackwell Scientific, London.

Isaacs, J. D. 1969. The nature of oceanic life. *Scientific American* 221:146–162.

Jahnke, R. A. 1998. Geochemical impacts of waste disposal on the abyssal seafloor. *Journal of Marine Systems* 14:355–375.

Jahnke, R. A. & G. A. Jackson. 1992. Spatial distribution of sea floor oxygen consumption in the Atlantic and Pacific Oceans, p. 295–307. *In* Rowe, G. T. & V. Pariente (eds.), Deep-Sea Food Chains and the Global Carbon Cycle. Kluwer Academic Publications.

Jarvis, I. & N. Higgs. 1987. Trace-element mobility during early diagenesis in distal turbidites: Late Quaternary of the Madeira Abyssal Plain, N Atlantic, p. 179–213. *In* Weaver, P. P. E. & J. Thomson (eds.), Geology and Geochemistry of Abyssal Plains. Geological Society Special Publication 31, Blackwell Scientific, London.

Johnson, R. G. 1972. Conceptual models of benthic marine communities, p. 148–159. *In* Schopf, T. J. M. (ed.), Models in Paleobiology. Freeman, Cooper & Co., San Francisco, California.

Jumars, P. A. & E. D. Gallagher. 1982. Deep–Sea community structure: Three plays on the benthic proscenium, p. 217–255. *In* Ernst, W. G. & J. G. Morin (eds.), The Environment of the Deep Sea. Prentice-Hall, Inc., New York.

Jumars, P. A. & R. A. Wheatcroft. 1989. Responses of benthos to changing food quality and quantity, with a focus on deposit feeding and bioturbation, p. 235–253. *In* Berger, W. H., V. S. Smetacek, & G. Wefer (eds.), Productivity of the Ocean: Present and Past. John Wiley & Sons Limited, New York.

Jumars, P. A. 1993. Concepts in Biological Oceanography. Oxford University Press, New York.

Kuenen, P. H. & C. I. Migliorini. 1950. Turbidity currents as a cause of graded bedding. *Journal of Geology* 58:91–127.

Larson, D. W. & D. C. Rhoads. 1983. The evolution of infaunal communities and sedimentary fabrics, p. 627–648. *In* Tevesz, M. J. S. & P. L. McCall (eds.), Biotic Interactions in Recent and Fossil Benthic Communities. Plenum Press, New York.

Li, W. Q., N. L. Guinasso, K. H. Cole, M. D. Richardson, J. W. Johnson, & D. R. Schink. 1985. Radionuclides as indicators of sedimentary processes in abyssal Caribbean sediments. *In* Young, D. K. & M. D. Richardson (eds.), Benthic Ecology and Sedimentary Processes of the Venezuela Basin: Past and Present. *Marine Geology* 68:187–204.

Lovell, J. P. B. 1969. Tyee Formation: Undeformed turbidites and lateral equivalents. *Geological Society of America Bulletin* 80:9–22.

Lutz, R. A., T. M. Shank, D. J. Fornari, R. M. Haymon, M. D. Lilley, K. L. VonDamm, & D. Desbruyeres. 1994. Rapid growth at deep-sea vents. *Nature* 371:663–664.

McBride, E. F. 1962. Flysch and associated beds of the Martinsburg Formation (Ordovician), central Appalachians. *Journal of Sedimentary Petrology* 32:39–91.

Menard, H. W. 1964. Marine Geology of the Pacific. McGraw-Hill Press, New York.

Menzies, R. J. & J. Imbrie. 1958. On the antiquity of the deep sea bottom fauna. *Oikos* 9:192–209.

Mills, E. L. 1983. Problems of deep-sea biology: an historical perspective, p. 1–79. *In* Rowe, G. T. (ed.), Deep-Sea Biology. The Sea, Vol. 8, Wiley-Interscience Publication. John Wiley & Sons, New York.

Moore, B. R. 1971. Paleographic and tectonic significance of diachronism in Siluro-Devonian Age Flysch Sediments, Melbourne Trough, Southeastern Australia. *Bulletin of the Geological Society America* Special Paper 107.

Mortensen, T. 1938. On the vegetarian diet of some deep-sea echinoids. *Annotationes Zoologicae Japonenses* 17:225–228.

Nuclear Energy Agency. 1984. Seabed Disposal of High-level Radioactive Waste: A Status Report of the NEA Coordinated Research Programme, Organization for Economic Co-operation and Development, Paris.

Pilkey, O. H. 1987. Sedimentology of basin plains, p.1–12. *In* Weaver, P. P. E. & J. Thomson (eds.), Geology and Geochemistry of Abyssal Plains. Geological Society Special Publication 31. Blackwell Scientific, London.

Piper, D. J. W. & W. R. Normark. 1983. Turbidite depositional patterns and flow characteristics, Navy submarine fan, California borderland. *Sedimentology* 30:681–694.

Piper, D. J. W. & D. A. V. Stow. 1991. Fine-grained turbidities, p. 360–376. *In* Einsele, G., W. Ricken and A. Seilacher (eds.), Cycles and Events in Stratigraphy. Springer-Verlag, Berlin, Germany.

Rhoads, D. C. 1974. Organism-sediment relations on the muddy sea floor. *Oceanography and Marine Biology, an Annual Review* 12:363–300.

Rhoads, D. C. 1970. Mass properties, stability, and ecology of marine muds related to burrowing activity, p. 391–406. *In* Crimes, T. P. & J. C. Harper (eds.), Trace Fossils. Seel House Press, Liverpool, United Kingdom.

Rhoads, D. C. & J. D. Germano. 1982. Characterization of organism–sediment relations using sediment profiling imaging: An efficient method of remote ecological monitoring of the seafloor (Remots™ System). *Marine Ecology Progress Series* 8:115–128.

Rhoads, D. C. & L. F. Boyer. 1982. The effects of marine benthos on physical properties of sediments: A successional perspective, p. 3–52. *In* McCall, P. L. & M. J. S. Tevesz (eds.), Animal-Sediment Relations. Plenum Press, New York.

Rhoads, D. C., P. L. McCall, & J. Y. Yingst. 1978. Disturbance and production on the estuarine seafloor. *American Scientist* 66:577–586.

Rhoads, D. C., I. G. Speden, & K. M. Waage. 1972. Trophic group analysis of Upper Cretaceous (Maastrichtian) bivalve assemblages from South Dakota. *American Association of Petroleum Geologists Bulletin* 56:1100–1113.

Rhoads, D. C. & J. W. Morse. 1971. Evolutionary and ecologic significance of oxygen-deficient marine basins. *Lethaia* 4:418–428.

Rice, A. L., M. H. Thurston, & B. J. Bett. 1994. The IOSDL DEEPSEAS programme: Introduction and photographic evidence for the presence and absence of a seasonal input of phytodetritus at contrasting abyssal sites in the northeastern Atlantic. *Deep-Sea Research* 41:1,305–1,320

Richardson, M. D. & D. K. Young. 1987. Abyssal benthos of the Venezuela Basin, Caribbean Sea: Standing stock considerations. *Deep-Sea Research* 34:145–164.

Richardson, M. D., K. B. Briggs, & D. K. Young. 1985. Effects of biological activity by abyssal benthic macroinvertebrates on sedimentary structure in the Venezuela Basin. *In* Young, D. K. & M. D. Richardson (eds.), Benthic Ecology and Sedimentary Processes of the Venezuela Basin: Past and Present. *Marine Geology* 68:243–267.

Rowe, G. T. 1983. Biomass and production of the deep-sea macrobenthos, p. 97–121. *In* Rowe, G. T. (ed.), The Sea, Vol. 8. Wiley, New York.

Rupke, N. A. & D. J. Stanley. 1974. Distinctive properties of turbiditic and hemipelagic mud layers in the Algero-Balearic Basin, Western Mediterranean Sea. Smithsonian Contributions Earth Sciences 13, Smithsonian Institution Press, Washington D.C.

Rusnak, G. A. & W. D. Nesteroff. 1964. Modern turbidites: Terrigenous abyssal plain versus bioclastic basin, p. 488–507. *In* Miller, R. L. (ed.), Papers in Marine Geology, Shepard Commemorative Volume. Macmillan Co., New York.

Sanders, H. L. 1968. Marine benthic diversity: A comparative study. *American Naturalist* 102:243–282.

Santos, V., D. S. M. Billett, A. L. Rice, & G. A. Wolff. 1994. Organic matter in deep-sea sediments from the Porcupine Abyssal Plain in the north-east Atlantic Ocean. I - Lipids. *Deep-Sea Research* 41:787–819.

Savrda, C. E. & D. J. Bottjer. 1991. Oxygen-related biofacies in marine strata: An overview and update, p. 201–219. *In* Tyson, R. V. & T. H. Pearson (eds.), Modern and Ancient Continental Shelf Anoxia. Geological Society Special Publication No. 58. Blackwell Scientific, London.

Seilacher, A. 1991. Events and their signatures—An overview, p. 222–226. *In* Einsele, G., W. Ricken, & A. Seilacher (eds.), Cycles and Events in Stratigraphy. Springer-Verlag, Berlin, Germany.

Sepkoski, J. J., R. K. Bambach, & M. L. Droser. 1991. Secular changes in Phanerozoic event bedding and the biological overprint, p. 298–312. *In* Einsele, G., W. Ricken, & A. Seilacher (eds.), Cycles and Events in Stratigraphy. Springer–Verlag, Berlin, Germany.

Shaw, P. M. & R. B. Johns. 1985. Comparison of lipid composition of sediments from three sites in the Venezuela Basin. *In* Young, D. K. & M. D. Richardson (eds.), Benthic Ecology and Sedimentary Processes of the Venezuela Basin: Past and Present. *Marine Geology* 68:205–216.

Smith, C. R. 1986. Nekton falls, low-intensity disturbance and community structure of infaunal benthos in the deep sea. *Journal of Marine Research* 44:567–600.

Smith, C. R. 1992. Factors controlling bioturbation in deep-sea sediments and their relation to models of carbon diagenesis, pp. 375–393. *In* Rowe, G. T. & V. Pariente (eds.), Deep-Sea Food Chains and the Global Carbon Cycle. Kluwer Academic Publications, Dordrecht.

Stordal, M. C., J. W. Johnson, N. L. Guinasso, & D. R. Schink. 1985. Quantitative evaluation of bioturbation rates in deep ocean sediments. II. Comparison of rates determined by 210Pb and 239,240Pu. *Marine Chemistry* 17:99–114.

Stow, D. A. V. 1985. Deep-sea clastics: Where are we and where are we going?, p. 67–93. *In* Brenchley, P. J. & P. J. Williams (eds.), Sedimentology: Recent developments and Applied Aspects. Geological Society, Blackwell Scientific Publications, Oxford, United Kingdom.

Sujkowski, Z. L. 1957. Flysch sedimentation. *Bulletin of the Geological Society of America* 68:543–554.

Thomson, C. W. 1878. The Voyage of the "Challenger". The Atlantic. A Preliminary Account of the General Results of the Exploring Expedition of H. M. S. "Challenger" During the Year 1873 and the Early Part of the Year 1876. Harper, New York.

Thomson, J. & P. P. E. Weaver. 1994. An AMS radiocarbon method to determine the emplacement time of recent deep-sea turbidites. *Sedimentary Geology* 89:1–7.

Thomson, J., N. C. Higgs, D. J. Hydes, T. R. S. Wilson, & J. Sorensen. 1987. Geochemical oxidation fronts in NE Atlantic distal turbidites and their effects in the sedimentary record, p. 167–177. *In* Weaver, P. P. E. & J. Thomson (eds.), Geology and Geochemistry of Abyssal Plains. Geological Society Special Publication 31, Blackwell Scientific, London.

Thomson, J., I. Jarvis, D. R. H. Green, & D. Green. 1997. Oxidation fronts in Madiera Abyssal Plain Turbidites: Persistence of early diagenetic trace–element enrichments during burial, ODP Site 950, p. 287–296. *In* Weaver, P. P. E., H. U. Schmincke, J. V. Firth & W. A. Duffield (eds.), Proceedings of the Ocean Drilling Program, Scientific Results, Vol. 157. College Station, Texas.

Thurston, M. H., B. J. Bett, A. L. Rice, & P. A. B. Jackson. 1994. Variations in the invertebrate abyssal megafauna in the North Atlantic Ocean. *Deep-Sea Research* 41:1321–1348.

Trauth, M. H. & M. Sarnthein. 1997. Bioturbational mixing depth and carbon flux at the seafloor. *Paleoceanography* 12:517–526.

Valent, P. J. & D. K. Young (eds.) 1995. Abyssal Seafloor Waste Isolation: Environmental Report. NRL/MR/7401-95-7576, Naval Research Laboratory, Stennis Space Center, Mississippi.

Waples, D. W. & J. R. Sloan. 1980. Carbon and nitrogen diagenesis in deep sea sediments. *Geochimica et Cosmochimica Acta* 44:1463–1470.

Weaver, P. P. E. & R. G. Rothwell. 1987. Sedimentation on the Madeira Abyssal Plain over the last 300,000 years, p. 71–86. *In* Weaver, P. P. E. & J. Thomson (eds.), Geology and Geochemistry of Abyssal Plains. Geological Society Special Publication 31, Blackwell Scientific, London.

Weaver, P. P. E., J. Thomson, & P. M. Hunter. 1987. Introduction, p. vii–xii. *In* Weaver, P. P. E. & J. Thomson (eds.), Geology and Geochemistry of Abyssal Plains. Geological Society Special Publication 31. Blackwell Scientific, London.

Wheatcroft, R. A. 1990. Preservation potential of sedimentary event layers. *Geology* 18:843–845.

Wheatcroft, R. A., P. A. Jumars, C. R. Smith, & A. R. M. Nowell. 1990. A mechanistic view of the particulate biodiffusion coefficient: Step lengths, rest periods and transport directions. *Journal of Marine Research* 36:71–91.

Wilson, T. T. S., M. D. Thomson, S. Colley, D. J. Hydes, N. C. Higgs, & J. Sorensen. 1985. Early organic diagenesis; the significance of progressive subsurface oxidation fronts in pelagic sediments. *Geochimica et Cosmochimica Acta* 49:811–822.

Wolff, G. A., D. Boardman, I. Horsfall, I. Sutton, N. Davis, R. Chester, M. Ripley, C. A. Lewis, S. J. Rowland, J. Patching, T. Ferrero, P. J. D. Lambshead, & A. L. Rice. 1995. The biogeochemistry of sediments from the Madeira Abyssal Plain–Preliminary results. *Internationale Revue der Gesamten Hydrobiologie* 80:333–349.

Young, D. K. & M. D. Richardson. 1998. Effects of waste disposal on benthic faunal succession on the abyssal seafloor. *Journal of Marine Systems* 14:319–336.

Young, D. K. & P. J. Valent. 1997. Is abyssal seafloor isolation an environmentally sound waste management option? *Underwater Technology* 22:155–165.

Young, D. K. & P. J. Valent. 1996. Environmental assessment of waste isolation on the abyssal seafloor. *Oceanology International 96* 1:317–326.

Young, D. K., W. H. Jahn, M. D. Richardson, & A. W. Lohanick. 1985. Photographs of deep-sea Lebensspuren: A comparison of sedimentary provinces in the Venezuela Basin, Caribbean Sea. *In* Young, D. K. & M. D. Richardson (eds.), Benthic Ecology and Sedimentary Processes of the Venezuela Basin: Past and Present. *Marine Geology* 68:269–301.

Unpublished Materials

Rice, A. L. 1997. Personal communication. Letter of July17, 1997, to D. K. Young.

Organism-Sediment Relations at Multiple Spatial Scales: Implications for Community Structure and Successional Dynamics

Roman N. Zajac

Abstract: *How organism-sediment relationships change over different spatial scales and the potential consequences for infaunal successional dynamics are addressed. Review and re-analysis of benthic surveys conducted in Long Island Sound indicate that benthic communities are shaped by a combination of fa[ctors] including the heterogeneity of sediments at [va]rying spatial scales and hydrographic, topog[raphic] ... large ... recent ... research taking a benthic landscape approach ... sedimentary conditions explain < 40% of the varia-tion in species abun[dance] ... however, when considered within the context of definable sea floor patches (10s–1,000s of [m]) sedimentary factors explained > 70% of the variation in sands but less in muds. Boundaries ... [signific]ant degree of population variability for some i[nfauna] ... ty of ambient community types composed of s[pecies with] ... ons and life history and population character[istics] ... [se]diment succession model is suggested in which ... h multiple successional pathways among them ...*

Organism-sedim[ent] [relationships are central to mari]ne soft-sediment ecology. Relation-ships between infauna [and sediments have been studied along] many different avenues. These include, for ex[ample,] ... [the] influence of sediment characteristics, how infauna affect sedimen[ts] and in turn shape community [structure], food resources, feeding patterns and trophic interactions, th[e effects of disturbance and subsequent] [re]colonization-succession dynamics, biogeochemical dynamics, and sediment-organism-flow interactions. Much of the literature on these areas of research are reviewed in Gray (1974), Rhoads (1974), Aller (1982), Rhoads and Boyer (1982), Probert (1984), Lopez et al. (1989), Hall et al. (1994), and Snelgrove and Butman (1994). Two of these areas: sediment characteristics and community structure, and disturbance-succession dynam-ics, are the focus of this paper. The main questions addressed are, How does benthic community structure change in relation to sediment characteristics at different spatial scales? and How do such relationships shape recolonization and succession? Specifically, I am interested in determining the patterns of population and community variation at different spatial scales as a basis for establishing the identity of "climax" or endpoint communities and predicting the successional pathways leading to the re-establishment of these communities following disturbance. By considering soft-sediment community variation at different spatial scales we might gain insights as to the conditions and factors that shape infaunal successional dynamics. Indeed, the nature of successional pathways and the factors shaping the responses may change as the spatial extent of the disturbance changes (e.g., Smith & Brumsickle 1989; Thrush et al. 1996; Zajac et al. 1998). Although little work has been done on this particular topic (Hall et al.1994), there is a growing body of research on the spatial characteristics and dynamics of soft-sediment communities (e.g., Jumars et al. 1977; Schnieder 1987; Pickney & Sandulli 1990; Thrush 1991; Robbins & Bell 1994; Thrush et al. 1997), in concert with similar efforts to understand spatial dynamics in ecology in general (e.g., Levin 1992; Hansson et al. 1995).

I address these possibilities by focusing on research conducted in Long Island Sound (LIS), where much of the seminal work on organism-sediment interactions has been done (e.g., Sanders 1956; Rhoads et al. 1978). My approach is to examine population and community responses to sediment characteristics in an area of LIS where sea floor imaging technologies were used to study benthic landscapes at multiple spatial scales and by re-analyzing information from several sound-wide surveys that were conducted during the 1970s and 1980s.

Patterns at the Benthoscape Scale

Assessing organism-sediment relations has profited greatly from advances in technology. Although basic survey designs and "blind" bottom grabs show varying levels of association between infauna and sediments (reviewed in Snelgrove & Butman 1994), they do not provide an a priori habitat context for analysis. Grab samples also generally destroy sediment structure and information on potentially important factors may be lost. The latter problem has been overcome to a large extent by the introduction of the sediment-profile camera (Rhoads & Cande 1971; Rhoads & Germano 1982, 1986), which has added greatly to our understanding of animal-sediment relations and how these may change under different environmental conditions (e.g., Rhoads & Boyer 1982; Schaffner et al. 1992; Rosenberg & Diaz 1993). The problem of having some understanding of the spatial distribution and extent of sedimentary habitat types prior to initiating studies of benthic community structure has been addressed by the use of side-scan sonar in conjunction with other techniques (Warwick & Uncles 1980; Menzie et al. 1982; Oliver & Kvitek 1984; Krost et al. 1990). In LIS, several areas have been studied intensively using side-scan sonar, video, and bottom grabs to understand how infaunal community structure varies in relation to the spatial characteristics of the seafloor (Zajac 1996, 1998; Twichell et al. 1997, 1998; Poppe & Poloni 1998). This landscape, or benthoscape, approach seeks to assess spatial structure and temporal dynamics in benthic communities among and within the various landscape features (or elements) composing the seafloor. The benthoscapes investigated range over spatial extents of 10s of km^2. The elements are distinct patches of different bottom types evident on the side-scan sonar images and can be related to the concept of ecological habitat patches.

To illustrate how animal-sediment relations may change at the population and community levels I will summarize some results from a side-scan-sonar study site located in the eastern end of LIS, an area with a relatively high degree of seafloor heterogeneity (Fig. 1). Based on the interpretation of the

Fig. 1. (Upper) Side-scan (100 kHz) mosaic of study area in eastern Long Island Sound (inset). The mosaic shown is approximately 2.5 x 8 km. (Lower) Sediment composition at sampling sites within the side-scan mosaic study area. Sites were grouped using cluster analysis (Canberra metric and unweighted pair-group with arithmetic averages, UPGMA), groups are shown in the dendrogram. A-E represent the subgroups in main clusters. Mean percent composition of sediment types for each cluster are given. G r = gravel, Sa = sand, Si = silt, CL = clay. Mud-Sand, Sand 1, and Sand 2 refer to three large habitat patches in the study area.

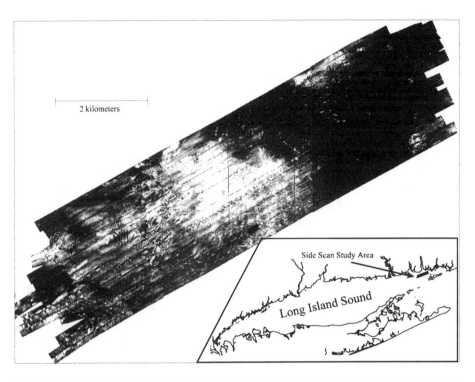

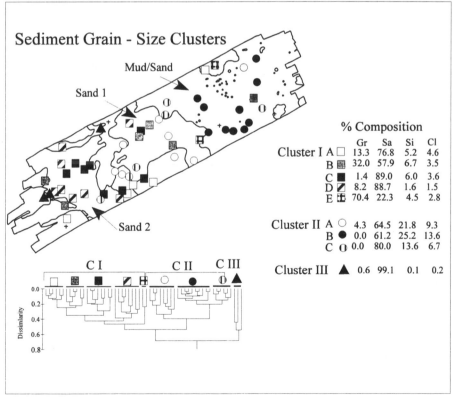

Sediment Grain - Size Clusters

Mud/Sand

Sand 1

Sand 2

% Composition

		Gr	Sa	Si	Cl
Cluster I	A	13.3	76.8	5.2	4.6
	B	32.0	57.9	6.7	3.5
	C	1.4	89.0	6.0	3.6
	D	8.2	88.7	1.6	1.5
	E	70.4	22.3	4.5	2.8
Cluster II	A	4.3	64.5	21.8	9.3
	B	0.0	61.2	25.2	13.6
	C	0.0	80.0	13.6	6.7
Cluster III		0.6	99.1	0.1	0.2

C I   C II   C III

Dissimilarity
0.0
0.2
0.4
0.6
0.8

side-scan-sonar mosaic, an array of 63 sampling sites were positioned in a nested, stratified random design to assess within and among element variation in infaunal populations (Zajac 1996, 2000). Strata and substrata corresponded to the large-scale element and within-element structure found in the side-scan-sonar mosaic.

Part of this work suggested that local sedimentary factors explained less than 40% of the variation in the abundances of the 16 numerically dominant species, in total abundance, and in species richness when considered across the entire benthic landscape (Table 1). Most of the best fit models comprised three or less of the seven sedimentary factors considered. These results agree with Snelgrove and Butman's (1994) conclusion that direct relationships between sedimentary factors and infaunal distributions are usually weak, and suggest that infaunal communities may only be loosely shaped by small sets of sedimentary factors when considered over a variety of patch types. However, this analysis did not take into account any information on the spatial location of the samples (i.e., the benthoscape context). When the analysis was performed by grouping the samples from the same benthoscape elements, a much higher degree of variation (often above 60%) was explained by sedimentary variables, and more factors were included in the best fit models (Table 1). Interestingly, higher amounts of variation were explained for sand areas relative to the mud area in the eastern portion of the study area (Table 1 & Fig. 1). Thus, within a specific seafloor element, infaunal abundances and distributions may be shaped by a wider range of sediment characteristics that affect all portions of the life cycle and associated activities. However, the relative mix of critical sediment factors may change with habitat type.

Cluster analysis indicated that while community groups tended to be distributed across the study area, there were distinct spatial clusters primarily within one of the benthoscape elements (Fig. 2). The communities identified tended to be dominated by species associated with surficial sediments, with the main differences being their relative abundances at particular sites (Table 2). However, the communities were a mix of species having different life histories, life modes, and potential effects on sedimentary characteristics. It is expected then that responses to disturbance may also be quite varied and successional dynamics be shaped by the mix of mesoscale (10s to 1,000s of m^2) features composing the patch. At this mesoscale, it is also likely that biotic interactions may be of some importance.

Another spatial component of the seafloor that was found to explain the abundance patterns of some species were transition zones among benthoscape elements, which can be identified visually on the side-scan sonar records and via image analysis. For example, the spionid polychaete *Prionospio steenstrupi* was the most abundant species at the time of sampling (June 1992), and the population included many newly settled juveniles. Analysis of variance based on a nested sampling design in the three largest landscape elements (Fig. 1) indicated no significant differences in abundance among these patches but significant mesoscale differences within the patches (Fig. 3a). Furthermore, when the samples were grouped based on whether they were located in transition zones or in the interior of patches, it was found that *Prionospio steenstrupi* had higher abundances in transition zones than in the interior of large patches (Fig. 3b).

The benthoscape study in eastern LIS suggested that sedimentary factors may play an important role in shaping infaunal communities at mesoscales that are associated with identifiable habitats and habitat transitions across the seafloor landscape. The habitats identified at the landscape level are formed and maintained by geomorphologic and hydrodynamic processes. These, in turn, likely shape the large-scale distributions of infauna by creating sedimentary environments that infauna are adapted

Table 1. Results of regression analyses relating species abundances to sediment characteristics (% composition of gravel, sand, silt, and clay [from sediment analyses]) and the incidence of shell hash, cobbles, and algae (from analysis of video data) at sampling sites in eastern Long Island Sound (Fig. 1). Multiple regressions were run for each combination of factors using Mallows Cp and minimization of the mean square error as selection criteria for the best fit model (Hintze 1997). Table entries are the R^2 values and the letters in parentheses indicating which factors composed the best model. "Landscape" indicates regressions that included 51 sites across the entire study area; Mud-Sand (n = 14), Sand 1 (n = 10), and Sand 2 (n = 10) regressions are for sampling sites located only within these landscape elements (Fig.1). Infaunal data based on 6-cm diameter cores collected from 0.1-m² Van Veen grabs and washed on a 300-μm sieve. Details given in Zajac (1996, 2000). Factors: A = shell hash, B = cobble, C = algae, D = gravel, E = sand, F = silt, G = clay, all = all factors included in model.

Taxon	Landscape	Mud-Sand	Sand 1	Sand 2
Nemertinea	0.222 (AB)	0.284 (CF)	0.670 (EG)	0.917 (ABCEG)
Nucula annulata	0.092 (DG)	0.318 (DEFG)	0.943 (BCDEFG)	0.890 (all)
Oligochaete sp. a	0.309 (ABCF)	0.905 (ABCDF)	0.961 (BCDEFG)	0.416 (CD)
Mediomastus ambiseta	0.271 (BG)	0.344 (BCDE)	0.708 (ABF)	0.800 (ACDEFG)
Tharyx dorsobranchialis	0.123 (BE)	0.299 (G)	0.118 (AG)	0.458 (DG)
Chaetozone sp. a	0.164 (AC)	0.079 (B)	0.939 (BCDEFG)	0.875 (BCFG)
Clymenella torquata	0.133 (BC)	0.140 (ACF)	0.834 (ABDEFG)	0.572 (AEG)
Nephtys incisa	0.155 (F)	0.112 (F)	0.790 (ACDEFG)	0.862 (ADEFG)
Aricidia catherinae	0.114 (D)	0.233 (CDF)	0.793 (BCE)	0.337 (ABG)
Prionospio steenstrupi	0.102 (F)	0.240 (F)	0.914 (BCDEFG)	0.923 (BDEFG)
Exogenes hebes	0.393 (ABEF)	0.072 (G)	0.739 (BEG)	0.715 (ABFG)
Polycirrus exumius	0.297 (ADEFG)	0.213 (DE)	0.843 (ADEFG)	0.919 (ABCE)
Ampilesca vadorum	0.561 (BEG)	0.212 (G)	0.988 (all)	0.576 (CEG)
Phoxocephalus holboli	0.111 (BE)	not found	0.694 (ABF)	0.889 (ABCFG)
Unicola irrorata	0.369 (BCF)	0.336 (ACF)	0.947 (BCDEFG)	0.903 (ABEG)
Microduetopus gryllotalpa	0.063 (AF)	0.571 (ADEFG)	0.777 (CDG)	0.961 (ABDEG)
Total Abundance	0.186 (DE)	0.242 (F)	0.930 (BCDEFG)	0.946 (BEFG)
Species Richness	0.249 (DG)	0.221 (EG)	0.784 (CDG)	0.716 (CEG)

Table 2. Composition of main community types determined by cluster analysis (communities III, IV, and V in Fig. 2) of samples from a study area in eastern Long Island Sound (Fig. 1). The species are given in the order they may appear in the sediments: from taxa found in the upper few centimeters of the sediment to deeper dwelling taxa. Abundances are given as number of individuals m^{-2}.

	III	IV	V	Feeding, Motility, Sediment Modification	Taxon
Prionospio steenstrupi	59,304	36,712	33,535	Surface deposit-feeding, filter feeding, discretely motile, tubiculous	Spionid
Tharyx dorsobranchialis	1,800	226	3,265	Surface deposit-feeding, discretely motile/motile, sediment bioturbating?	Cirratulid
Chaetozone sp. a	106	353	501	Surface deposit feeder, discretely motile	Cirratulid
Ampelisca vadorum	4,236	385	116	Surface deposit/suspension feeder, tubiculous	Amphipod
Unicola irrorata	1,800	159	205	Surface deposit/suspension feeder, tubiculous	Amphipod
Microduetopus gryllotalpa	212	64	237	Surface deposit/suspension feeder, tubiculous	Amphipod
Phoxocephalus holboli			469	Surface deposit/suspension feeder, tubiculous	Amphipod
Aricidea catherinae	247	385	413	Herbivore, surface deposit feeder, motile, burrower	Paraonid
Polycirrus exumis	247	4	205	Surface deposit feeder, discretely motile	Terrebellid
Exogenes hebes			794	Herbivore, surface deposit feeder, carnivore, motile, burrower, nontubiculous	Syllid
Oligochaete sp. a	2,965	2,330	2,577	Burrowing deposit feeder, motile,	
Mediomastus ambiseta	247	247	116	Burrowing deposit feeder, motile, pelletization	Capitellid
Nucula annulata	318	212	28	Subsurface deposit feeder, discretely motile	Protobranch
Nemertean	141	642	88	Carnivore, burrowing	
Nephtys sp.	353	226	381	Carnivorous, burrowing deposit feeder, motile, burrows	Nephtyid
Clymenella torquata	353	159	148	Subsurface deposit feeder, sessile, tubiculous, bioturbation, oxygenation	Maldanid

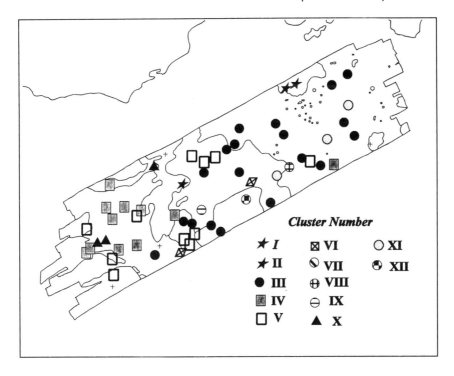

Fig. 2. Spatial distribution of community types in the side-scan mosaic study area in eastern Long Island Sound as determined by cluster analysis. Analysis was conducted on the 16 most abundant taxa (see Table 1) using Moritsita's resemblance function and an UPGMA averages clustering algorithm (Rohlf 1993).

to at the individual level and/or flow conditions that distribute larvae and/or food resources. However, at mesoscales more recognizable patterns are evident as the specifics of species adaptations are played out and the effects of any biotic interactions emerge. The communities comprised a mix of species types spanning the "opportunist-type" to "climax-type" spectrum (see below).

Patterns at the Regional Scale

Are patterns of animal-sediment relationships evident at the landscape level (e.g., the study area in eastern LIS) repeated over a larger geographic region (e.g., the whole of LIS)? We can attempt to answer this question by reassessing the results of large-scale surveys and exploring patterns of community structure relative to the regional distribution of seafloor landscapes, their characteristics, and other factors that may shape distributions and dynamics at this spatial scale. I do so here by considering the results of several benthic surveys conducted in LIS.

Although his work was focused in the central portion of LIS, Sanders (1956) provided key initial insights as to relationships between sediment characteristics, community structure, and feeding modes, and the classification of community types. With respect to the latter, he was able to recognize one community type at four of the eight sites he sampled in the central portion of LIS: the *Nephtys incisa-Yoldia limatula* community, which was found at 4–30 m depths in sediments of > 25% silt-clay content. Perhaps more importantly however, was that Sanders (1956) recognized that variation in

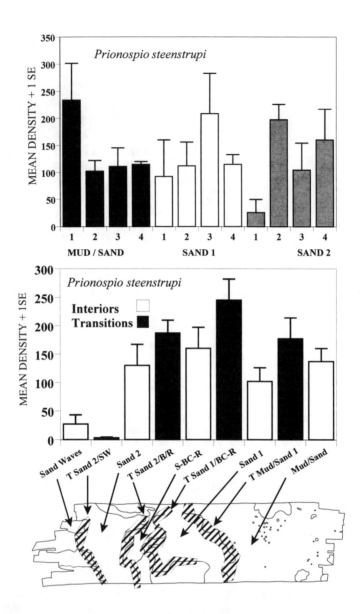

Fig. 3. (Upper) Differences in the mean density per 6-cm diameter core of the spionid polychaete *Prionospio steenstrupi* within subdivisions (noted as 1–4) of, and among, the largest benthic landscape elements (Mud-Sand, Sand 1, and Sand 2) of the study area shown in Fig. 1. (Lower) Differences in the density per 6-cm diameter core of the spionid polychaete *Prionospio steenstrupi* among transition zones (solid bars indicated by a T) and the interior portions (open bars) of landscape elements in the side-scan study area shown in the map below the graph. Hatched areas show the transition areas. SW = sand wave patches; BC, B = boulder and cobble patches; R = rubble and mixed sediment patch; S = sand/mud patch; 'T Sand/SW' indicates, for example, the transition area between sand wave patches in the western portion of the study area and the Sand 2 landscape element.

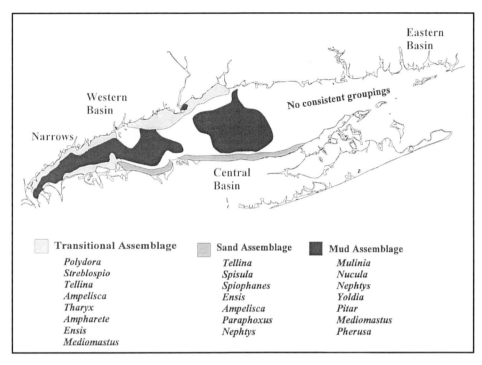

Fig. 4. Infaunal community types recognized by Reid et al. (1979) in Long Island Sound. Below each assemblage type are the dominant species in decreasing order of abundance. Data are from 0.1-m² grab samples washed on a 1-mm sieve.

habitat characteristics both within and among habitats was a potentially important determinant of sound-wide trends in benthic community structure. He noted that while there were predominant sediment grain-size classes at each station, the sediments were often heterogeneous. For example, in some coarse sand areas there were also significant amounts of clay and silt. He also noted the potential effects of other habitat characteristics on infaunal communities, such as the presence of shell hash.

The first sound-wide benthic survey in LIS was conducted between 1972 and 1973 by Reid at al. (1979). Using data from 142 samples collected during the summer of 1972 (located every 3–5 km on north-south transects spaced 8.7 km apart), Reid et al. (1979) used cluster analysis to determine the relative similarity of infaunal community structure throughout LIS. They recognized three faunal groups in the central and western basins of the sound; no consistent groups were identified for the eastern basin (Fig. 4). The three faunal groups consisted of a muddy, deep-water assemblage distributed throughout much of the central and western basins, a shallow sandy assemblage along much of the north shore of Long Island, New York, except in western portions of the sound, and a transitional shallow water assemblage in the western portion of the sound and along the Connecticut shore roughly between Fairfield and Branford. The three groups (Fig. 4) each comprised a mixture of species with varying life modes and life histories. The muddy deep-water and the shallow sandy groups were dominated by bivalves, whereas several polychaetes dominated the transitional group.

Table 3. Dominant species based on mean abundance in the stations composing each of the community types shown in Fig. 5. Shown in parentheses are the number of species in the community out of the 35 used in the analysis.

Community Type	Dominant Species
A	*Mulinia, Nephtys, Pectinaria* (6)
B	*Mulinia, Nucula, Nephtys* (16)
C1	*Nucula, Mulinia, Nephtys* (17)
C2	*Mulinia, Nucula, Nephtys* (13)
D	*Pectinaria, Corophium, Mulinia* (22)
E	*Nucula, Nephtys, Paraonis, Yoldia* (9)
F	*Mulinia, Clymenella, Mediomastus* (21)
G	*Pectinaria, Clymenella, Pitar, Asabellides* (10)
H1	*Asabellides, Tellina, Spiophanes* (29)
H2	*Ampelisca, Corophium, Spiophanes* (28)
H3	*Unicola, Aricidea, Capitella* (14)
I	*Cirratulis, Corophium, Prionospio* (21)
J	*Assabelides, Polydora, Spiophanes,Leptocherius* (15)
K	*Protohaustorius, Acanthohaustorius* (15)

Species richness was lower in the muddy, deep-water and shallow sandy groups than in the transitional group. Reid et al. (1979) suggested that the overlap in community composition among the three groups indicated that benthic infauna in LIS are not distributed as discrete, well-defined communities but rather form a faunal continuum from one area to another.

Pellegrino and Hubbard (1983) conducted a benthic survey (413 stations) in Connecticut waters of LIS (Fig. 5). They found that community composition varied considerably throughout the Sound. In the western basin, the communities were dominated by *Mulinia lateralis, Nucula annulata, Pitar morrhuana*, and *Cestinoides gouldii*; in the central basin *Nephtys incisa, Nucula annulata, Yoldia limatula*, and *Mulinia lateralis* were dominant; in the eastern basin a variety of polychaetes and amphipods (different from the western and central basins) were the dominant species (Pellegrino & Hubbard 1983). To assess organism-sediment relations at the community level in more detail across the spatial extent of LIS, I re-analyzed Pellegrino and Hubbard's (1983) data and compared the results with a sound-wide sediment distribution map developed by Freidrich et al. (1986).

Cluster analysis, based on the 35 most abundant species throughout the sound, identified 10 main community types (Fig. 5). Similarity among the communities ranged between 5% and 30% (Zajac 2000). This analysis indicates that community structure varies in LIS on several spatial scales. At the regional scale, sound-wide trends are evident in the distributions of the communities but there is also a fair degree of variation at smaller spatial scales (i.e., within the benthoscapes of the Sound). Community types A-G were primarily found in the central and western basins of LIS, whereas community types H-K were generally found in the eastern end of the Sound (Fig. 5). However, several community types, for example D, F, and H1, were found to be distributed throughout the Sound.

Table 4. Frequency of community types identified in Fig. 5 in different types of sediments in Long Island Sound. Sediment distributions are also shown in Fig. 5.

Community	Sediment Type						Grand Total
	Gravel	Gravelly Sand	Coarse Sand	Medium Sand	Fine Sand	Silt	
A				1	6	3	10
B			3	6	12	45	66
C1			3	1	7	61	72
C2				1	16	22	39
D			4		9	3	16
E				2	1	2	5
F			5	3	14	8	30
G				1	5		6
H1			14	8	44	5	71
H2		1	2	5	12	2	22
H3			9	1	2		12
I	1	2	18	5	8		34
J			2	1		1	4
K			8	4	8		20
No Data			1		2	1	4
Grand Total	1	3	69	39	146	153	411

In the western and central basins, most stations were composed of community types B, C1, or C2. The species in these communities included *Nephtys incisa*, *Cistenoides gouldii*, *Mulinia lateralis*, *Nucula annulata*, and *Pitar morrhuana* (Table 3). The communities differed primarily based on the relative number of these species, with higher numbers of *Mulinia* and *Nucula* in community C1 and C2 than in community B stations. In the eastern portion of the Sound, seven community types were found, and these were relatively more distinct with respect to species composition and relative abundance. For example, community type F, which was primarily found at the transition from the central basin to the eastern Sound (Fig. 5 & Table 3), was dominated by *Clymenella zonalis* but also included *Mulinia lateralis* and *Mediomastus ambiseta*. Community type I, which was found predominantly in the eastern end of the Sound, was dominated by the polychaetes *Cirratulis grandis*, *Cirratuli cirratus*, *Prionospio heterobranchia*, and *Prionospio tenuis*, and the amphipod *Aeginnia longicornis*. The mix of community types in the far western, or Narrows, section of the Sound was also heterogenous (Fig. 5 & Table 3).

This analysis suggests that landscape-scale variation in community structure is lower for large areas of the western and central basins compared with that found in the Narrows and eastern portion of LIS. Each community type was found in most types of sediments, but particular communities tended to be associated with certain sediment types (Fig. 5 & Table 4). For example, community types B, C1, and C2 were found in varying combinations in primarily silty sediments of the western and central basin of the Sound. The other communities were also found in a variety of sediment types but were primarily associated with coarser sediments (Table 4).

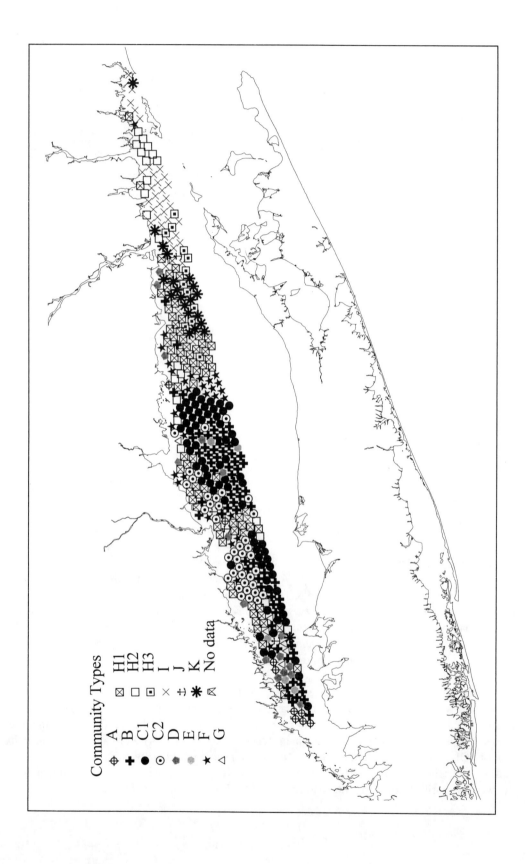

Community Types

A	⊕	H1 ⊠
B	+	H2 □
C1	●	H3 ⊡
C2	⊙	I ×
D	⬣	J ⊥
E	⬣	K ✳
F	★	No data ⊠
G	△	

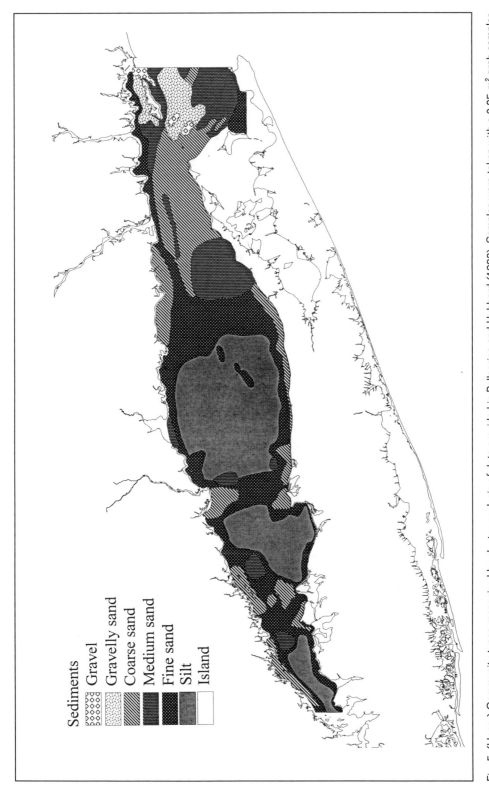

Fig. 5. (Upper) Community types recognized by cluster analysis of data provided in Pellegrino and Hubbard (1983). Samples were taken with a 0.25-m² grab samples washed on a 1-mm sieve. The cluster analysis used a Bray-Curtis resemblance function and an UPGMA clustering algorithm (Rohlf 1993). Community characteristics are given in Table 3. (Lower) Distribution of general sediment types in Long Island Sound based on Freidrich et al. (1986).

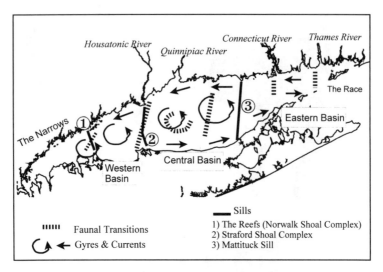

Fig. 6. Transitions in soft-sediment infaunal community structure relative to several physical features in Long Island Sound. The transitions are based on the distribution of community assemblages shown in Fig. 5. The position of hypothesized gyres and current patterns are taken from Welsh (1993). Other characteristics of each basin are given in Zajac (1996) and Poppe and Poloni (1998).

Across large, regional scales, the relationship between sediment and community type is not absolute, but clearly there are strong associations. Overall spatial variation of infaunal communities in LIS is likely related to several factors, including sediment grain-size, the geographic location of the habitat, topography, and hydrodynamic patterns. Some initial idea of how these factors may interact to shape infaunal communities can be derived by looking at the transitions in community types along the length of LIS relative to changes in its geophysical characteristics (Fig. 6). The boundaries among regions, landscapes, and landscape elements can be important foci of ecologic activity, reflecting different scales of biological and geophysical structure and dynamics (e.g., Ebert & Russell 1988; Holland et al. 1991; Posey & Ambrose 1994; Forman 1995). Benthic community structure and infaunal species populations (Fig. 3) can exhibit significant changes along such transition zones.

There appear to be several distinct boundaries among benthic communities in LIS (Fig. 6) as determined by changes in benthic community structure (Fig. 5). What is interesting to note is that these faunal boundaries are associated with different sets of geophysical features. In the western portion of LIS, relatively sharp breaks in benthic community structure occur in the areas of bathymetric highs, specifically the Reefs and Stratford Shoals. These two boundaries do differ, however. Changes in community structure along Stratford Shoal coincide with local changes in bathymetry and sediment composition. The mix of community types in the western and central basins, which the shoal separates, are similar, and therefore this boundary does not mark a distinct shift in biotic assemblages between these two regions of the Sound. In contrast, the boundary associated with the Reefs area does mark a distinct change in the mix of community types between the Narrows and the western basin. This boundary may also be associated with a small gyre (Fig. 6).

In the central basin, there are two faunal breaks that appear to be associated with mesoscale circulation patterns, specifically a system of gyres that may occur in this basin (see Welsh 1993). In the western portion of the central basin there is a large area primarily composed of one community type (Type B in Fig. 5). This is in the vicinity of a smaller gyre as shown in Fig. 6. The other faunal break in the central basin is to the east (Fig. 6) and is more extensive (the change from community type C1 to F in Fig. 5). This boundary may be associated with a large gyre as shown in Fig. 6. Neither of these two boundaries appear to be associated with a sharp change in depth. The larger transition to the east is also associated with a change in sediment type (Fig. 5), but the smaller, circular faunal transition is not.

No strong faunal break appears to be associated with the Mattituck Sill, and there is a relatively high degree of community variation over a broader spatial scale in the eastern basin. Distinct faunal breaks do occur, however, to the west of the Connecticut and Thames rivers (Fig. 6). Bottom substrates in these areas are complex and are characterized by a mixture of coarse- to fine-grained habitats.

Although qualitative in nature, these patterns suggest that benthic community structure in LIS may, in addition to depth and substrate composition, be controlled by a number of hydrodynamic factors, including gyres and river plumes. Features associated with the breaks, and potentially influencing faunal transitions, change with geographic position in the estuary. River plumes and varying sediment composition appear critical in the east, and gyres in the central basin. In the western basin and Narrows, hydrodynamic features appear less critical and faunal changes are associated primarily with bathymetric features.

Implications for Successional Dynamics

The studies reviewed above provide a basic understanding of how soft-sediment communities in LIS vary over different spatial scales. Several community types are recognized, which were found in varied sedimentary environments but associated most frequently with one type of sediment. Communities in mud environments are more similar to one another over larger areas. Sand associated communities are more complex and change significantly over smaller spatial scales. Many communities are not continuous but are found over mesoscale (2–3 km^2) areas of the seafloor. What implications do these patterns have for recolonization and succession following disturbance? The most critical aspect of this question is perhaps establishing the identity of the climax or endpoint community, because the characteristics of species composing these communities will set the patterns and tempo of succession via their interactions with changing physical, chemical, and biological factors following disturbance.

There has been extensive work on the dynamics of soft-sediment communities in the central basin of LIS that has led to key insights about the effects of disturbance on infaunal communities and successional dynamics. Much of this work is integrated into a model of infaunal succession presented by Rhoads et al. (1978). The key element of the model is the recognition of specific stages of benthic community development that differ based on the life histories and sediment interactions of the species (Fig. 7). The stages can also represent different states of degradation in response to a pollution gradient (Pearson & Rosenberg 1978). The model recognizes three phases in community recovery. Following disturbance, the communities will be dominated by opportunistic species (small size, short life spans, high population growth rates), the infauna will be concentrated in the upper portions of the sediment, sediment binding may be high, and there will be a shallow redox potential discontinuity (RPD) level. This is referred to as a Stage I community. In contrast, climax communities will be composed of more equilibrium-type species (larger size, longer life spans, lower population growth

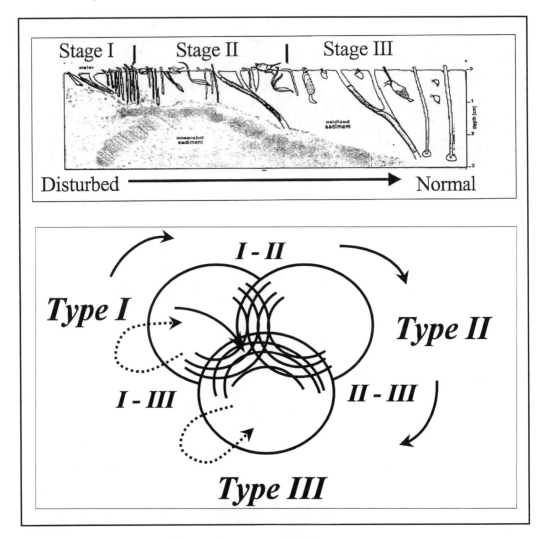

Fig. 7. (Upper) Succession model proposed by Rhoads et al. (1978). Figure is adapted from Rhoads and Germano (1982). (Lower) Proposed modification of the Rhoads et al. (1978) model in which a variety of successional endpoints are possible (e.g., community types I, II, and III, and combinations of each as represented by the arcs) with animal-sediment characteristics as depicted in the upper model. Arrows indicate examples of potential successional transitions among the community types. See text for details.

rates), which live deeper in the sediments. Their activities will tend to mobilize the sediments and also to create a deeper RPD layer. This is referred to as a Stage III community. Stage II communities exhibit intermediate characteristics. Other aspects of this model are presented in Rhoads and Boyer (1982). This model has been used to map and interpret community changes and sediment-organism relationships in a variety of environments in response to human impacts such as dredge spoil disposal and hypoxia-anoxia (e.g., Rhoads & Germano 1986).

Based on the studies reviewed above, no consistent Stage III assemblage could be easily identified which was distributed across LIS. Both Reid et al.'s (1979) and Pellegrino and Hubbard's (1983) studies point to a diverse suite of community types, many of which are composed of species that would exhibit the organism-sediment relations predicted by each stage in the Rhoads et al. (1978) model (Fig. 4 & Table 3). If we interpret the model strictly over the regional scale of LIS, the data suggest that there is high spatial variability in the distribution of Type I, II, and III communities and that most areas are at some intermediate stage of succession. Although there are many types of disturbances that affect the infauna of the Sound, and some occur frequently (such as dredge disposal and trawling), the "temporal mosaic" (sensu Johnson 1973) should include areas that are not undergoing succession.

An alternative view is that some subset of the variety of infaunal assemblages in LIS represent different successional endpoints, with distinct sets of community characteristics and dynamics. A simple graphical model of this alternative view is shown in Fig. 7. Three general community types (Types I, II, and III) are suggested which are equivalent in terms of their ecologic characteristics to Stage I, II, and III communities, respectively, in the Rhoads et al. (1978) model. The difference is that each represents a potential successional endpoint (or climax community) that can persist over time as a recognizable community type within the domain of prevailing environmental conditions. However, the sediment-organism relations predicted by the Rhoads et al. (1978) model (Fig. 7) for each community type still hold. Due to environmental heterogeneity, variable responses to sediment types, and other factors, combinations of community types (e.g., a type I-III community) will be found. Rhoads and Germano (1986) have noted mixtures of Stage I and Stage III communities. When these community types are disturbed, the successional pathways also differ. For example, for a Type I community, composed of primarily short-lived species living in the upper layers of the sediment, succession may proceed relatively rapidly and without easily recognizable seral stages (Zajac & Whitlatch 1982a). In contrast, in Type III communities, succession may proceed more slowly along the pathway predicted by the Rhoads et al. (1978) model. Successional pathways for each community type will be shaped by factors such as seasonality and species life histories.

The successional dynamics of each proposed community type may also be shaped by interactions between sediment type and the spatial extent of the disturbance. For example, in the large central basin, which is primarily composed of muds, responses to disturbances of different spatial extent may be quite similar, as the variation in community structure is not large over this region of LIS. Large-scale hydrodynamic patterns such as gyres may also provide a relatively less variable supply of colonists to disturbed areas. In contrast, succession in sandier environments may be more variable, as smaller scale sedimentary differences and species responses affect succession. When disturbance changes sediment characteristics in sand environments, more variable successional trajectories may be elicited, as infauna may respond with more specificity based on sediment-organism relationships at the species level in sand habitats (Table 1). Hydrodynamic patterns in sand environments may also be more variable.

Conclusions

In a recent and insightful review, Snelgrove and Butman (1994) argued that, although many studies have revealed some relationship between sediment type and infaunal community structure, there is considerable variability in species responses to specific sediment characteristics. They suggested that the factors ultimately controlling infaunal distributions may not be sediment grain-size per se or correlates such as organic content, but rather interactions between hydrodynamics, sediments, and infauna and how these affect sediment distribution, larval supply, particle flux, and porewater chemistry (Snelgrove & Butman 1994). Such interactions are undoubtably complex, and it will be difficult to tease them apart and their various outcomes. The extent to which we are able to generalize may be related to the spatial and temporal scales at which we consider each factor or combination of factors (e.g., Zajac et al. 1998) and which ecological characteristics of soft-sediment infauna are considered (i.e., individual, population, community, or ecosystem-level characteristics and dynamics). At small spatial scales (≤ 5 m^2), individual species adaptations to sediment type may be obscured as distributions are determined by responses to a complex set of prevailing local abiotic and biotic conditions. Therefore, successional patterns and dynamics at this scale may not differ with respect to sediment type as they may at larger spatial scales. At mesoscales (10s to 1,000s of m^2), composite responses may have a distinguishable sediment signal as individual- and population-level responses are amplified across defined sedimentary environments. At these scales the effects of post-settlement processes in determining community structure may also be most evident, as opposed to the direct effects of hydrodynamics (Olafsson et al.1994; Hewitt et al. 1997) Large-scale habitat patterns are generated by interactions among sediment, flow, and geomorphology, and the nature and distribution of community types found in these habitats are likewise shaped by these factors.

Within this context there may exist a variety of persistent community types which likely exhibit relatively distinct successional dynamics. The details of how the factors noted above interact to shape infaunal successional dynamics relative to organism-sediment interactions over different spatial and temporal scales remain to be fully understood. For example, only one controlled successional study has been conducted in the offshore waters of LIS (McCall 1977). This was in the mud habitats of the central basin. In addition to conducting more succession studies in different types of sedimentary environments, it is critical that detailed population studies be conducted focusing on the key species found in different types of infaunal communities, and, in particular, studies that link benthic and pelagic phases of the life cycle (Eckman 1996) Such studies can make use of previous surveys to identify the pattern and extent of differences in population abundances over varying spatial scales in relation to benthic landscape structure and prevailing hydrodynamic patterns. This may be especially insightful as detailed knowledge on more benthic landscapes becomes available from the use of technologies such as side-scan sonar. Disturbance is an important determinant of soft-sediment community dynamics. Elucidating how various factors interact to shape climax communities and the various stages of succession that may occur in a particular type of sedimentary habitat and/or within a particular region (e.g., Rumohr et al. 1996) is a key component of understanding organism-sediment interactions.

Acknowledgments

This work was supported by Long Island Sound Research Fund grant CWF-221-R, Connecticut Department of Environmental Protection. I worked with several colleagues to develop the data presented for the New London study area: Ralph Lewis, who developed the side-scan mosaic, Larry Poppe, who collected and prepared the sediment data, and Joe Vozarik, who provided the taxonomic expertise. Jana Newman, Kevin Zawoy, Ann Asterista, Ed King, Kristin Cramer, Yuan Chen, Joanna Dowgialo, Sue Wilson, Lisa Vasiloskos, and Matt Robinson provided technical assistance during various phases of the work. Simon Thrush and an anonymous reviewer provided a number of suggestions that greatly enhanced the manuscript. To all my deepest thanks. I am especially grateful to Peter Pellegrino and William Hubbard for providing their survey data. Also, many thanks to Fran, Katya, and Julia.

Literature Cited

Aller, R. C. 1982. The effects of macrobenthos on chemical properties of marine sediments and overlying waters, p. 53–102. *In* McCall, P. L. & M. J. S. Tevesv (eds.), Animal-Sediment Relations. Plenum Publishing, New York.

Ebert, T. E. & M. P. Russell. 1988. Latitudinal variation in size structure of the west coast purple sea urchin: A correlation with headlands. *Limnology and Oceanography* 33:286–294.

Eckman, J. E. 1996. Closing the larval loop: Linking larval ecology to the population dynamics of marine benthic invertebrates. *Journal of Experimental Marine Biology and Ecology* 200:207–237.

Freidrich, N. E., R. L. McMaster, H. F. Thomas, & R. S Lewis. 1986. Non-energy resources: Connecticut and Rhode Island coastal waters. U.S. Department of the Interior, Minerals Management Service Cooperative Agreement No. 14-12-0001-30115, Final Report FY 1984.

Gray, J. S. 1974. Animal sediment relationships. *Oceanography and Marine Biology Annual Review* 12:223–261.

Forman, R. T. T. 1995. Land Mosaics: The Ecology of Landscapes and Regions. Cambridge University Press, Cambridge.

Hall, S. J., D. Raffaelli, & S. F. Thrush. 1994. Patchiness and disturbance in shallow water benthic assemblages, p.333–375. *In* Giller, P. S., A. G. Hildrew, & D. G. Rafaelli (eds.), Aquatic Ecology: Scale, Pattern, and Process. Blackwell Scientific Publications, London.

Hansson, L. Fahrig, & G. Merriam (eds.), 1995. Mosaic Landscapes and Ecological Processes. Chapman and Hall, London.

Hewitt, J. E., R. D. Pridmore, S. F. Thrush, & V. J. Cummings. 1997. Assessing the short-term stability of spatial patterns of macrobenthos in a dynamic estuarine system. *Limnology and Oceanography* 42:282–288.

Hintze, J. L. 1997. NCSS Statistical Software. NCSS, Kaysville, Utah.

Holland, M. M., R. J. Naiman, & P. G. Risser. 1991. Role of landscape boundaries in the management and restoration of changing environments. Chapman and Hall, New York

Johnson, R. G. 1973. Conceptual models of benthic communities, p. 148–159. *In* Schopf, T. J. M. (ed.), Models in Paleobiology. Freeman and Cooper, San Francisco.

Jumars, P. A., D. Thistle, & M. L. Jones. 1977. Detecting two-dimensional spatial structure in biological data. *Oecologia* 28:109–123.

Krost, P., M. Bernhard, F. Werner, & W. Hukriede. 1990. Otter tracks in Kiel Bay (Western Baltic) mapped by side-scan sonar. *Meeresforschung* 32:344–353.

Levin, S. A. 1992. The problem of pattern and scale in ecology. *Ecology* 73:1,943–1,983.

Lopez, G. R., G. Taghon, & J. Levinton (eds.). 1989. Ecology of marine deposit feeders. Lecture Notes on Coastal and Estuarine Studies 31. Springer, New York.

McCall, P. L. 1977. Community patterns and adaptive strategies of the infaunal benthos of Long Island Sound. *Journal of Marine Research* 35:221–266.

Menzie, C. A., J. Ryther, L. F. Boyer, J. D. Germano, & D. C. Rhoads. 1982. Remote methods of mapping seafloor topography, sediment type, bedforms, and benthic biology, p. 1046–1051. In *Oceans '82* Conference Record. IEEE publication number 82CH1827-5. IEEE, Piscataway, New Jersey.

Olafsson, E. B., C. H. Peterson, & W. G. Ambrose, Jr. 1994. Does recruitment limitation structure populations and communities of macro-invertebrates in marine soft sediments: The relative significance of pre- and post-settlement processes. *Oceanography and Marine Biology Annual Review* 32:65–109.

Oliver, J. S. & R. G. Kvitek. 1984. Side-scan sonar records and diver observations of the Gray Whale (*Eschrichtius robustus*) feeding grounds. *Biological Bulletin* 167:264–269.

Pearson, T. H. & R. Rosenberg. 1978. Macrobenthic succession in relation to organic enrichment and pollution of the marine environment. *Oceanography and Marine Biology Annual Review* 16:229–311.

Pellegrino, P. & W. Hubbard. 1983. Baseline shellfish data for the assessment of potential environmental impacts associated with energy activities in Connecticut's coastal zone. Volumes I & II. Report to the State of Connecticut Department of Agriculture, Aquaculture Division. Hartford, Connnecticut.

Pinckney, J. & R. Sandulli. 1990. Spatial autocorrelation analysis of meiofaunal and microalgal populations on an intertidal sandflat: Scale linkage between consumers and resources. *Estuarine, Coastal and Shelf Science* 30:341–353.

Poppe, L. J. & C. Polloni. 1998. Long Island Sound Environmental Studies. U.S. Department of the Interior, U.S. Geological Survey, Atlantic Division, Woods Hole, Massachusetts. Open File Report 98–502.

Posey, M. H. & W. G. Ambrose. 1994. Effects of proximity to an offshore hard-bottom reef on infaunal abundances. *Marine Biology* 118:745–753.

Probert, P. K. 1984. Disturbance, sediment stability, and trophic structure of soft-bottom communities. *Journal of Marine Research* 42:893–921.

Reid, R. N., A. B. Frame, & A. F. Draxler. 1979. Environmental baselines in Long Island Sound, 1972–1973. U.S. Department of Commerce, National Oceanographic and Atmospheric Administration (NOAA), Technical Report NMFS SSRF-738.

Rhoads, D. C. 1974. Organism-sediment relations on the muddy sea floor. *Oceanography and Marine Biology Annual Review* 12:263–300.

Rhoads, D. C. & L. F. Boyer. 1982. The effects of marine benthos on physical properties of sediments: A successional perspective, p. 3–52. In McCall, P. L. & M. J. S. Tevesv (eds.), Animal-Sediment Relations. Plenum Publishing, New York.

Rhoads, D. C. & S. Cande. 1971. Sediment profile camera for in situ study of organism-sediment relations. *Limnology and Oceanography* 16:110–114.

Rhoads, D. C. & J. D. Germano. 1982. Characterization of organism-sediment relations using sediment profile imaging: An efficient method of remote ecological monitoring of the seafloor (REMOTS System). *Marine Ecology Progress Series* 8:115–128.

Rhoads, D. C. & J. D. Germano. 1986. Interpreting long-term changes in benthic community structure: A new protocol. *Hydrobiologia* 142:291–308.

Rhoads, D. C., P. L. McCall, & J. Y. Yingst. 1978. Production and disturbance on the estuarine seafloor. *American Scientist* 66:577–586.

Rhoads, D. C. & D. K.Young. 1970. The influence of deposit-feeding organisms on sediment stability and community trophic structure. *Journal of Marine Research* 28:150–78.

Robbins, B. D. & S. S. Bell. 1994. Seagrass landscapes: A terrestrial approach to the marine subtidal environment. *Trends in Ecology and Evolution* 9:301–303.

Rohlf, F. J. 1993. NTSYS-pc: Numerical Taxonomy and Multivariate Analysis System. Exeter Sofware, Setauket, New York.

Rosenberg, R. & R. J. Diaz. 1993. Sulfur bacteria (*Beggiatoa* spp.) mats indicate hypoxic conditions in the inner Stockholm Archipelago. *Ambio* 22:32–36.

Rumohr, H., E. Bonsdorff, & T. H. Pearson. 1996. Zoobenthic succession in Baltic sedimentary habitats. *Archives of Fisheries and Marine Research* 44:179–214.

Sanders, H. L. 1956. Oceanography of Long Island Sound. X. The biology of marine bottom communities. *Bulletin of the Bingham Oceanography Collection* 15:245–258.

Schaffner, L., P. Jonsson, R. J. Diaz, R. Rosenberg, & P. Gapcynski. 1992. Benthic communities and bioturbation history of estuarine and coastal systems: Effects of hypoxia and anoxia. *Science of the Total Environment*, Supplement p. 1,001–1,016.

Schneider, D. C., J. Gagnon, & K. D. Gilkinson. 1987. Patchiness of epibenthic megafauna on the outer Grand Banks of New Foundland. *Marine Ecology Progress Series* 39:1–13.

Smith, C. R. & S. J. Brumsickle. 1989. The effect of patch size and substrate isolation on colonization modes and rate in an intertidal sediment. *Limnology and Oceanography* 34:1,263–1,277.

Snelgrove, P. V. & C. A. Butman. 1994. Animal-sediment relationships revisited: Cause versus effect. *Oceanography and Marine Biology Annual Review* 32:111–177.

Thrush, S. F. 1991. Spatial patterns in soft-bottom communities. *Trends in Ecology and Evolution* 6:75–79.

Thrush, S. F., R. B. Whitlatch, R. D. Pridmore, J. E. Hewitt, V. J. Cummings, & M. R. Wilkinson. 1996. Scale-dependent recolonization: The role of sediment stability in a dynamic sandflat habitat. *Ecology* 77:2,471–2,487.

Thrush, S. F., R. D. Pridmore, R. G. Bell, V. J. Cummings, P. K. Dayton, R. Ford, J. Grant, M. O. Green, J. E. Hewitt, A. H. Hines, T. M. Hume, S. M. Laurie, P. Legendre, B. H. McCardle, D. Morrisey, D. C. Schneider, S. J. Turner, R. A. Walters, R. B. Whitlatch, & M. R. Wilkinson. 1997. The ecology of soft-bottom habitats: Matching spatial patterns with dynamic processes. *Journal of Experimental Marine Biology and Ecology*, Special Issue 216:1–254.

Twichell, D. C., R. N. Zajac, L. J. Poppe, R. S. Lewis, V. Cross, D. Nichols, & M. DiGiacomo-Cohen. 1997. Sidescan sonar image, surficial geologic interpretation and bathymetry of the Long Island Sound seafloor off Norwalk, Connecticut. U.S. Geological Survey, Geologic Investigations Series Map I-2589.

Twichell, D. C., R. N. Zajac, L. J. Poppe, R. S. Lewis, V. Cross, D. Nichols, & M. DiGiacomo-Cohen. 1998. Sidescan sonar image, surficial geologic interpretation and bathymetry of the Long Island Sound seafloor off Milford, Connecticut. U.S. Geological Survey, Geologic Investigations Series Map I-2632.

Warwick, R. M. & R. J. Uncles. 1980. Distribution of benthic macrofauna associations in the Bristol Channel in relation to tidal stress. *Marine Ecology Progress Series* 3:97–103.

Welsh, B. L. 1993. Physical oceanography of Long Island Sound: An ecological perspective, p. 23–33. *In* Van Patten, M. S. (ed.), Long Island Sound Research Conference Proceedings. publication no. CT-SG-93-03. Connecticut Sea Grant Program, University of Connecticut, Groton, Connecticut.

Zajac, R. N. 1996. Ecologic mapping and management-based analyses of benthic habitats and communities in Long Island Sound. Final Report. Office of Long Island Sound Programs, State of Connecticut, Department of Environmental Protection, Hartford, Connecticut.

Zajac, R. N. 1998. Spatial and temporal characteristics of selected benthic communities in Long Island Sound and management implications. Final Report. Office of Long Island Sound Programs, State of Connecticut, Department of Environmental Protection, Hartford, Connecticut.

Zajac, R. N. , R. S. Lewis, L. J. Poppe, D. C. Twichell, J. Vozarik, & M. L. DiGiacomo-Cohen. 2000. Relationships among sea-floor structure and benthic communities in Long Island Sound at regional and benthoscape scales. *Journal of Coastal Research* 16:627–640.

Zajac, R. N. & R. B. Whitlatch. 1982. Responses of estuarine infauna to disturbance. I. Spatial and temporal variation of initial recolonization. *Marine Ecology Progress Series* 10:1–14.

Zajac, R. N., R. B. Whitlatch, & S. F. Thrush. 1998. Recolonization and succession in soft-sediment infaunal communities: The spatial scale of controlling factors. *Hydrobiologia* 375/376:227–240.

Population Consequences of Intermediate Disturbance: Recruitment, Browsing Predation, and Geochemistry

David S. Wethey*, Sara M. Lindsay, Sarah A. Woodin, and
Roberta L. Marinelli

Abstract: *Biogenic disturbance is a dominant feature of marine soft-sediments. Sediment dwellers continuously feed and defecate on the surface, thereby causing mortality of larvae and juveniles. Browsing predation on adults can reduce disturbance imposed by adults because individuals feed and defecate less while they regenerate. While consuming adult body parts, browsers themselves disturb the sediments, resulting in changes in the surficial chemical properties, as well as ingestion of larvae. Previous studies suggest that changes in surficial sediment chemistry may result in changes in the acceptability of a site to larvae. Thus there is a potential balance between browsing predation, sediment disturbance, and larval recruitment in marine sedimentary habitats. We constructed a population model consisting of four coupled differential equations to examine the relationship between predation, disturbance, sediment chemistry, and their interactive effects on the population dynamics of adults, recovering individuals, and settlers. We parameterized the model using demographic and predation data on* Macoma balthica *for the Wadden Sea. In the model,* Macoma *shows enhanced recruitment at intermediate browsing intensities in sediments ranging from sands to muds. The enhancement of recruitment by browsing is a function of the ratio of the recovery rate to the browsing rate, and the ratio of the rate of disturbance by intact adults to the rate of consumption of larvae by browsers. This phenomenon is a population-level analog of Connell's enhancement of species diversity at intermediate disturbance rates. The steepness of the concentration gradient of porewater solutes in the sediment determines the equilibrium population densities through its effects on the rate of recovery of disturbed sediments and their acceptability to larvae. If larvae respond positively to the chemical signature of disturbed surficial sediments, then population maxima are predicted to occur in muds. In contrast, if larvae respond negatively to disturbed surficial sediments, then population maxima are predicted to occur in less reactive sediments. The data on* Macoma *densities suggest the recruits respond positively to geochemical signals associated with disturbance.*

Introduction

Sediments are dynamic habitats where disturbances, both biotic and physical, are common. Such disturbances include removal of surficial sediments by erosion, deposit feeding, and digging or biting by predators as well as deposition onto the surface by defecation, sediment transport, or sediment bulldozing by epifauna (see references in Hall 1994 and Wilson 1991). All of these disturbances alter the chemistry of the surficial sediments as well as the physical setting. Such disturbed surfaces, whether biologically or physically generated, are likely to be chemically similar to those under suboxic or anoxic conditions. For example, removal of sediments by erosion or surface-deposit feeding exposes subsurface sediments which may have low or no oxygen, and concomitantly high concentrations of reduced compounds (Aller & Yingst 1978, 1985; Jorgensen & Revsbech 1983; Marinelli & Boudreau 1996). Similarly, deposition of feces onto the sediment surface imposes a strong suboxic signature to the sediment surface, due largely to the anoxic conditions in infaunal guts (Plante & Jumars 1992), the intense microbial activity of hindguts and fecal pellets (Lopez & Levinton 1987; Plante & Mayer 1995), and upward advection of deep anoxic sediment by subsurface-feeding deposit feeders (Clough & Lopez 1993). Even the deposition of previously eroded nearshore sediments can impose seemingly "reduced" conditions on the sediment surface due to desorption of ammonium and possibly, desorption and decomposition of newly exposed organic material (Rosenfeld 1979; Aller 1994; Keil et al. 1994).

Many of these disturbances are dependent upon the feeding and defecation activities of infaunal adults and the activities of predators. For example, rates of defecation and feeding are a function of adult size (Brey 1991; Krager & Woodin 1993), and both rates are reduced following tissue loss to browsing predators (Woodin 1984; Lindsay & Woodin 1992, 1996; Kamermans & Huitema 1994). The degree of biotic sediment disruption is thus affected by the densities of infauna, their mortality rates, the rate of browsing predation, the recovery rates of infauna following tissue loss, the size distribution of infauna, and the rates of sediment disruption by other forces such as digging predators and erosion. The recovery rate of the solute concentrations in the surficial sediments depends in part on the nature of the sediment disturbance. Properties such as the frequency, severity, and spatial persistence of a disturbance all will contribute to the maintenance of the chemical signature associated with a disruption. At the same time, the recovery rate is determined by the transport-reaction properties inherent to the habitat. For example, highly organic muds with steep chemical gradients will take longer to be restored to a predisturbance chemical state than sandy sediments with shallow gradients (Woodin et al. 1998).

A number of studies have shown that sediment disruption can significantly alter population dynamics by influencing growth and mortality rates of settling larvae and juveniles (Wilson 1980, 1981; Levin 1981; Brenchley 1982; Elmgren et al. 1986; Posey 1986; Hines et al. 1989; Ólafsson 1989; Turner & Miller 1991; Flach 1992). We have suggested that surface chemistry associated with sediment disruption can convey significant information about the probability of survivorship to recruits. Consistent with this hypothesis, recruits of some species are known to respond positively to elevated hydrogen sulfide in surficial sediments (Cuomo 1985 but see Dubilier 1988), while others respond negatively to elevated concentrations of ammonium typical of subsurface habitats (Woodin et al. 1998) and reduced surficial oxygen following disturbance (Marinelli & Woodin unpublished data).

Given that recruitment is affected by the chemical signal of surficial sediments and that the chemical signal is determined by the disturbance history and governing transport-reaction processes inherent to sediments, the dynamics of the community may be driven by a complex interaction among the input of recruits, the densities and activities of the resident infauna, the activities of predators, and sediment chemical processes. Below we present coupled numerical models of the chemical recovery of surficial sediments, the impact of browsing predators on infaunal activities, and the recovery of the affected adult infauna. The goal is to examine how the interaction of these components affects the chemical and population dynamics of the assemblage and whether recruitment will be affected by the inhibition of adult activities due to tissue loss. The organism of interest in this modeling effort is *Macoma balthica*, a tellinid bivalve with a cosmopolitan distribution for which there are considerable data on recruitment intensity, population density, tissue losses to predators, feeding rates, and regeneration rates.

Our hypothesis is that browsing predators at intermediate densities should increase recruitment success by reducing the rate of larval mortality due to ingestion and disturbance by adults. In addition, if the recruits respond negatively to the chemical signature of disturbed sediments, then recruitment and population densities will be greatest in less reactive sediments such as sands. In contrast, if recruits respond positively to the chemical signature of disturbed sediments, then recruitment and population densities will be greatest in more reactive sediments such as muds. Our hypothesis is analogous to the intermediate disturbance hypothesis in which disturbance is predicted to increase diversity by reducing mortality from competition (Connell 1971).

Model Description

The population model assumes an open population in which local reproduction has no influence on the rate of recruitment (e.g., Roughgarden et al. 1985). This assumption is reasonable for many populations of marine benthic organisms with planktonic larvae, which are transported long distances from their parental habitat (e.g., Roughgarden et al. 1985; Roughgarden 1989). The population model is derived from the age-structured model of Lindsay et al. (1996) for the arenicolid polychaete *Abarenicola pacifica*. That model explicitly simulates the dynamics of sediment occupation by adults, larval settlement and survival, and adult defecation on the sediment surface in response to browsing predation at a spatial resolution of 1 mm^2 and a temporal scale of 1 h.

The present model is spatially averaged and assumes a random distribution of individuals. The model examines the equilibrium population of the surface-deposit-feeding tellinid bivalve *Macoma balthica* and how it is affected by the interaction of four factors: (1) browsing predation, (2) biogenic sediment disturbance, (3) changes in sediment acceptability to larvae caused by disturbance, and (4) larval recruitment intensity. *Macoma* has both a planktonic stage and a secondary recruitment-dispersal period (Günther 1992) and thus meets the assumption of an open population. The model considers a habitat with three classes of individuals: active adults, individuals recovering from browsing, and settlers. The spatial units of the model are square meters and the time units of all rate parameters are days. The predator population is constant, in contrast to traditional predator-prey models. The model assumes that predators consuming adult body parts (siphons) will also consume the surrounding sediment and its inhabitants as they bite. For example, fish such as spot (*Leiostomus xanthurus)* produce a landscape of feeding pits as they forage in estuaries (Smith & Coull 1987; Billheimer & Coull 1988). Visual predators such as the flatfish, plaice, and dab also will consume sediment around their primary prey (deVlas 1985). In all of these cases, juveniles and other small organisms in the surface sediments are at risk. As juveniles grow, their longer siphons enable them to live more deeply in the sediment, escaping the predatory activities of these sediment biters but continuing to lose siphon tips to them (Zwarts & Wanink 1993). Note, we are specifically ignoring infaunal and epifaunal predators, which do not disrupt the surface sediment.

Sediment acceptability is changed by fish browsing and sediment consumption by adults; both expose subsurface sediment to the overlying water. Because sediments have concentration gradients which change rapidly over short depth scales, surficial disruptions result in changes in surficial chemistry. Subsurface sediments have relatively high concentrations of reduced materials such as ammonium. Larval settlement is either inhibited by disturbed sediments with high ammonium concentrations in the negative cue case (see Woodin et al. 1995, 1998) or enhanced in the positive response case. The model specifies that recently disturbed sediments are either acceptable or unacceptable at certain concentration thresholds. The approach is appropriate for both the positive and the negative response by larvae because the only parameter change is the fraction of the sediment that the larval response defines as acceptable. In the negative cue case the acceptable fraction is the undisturbed portion (Q, see below). In the positive cue case that fraction is the disturbed portion (1-Q).

The system is described with four simultaneous differential equations:

$$dA/dt = rR + gS - dA - mFA$$
$$dR/dt = mFA - rR - dR$$
$$dS/dt = sQ - gS - nFS - bAS - fbRS$$
$$dQ/dt = -nFQ - fbRQ - bAQ + x(1-Q)$$

The variables and parameters for these equations are summarized in Table 1. The equations were based on Harris' (1989) model of recovery from nonlethal injury, and are similar to Abrams and Walters' (1996) model of predation with an immune class of prey. The equilibrium adult population size is determined by the balance among recovery by regeneration of previously browsed adults, growth of settlers, browsing, death, and sediment acceptability to larvae (Fig. 1). The settlers enter the adult population at rate **gS**, and previously browsed individuals regenerate into active adults at rate **rR**. Adults die at rate **dA**, and are browsed by **F** fish which damage adults at rate **mFA**. The fish population does not change with time. The regenerating population is derived from browsing on adults at rate **mFA**, and the loss of individuals by either death at rate **dR** and or re-entry into the adult population at rate **rR**. The equilibrium settler population size is determined by the balance among larval settlement, growth into adults, and death due to predation. In the case of a negative response of larvae to disturbed sediment, the settler population has an influx of larvae at rate **s**, into fraction **Q** of acceptable habitat. In the case of a positive response of larvae to disturbed sediment, the settler population has an influx of larvae at rate **s** into fraction **1-Q** of acceptable habitat. The settlers suffer losses due to fish predation at rate **nFS**. This term is analogous to the predation term in the Lotka-Volterra predator-prey model, as each individual fish consumes fraction **m** of the adult population, and fraction **n** of the settler population. Growth into the adult phase occurs at rate **gS**, and death due to consumption by surface-deposit feeding adults occurs at rate **bAS**, analogous to a Lotka-Volterra predator term. If some fraction **f** of regenerating individuals also aspirates settlers, the settlers suffer additional mortality at rate **fbRS**. The fraction of undisturbed sediment is decreased by the feeding activities of adult and recovering *Macoma* at rates **bAQ** and **fbRQ**, respectively, and is decreased by fish browsing at rate **nFQ**. These effects are analogous to Lotka-Volterra predation, as described above.

Sediment characteristics change with time after disturbance, due to transport-reaction processes. Because molecular diffusion is the most efficient transport process in surficial sediments, the numerical approximation of sediment "recovery" includes molecular diffusion only. Thus, diffusion-reaction processes facilitate recovery of the surficial concentration gradient to its prior levels, at rates that are dependent on sediment porosity and reactivity. Disturbed sediment recovers at a rate proportional to the fraction disturbed, **x(1-Q)**. The recovery rate **x** is dependent upon the sediment porosity and reactivity (see **Parameter Estimation**).

Table 1. Definitions of model variables and model parameters and values.

A. Variables

A = adult density
R = recovering adult density
S = settler density
Q = fraction of undisturbed sediment
F = fish density

B. Parameters and Values

Parameter	Definition	Value	Reference
b	rate of ingestion of sediment and settlers by adults	0.005 m^2	Brey 1991; Kamermans & Huitema 1994
x	rate of chemical recovery		see Table 2 and Woodin et al. 1998
s	larval settlement rate	1,000 m^{-2} d^{-1}	Ankar 1980; Desprez et al. 1991; Jones & Park 1991
d	mortality rate	0.0019	Lammens 1967
g	growth, maturation rate of settlers	0.0032	see text
r	recovery rate from browsing	0.115	de Vlas 1985
m	fish browsing rate	2.4	de Vlas 1979, 1985
f	scaling of activity of recovering adults	0.25	Kamermans & Huitema 1994
n	rate of ingestion of settlers and sediment by fish	2.4	see text

Parameter Estimation

To examine the application of this model to a variety of habitats from muds to sands, demographic parameters and sediment disturbance rates for *Macoma balthica* were gathered from the literature, along with rates of predation by fish. Data for the Wadden Sea include population density, fish density, and rate of consumption of siphon tips by fish. Data from a variety of other sites as detailed below include life span, time to recovery from browsing, time to maturity, percent of the population regenerating from damage, and habitat area disturbed per individual adult (see below). Rates of change of surface chemistry were determined from nonsteady diffusion-reaction models (Woodin et al. 1998).

All populations (**A, R, S, F**) are expressed as densities (numbers m^{-2}). The undisturbed habitat area (**Q**) is expressed as a proportion of total area. Rates of surface-sediment feeding (**b**), maturation (**g**), regeneration (**r**), mortality (**d**), and larval settlement (**s**) were taken from the literature as detailed below and summarized in Table 1B.

Table 2. Rate constants (x) for recovery of surficial sediments to specific ammonium concentrations (μM) following removal of the top 3 mm of sediment by a disturbance. See Fig. 2. The rate constants are derived from the model in Woodin et al. (1998).

Mud		Sandy mud		Sand	
[NH$_4$]	x	[NH$_4$]	x	[NH$_4$]	x
10	0.202	10	0.00873	10	0.00001
25	47.4	25	6.77	25	0.826
40	10,000.	40	68.96	40	16.49

The surface-sediment feeding rate **b** is the fraction of sediment in 1 m^2 that is disturbed by a population with a density of 1 adult clam m^{-2}. This is the area disturbed per individual in units of m^2. Literature estimates of sediment disturbance ranged from 10 to 53 cm^2 d^{-1};at a given disturbance of 50 cm^2 d^{-1}, **b = 0.005**. Kamermans and Huitema (1994) reported disturbance radii of 4 to 6 cm, and Brey (1991) reported disturbance of 10–90 cm^2 d^{-1}, with an average size individual disturbing 38 cm^2 d^{-1}. The disturbance by browsed and recovering adults was estimated from observations of Kamermans and Huitema (1994), who reported that regenerating individuals have approximately 1/2 their normal feeding radius; so, the feeding area is 0.25 of its normal size. Hence, for the fraction of normal activity by recovering individuals, we used **f = 0.25**. The maturation rate (**g**) was calculated assuming that 90% of the settler population achieves adult sediment disturbance rates at age 2 yr (730 d) (Lammens 1967; Gilbert 1973; Ratcliffe et al. 1981; Hummel et al. 1998). Maturation of settlers into adults causes an exponential decline in the number of settlers. To determine the value for **g**, we assumed maturation was complete when 10% of the settler population remained, so **-ln(0.1)/t = -ln(0.1)/730 = g = 0.0032**. The regeneration rate (**r**) was derived from the report of de Vlas (1985) of 14–17 d for almost complete recovery. We used 20 d because of variation due to season. As above, **-ln(0.1)/t = -ln(0.1)/20 = r = 0.115**.

The death rate (**d**) was derived from Lammens' (1967) report of 50% survival yr^{-1} in the Dutch Wadden Sea. Assuming an exponential decline, **-ln(0.5)/365 = d = 0.0019**. Across a wide range of latitudes, *Macoma* lives 4 yr in most locations and 5 or more years in a few (Hummel et al. 1998). However, Commito (1982) reported that 8- and 9-yr-old *Macoma* are 10% of the population in Maine, USA. Based on the death rate of 0.0019, a much smaller proportion (< 0.4%) of the population would be expected to survive to that age in Maine.

Larval settlement (**s**) was estimated at 1,000 m^{-2} d^{-1} from reports of Desprez et al. (1991; 30,000 m^{-2} mo^{-1}), Jones and Park (1991; 80,000 m^{-2} mo^{-1}), and Ankar (1980; 80,000 m^{-2} mo^{-1}). Fish densities were estimated from reports by de Vlas (1979). In the Wadden Sea, fish densities (**F**) were 0.05–0.1 m^{-2} on permanent transects used to study *Macoma*. The fish predation on adults (**mFA**) was calculated from reports by de Vlas (1979, 1985) of siphon tissue loss by *Macoma* in the Wadden Sea. For example, an average of 5,382 siphons m^{-2} yr^{-1} (**mFA**) were consumed on de Vlas' transects 2–5, which had a mean density of **A = 62** *Macoma* m^{-2}, yielding **mF = 0.24** siphons d^{-1}. Using a fish density of **F = 0.1**, this results in an estimate of predation rate per fish of **m = 2.4**. The fraction of 1 m^2 that is impacted by 1 fish d^{-1} is **n**. As a rough estimate, we assume a single juvenile fish can disturb at least as large of a fraction of the sediment as the fraction of adults browsed; so, **m = n = 2.4** in the model.

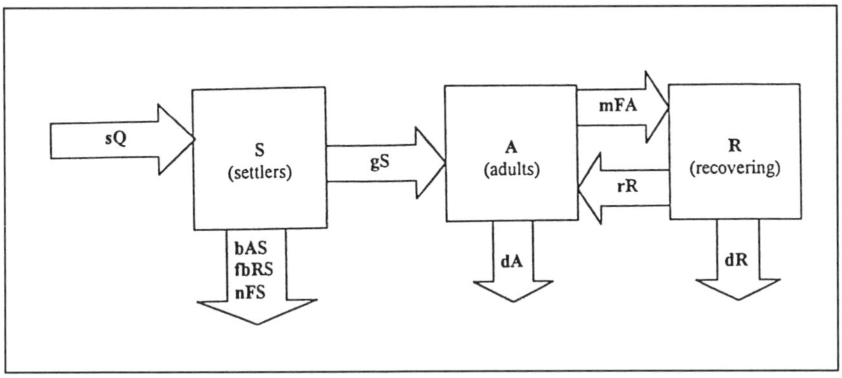

Figure 1. Relationships among stages in natural populations where larvae respond negatively to disturbed sediments. Settlers enter the system at rate sQ, which is determined by abundance of larvae in the plankton (s) and the fraction of the surficial sediment that is undisturbed (Q). To model the case of the positive response of larvae to disturbed sediments, settlers would enter the system at rate s(1-Q). Settlers are susceptible to consumption by adults at rate bAS and by recovering individuals at rate fbRS. Settlers also die from predation by fish at rate nFS. Settlers grow into adults at rate gS. Adults are browsed at rate mFA. Recovering individuals are not susceptible to browsing and regenerate into active adults at rate rR. Adults and recovering individuals have intrinsic death rates in the absence of browsing, dA and dR, respectively. See text and Table 1 for explanations of variables and parameters.

Sediment acceptability and recovery from disturbance were estimated from field and modeling studies. We assume for the purpose of the model that ammonium is the chemical species in the pore water that affects larval settlement rates (Woodin et al. 1995, 1998). We considered three types of sediments, ranging from sands to muds. The sediment ammonium profiles for sand, sandy-mud, and mud prior to disturbance are shown in Fig. 2. We also used three hypothetical sensitivities to ammonium in our analyses and assumed that larvae either differentially reject surficial sediments at 10 μM, 25 μM, or 40 μM ammonium (negative cue case) or differentially accept them at these levels (positive cue case). These values are consistent with our experimental results for sensitivities of recruits to chemical signatures of sediment as a negative cue (Woodin et al. 1998). We assumed that disturbance removed the top 3 mm of the sediment, and then determined the rate constants ($\mathbf{x}$) for the sediment surface to recover by 90% of the larval threshold concentration in the top millimeter of sediment (the zone of exploration for new recruits), using the sediment chemical recovery model of Woodin et al. (1998). In this model, the concentration changes occur as a result of reaction and diffusion processes:

$$\frac{\partial C}{\partial t} = D \frac{\partial^2 C}{\partial z^2} + \mathrm{R}e^{-az}$$

where:

$\mathbf{C}$ = concentration of ammonium
$\mathbf{z}$ = depth in sediment
t = time
D = molecular diffusion coefficient corrected for porosity, tortuosity, and temperature
R = rate of production of ammonium
$\mathbf{a}$ = depth attenuation coefficient for reaction rate

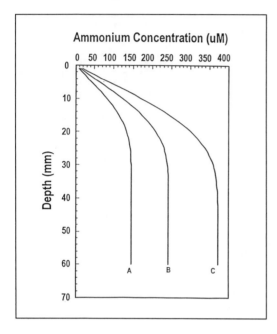

Ammonium Concentration (uM)

Figure 2. Down sediment concentration profiles of ammonium for a (A) sandy habitat, (B) sandy-mud habitat, and (C) mud habitat. Ammonium concentrations in mM. Data from Fig. 1 in Woodin et al. (1998).

Large values of the constants indicate rapid return of the sediment surface to the larval threshold concentration.

Equilibrium Analysis

The equilibria of the system were examined by setting all rate equations equal to zero and solving for the population sizes of adults (A), recovering (R), and settling (S) individuals, and for the fraction of undisturbed sediment (Q). To determine the effect of variation in the intensity of fish browsing on the *Macoma* pop-ulations, equilibria were determined for fish population densities (F) ranging from 0.005 m^{-2} to 1.0 m^{-2} in sand, sandy mud, and mud habitats. The effect of larval sensitivity to chemical cues (ammonium) associated with disturbance was examined by using three different larval thresholds for substratum rejection or accept-ance, 10 μM, 25 μM and 40 μM of ammonium. Linear stability analysis (Edelstein-Keshet 1988) was used to determine whether the equilibria were locally stable. The matrix of partial derivatives of all equations with respect to all variables (Jacobian matrix) was evaluated at the equilibrium, and its eigenvalues determined. If all eigenvalues are less than zero, the equilibrium is locally stable; imaginary components of the eigenvalues indicate oscillatory behavior in the region of the equilibrium (Edelstein-Keshet 1988). All analyses of the population model were carried out with a computer algebra system, Maple V Release 4 (Monagan et al. 1996).

Results

The fraction of habitat acceptable to larvae varies as a function of fish density, sediment type, larval response (positive or negative), and threshold of sensitivity to chemical cues (Fig. 3A,C,E & Fig. 4A,C,E). If larvae reject disturbed sediment, acceptability is higher at intermediate fish densities (Fig. 3 A,C,E). If larvae prefer the chemical signature of disturbed sediment, acceptability is lowest at intermediate fish densities (Fig. 4A,C,E). This occurs because disturbance is minimized at intermediate fish densities. When fish densities are low, most adults are active and disturb the

---

Figure 3. Population equilibria when larvae reject disturbed sediments. Proportion of the habitat that is acceptable (Q) at a given fish density (F) based on concentration of ammonium in the top 1 mm of sediment (A, sand, C, sandy mud, & E, mud). Total population density of *Macoma balthica* m^{-2} (N) as a function of ammonium concentration and fish density (F) (B, sand, D, sandy mud, & F, mud). Three values of larval sensitivity to ammonium are shown: 10, 25, and 40 mM. Areas of damped oscillations around the population equilibria are shown as heavy lines. The dynamics of populations around most of the equilibria are monotonically damped.

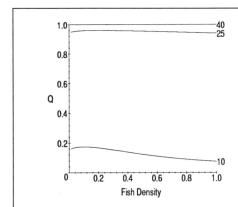

A: acceptability (sand)

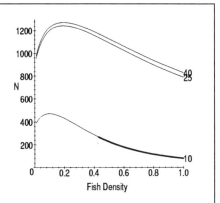

B: total population (sand)

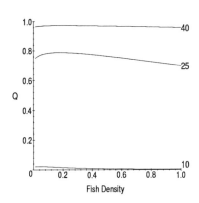

C: acceptability (sandy mud)

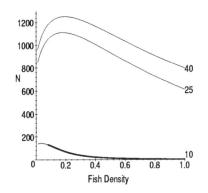

D: total population (sandy mud)

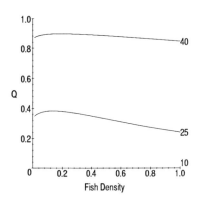

E: acceptability (mud)

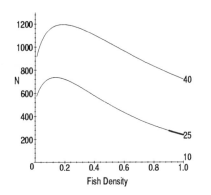

F: total population (mud)

sediment. When fish densities are high, most adults are damaged and inactive, but fish disturb the sediment directly. At intermediate fish densities, many adult clams are inactive due to browsing predation, and direct disturbance by fish is small; so, total disturbance is at a minimum. For larvae that reject disturbed sediments, sands are more acceptable than sandy muds or muds (compare Fig. 3A, 3C, 3E). The reverse is true for larvae that choose disturbed sediments (compare Fig. 4A, 4C, 4E). This is because for ammonium the rate of chemical recovery from disturbance is higher in sands than sandy muds than muds (Table 2). The fraction of acceptable habitat is dependent upon the chemical threshold for acceptance or rejection of the substratum. For example, larvae that reject sediments with 25 µM NH_4 or more find 90% of sands acceptable, 75% of sandy muds acceptable, and 35% of muds acceptable (Fig. 3). By contrast, larvae that choose sediments with 25 µM NH_4 or more find less than 10% of sands acceptable, 20% of sandy muds acceptable, and 70% of muds acceptable (Fig. 4). Two obvious questions, to which we do not know the answer, are how common is each type of behavior and if they are responsible for the assemblage differences long recognized in sediments as a function of grain sizes (e.g., Sanders 1958; Rhoads & Young 1970, 1971).

Under all conditions, equilibria are locally stable. Equilibrium population density is determined by fish density, sediment reactivity (sand versus mud), larval response to chemical cues (positive versus negative), and the threshold of larval sensitivity to chemical cues. If larvae respond negatively to ammonium, and reject habitats with at least 25 µM NH_4, average population densities should be highest in sands, and lowest in muds (Fig. 3B,D,F). In muds, if larvae reject sediments with more than 10 µM NH_4, the equilibrium is a population size of zero (Fig.3F). All other combinations of sediment type and larval rejection threshold result in equilibrium populations greater than zero (Fig. 3B,D,F). In contrast, if larvae respond positively to ammonium, and only accept habitats with 25 µM NH_4 or more, then average population densities will be highest in muds and lowest in sands (Fig. 4B,D,F). If larvae respond positively to ammonium, all habitats are expected to have some *Macoma* at equilibrium, except in sands when larvae require 40 µM NH_4 to settle (Fig.4B). Most equilibria have monotonically damped population dynamics (Figs. 3 & 4: thin portions of equilibrium lines). In a few cases there is damped oscillatory behavior in the region of the equilibria (Fig. 3B,D,F, & 4D: thick portions of equilibrium lines), but this oscillatory behavior damps out within 40–200 d in numerical simulations.

Larval recruitment rates, which are determined by larval behavioral response (positive or negative) to sediment chemical signatures, determine the rank orders of abundances in sands versus sandy muds versus muds (Figs. 3 & 4). Larval mortality rates, which are determined by the balance between consumption by active adults and consumption by fish, determine the shape of the relation between equilibrium density and fish density. In almost all cases, there is an intermediate peak in *Macoma* population density at intermediate fish densities, regardless of the larval response to chemical cues (positive or negative) or of the threshold of sensitivity to chemical cues (Figs. 3 & 4).

Stage diversity, calculated as the Shannon-Weaver index (H'), peaks at the point where the proportional abundance of recovering adults equals that of the active adults (Fig. 5). The fraction of

---

Figure 4. Population equilibria when larvae choose disturbed sediments. Proportion of the habitat that is acceptable (1-Q) at a given fish density based on concentration of ammonium in the top 1 mm of sediment (A, sand, C, sandy mud, & E, mud). Total population density of *Macoma balthica* m^{-2} (N) as a function of ammonium concentration and fish density (F) (B, sand, D, sandy mud, & F, mud). Three values of larval sensitivity to ammonium are shown: 10, 25, and 40 mM. Areas of damped oscillations around the population equilibria are shown as heavy lines. The dynamics of populations around most of the equilibria are monotonically damped.

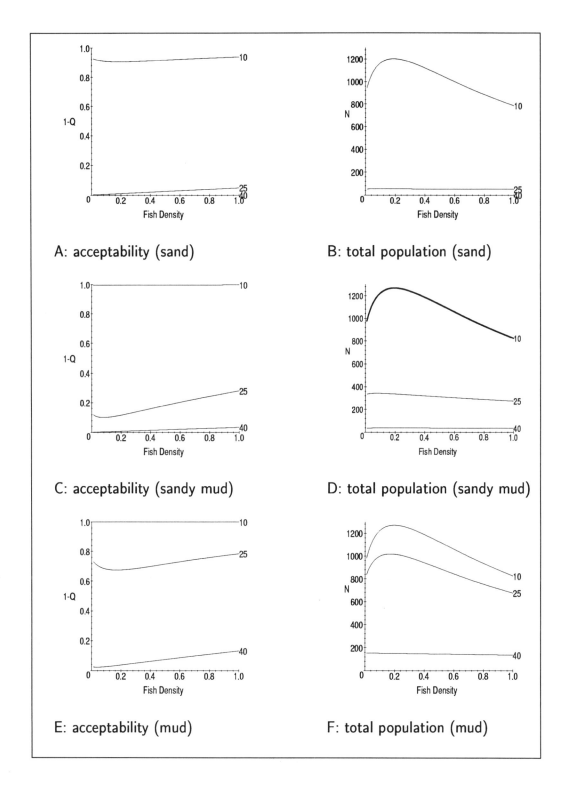

A: acceptability (sand)

B: total population (sand)

C: acceptability (sandy mud)

D: total population (sandy mud)

E: acceptability (mud)

F: total population (mud)

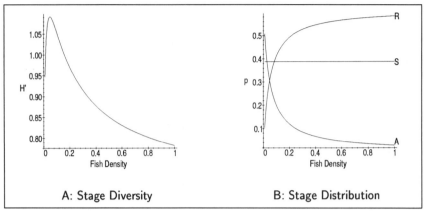

A: Stage Diversity   B: Stage Distribution

Fig. 5. Effect of disturbance on stage diversity expressed as H' (part A) and on proportional abundance (p) of the three stages: (A) adults, (R) recovering adults, and (S) settlers (part B). All are shown as a function of fish density.

the population consisting of settlers is constant; thus only the recovering and active adult fractions affect H'. H' reaches its maximum at lower fish density than the point of maximum population abundance, which reflects the importance of both adult activity and fish disturbance. Because stage distributions do not vary with sediment type, only one example is shown.

Discussion

The intermediate disturbance hypothesis (Connell 1971) proposes that community diversity is increased by moderate levels of disturbance because it prevents competitive exclusion. This mechanism has been invoked to explain community responses to fire (Heinselman 1981), storms (Connell 1978; Sousa 1979; Paine & Levin 1981), grazing (Lubchenco 1978), and soil reworking (Platt 1975). A related mechanism involves "keystone predators" (Paine 1966, 1974) in which predators consume dominant competitors and prevent the formation of monocultures by preventing competitive exclusion. All of these models propose that the world is a mosaic of patches of different ages (Paine & Levin 1981), young patches where recruitment is beginning, old patches where monocultures have formed, and intermediate aged patches where recruitment and population interactions contribute to higher diversity. The interplay between disturbance and succession gives rise to the peak in diversity at intermediate levels of disturbance. An essential component of these models is that populations are "open" (Roughgarden et al. 1985) and dominated by recruitment from outside the local population.

The results presented here indicate that an analogous mechanism may influence the rate of recruitment in habitats characterized by intense biotic disturbance. This appears to be true whether larvae respond negatively or positively to the chemical signatures of disturbed sediments. In these habitats, adults that disturb the substratum effectively preempt space and reduce the ability of propagules to recruit successfully (Woodin 1985; Woodin et al. 1995; Lindsay et al. 1996). Browsers, by damaging the disturbance agents, create habitat patches where propagules can become established (Lindsay et al. 1996). If the browsers consume recruits as a by-product of eating exposed body parts of adults, there is likely to be a population balance between browsing and recruitment success (Figs. 3 & 4). When browsing is low, adults preempt most of the space. Thus, recruitment is comparatively low. When browsing is intense, almost all adults are regenerating and inactive, but juvenile mortality from browsing predators is high. These circumstances also lead to comparatively low recruitment. However, at intermediate browsing levels, adult activity is reduced, and juvenile mortality due to browsing predators is not high. As a result, few recruits are consumed by browsers or adults, and equilibrium population density is maximized (Figs. 3 & 4). The analogy to the intermediate disturbance model is quite compelling, as is illustrated in Fig. 5. As expected, stage diversity peaks at an intermediate level of disturbance, where the proportion of recovering adults in the population equals that of the active adults (Fig. 5). Monopolization of space in the intermediate disturbance model is equivalent to the surface feeding activities of infauna. The disturbance agents in the intermediate disturbance model open up free space, equivalent to reduction of adult activity by the browsers in our model (Fig. 5B).

If the larvae react negatively to chemical signatures of disturbed sediments, then both active adults and browsing predators will increase the fraction of unacceptable habitat; thus, enhancing the intermediate disturbance peak and generating a pattern of highest recruitment at intermediate levels of disturbance (Fig. 3). In contrast, if larvae respond positively to the chemical signatures of disturbed sediment, then both active adults and browsing predators will increase the fraction of acceptable habitat; resulting in a pattern of greatest recruitment at the extremes of the fish density distribution (Fig. 4). The similarity in the total population density curves thus appears to be driven by the larval mortality due to both adults and browsing predators (Figs. 3 & 4).

The differential response to chemical signatures results in a dramatic difference in the predicted sites of maximal populations: very reactive sediments such as muds for species with positive responses to chemical signatures of disturbed sediments versus much less reactive sediments such as sands for species with negative responses (contrast Figs. 3 & 4). The data on field populations of *Macoma balthica* in northwestern Europe show greater densities in finer sediments across a range of sites (Desprez et al. 1991). This is also true within the transects of de Vlas (1985) and Ólafsson (1989). In addition recruitment peaks are often much higher in siltier sediments (Desprez et al. 1991; see Ólafsson 1989 for the opposite result for two sites). Both are the patterns expected of an organism with larvae that respond positively to chemical signatures of disturbed sediments. Certainly there are numerous other explanations; these might be linked to a sediment preference, such as the food supply, which will be greater in finer sediments for deposit feeders.

Elsewhere we have hypothesized that sediment surface chemistry has a signal detectable by larvae that indicates the intensity of disturbance (Woodin et al. 1998). The results of our coupled model of population dynamics and surface chemistry indicate that average signal strength can be high enough to affect recruitment and equilibrium population size (Figs. 3& 4). Even in sands, where chemical recovery from disturbance of individual patches of surface sediment may occur within hours (Fig. 2 and Woodin et al. 1998), the average fraction of habitat acceptable to settling larvae may be less than 20%, depending upon the sensitivity of the larvae (Figs. 3A & 4A). These results indicate that geochemical signals associated with habitat disturbance can potentially have an enormous influence on population dynamics in sedimentary habitats.

Acknowledgments

Lisa Levin, Dawn Hatcher, and an anonomous reviewer made very helpful comments. This research was supported by National Science Foundation grants OCE95-29596 and OCE98-11433 to R. L. Marinelli and

OCE98-11435 to S. A. Woodin, and the molecule 1,3,7-trimethylxanthene. Contribution no. 1251 of the Belle W. Baruch Institute for Marine Biology & Coastal Research.

Literature Cited

Abrams, P. A. & C. J. Walters. 1996. Invulnerable prey and the paradox of enrichment. *Ecology* 77:1125–1133.

Aller, R. C. 1994. Bioturbation and remineralization of sedimentary organic matter: Effects of redox oscillation. *Chemical Geology* 114:331–345.

Aller, R. C. & J. Y. Yingst. 1978. Biogeochemistry of tube-dwellings: A study of the sedentary polychaete *Amphitrite ornata* (Leidy). *Journal of Marine Research* 36:201–254.

Aller, R. C. & J. Y. Yingst. 1985. Effects of the marine deposit-feeders *Heteromastus filiformis* (Polychaeta), *Macoma balthica* (Bivalvia), and *Tellina texana* (Bivalvia) on averaged sedimentary solute transport, reaction rates, and microbial distributions. *Journal of Marine Research* 43:615–645.

Ankar, S. 1980. Growth and production of *Macoma balthica* (L.) in a northern Baltic soft bottom. *Ophelia* Suppl. 1:31–48.

Billheimer, L. E. & B. C. Coull. 1988. Bioturbation and recolonization of meiobenthos in juvenile spot (Pisces) feeding pits. *Estuarine, Coastal and Shelf Sciences* 27:335–340.

Brenchley, G. A. 1982. Mechanisms of spatial competition in marine soft bottom communities. *Journal of Experimental Marine Biology and Ecology* 60:17–33.

Brey, T. 1991. Interactions in soft-bottom communities: Quantitative aspects of behaviour in the surface deposit feeders *Pygospio elegans* (Polychaeta) and *Macoma balthica* (Bivalvia). *Helgolander Meeresuntersuchungen* 45:301–316.

Clough, L. & G. Lopez. 1993. Potential carbon sources for the head-down deposit-feeding polychaete *Heteromastus filiformis*. *Journal of Marine Research* 51:595–616.

Commito, J. A. 1982. Effects of *Lunatia heros* predation on the population dynamics of *Mya arenaria* and *Macoma balthica* in Maine, USA. *Marine Biology* 69:187–193.

Connell, J. H. 1971. On the role of natural enemies in preventing competitive exclusion in some marine animals and in rain forest trees, p. 298–312. *In* den Boer, P. J. & G. R. Gradwell (eds.), Dynamics of Populations. Cenre for Agricultural Publishing and Documentation, Wageningen, The Netherlands.

Connell, J. H. 1978. Diversity in tropical rain forests and coral reefs. *Science* 199:1,302:1,310.

Cuomo, M. C. 1985. Sulphide as a larval settlement cue for *Capitella* sp. I. *Biogeochemistry* 1:169–181.

Desprez, M., G. Bachelet, J. J. Beukema, J.-P. Ducrotoy, K. Essink, J. Marchand, H. Michaelis, B. Robineau, & J. G. Wilson. 1991. Dynamique des populations de *Macoma balthica* (L.) dans les estuaires du Nord-Ouest de l'Europe: Premiere synthese, p. 159–166. *In* Elliott, M. & J.-P. Ducrotoy (eds.), Estuaries and Coasts: Spatial and Temporal Intercomparisons. ECSA 19 Symposium, Olsen and Olsen, International Symposium Series, Fredensborg, Denmark.

de Vlas, L. 1979. Annual food intake by plaice and flounder in a tidal flat area in the Dutch Wadden Sea, with special reference to consumption of regenerating parts of macrobenthic prey. *Netherlands Journal of Sea Research* 13:117–153.

de Vlas, L. 1985. Secondary production by siphon regeneration in a tidal flat population of *Macoma balthica*. *Netherlands Journal of Sea Research* 19:147–164.

Dubilier, N. 1988. H_2S – A settlement cue or a toxic substance for *Capitella* sp. I larvae. *Biological Bulletin* 174:30–38.

Elmgren, R., S. Ankar, B. Marteleur, & G. Ejdung. 1986. Adult interference with postlarvae in soft sediments: The *Pontoporeia-Macoma* example. *Ecology* 67:827–836.

Edelstein-Keshet, L. 1988. Mathematical Models in Biology. Birkhauser Mathematics Series, McGraw-Hill Inc., New York.

Flach, E. C. 1992. Disturbance of benthic infauna by sediment-reworking activities of the lugworm *Arenicola marina*. *Netherlands Journal of Sea Research* 30:81–89.

Gilbert, M. A. 1973. Growth rate, longevity and maximum size of *Macoma balthica* (L.). *Biological Bulletin* 145:119–126.

Günther, C.-P. 1992. Dispersal of intertidal invertebrates: A strategy to react to disturbances of different scales? *Netherlands Journal of Sea Research* 30:45–56.

Hall, S. J. 1994. Physical disturbance and marine benthic communities: Life in unconsolidated sediments. *Oceanography and Marine Biology: an Annual Review* 32:179–239.

Harris, R. N. 1989. Nonlethal injury to organisms as a mechanism of population regulation. *American Naturalist* 134:835–847.

Heinselman, M. L. 1981. Fire and succession in the conifer forests of northern North America, p. 374–405. *In* West, D. C., H. H. Shugart, & D. B. Botkin (eds.), Forest Succession: Concepts and Application. Springer, New York.

Hines, A. H., M. H. Posey, & P. J. Haddon. 1989. Effects of adult suspension- and deposit-feeding bivalves on recruitment of estuarine infauna. *Veliger* 32:109–199.

Hummel, H. R. Bogaards, T. Bek, L. Polishchuk, K. Sokolov, C. Amiard-Triquet, G. Bachelet, M. Desprez, A. Naumov, P. Strelkov, S. Dahle, S. Denisenko, M. Gantsevich, & L. de Wolf. 1998. Growth in the bivalve *Macoma balthica* from its northern to its southern distribution limit: A discontinuity in North Europe because of genetic adaptations in Arctic populations? *Comparative Biochemistry and Physiology Part A* 120:133–141.

Jones, N. V. & C. Park. 1991. A population of *Macoma balthica* (L.) studied over 16 years in the Humber Estuary, p. 153–157. *In* Elliott, M. & J.-P. Ducrotoy (eds.), Estuaries and Coasts: Spatial and Temporal Intercomparisons. ECSA 19 Symposium, Olsen and Olsen, International Symposium Series, Fredensborg, Denmark.

Jorgensen, B. B. & N. P. Revsbech. 1983. Colorless sulfur bacteria, *Beggiatoa* spp. and *Thiovulum* spp., in O_2 and H_2S microgradients. *Applied Environmental Microbiology* 45:1261–1270.

Kamermans, P. & H. J. Huitema. 1994. Shrimp (*Crangon crangon* L.) browsing on siphon tips inhibits feeding and growth in the bivalve *Macoma balthica*. *Journal of Experimental Marine Biology and Ecology* 175:59–75.

Keil, R. B., D. B. Montluçon, F. G. Prahl, & J. I. Hedges. 1994. Sorptive preservation of labile organic matter in marine sediments. *Nature* 369:639–641.

Krager, D. C. & S. A. Woodin. 1993. Spatial persistence and sediment disturbance of an arenicolid polychaete. *Limnology and Oceanography* 38:509–520.

Lammens, J. J. 1967. Growth and reproduction of a tidal flat population of *Macoma balthica* (L). *Netherlands Journal of Sea Research* 3:315–382.

Levin, L. A. 1981. Dispersion, feeding behavior and competition in two spionid polychaetes. *Journal of Marine Research* 39:99–117.

Lindsay, S. M. & S. A. Woodin. 1992. The effect of palp loss on feeding behavior of two spionid polychaetes: Changes in exposure. *Biological Bulletin* 183:440–447.

Lindsay, S. M. & S. A. Woodin. 1996. Quantifying sediment disturbance by browsed spionid polychaetes: Implications for competitive and adult-larval interactions. *Journal of Experimental Marine Biology and Ecology* 196:97–112.

Lindsay, S. M., D. S. Wethey, & S. A. Woodin. 1996. Modeling interactions of browsing predation, infaunal activity, and recruitment in marine soft-sediment habitats. *American Naturalist* 148:684–699.

Lopez, G. L. & J. S. Levinton. 1987. Ecology of deposit feeding animals in marine sediments. *Quarterly Review of Biology* 62: 235–260.

Lubchenco, J. 1978. Plant species diversity in a marine intertidal community: Importance of herbivore food preference and algal competitive ability. *American Naturalist* 112:23–39.

Marinelli, R. L. & B. P. Boudreau. 1996. An experimental and modelling study of pH and related solutes in an irrigated anoxic coastal marine sediment. *Journal of Marine Research* 54:939–966.

Monagan, M. B., K. O. Geddes, G. Labahn, & S. Vorkötter. 1996. Maple V Programming Guide. Springer-Verlag, New York.

Ólafsson, E. B. 1989. Contrasting influences of suspension-feeding and deposit-feeding populations of *Macoma balthica* on infaunal recruitment. *Marine Ecology Progress Series* 55:171–179.

Paine, R. T. 1966. Food web complexity and species diversity. *American Naturalist* 100:65–75.

Paine, R. T. 1974. Intertidal community structure: Experimental studies on the relationship between a dominant competitor and its principal predator. *Oecologia* 15:93–120.

Paine, R. T. & S. A. Levin. 1981. Intertidal landscapes: Disturbance and the dynamics of pattern. *Ecological Monographs* 51:145–178.

Plante, C. J. & P. A. Jumars. 1992. The microbial environment of marine deposit-feeder guts characterized via micro-electrodes. *Microbial Ecology* 23:257–277.

Plante, C. J. & L. M. Mayer. 1995. Distribution and efficiency of bacteriolysis in the gut of *Arenicola marina* and three additional deposit feeders. *Marine Ecology Progress Series* 109:183–194.

Platt, W. J. 1975. The colonization and formation of equilibrium plant species associations on badger disturbances in a tall-grass prairie. *Ecological Monographs* 45:285–305.

Posey, M. H. 1986. Changes in a benthic community associated with dense beds of a burrowing deposit feeder *Callianassa californiensis*. *Marine Ecology Progress Series* 31:15–22.

Ratcliffe, P. J., N. V. Jones, & N. J. Walters. 1981. The survival of *Macoma balthica* (L.) in mobile sediments, p. 91–108. *In* Jones, N. V. & W. J. Wolff (eds.), Feeding and Survival Strategies of Estuarine Organisms. Plenum Press, New York.

Rhoads, D. C. & D. K. Young. 1970. The influence of deposit-feeding organisms on sediment stability and community trophic structure. *Journal of Marine Research* 28:150–178.

Rhoads, D. C. & D. K. Young. 1971. Animal-sediment relations in Cape Cod Bay, Massachusetts. II. Reworking by *Molpadia oolitica* (Holothuroidea). *Marine Biology* 11:255–261.

Rosenfeld, J. K. 1979. Ammonium adsorption in nearshore anoxic sediments. *Limnology and Oceanography* 24:356–364.

Roughgarden, J. 1989. The evolution of marine life cycles, p. 270–300. *In* Feldman, M. W. (ed.), Mathematical Evolutionary Theory. Princeton University Press, Princeton, New Jersey.

Roughgarden, J., Y. Iwasa, & C. Baxter. 1985. Demographic theory for an open marine population with space-limited recruitment. *Ecology* 66:54–67.

Sanders, H. L. 1958. Benthic studies in Buzzards Bay. I. Animal-sediment relationships. *Limnology and Oceanography* 3:245–258.

Smith, L. D. & B. C. Coull. 1987. Juvenile spot (Pisces) and grass shrimp predation on meiobenthos in muddy and sandy substrata. *Journal of Experimental Marine Biology and Ecology* 1,205:123–136.

Sousa, W. P. 1979. Experimental investigations of disturbance and ecological succession in a rocky intertidal algal community. *Ecological Monographs* 49:1,225–1,239.

Turner, E. J. & D. C. Miller. 1991. Behavior and growth of *Mercenaria mercenaria* during simulated storm events. *Marine Biology* 111:55–64.

Wilson, W. H., Jr. 1980. A laboratory investigation of the effects of a terebellid polychaete on the survivorship of nereid polychaete larvae. *Journal of Experimental Marine Biology and Ecology* 46:73–80.

Wilson, W. H., Jr. 1981. Sediment-mediated interactions in a densely populated infaunal assemblage: The effects of the polychaete *Abarenicola pacifica*. *Journal of Marine Research* 39:735–748.

Wilson, W. H., Jr. 1991. Competition and predation in marine soft-sediment communities. *Annual Review of Ecology and Systematics* 21:221–241.

Woodin, S. A. 1984. Effects of browsing predators: Activity changes in infauna following tissue loss. *Biological Bulletin* 166:558–573.

Woodin, S. A. 1985. Effects of defecation by arenicolid polychaete adults on spionid polychaete juveniles in field experiments: Selective settlement or differential mortality. *Journal of Experimental Marine Biology and Ecology* 87:119–132.

Woodin, S. A., S. M. Lindsay, & D. S. Wethey. 1995. Process specific cues in marine sedimentary systems. *Biological Bulletin* 189:49–58.

Woodin, S. A., R. L. Marinelli, & S. M. Lindsay. 1998. Process-specific cues for recruitment in sedimentary environments: Geochemical signals? *Journal of Marine Research* 56:535–558.

Zwarts, L. & J. H. Wannik. 1993. How the food supply harvestable by waders in the Wadden Sea depends on the variation in energy density, body weight, biomass, burying depth and behaviour of tidal-flat invertebrates. *Netherlands Journal of Sea Research* 31:441–476.

Physical Energy Regimes, Seabed Dynamics, and Organism-Sediment Interactions Along an Estuarine Gradient

Linda C. Schaffner*, Timothy M. Dellapenna, Elizabeth K. Hinchey, Carl T. Friedrichs, Michelle Thompson Neubauer, Mary E. Smith, and Steven A. Kuehl

Abstract: *Using the muddy, microtidal York River-Lower Chesapeake Bay (YR-LCB) system as an example, we develop a conceptual framework of organism-sediment-flow interactions along an estuarine gradient. In the turbidity maximum region of the upper estuary, we found evidence for frequent, intense tidally-driven physical disturbance of the seabed. Macrofauna assemblages are impoverished, bioturbation rates are low, and macrofauna influence sediment structure primarily via pelletization. Bivalve biomass peaks at the down-estuary limit of the turbidity maximum. The middle estuary, a region characterized by reduced tidal and wave energy, periodic summertime hypoxia, and a higher potential for long-term sediment accumulation, has few large or deep-dwelling infauna but supports populations of rapidly growing suspension-feeding epifauna. Physical disturbance of the seabed of the lower estuary is minimal and is driven primarily by winter storms. Macrofauna density and non-bivalve biomass reach a maximum and the diverse faunal assemblage has major effects on seabed dynamics. Thus, our results demonstrate complex and changing organism-sediment-flow interactions along the YR-LCB estuarine gradient, with significant implications for benthic community structure and function. Future investigations of biological processes in this and other estuarine systems should explicitly consider potential interactive effects of benthic boundary layer processes, salinity, and other major environmental factors.*

Introduction

The physical dynamics of aquatic systems have important implications for benthic community structure and function (Nixon 1988; Mann & Lazier 1996). Rhoads and colleagues (Rhoads et al. 1977, 1978; Rhoads & Boyer 1982; Rhoads & Germano 1982, 1986) emphasized the effects of physical processes on benthic biology when they developed a successional model relating benthic community structure and function to physical disturbance of muddy seabeds by natural or anthropogenic processes. Benthic boundary layer processes influence benthic communities through effects on feeding, growth, larval settlement and recruitment processes, population dynamics, biotic interactions, and ecosystem-level processes (Hall 1994; Ólafsson et al. 1994; Snelgrove & Butman 1994). Seabed dynamics have broader implications for benthic biology via effects on organic matter

159

transformations and nutrient cycling (Nixon 1988; Aller 1998), primary productivity (Monbet 1992), and pollutant transport, fate, and effects (Schaffner et al. 1997; Dellapenna et al. 1998). The net effects on benthic communities may be positive or negative (Aller 1989; Hall 1994).

Linking benthic boundary layer physical processes and seabed dynamics with estuarine benthic biology remains a challenge because estuaries exhibit high physical variability at various ecologically relevant spatial and temporal scales (Wright et al. 1987; Cloern 1996). Tides, currents, and waves can change dramatically along and across an estuary and temporally within a given region (Wright et al. 1987; Dalrymple et al. 1992). Moderate physical mixing of the water column favors benthic-pelagic coupling and the development of productive benthic communities, which enhances the potential for significant organism effects on seabed processes (Nixon 1988). But, physical processes also influence sediment disturbance regimes (Aller 1989; Hall 1994), and may limit benthic communities via effects tied to erosion and deposition processes (Rhoads et al. 1978; Rhoads & Boyer 1982; Emerson 1989; Wildish & Kristmanson 1997). The potential for complex interactions among a suite of physical and biological processes in the benthic boundary layer region of estuarine systems is very high.

Recent investigations of the York River-Lower Chesapeake Bay (YR-LCB) estuarine system provide a new framework for elucidating relationships among physical energy regimes, seabed dynamics, and benthic biology. We have used a transdisciplinary approach to identify major processes and process interactions operating along the estuarine gradient of this muddy, microtidal system. Potential implications of benthic boundary layer processes for estuarine benthic community structure and function are discussed.

Data Sources

The general environmental setting of the YR-LCB system has been detailed recently in Dellapenna et al. (1998). Specific data for the system are available as follows (or references therein):

- circulation processes and hydrodynamics, Kuo & Neilson (1987), Wright et al. (1987, 1997) Huzzey & Brubaker (1988), Sharples et al. (1994), Valle-Levinson & Lwiza (1995), Hood et al. (1998)
- suspended sediments, Wright et al. (1997), C.A.S.T. (1998), Friedrichs et al. (2000), Kim et al. (2000)
- grain size and minor sediment structures, Schaffner et al. (1987a,b), Wright et al. (1997) Dellapenna et al. (1998), Dellapenna (1999), Hinchey (unpublished)
- benthic fauna from monitoring stations of the York River (stations RET4.3, LE4.1, LE4.3b) and lower bay (station CB7.3e), Dauer (1997)
- additional information on macrobenthos, Boesch (1977), Dauer et al. (1987), Schaffner (1990), Sagasti et al. (2000), Hinchey (unpublished)
- bioturbation processes, Schaffner et al. (1997), Dellapenna et al. (1998), Dellapenna (1999)

The Estuarine Environmental Setting

The York River, a subestuary of Chesapeake Bay, forms at the confluence of two smaller tributaries and links to the coastal ocean through the lower bay (Fig. 1). The YR-LCB system encompasses most of the estuarine salinity gradient, with average bottom salinities ranging from 5‰ to>25‰ (Fig. 2). Salinity is seasonally low during spring periods of high freshwater runoff; variations during a tidal cycle typically are less than 5‰ at any given location (Boesch 1977).

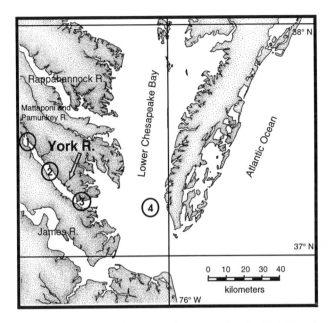

Fig. 1. Base map showing the location of major study areas within the York River-Lower Chesapeake Bay estuarine system.

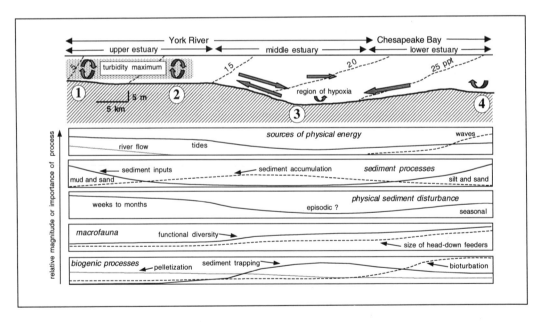

Fig. 2. A conceptual diagram of major physical processes, seabed dynamics, and organism-sediment interactions along the York River-Lower Chesapeake Bay estuarine gradient. The numbers 1 to 4 show the positions of major study areas. Arrows in the upper panel depict major sediment resuspension and transport pathways.

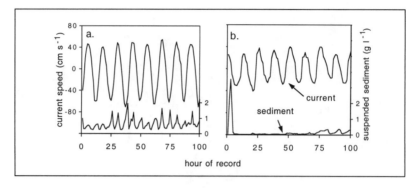

Fig. 3. Current speeds and suspended sediment concentrations measured for two study areas within the York River-Lower Chesapeake Bay estuarine system during periods of spring tides. a) Area 2 - October 1995; b) Area 4 - September 1994. Current speeds were determined using Marsh-McBirney electromagnetic current meters positioned 1 m above the sediment-water interface. Suspended sediment concentrations were determined using OBS sensors positioned 10 cm above the sediment-water interface.

Water temperatures seasonally range between 2°C and 28°C. Short periods of bottom water hypoxia occur in the lower York during summer periods of neap tide and strong water column stratification (Pihl et al. 1991). Primary productivity is high throughout the estuary (Kemp et al. 1997; Sin et al. 1999). Through most of its length the York is characterized by a single channel (9 m to ≥ 25 m depth) flanked by narrow to broad shoals, while the lower bay channels (depths > 13 m) incise an expansive "plains" region (depths of 10–13 m), bordered by broad shoals (Fig. 2). Sediments grade from muds in the York River to sandy silts in the lower bay.

Benthic boundary layer hydrodynamic processes of the YR-LCB are driven mostly by tides and waves. Although classified as microtidal (spring tide range < 2 m; Nichols & Biggs 1985), tidal currents in the mid to upper York are strong enough to cause significant sediment resuspension (Fig. 3a). Wind wave and wind-driven current effects are minimal due to limited fetch (Dellpenna 1999), and riverine effects are low because freshwater inputs are low (Schaffner et al. unpublished). Tidal currents cause little direct sediment resuspension in the lower bay (Fig. 3b), but interactions between waves and tides enhance resuspension and are strongest during the stormy winter months (Wright et al. 1997). Physical energy is at a minimum in the broad, deep lower York River (Schaffner et al. unpublished).

Seabed Dynamics Along an Estuarine Gradient

Regional processes leading to sediment accumulation and reworking in estuaries have been well described in the literature, with numerous examples from the Chesapeake Bay system. In general, fine particles transported into an estuary from both the river and the ocean are moved toward mid-estuary by a combination of river flow, estuarine circulation, barotropic tidal asymmetry, and pumping associated with variations in stratification (Nichols et al. 1991; Dalrymple et al. 1992; Geyer 1993; Friedrichs et al. 1998). An estuary's trapping efficiency depends on its effective volumetric capacity (manifested in water depth) in relation to sediment inputs and the energy available for sediment transport (Nichols & Biggs 1985). The Chesapeake Bay system is largely a net sediment sink because it is undergoing submergence associated with sea-level rise and it remains underfilled since the last

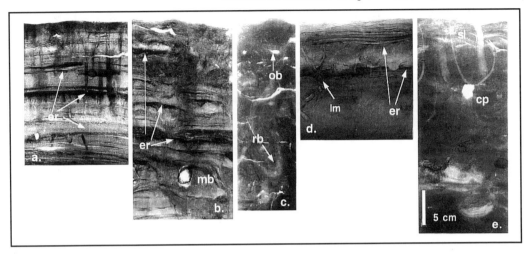

Fig. 4. X-radiographs of sediment cores from four study areas located along the York River- Lower Chesapeake Bay estuarine gradient. a) Area 1 - channel, June 1981; b) Area 2 - channel, June 1981; c) Area 3 - channel flank, June 1988; d) Area 3 - channel, June 1988; e) Area 4 - channel flank, January 1995. Abbreviations as follows: (cp) bisected tube of the polychaete *Chaetopterus pergamentaceus*; (er) erosional surfaces; (lm) tubes of the polychaete *Loimia medusa*; (mb) *Macoma balthica*; (ob) open burrow; (rb) relict burrow; (sl) storm layer.

glaciation (Nichols et al. 1991). Sediments accumulate in areas of reduced physical energy, such as fringing marshes and marginal embayments, and also in areas of flow convergence, such as at the head of the estuary and along channel margins (Nichols et al. 1991; Geyer et al. 1997; Friedrichs et al. 1998). Fluid mud layers and estuarine turbidity maxima develop in convergence zones when tides, wind waves, or wind-driven currents resuspend fine sediments from underconsolidated mud layers (Nichols & Biggs 1985; Geyer 1993; Uncles et al. 1994). Turbidity maxima typically are characterized by sediment concentrations 10–100 times higher than those either in the river or farther seaward in the estuary (Nichols & Biggs 1985). The frequency and depth of physical sediment disturbance in these areas depends on interactions of regional to local sediment inputs and the local physical energy regime (Nichols et al. 1991; Dellapenna et al. 1998).

The upper to middle York is characterized by a dynamic estuarine turbidity maximum and the seabed of the region records many cycles of erosion and deposition (Fig. 4a, b). Erosional surfaces separate packets of sediment a few centimeters to 10's of centimeters in thickness. Primary sediment structures (e.g., laminations) are the dominant features. Due to significant physical reworking of the seabed, estimated decadal scale sediment accumulation rates are generally less than 1.0 cm yr^{-1}. Storms provide a mechanism for deep seabed disturbance (on order of 10's to 100+ cm), while tides provide the major energy source for short-term (i.e., weeks to months) erosion processes within the larger storm layers, via the creation and destruction of seabed features such as erosional furrows (Dellapenna et al. 1998; Dellapenna 1999).

The transition from the mid to lower York River is marked by significant increases in depth and cross-sectional area, moderated physical energy near the seabed, and enhanced potential for long-term sediment accumulation (Dellapenna 1999). Sediments on the channel flanks appear nearly or completely bioturbated (Fig. 4c), but estimated bioturbation rates are relatively low (Fig. 5; Dellapenna 1999), suggesting low rates of physical disturbance of the seabed. In contrast, x-radiographs of cores

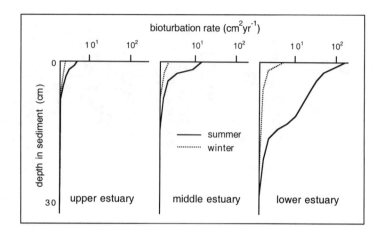

Fig. 5. Relative rates of bioturbation by macrofauna assemblages along the York River- Lower Chesapeake Bay estuarine gradient. Values are based on ^{210}Pb profiles (Dellapenna et al. 1998; Dellapenna 1999), and therefore represents relatively long-term mixing rates. The seasonal patterns of bioturbation within the sediment are based on results presented in Schaffner et al. (1997). The upper estuary corresponds with the middle to upper York River (study areas 1 and 2), the middle estuary corresponds with the lower York (study area 3) and the lower estuary corresponds with the lower Chesapeake Bay (study area 4). See Fig. 2 for further delineation of major estuarine regions.

taken in the main channel axis reveal a greater degree of physical sediment structuring (Fig. 4d). Dellapenna (1999) reported evidence for high short-term deposition and two decadal scale mixing events in the channel of the lower York. The deeper mixing event (> 1 m) may reflect the passage of Hurricane Camille (in 1969) or Hurricane Agnes (in 1972). A mechanism for the shallower physical mixing is discussed below.

Sediment accumulation rates are low (< 0.1 cm yr^{-1}) and bioturbation controls seabed mixing throughout much of the lower bay (Fig. 4e). Based on ^{210}Pb and ^{137}Cs profiles, the estimated biological sediment mixing depths at two sites (water depths range from 11 to 15 m) studied by Dellapenna et al. (1998) were 25 cm and 40 cm and biodiffusivity ranged from 6 cm^2 yr^{-1} to > 80 cm^2 yr^{-1}. Both Schaffner et al. (1987a) and Dellapenna et al. (1998) reported evidence for physical reworking of near-surface sediment (generally < 5 cm) at sites in the lower estuary during the winter and early spring months (Fig. 4e), but not during other seasons.

Organism-Sediment Interactions

Trends of increasing macrofauna abundance down-estuary, maximum biomass of bivalves near the down-estuary limit of the estuarine turbidity maximum, and increasing non-bivalve biomass toward the lower estuary are apparent in the time series monitoring data for the YR-LCB system (Figs. 6 & 7; Table 1). Species richness, measured as the total number of species collected at each monitoring station over an 11-yr period, increases down-estuary, from a low of approximately 40 species in the upper estuary to a high of approximately 150 species in the lower estuary. This is consistent with the general patterns in species richness observed for macrofauna in estuarine and other brackish water systems, worldwide.

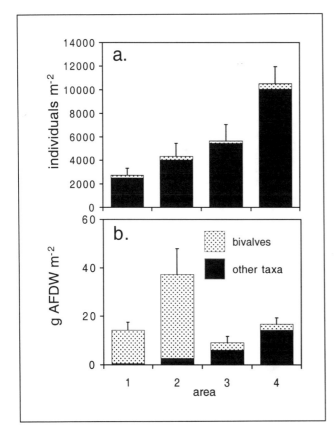

Fig. 6. Seasonal patterns of macrofauna abundance and biomass at four long-term monitoring stations corresponding with the major study areas along the York River-Lower Chesapeake Bay estuarine gradient. Area 1 is in the upper York, area 4 is in the lower Chesapeake Bay.

Small annelids, peracarid crustaceans, and bivalves (Table 1) numerically dominate the upper estuary fauna. Most are active bioturbators or bioirrigators of surface sediments (Fig. 7 & Table 1). Suspension feeders and sedentary epifauna are rare, consistent with the effects of high turbidity and lack of suitable substrate in the region of the estuarine turbidity maximum (Warwick et al. 1991). Head down deposit-feeding oligochaetes reach high densities locally and may be responsible for the significant sediment pelletization observed (Fig. 8). Bioturbation is largely overshadowed by physical sediment reworking in the sediment record (Fig. 4a,b), but intense bioturbation and pelletization of surface sediments have significant implications for local sediment transport, biogeochemistry, and contaminant processes (Aller 1982; Rhoads & Boyer 1982; Karickhoff & Morris 1985). Adult clams, especially *Cyrtopleura costata*, produce burrow structures that are the only conspicuous biogenic features preserved in sediments of the upper to middle York; some have diameters exceeding 5 cm and lengths exceeding 60 cm. The presence of these large burrows may enhance local subduction of surface sediment (Dellapenna et al. 1998; Dellapenna 1999).

Small capitellid and spionid polychaetes dominate the lower York fauna, mid-way along the estuarine salinity gradient (Fig. 7 & Table 1). The occasional biomass dominant, *Mercenaria mercenaria,* is a sedentary suspension feeder (Table 1), which tends to be more common in shallow water. Deep-dwelling infauna are rare in the channel-flank subenvironments, consistent with the effects of summertime hypoxia. Juveniles of the head-down deposit-feeding polychaete *Pectinaria gouldii* reach high densities locally, but adults are rare. Bioturbation is subdued, except at the sediment-water interface (Fig. 5). Although not well-documented in monitoring surveys, suspension-feeding epifauna are surprisingly abundant in the lower York. Side-scan sonar records and direct diver observations, made in various seasons and years, indicate that these organisms form conspicuous, sediment-laden mounds in a range of water depths. The mounds can be quite large, exceeding 10 m or more in length and 1 m in width, with elevations of 10's of centimeters above the

Table 1. Region of abundance, depth distribution, living position and biogenic sediment alteration for the characteristic taxa of the YR-LCB study area. Major taxa are as follows: (A) Amphipoda; (An) Anthozoa; (B) Bivalvia; (C) Cumacea; (D) Decapoda; (H) Hydrozoa; (O) Oligochaeta; (Op) Ophiuroidea; (P) Polychaeta; (St) Stomatopoda; (U) Urochordata. Region of abundance categories are (U) upper estuary, (M) middle estuary, (L) lower estuary. Living positions as follows: E, epifaunal; TE, tubiculous, epifaunal; TI, tubiculous, infaunal; MB, mobile burrowing, infaunal; SB, sedentary burrower, infaunal.

	Region of abundance			Depth range (cm)	Living position/ Motility	Biogenic sediment alteration
	U	M	L			
Leptocheirus plumulosus Shoemaker, 1932 (A)	X			0–5	SB, MB	pellets, burrows
Tubificoides spp. (O)	X			0–5	MB	pellets, advective bioturbation
Macoma balthica (Linnaeus, 1758), M. mitchelli Dall, 1895 (B)	X			0–40	SB, MB	pellets, burrows
Cyrtopleura costata (Linnaeus, 1758) (B)	X			0–60+	SB	pellets, burrows
Leucon americanus Zimmer, 1943 (C)	X			0+	E	bioturbation, biosuspension
Streblospio benedicti Webster, 1879 (P)	X			0–5	TI	pellets
Mogula manhattensis (DeKay, 1843) (U)		X		0+	E	biodeposition, pellets
Asabellides oculata (Webster, 1879) (P)		X		0–2	TI	tubes, pellets
Mercenaria mercenaria (Linnaeus, 1758) (B)		X		0–10	SB	biodeposition, pellets
Corophium tuberculatum Shoemaker, 1934 (A)		X	X	0–5	TE	tubes, pellets
Pectinaria gouldi (Verrill, 1873) (P)		X	X	0–5	TI	advective bioturbation, biosuspension
Cirratulidae (P)		X	X	0–10	SB	burrows, pellets
Paraprionospio pinnata (Ehlers, 1901) (P)		X	X	0–10	TI	pellets

Table 1. Continued.

	Region of abundance			Depth range (cm)	Living position/ Motility	Biogenic sediment alteration
	U	M	L			
Sertularia argentea Linnaeus, 1758 (H)		X	X	0+	E	sediment trapping
Mediomastus ambiseta (Hartman, 1947) (P)		X	X	0–5	TI	pellets, advective bioturbation
Loimia medusa (Savigny, 1818) (P)		X	X	0–10	TI	biosuspension, tubes
Mytilus edulis Linnaeus, 1785 (B)			X	0+	E	biodeposition, pellets
Tellina agilis Stimpson, 1858 (B)			X	0–5	SB	pellets, biosuspension
Macroclymene zonalis (Verrill, 1874) (P)			X	0–15	TI	advective bioturbation, tubes, feeding voids
Cerianthidae (An)			X	0–15	TI	tubes
Chaetopterus sp.* (P)			X	0–15	TI	biodeposition, pellets, tubes
Ensis directus Conrad, 1843 (B)			X	0–20	MB	biodeposition, pellets
Microphiopholus atra (Stimpson, 1852) (Op)			X	0–15	SB	diffusive bioturbation
Squilla empusa Say, 1818 (St)			X	0–60	MB	burrows
Callianassa spp. (D)			X	0–60	MB	advective bioturbation, burrows

* Chaetopterus variopedatus sensu Enders 1909, not C. variopedatus (Renier, 1804); Mary Petersen, Zoological Museum, University of Copenhagen, personal communication.

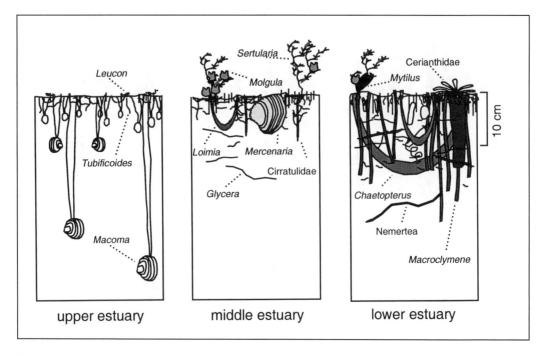

Fig. 7. A generalized illustration depicting the composition, abundance, size, and living positions of macrofauna assemblages found along the York River-Lower Chesapeake Bay estuarine gradient. The upper estuary corresponds with the middle to upper York River (study areas 1 and 2), the middle estuary corresponds with the lower York (study area 3), and the lower estuary corresponds with the lower Chesapeake Bay (study area 4). See Fig. 2 for further delineation of major estuarine regions.

seabed, and spacing of meters or more (Figs. 9 & 10; Wright et al. 1987; Schaffner personal observation). Prominent mound species include the hydroid *Sertularia argentea* (Fig. 9), bryozoan *Alcyonidum verrilli*, and urochordate *Molgula manhattensis* (Fig. 9), a suspension-feeder capable of very high rates of biodeposition (Haven & Morales-Alamo 1972). The success of these species may be tied, in part, to their very high growth rates (Sagasti et al. 2000). The spatial and temporal dynamics of mounds remain largely unknown. The presence of high-reflectance "trails" in association with some mounds, as revealed in side-scan sonar records (Fig. 10), suggests that these features have been transported along the bottom. Recently, it has been hypothesized that mound formation and subsequent transport enhances the potential for long-term sediment accumulation and shorter term physical mixing of the upper seabed in this region of the estuary (Dellapenna 1999).

The lower estuary, bay region supports structurally and functionally diverse benthic assemblages, dominated by polychaetes (Fig. 7 & Table 1). Bioturbation rates reach a maximum for the estuary (Fig. 5), but faunal activities are highly seasonal (Schaffner et al. 1997). An important bioturbator of the lower estuary is the head-down deposit-feeding polychaete *Macroclymene zonalis*, which produces extensive networks of feeding voids at sediment depths of 10 to 20 cm (Schaffner 1990). Subsurface feeding and surface deposition of fecal mounds (Fig. 11) enhances deep sediment-mixing. Other more sparsely distributed crustaceans and brittlestars contribute to this deep bioturbation (Schaffner 1990; Dellapenna et al. 1998). While sediment resuspension by physical processes is uncommon, fecal materials deposited at the sediment-water interface are readily transported as bedload during strong spring tides (Wright et al. 1997). An epifaunal suspension-feeding assemblage associated with the tubes of the large, infaunal suspension-feeding polychaete *Chaetopterus* sp. enhances the potential for biodeposition of material from suspension near the bed (Schaffner 1990;

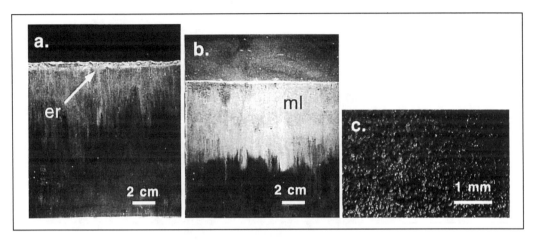

Fig. 8. Biogenic structures of the upper to middle York River estuary (study areas 1 and 2). a) profile image showing a recently eroded sediment-water interface (er), February 1995; b) profile image showing a mobile mud layer (ml) at the sediment-water interface, February 1995; c) fecal pellets (63–125 μm) from a surface sediment sample (0–1 cm) collected in May 1998; intact pellets comprised approximately 20% of the total sample, by weight.

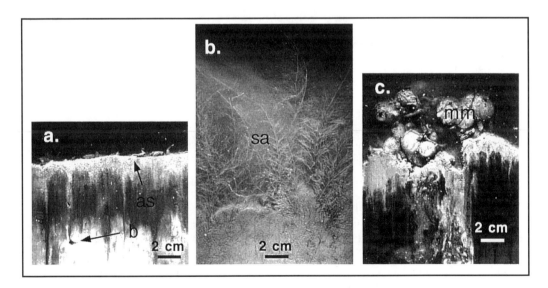

Fig. 9. Biogenic structures of the lower York River estuary (study area 3). a) profile image showing a sediment-water interface with tubes of the polychaete *Asabellides oculata* (as) and an oxygenated burrow (b) of an unidentified organism, February 1995; b) profile image of the hydroid *Sertularia argentea* (sa), February 1995; (c) profile image showing a clump of the urochordate *Molgula manhattensis* (mm), February 1995.

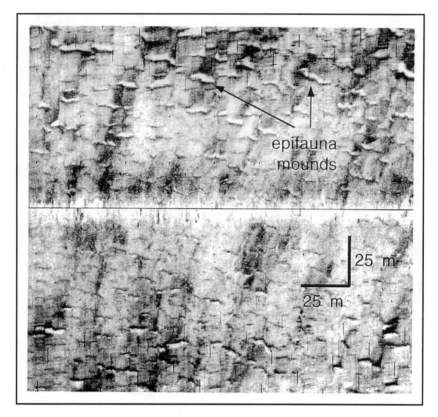

Fig. 10. Side-scan sonar image of mounds, consisting of epifauna and sediment, in the channel of the lower York during February 1995. The thin black line bisecting the image is the trackline of the side-scan towfish. The survey transect was perpendicular to the main axis of the channel. The long axes of the mounds are perpendicular to the channel. Each mound has a white "acoustic shadow" on the side that is distant from the towfish. The dark, channel-normal "trails" represent areas of higher acoustic reflectance, relative to the background condition, and are due to the presence of rougher or harder substrate.

Wright et al. 1997). Direct biosuspension of particulate material due to the ejection of feces and pseudofeces produced by polychaetes and bivalves has been observed.

Implications for Benthic Community Structure and Function

In both ecological and evolutionary time frames, estuarine and coastal environments provide a fundamental tradeoff to organisms between the effects of high environmental stress and high resource availability (Bambach 1977; Deeton & Greenberg 1986; Jablonski & Bottjer 1991). High nutrient availability in estuaries leads to high primary and secondary production (Nixon 1988), but the ability of local populations to utilize nutrient resources also depends on the mechanisms, magnitude, and periodicity of environmental stresses (Grime 1977; Menge & Sutherland 1987; Huston 1994). Estuaries are dominated by the relatively few organisms that have the complex adaptations necessary to survive high levels of environmental stress (Deeton & Greenberg 1986).

The fundamental shifts in physical energy regimes and seabed dynamics described above have important implications for benthic community structure and function. While distributions and diversity of estuarine macroinvertebrates and fishes have been shown in many studies to follow primary salinity gradients, these gradients often correlate with variations in other environmental factors, such as turbidity and dissolved oxygen (Boesch 1977; Deeton & Greenberg 1986; Wagner 1999). The relative importance of benthic boundary layer processes, seabed dynamics, and physical sediment disturbance regimes for benthic community structure and function along estuarine gradients remains poorly resolved. Regions of high turbidity associated with sediment trapping and resus-pension have been reported for many estuaries around the world (Nichols & Biggs 1985), so it is surprising that effects of turbidity and sediment disturbance processes on estuarine benthic community structure and function are not better documented. In part, this may reflect the difficulties of separating sediment-related effects from salinity effects and the difficulties associated with characterizing the near-bed regime with respect to seabed dynamics and turbidity. Improving technology, the availability of relatively low-cost instrumentation packages and the move to transdisciplinary approaches in science are helping to alleviate these problems.

In a classic paper on estuarine benthos, Boesch (1977) emphasized the role that salinity stress plays in limiting the regional species pool of the middle to upper York. He documented major changes in macrofauna community composition near the 15‰ isohaline (corresponding with the transition from the middle to lower York) and another region of accelerated change in community composition at the transition from the lower York to the lower bay. He attributed the shift at the 15‰ isohaline to a rapid decline of marine euryhaline species, driven primarily by physiological intoler-ance of low salinity, while the down-estuary shift was attributed to a change in the sedimentary environment. We now know that both regions are characterized by major shifts in physical energy regimes and seabed dynamics. Salinity certainly plays a major role in determining species distribution patterns along the YR-LCB estuarine gradient, but the potential role of benthic boundary layer physi-cal processes and potential interactions with the salinity gradient are now readily apparent. For the YR-LCB system, physical disturbance of the seabed increases concordant with decreases in salinity, making the effects of each stress more difficult to separate, especially without a priori knowledge of benthic boundary layer processes.

In previous studies of other estuarine systems, comparisons of relatively quiescent versus highly energetic subenvironments within a salinity regime have provided some insights into the effects of physical sediment disturbance processes while controlling (somewhat) for salinity effects. For example, in the highly dynamic Columbia River system, the estuarine zone is characterized by both a turbidity maximum and a highly variable salinity regime. Density and standing stocks of benthic infauna are high in protected tidal flat habitats but are comparatively impoverished in the energetic channels (Jones et al. 1990). McLusky et al. (1993) documented a relative impoverishment of subtidal faunas in a comparison of intertidal and subtidal benthic communities in the upper Forth estuary, eastern Scotland, UK. They attributed this to a much greater degree of sediment instability in the estuary's main channel. Negative effects of strong tidal currents on benthic community structure and function also have been noted by comparing subenvironments within high salinity estuarine and coastal regions. In the macro (hyper) tidal Bay of Fundy estuarine system, low biomass and production, and a predominance of specialized deposit-feeders, usually deep burrowers, characterize the impoverished macrofauna assemblages found in the highest energy areas, where mean current speeds at 1–2 m above the bed exceed 30 cm s^{-1} (Wildish & Kristmanson 1997). For this system, suspension-feeder biomass peaks in areas characterized by mean current speeds in the range of

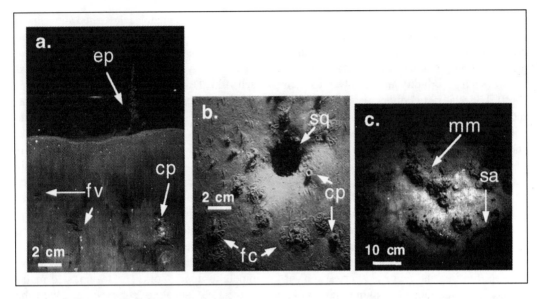

Fig. 11. Biogenic structures of the lower bay (study area 4). a) a profile image showing well-bioturbated sediment with feeding voids (fv) of the polychaete *Macroclymene zonalis*, a bisected *Chaetopterus* sp. (cp), and unidentified epifauna (ep), September 1994; b) a plan image showing a burrow of the stomatopod *Squilla empusa* (sq), tube tops of *Chaetopterus* sp. (cp), and fecal coils (fc) of *Macroclymene zonalis*, May 1985; c) a plan image showing clumps of the urochordate *Molgula manhattensis* (mm) and a colony of the hydroid *Sertularia argentea* (sa), all attached to tube tops of *Chaetopterus* sp., April 1985.

10–30 cm s^{-1}, while deposit feeder biomass increases steadily with decreasing current speeds to a maximum in areas where current speeds drop to about 10 cm s^{-1}. Warwick & Uncles (1980) found impoverished benthic communities associated with regions of very high bed stress in the Bristol Channel.

Warwick et al. (1991) argued for substantial effects of sediment instability and high turbidity on estuarine biota by comparing macrobenthic communities among estuaries of southwest Britain. They found clear differences in the faunal composition and the abundance and biomass of individual invertebrate species between the Severn River estuary and five other estuaries studied. The Severn is a large macro (hyper) tidal estuary (tide range > 4 m) characterized by high levels of sediment instability and turbidity (values typically exceed 500 mg l^{-1}) throughout its length, while the other estuaries studied are smaller, have restricted to moderate tidal ranges, and are generally far less turbid. High levels of turbidity occur in the smaller estuaries only at the estuarine turbidity maximum and for only brief periods during summer months when run-off is low. Suspension-feeding invertebrates are absent from the Severn and the dominant species are primarily motile forms, whereas suspension feeders and sedentary species are abundant in the less dynamic estuaries.

Emerson (1989) found negative relationships between benthic secondary production and wind stress for 201 widely distributed coastal sites. Multiple regression analysis demonstrated that wind stress, tidal height, shelter indices, and water temperature explained approximately 90% of the variance in total benthic secondary production (including micro-, meio-, and macrobenthos), and that neither benthic or pelagic primary production contributed to a significant reduction in the variance in benthic secondary production. Emerson argued that the effects of wind stress on benthic secondary production are mediated largely by sediment transport and that smaller organisms living near the sediment-water interface are likely to be the most susceptible. Removal of food supply, inhibition of feeding, injury from abrasion, and direct mortality will translate a higher susceptibility to sediment transport into lower secondary production.

Some of the most dramatic examples of faunal impoverishment due to high physical disturbance of muddy seabeds have been recorded for the inner shelf regions off the mouths of large rivers such as the Changjiang and Amazon, where both physical energy and rates of delivery of fine sediments are very high (Rhoads et al. 1985; Aller & Aller 1986; Aller & Stupakoff 1996). These regions are characterized by high turbidity and dynamic seabeds. Areas of highest physical disturbance exhibit reductions in the size, abundance, and reproductive output of macrofauna, increases in the prevalence of shallow-dwelling opportunistic species, decreases in macrofauna diversity, and enhancement of microbial over metazoan productivity. Given that these regions are associated with large river plumes, the faunas may also be influenced to some degree by salinity stress. Benthic habitats in these regions tend to develop more persistent and diverse populations of macrofauna as physical disturbance decreases in space or time.

The success of a few species of bivalves, especially *Macoma balthica* and *Cyrtopleura costata*, in the upper to middle York River estuary is interesting, given the high degree of physical disturbance documented for the seabed. Juvenile *M. balthica* are highly motile and use physical transport processes to facilitate dispersal (Beukema & de Vlas 1989). In both the York and James River subestuaries of lower Chesapeake Bay, the burrowing depths of *M. balthica* increase as they grow; juveniles of 1 cm width are found at ~ 10 cm and adults of 4 cm width can be found as much as 40 cm below the sediment-water interface (Schaffner et al. 1987a; Hinchey unpublished). Deep burrowing provides a predator refuge for adult clams (Blundon & Kennedy 1982), but in the YR-LCB system it also makes them far less susceptible to physical disturbance, which is most intense in surface sediments.

While a paucity of large or deep-dwelling infauna in the lower York is consistent with effects of physiological stresses due to hypoxia (Diaz & Rosenberg 1995), sediment trapping associated with mounds of epifauna and subsequent transport of these mounds along the seabed might also limit infauna. Given their abundance, the dynamics of successful epifauna and the mounds they form warrant further investigation. Our results demonstrate that side-scan sonar and underwater profile cameras provide rapid and effective means to identify and map the distributions of epifauna assemblages in turbid estuarine waters.

Resident biota of the lower estuary, bay environment experience minimal physical disturbance of the seabed and relatively low physiological stress. This is the region of the estuary where the structural and functional diversity of benthic macrofauna assemblages are highest (Schaffner 1990). Many of the dominant species have maximum life spans that exceed a year and their high standing stocks and relatively high growth rates lead to high secondary production (Diaz & Schaffner 1990; Thompson & Schaffner in press). During the spring and early summer months, following the period when the upper few centimeters of the seabed are disturbed by physical processes, the lower estuary also supports high densities of small, fast-growing species (Schaffner 1990). Both the high standing stock of suspension feeders and high productivity of the macrofauna in this region of the estuary argue for efficient benthic-pelagic coupling (Thompson-Neubauer 2000). The moderate tidal currents observed near the bed in the lower bay can be expected to affect efficient delivery of primary production and removal of waste materials, thereby serving as a source of auxiliary energy (sensu

Mann & Lazier 1996; see also Nixon 1988; Wildish & Kristmanson 1997) to enhance secondary productivity.

Conclusions and Recommendations for Future Work

How representative is the YR-LCB estuarine gradient, and how useful are the results presented here for understanding processes in other estuaries or other marine systems? It can always be argued that, to some extent, each estuary is unique. Our work-in-progress suggests that even the subestuaries of the Chesapeake Bay system vary with respect to physical energy regimes (Schaffner et al. unpublished). Nonetheless, we are encouraged that the organism-sediment-flow interactions we report for the YR-LCB system are consistent with the results reported previously for coastal systems as diverse as the Severn River estuary, Bay of Fundy, and Amazon River shelf environment. Benthic macrofauna assemblages become impoverished when physical energy regimes are sufficient to cause frequent, intense physical disturbance of muddy seabeds. As in many of the studies discussed above, we observed increases in the structural and functional diversity of the macrofauna, increased densities of macrofauna and increased standing stocks of suspension feeders as physical disturbance of the seabed declined. Given that the Chesapeake Bay and its subestuaries are microtidal, we were somewhat surprised by our findings regarding the dynamic nature of seabed processes in the upper York River. Thus, we offer a caveat to those working in muddy, coastal environments. Convergent transport processes operate to enhance the trapping of fine sediment in many estuarine environments. Thus, it can not be assumed that sediment grain size and physical energy regime will correlate in these systems. Underconsolidated muddy sediments may be found in relatively high energy areas where they will be easily mobilized by physical processes.

For the YR-LCB system, the physical sediment disturbance gradient is correlated with the estuarine salinity gradient. This is significant because the size of the regional species pool in estuaries and other bodies of brackish water declines with decreasing salinity, usually reaching a minimum at 5-8‰ (Deeton & Greenberg 1986). Most estuarine organisms are derived from marine species and their up-estuary limits often are set by physiological tolerances of low salinity (Vernberg & Vernberg 1972). The potential for synergistic effects of salinity stress and physical disturbance on macrofauna seems high based on studies of the responses of individuals to multiple physiological stresses (Vernberg & Vernberg 1972). Physical sediment disturbance processes imposed in addition to salinity stress might have greater effects on individuals or populations living near their physiological tolerance limits in the upper estuary than those observed for individuals or populations residing at higher salinities in the middle to lower estuary.

A variety of approaches will be needed to separate the effects of near-bed flow regimes and sediment disturbance processes from the effects of other environmental factors, such as salinity, oxygen, and food availability, that influence the structure and function of estuarine benthic communities. Experimental manipulations of bed disturbance regimes offer the potential for significant control over some variables; however, it will be difficult to reproduce fully the disturbance environment experienced by resident biota, and scaling issues will be a concern. "Natural" physical disturbance events, such as the formation of furrows in the upper to middle York, may provide significant insights if effects can be followed in time and space. Comparisons of distinct estuarine subenvironments characterized by different physical energy and salinity regimes (e.g., protected versus exposed areas, high salinity versus low salinity) within single estuaries will provide control over regional differences in the environmental setting (e.g., nutrient or sediment loadings). Following faunal responses to "natural" disturbance events or comparing subenvironments within a single system offers no control over the regional disturbance regime. The regional disturbance regime has important implications for community dynamics following disturbance events because it influences the size and characteristics of the regional species pool available as potential colonizers (Huston 1994; Palmer et al. 1996). Comparisons among a variety of estuarine systems, particularly those that represent "endmembers" along a gradient of physical energy regimes, can be expected to provide insights regarding the relative importance of disturbance versus other factors that vary among the systems. This approach has been successfully employed to examine the relationships among nutrient loading, tidal energy regime, and primary production in microtidal versus macrotidal estuaries (Monbet 1992).

We urge further investigations of the linkages between physical and biological processes in the benthic boundary layer of estuarine environments using the approaches outlined above. We can envision many interesting and fruitful comparisons with other physically dynamic environments. Estuaries provide a sharp contrast to rocky intertidal environments because of the differences in the nature of the substrates, the mechanisms by which physical disturbance influences the substrate, the nature of the geochemical setting, and differences in taxonomic composition of dominant resident organisms. Characteristics of estuaries, such as high physical variability, high productivity, relatively small size, and ephemeral nature, could make for interesting comparisons with deep-sea vent habitats. Finally, the well-documented effects of physical seabed disturbance processes on biology and chemistry of continental shelf benthic habitats off large river systems have provided important insights for the interpretation of processes operating in a much smaller estuarine system. We predict that a better understanding of processes operating in the benthic boundary layers of estuaries will enhance general understanding of muddy seabed dynamics and help us to understand the ramifications of seabed processes for ecosystem structure and function in a wide variety of marine settings.

Acknowledgments

The Office of Naval Research, Harbor Processes Program (grant N00014-93-1-0986) and the Virginia Institute of Marine Science funded this study. Two anonymous reviewers provided very useful comments that helped us improve the manuscript. We thank Dr. Joseph Kravitz of the Office of Naval Research and Dr. Don Wright of Virginia Institute of Marine Science for their support and guidance during our investigations. The field work that formed the basis of our insights could not have been accomplished without the untiring efforts (and strong backs) of many students, technicians, and vessel operations support staff at the Virginia Institute of Marine Science. We thank them all! Contribution number 2231 of the Virginia Institute of Marine Science.

Literature Cited

Aller, J. Y. 1989. Quantifying sediment disturbance by bottom currents and its effects on benthic communities in a deep-sea western boundary zone. *Deep-Sea Research* 36:901–943.

Aller, J. Y. & I. Stupakoff. 1996. The distribution and seasonal characteristics of benthic communities on the Amazon shelf as indicators of physical processes. *Continental Shelf Research* 16:717–751.

Aller, R. C. 1982. The effects of macrobenthos on chemical properties of marine sediments and overlying water, p. 53–102. *In* McCall, P. L. & M. J. S. Tevesz (eds.) Animal-Sediment Relations. Plenum Press, New York.

Aller, R. C. 1998. Mobile deltaic and continental shelf muds as suboxic, fluidized bed reactors. *Marine Chemistry* 61:143–155.

Aller, R. C. & J. Y. Aller. 1986. General characteristics of benthic faunas of the Amazon inner continental shelf with comparisons to the shelf off the Changjiang River, East China Sea. *Continental Shelf Research* 6:291–310.

Bambach, R. K. 1977. Species richness in marine benthic habitats through the Phanerozoic. *Paleobiology* 3:152–167.

Beukema, J. J. & J. de Vlas 1989. Tidal-current transport of thread-drifting postlarval juveniles of the bivalve *Macoma balthica* from the Wadden Sea to the North Sea. *Marine Ecology Progress Series* 52:193–200.

Blundon, J. A. & V. S. Kennedy. 1982. Refuges for infaunal bivalves from blue crab, *Callinectes sapidus* (Rathburn), predation in Chesapeake Bay. *Journal of Experimental Marine Biology and Ecology* 65:67–81.

Boesch, D. F. 1977. A new look at the zonation of benthos along the estuarine gradient, p. 245–266. *In* Coull, B. C. (ed.), Ecology of Marine Benthos. University of South Carolina Press, Columbia, South Carolina

C.A.S.T. 1998. Contaminant and Sediment Transport Project. http://www.vims.edu/physical/castnew/yorkmap.html

Cloern, J. E. 1996. Phytoplankton bloom dynamics in coastal ecosystems: A review with some general lessons from sustained investigation of San Francisco Bay, California. *Reviews of Geophysics* 34:127–168.

Dalrymple, R. W., B. A. Zaitlin, & R. Boyd. 1992. Estuarine facies models: Conceptual basis and stratigraphic implications. *Journal of Sedimentary Petrology* 62:1,130–1,146.

Dauer, D. M. 1997. Chesapeake Bay Benthic Biological Monitoring Program, 1985–1996. Chesapeake Bay Program, Annapolis, Maryland. Rpt. No. CBP/TRS191/97.

Dauer, D. M., R. M. Ewing, & A. J. Rodi, Jr. 1987. Macrobenthic distribution within the sediment along an estuarine salinity gradient. *Internationale Revue der Gesamten Hydrobiologie* 72:529–538.

Deeton, L. E. & M. J. Greenberg. 1986. There is no horohalinicum. *Estuaries* 9:20–30.

Dellapenna, T. M. 1999. Fine-scale strata formation in biologically and physically dominated estuarine systems within the lower Chesapeake and York River subestuary. Ph.D. Dissertation, School of Marine Science, The College of William and Mary, Virginia Institute of Marine Science, Gloucester Point, Virginia.

Dellapenna, T. M., S. A. Kuehl, & L. C. Schaffner. 1998. Seabed mixing and particle residence times in biologically and physically dominated estuarine systems: A comparison of lower Chesapeake Bay and the York River subestuary. *Estuarine, Coastal and Shelf Science* 46:777–795.

Diaz, R. J. & R. Rosenberg. 1995. Marine benthic hypoxia: A review of its ecological effects and the behavioural responses of benthic macrofauna. *Oceanography and Marine Biology Annual Review* 33:245–303.

Diaz, R. J. & L. C. Schaffner. 1990. The functional role of estuarine benthos, p. 25–56. *In* Haire, M. & E. C. Krome (eds.), Perspectives on the Chesapeake Bay, 1990. Advances in Estuarine Sciences. Chesapeake Research Consortium, Gloucester Point, Virginia. Rpt. No. CBP/TRS41/90.

Emerson, C. W. 1989. Wind stress limitation of benthic secondary production in shallow, soft-sediment communities. *Marine Ecology Progress Series* 53:65–77.

Friedrichs, C. T., B. A. Armbrust, & H. E. de Swart. 1998. Hydrodynamics and sediment dynamics of shallow, funnel-shaped tidal estuaries, p. 315–328. *In* Dronkers, J. and M. Scheffers (eds.), Physics of Estuaries and Coastal Seas. Balkema Press, Rotterdam, The Netherlands.

Friedrichs, C. T., L. D. Wright, D. A. Hepworth, & S.-C. Kim. 2000. Bottom boundary layer processes associated with fine sediment accumulation in coastal seas and estuaries. *Continental Shelf Research* 20:807–841.

Geyer, W. R. 1993. The importance of suppression of turbulence by stratification on the estuarine turbidity maximum. *Estuaries* 16:113–125.

Geyer, W. R., R. P. Signell, & G. C. Kineke. 1997. Lateral trapping of sediment in a partially mixed estuary, p. 115–124. *In* Dronkers, J. & M. Scheffers (eds.), Physics of Estuaries and Coastal Seas. Balkema Press, Rotterdam, The Netherlands.

Grime, J. P. 1977. Evidence for the existence of three primary strategies in plants and its relevance to ecological and evolutionary theory. *American Naturalist* 111:1,169–1,194.

Hall, S. J. 1994. Physical disturbance and marine benthic communities: Life in unconsolidated sediments. *Oceanography and Marine Biology Annual Review* 32:179–239.

Haven, D. S. & R. Morales-Alamo. 1972. Biodeposition as a factor in sedimentation of fine suspended solids in estuaries. *Geological Society of America, Memoir* 133:121–130.

Hood, R. R., H. V. Wang, J. E. Purcell, E. D. Houde, & L. W. Harding, Jr. 1998. Modeling particles and pelagic organisms in Chesapeake Bay: Convergent features control plankton distributions. *Journal of Geophysical Research* 104(C1):1223–1243 (correction in 104(C2):3289–3290).

Huston, M. A. 1994. Biological Diversity: The Coexistence of Species on Changing Landscapes. Cambridge University Press, New York.

Huzzey, L. M. & J. M. Brubaker. 1988. The formation of longitudinal fronts in a coastal plain estuary. *Journal of Geophysical Research* 93:1,329–1,334.

Jablonski, D. & D. J. Bottjer. 1991. Environmental patterns in the origins of higher taxa: The post-Paleozoic fossil record. *Science* 252:1,831–1,883.

Jones, K. K., C. A. Simenstad, D. L. Higley, & D. L. Bottom. 1990. Community structure, distribution, and standing stock of benthos, epibenthos and plankton in the Columbia River Estuary. *Progress in Oceanography* 25:211–241.

Karickhoff, S. W. & K. R. Morris. 1985. Impact of tubificid oligochaetes on pollutant transport and in bottom sediments. *Environmental Science and Technology* 19:51–56.

Kemp, W. M., E. M. Smith, M. Marvin-DiPasquale, & W.R. Boynton. 1997. Organic carbon balance and net ecosystem metabolism in Chesapeake Bay. *Marine Ecology Progress Series* 150:229–248.

Kim, S.-C., C. T. Friedrichs, J. P.-Y. Maa, & L. D. Wright. 2000. Estimating bottom stress in a tidal boundary layer from Acoustic Doppler Velocimeter Data. *ASCE Journal of Hydraulic Engineering* 126:399–406.

Kuo, A. Y. & B. J. Neilson. 1987. Hypoxia and salinity in Virginia estuaries. *Estuaries* 10:277–283.

Mann, K. H. & J. R. N. Lazier. 1996. Dynamics of Marine Ecosystems. Blackwell Science Inc., Cambridge, Massachusetts.

Menge, B. A. & J. P. Sutherland. 1987. Community regulation: Variation in disturbance, competition, and predation in relation to environmental stress and recruitment. *American Naturalist* 130:730–757.

McLusky, D. S., S. C. Hull, & M. Elliott. 1993. Variations in the intertidal and subtidal macrofauna and sediments along a salinity gradient in the upper Forth Estuary. *Netherlands Journal of Aquatic Ecology* 27:101–109.

Monbet, Y. 1992. Control of phytoplankton biomass in estuaries: A comparative analysis of microtidal and macrotidal estuaries. *Estuaries* 15:563–571.

Nichols, M. M. & R. B. Biggs. 1985. Estuaries, p. 77–186. *In* Davis, R. A. (ed.), Coastal Sedimentary Processes. Springer-Verlag, New York.

Nichols, M. N., G. H. Johnson, & P. C. Peebles. 1991. Modern sediments and facies model for a microtidal coastal plain estuary, the James Estuary, Virginia. *Journal of Sedimentary Petrology* 61:883–899.

Nixon, S. W. 1988. Physical energy inputs and the comparative ecology of lake and marine systems. *Limnology and Oceanography* 33:1,005–1,025.

Ólafsson, E. B., C. H. Peterson, & W. G. Ambrose. 1994. Does recruitment limitation structure populations and communities of macro-invertebrates in marine soft sediments: The relative significance of pre- and post-settlement processes. *Oceanography and Marine Biology Annual Review* 32:65–109.

Palmer, M. A., J. D. Allan, & C. A. Butman. 1996. Dispersal as a regional process affecting the local dynamics of marine and stream benthic invertebrates. *Trends in Ecology and Evolution* 11:322–326.

Pihl, L., S. P. Baden, & R. J. Diaz. 1991. Effects of periodic hypoxia on distribution of demersal fish and crustaceans. *Marine Biology* 108:349–360.

Rhoads, D. C., R. C. Aller, & M. Goldhaber. 1977. The influence of colonizing macrobenthos on physical properties and chemical diagenesis of the estuarine seafloor, p. 113–138. *In* B. C. Coull (ed.), Ecology of Marine Benthos. University of South Carolina Press, Columbia, South Carolina.

Rhoads, D. C., D. F. Boesch, T. Zhican, X. Fengshan, H. Liqiang, & K. J. Nilsen. 1985. Macrobenthos and sedimentary facies on the Changjiang delta platform and adjacent continental shelf, East China Sea. *Continental Shelf Research* 4:189–213.

Rhoads, D. C. & L. F. Boyer. 1982. The effects of marine benthos on physical properties of sediments: A successional perspective, p. 3–52. *In* McCall, P. L. & M. J. S. Tevesz (eds.), Animal-Sediment Relations. Plenum Press, New York.

Rhoads, D. C. & J. C. Germano. 1982. Characterization of organism-sediment relations using sediment profile imaging: An efficient method of remote ecological monitoring of the seafloor (Remots™ System). *Marine Ecology Progress Series* 8:113–128.

Rhoads, D. C. & J. C. Germano. 1986. Interpreting long-term changes in benthic community structure: A new protocol. *Hydrobiologia* 142:110–114.

Rhoads, D. C., P. L. McCall, & J. Y. Yingst. 1978. Disturbance and production on the estuarine seafloor. *American Scientist* 66:577–586.

Sagasti, A., L. C. Schaffner, & J. E. Duffy. 2000. Epifaunal communities thrive in an estuary with hypoxic episodes. *Estuaries* 23:474–487.

Schaffner, L. C. 1990. Small-scale organism distributions and patterns of species diversity: Evidence for positive interactions in an estuarine benthic community. *Marine Ecology Progress Series* 61:107–117.

Schaffner, L. C., R. J. Diaz, C. R. Olsen, & I. L. Larsen. 1987a. Faunal characteristics and sediment accumulation processes in the James River Estuary, Virginia. *Estuarine, Coastal and Shelf Science* 25:211–226.

Schaffner, L. C., R. J. Diaz, & R. J. Byrne. 1987b. Processes affecting recent estuarine stratigraphy, p. 584–599. *In* Kraus, N. C. (ed.), Coastal Sediments '87, Proceedings of a Specialty Conference on Advances in Understanding of Coastal Sediment Processes. Vol. I. American Society of Civil Engineers, New York.

Schaffner, L. C., R. M. Dickhut, S. Mitra, P. W. Lay, & C. Brouwer-Riel. 1997. Effects of physical chemistry and bioturbation by estuarine macrofauna on the transport of hydrophobic organic contaminants in the benthos. *Environmental Science and Technology* 31:3,120–3,125.

Sharples, J., J. H. Simpson, & J. M. Brubaker. 1994. Observations and modelling of periodic stratification in the upper York River estuary, Virginia. *Estuarine, Coastal and Shelf Science* 38:301–312.

Sin, Y., R. L. Wetzel, & I. C. Anderson. 1999. Spatial and temporal characteristics of nutrient and phytoplankton dynamics in the York River estuary, Virginia: Analysis of long-term data. *Estuaries* 22:260–275.

Snelgrove, P. V. R. & C. A. Butman. 1994. Animal-sediment relationships revisited: Cause versus effect. *Oceanography and Marine Biology Annual Review* 32:111–177.

Thompson, M. L. & L. C. Schaffner. In press. Demography of the polychaete *Chaetopterus pergamentaceus* within the lower Chesapeake Bay and relationships with environmental gradients. *Bulletin of Marine Science*

Uncles, R. J., M. L. Barton, & J. A. Stephens. 1994. Seasonal variability of mobile mud deposits in the Tamar estuary, p. 374–387. *In* Pattiaratchi, C. (ed.), Mixing in Estuaries and Coastal Seas. American Geophysical Union, Washington, D.C.

Valle-Levinson, A. & M. M. Lwiza. 1995. The effects of channels and shoals on exchange between the Chesapeake Bay and the adjacent ocean. *Journal of Geophysical Research* 100:18,551–18,563.

Vernberg, W. B. & F. J. Vernberg. 1972. Environmental Physiology of Marine Animals. Springer-Verlag, New York.

Wagner, C.M. 1999. Expression of the estuarine species minimum in littoral fish assemblages of the lower Chesapeake Bay tributaries. *Estuaries* 22:304–312.

Warwick, R. M., J. D.Goss-Custard, R. Kirby, C. L. George, N. D. Pope, & A. A. Rowden. 1991. Static and dynamic environmental factors determining the community structure of estuarine macrobenthos in SW Britain: Why is the Severn Estuary different? *Journal of Applied Ecology* 28:329–345.

Warwick, R. M. & R. J. Uncles. 1980. Distribution of benthic macrofauna associations in the Bristol Channel in relation to tidal stress. *Marine Ecology Progress Series* 3:97–103.

Wildish, D. J. & D. D. Kristmanson. 1997. Benthic Suspension Feeders and Flow. Cambridge University Press, Cambridge, United Kingdom.

Wright, L. D., D. B. Prior, C. H. Hobbs, R. J. Byrne, J. D. Boon, L. C. Schaffner, & M. O. Green. 1987. Spatial variability of bottom types in the Lower Chesapeake Bay and adjoining estuaries and inner shelf. *Estuarine, Coastal and Shelf Science* 24:765–784.

Wright, L. D., L. C. Schaffner, & J. P.-Y. Maa. 1997. Biological mediation of bottom boundary layer processes and sediment suspension in the lower Chesapeake Bay. *Marine Geology* 141:27–50.

Sources of Unpublished Materials

Hinchey, E. K. Effects of physical disturbance processes on estuarine benthic communities. Ph.D. dissertation in progress, School of Marine Science, The College of William and Mary, Virginia Institute of Marine Science, Gloucester Point, Virginia.

Schaffner, L. C., C. F. Friedrichs, E. K. Hinchey, T. A. Dellapenna, K. Dorgan, & S. A. Kuehl. Physical energy regimes and benthic subenvironments: A comparison of subestuaries of Chesapeake Bay. (manuscript).

Scale-Dependent Recovery of the Benthos: Effects of Larval and Post-Larval Life Stages

Robert B. Whitlatch*, Andrew M. Lohrer, and Simon F. Thrush

Abstract: *Marine soft-sediment recolonization following disturbance is dependent upon two, often independent, sources of colonists: larvae settling from the water column and post-settlement life stages (juveniles and adults) laterally advected across the seabed. The relative importance of these two colonist pools should vary with spatial scale of disturbance; smaller disturbances with greater edge-to-surface area ratios being more affected by post-settlement immigration than larger disturbances. While literature indicates rates of post-settlement immigration can rival larval recruitment rates, a review of experimental studies reveals no clear relationship between scale of disturbance and recovery time. The relative importance of the two colonist pools to the recolonization process is likely a function of both intrinsic (e.g., species' life history traits) and extrinsic (e.g., habitat conditions) factors. To illustrate how differences in life history characteristics can influence the contribution of the two colonist pools, a scale-dependent recolonization model was developed. Model simulations indicate that recovery rates of species possessing "opportunistic" life-history traits were less sensitive to the effects of post-settlement immigration than species possessing life history features characteristic of later successional stages. To examine how extrinsic factors (hydrodynamic regime) influence the relative contribution of the two colonist pools, experiments were conducted using an automated time-series recruitment sampler. Recruitment of the polychaete Polydora cornuta to defaunated sediment was most pronounced during slack water periods (particularly at low tide) and most of the colonists during this period were recently settled larvae. In contrast, during flood and ebb tidal periods most of the colonists were juvenile and adult stages. Future studies on scale-dependent recolonization dynamics need to measure both sources of immigrants simultaneously in order to assess the generality of this finding for species found in the entire recolonization sequence.*

Introduction

The seminal studies on disturbance-recovery dynamics of soft-sediment habitats by D. C. Rhoads and colleagues (Rhoads et al. 1977, 1978; Rhoads & Boyer 1982) have provided a fundamental conceptual framework for examining how the sedimentary milieu sets the boundary conditions for the benthic recolonization process (Zajac & Whitlatch 1985). The model has also been instrumental in linking disturbance-recovery processes and successional dynamics based on species' adaptations to the abiotic environment. The prevailing view of soft-sediment successional dynamics is that some species ("opportunists" or "early colonists") possess life history characteristics that facilitate rapid responses to recently perturbed areas, while other ("equilibrium" or "late successional") species are more constrained in their population responses and have much slower rates of recolonization of disturbed patches. While there remains some debate regarding the specific mechanism(s) associated with the replacement of early colonizing species with species found in the latter parts of the successional sequence (e.g., Gallagher et al. 1983; Chesney 1985; Whitlatch & Zajac 1985), the identification of "indicator" species as sentinels of habitat disturbance and recovery and environmental degradation remains a cornerstone of both basic and applied benthic recolonization studies.

While the Rhoads et al. (1978) disturbance-recovery model has been widely adopted, it is important to recognize that disturbances can generate variable abiotic and biotic conditions, which in turn may lead to multifaceted population- and community-level responses (Zajac & Whitlatch 1991). For example, the timing of disturbance and differences in habitat type may result in the lack of clearly defined opportunistic species responses (e.g., Bonsdorff 1980; Zajac & Whitlatch 1982a; Thrush et al. 1996) or lack of replacement of opportunistic forms by species hypothesized to be better resource competitors (e.g., Zajac & Whitlatch 1982b; Flemer et al. 1997). In addition, the colonization modes species use to respond to disturbances depend on life history features, habitat conditions, and life-stage-specific mobility patterns (e.g., Commito et al. 1995; Shull 1997).

In addition to the above, the soft-sediment recolonization process is now recognized to be influenced by two, often independent, sources of colonists: larvae and post-settlement life stages (Gunther 1992 for a recent review). The relative contributions of these colonist pools to the recolonization process are likely dependent upon a number of intrinsic and extrinsic factors. For example, life history features (e.g., organism size, fecundity, temporal patterns of larval release, organism mobility) of component species may lead to differences in the supply of larval and post-settlement life-stages, while habitat features (e.g., size of the disturbance, sediment mobility) may also add to variations in the supply of the two colonist pools. All the factors have the potential to result in substantive deviations in the predictions set forth by the classical disturbance-recovery model.

The objectives of this paper are to: (a) provide a general overview of our present understanding of how the two colonist pools might influence recovery of marine benthic disturbances, (b) present a spatially varying recolonization model that illustrates how the effects of post-settlement immigration can differ in early and late successional stage species, and (c) describe preliminary data examining how an extrinsic factor, namely the hydrodynamic regime, can influence the relative contributions of the two colonist pools in the recolonization process for the opportunistic polychaete *Polydora cornuta* (Bosc).

Background

As in most habitats, both natural and anthropogenic disturbances are recognized as important forces affecting soft-sediment population and community dynamics. These disturbances can vary in spatial scale and range from millimeters to kilometers depending upon the specific disturbance agent (see Zajac et al. 1998 for a recent review). Soft-sediment recolonization is often thought to be related to spatial scale, whereby recovery time is positively related to size of the disturbance (e.g., Hall et al. 1994). While this notion seems intuitively appealing, the existence of scale-dependent recovery is likely to be influenced by the relative contributions of different colonist pools (larvae, post-settlement juveniles, and adult life stages). Traditionally, recolonization was thought to be primarily from a pool of invertebrate larvae that settle into disturbed areas of the seafloor. This view was first articulated by Thorson (1950, 1966), who stated that most benthic invertebrate species (in low and middle latitudes) have pelagic larvae and recruitment occurs via larval settlement. Given the wide-scale dispersal ability of many species, and assuming sufficient larval supply and similar conditions in disturbances of different size, the density of colonists supplied from the water column should be independent of the size of the disturbance. Recolonization rates, therefore, are not necessarily predicted to vary directly with disturbance size.

The second colonist pool (juvenile and adult life stages) has been increasingly recognized in the recolonization process. For example, a variety of studies have found that many invertebrate colonists to recently disturbed patches of sediment are dominated by juvenile and adult life stages (e.g., Van Blaricom 1982; Thrush 1986; Zajac & Whitlatch 1989; Whitlatch et al. 1998). If lateral advection of post-settlement life stages is important to the recovery process, then recolonization should depend upon surface-to-edge ratios of a given patch (e.g., Smith & Brumsickle 1989). Larger disturbed patches with lower edge-to-surface area ratios should have proportionately lower densities of post-larval life-stage colonists than smaller patches possessing higher edge-to-surface area ratios. One would predict, therefore, the presence of scale-dependent recovery if the post-settlement colonist pool contributes substantively to the recovery process.

A limited literature review of shallow-water studies that have experimentally defaunated patches of sediment and then followed the recolonization process indicates little or no evidence of a clear relationship between the size of the disturbed patch and time for the patch to return to predisturbance or ambient conditions (Table 1). This result is not surprising for a number of reasons. By necessity, disturbance-recovery experiments must be conducted over relatively small spatial scales (e.g., 2–4 orders of magnitude). In addition, these studies have been conducted using a variety of experimental procedures (e.g., different methods of defaunating sediment, placement of sediments in the field, etc.) and have been done in a wide variety of habitat types. Despite these limitations, however, the data do indicate considerable variability in recovery time (a few days to more than 1 year) even over small spatial scales of disturbance.

The data presented in Table 1 clearly demonstrate the dynamic nature of the recovery process and the need for better assessment of the relative contributions of the two colonist pools on recolonization dynamics. A central issue, therefore, is the relative magnitude in the recruitment rates of the two colonist pools. Data from a number of studies indicate the presence of both larval and post-settlement life stages in recolonization of small-scale disturbed patches of sediment (e.g., Grant 1981; Zajac & Whitlatch 1982a; Thrush et al. 1996) and the colonization rates of the two colonist pools can be reasonably similar in magnitude for many species studied (e.g., Crowe et al. 1987; Shull 1997; Whitlatch et al. 1998). While differences in experimental procedures and habitat conditions result in considerable variability in these data sets, they do demonstrate the recovery process is not driven solely by a limited number of factors.

Recovery of benthic communities at larger scales of disturbance is equally variable (Zajac 1999). Examining the importance of the two colonist pools at larger spatial scales is more difficult, since many studies do not have a well-defined knowledge of predisturbance conditions or fail to monitor undisturbed reference sites in conjunction with monitoring the recovery of the disturbed sites. Also, the spatial extent of disturbance is often not always known, and many of the large-scale disturbances are associated with contaminated sediments or sediments that differ from surrounding ambient sediments. There are relatively few studies that have sampled large-scale disturbances with sufficient spatial and temporal resolution to discern the time-course of the recovery process adequately. Lastly, few studies have explicitly (e.g., Dauer & Simon 1976; Santos & Simon 1980; Zajac & Whitlatch 1989) or implicitly (e.g., Oliver & Slattery 1985; Oliver et al. 1985) examined the

Table 1. A limited review of field studies that have followed the recovery of experimentally defaunated patches of sediment. List is arranged in descending size of the disturbance. Recovery time refers to the authors' published estimates or when densities of infauna in disturbed plots resembled ambient or reference conditions (+ = no recovery observed by the end of the sampling period).

Habitat	Sediment Type	Disturbance Size (cm^2)	Recovery Time (days)	Reference
Intertidal	Sand	32,400	270+	Thrush et al. 1996
Subtidal	Sand	15,000	720	Arntz & Rumohr 1982
Intertidal	Sand	8,100	270+	Thrush et al. 1996
Intertidal	Sand	5,024	1–3	Thrush et al. 1991
Intertidal	Sand	4,000	40+	Levin 1984
Intertidal	Sand	2,030	270+	Thrush et al. 1996
Subtidal	Mud	2,000	260	Berge 1990
Subtidal		1,200	56–77	Bonsdorff & Osterman 1984
Subtidal	Mud	1,000	175–223	McCall 1977
Subtidal	Mud	1,000	90–120	Diaz-Castaneda et al. 1989
Intertidal	Sand	876	28–30	Grant 1981
Subtidal		706	80+	Oliver et al. 1985
Intertidal	Mud	176	20	Thrush & Roper 1988
Intertidal	Sand	100	5–7	Savage & Taghon 1988
Intertidal	Sand	78	2	Neto & Lana 1994

size-structure of organisms recruiting to large-scale disturbances. In all but one study (Santos & Simon 1980), juvenile and adult life stages are the most common contributors to the early stages of the recolonization process.

Life-Stage-Based Recolonization Model

One of the important consequences of life stage differences in dispersal ability is the prediction that small-scale disturbances should contain a wider representation of different life stages of a given species than those inhabiting large-scale disturbances. Differences in dispersal ability should also influence the within-disturbed patch population dynamics, resulting in variations in the population structure of organisms residing in disturbances of different sizes. Even if adult immigration is only a fraction of larval recruitment, a substantial fraction of those individuals may arrive as sexually mature (or nearly so) colonists and potentially could accelerate the recovery process. In addition, larger size colonists can escape larval predators (e.g., Watzin 1986) and enjoy lower risks of mortality relative to larval colonists.

In order to assess the effect of immigration by post-settlement life stages on the population dynamics and recovery of disturbed patches, we constructed a life-stage-specific recolonization model. The dynamics of patch recovery were modeled using STELLA Research software (version

5). Modeling in STELLA is a graphical programming process; variables influencing the system are classified and arranged in graphical ways using a series of icons. Mathematical relationships between variables are defined explicitly by the modeler.

Species life history traits used in the recruitment model simulations were chosen for their contrasting features, representing both an "early" (*Polydora cornuta*) and "late" (*Nephtys incisa* (Malmgren)) successional stage species. Demographic data (stage-specific survivorship, fecundity, maturation time, life span) are available for both polychaetes, making model parameterization less subjective. While *Polydora* populations normally consist of six to eight separate life stages (larval, two juvenile, and three to five adult stages) (Zajac 1991) and *Nepthys* populations commonly have six life stages (larval, juvenile, and four adult stages) (Zajac & Whitlatch 1989), both species were modeled using a collapsed life cycle containing three distinct life stages: larvae, juveniles, and adults, in order to simplify the modeling procedure. The model (Fig. 1) is loosely based on a three-stage insect described in Hannon and Ruth (1997) and on a previously developed patch recovery model

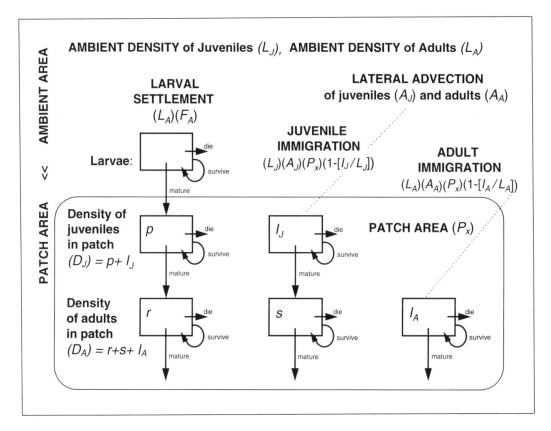

Fig. 1. Simplified schematic of the STELLA recolonization model. Lateral advection rates (A_x) and patch areas (P_x) were fixed parameters during model runs, but their effects were observed via alterations of them between model runs. Density pools (p, I_J, r, s, and I_A) all had initial densities of 0 and were dynamic variables that changed over time. Model runs were terminated when the patch had recovered; patch recovery was defined as the time step when the patch population abundance mimicked ambient population abundance. The density pools I_J and I_A are directly dependent on immigration, pool s is an indirect result of it and p and r are independent of immigration. (See Lohrer & Whitlatch In Press for a complete model description.)

(Whitlatch et al. 1998). A complete description and CD-ROM version of the model can be found in Lohrer and Whitlatch (in press).

At time zero, recently disturbed (completely defaunated) patches became available for recolonization. The circular defaunated patches were relatively small and were embedded within a large ambient area (total area: 10 km^2) . The effects of patch size on recovery time of a defaunated patch were modeled using patches with radii of 200 m, 20 m, and 2 m. The largest defaunated patch modeled was still quite small relative to the ambient area, and therefore, within-patch dynamics were assumed not to change the densities of worms in the ambient area (i.e., the ambient density of each life stage of each species did not change as time progressed in a simulation). We also assumed that the density of larvae that settled into a patch from the water column was unaffected by patch size. Immigration of post-settlers (both juveniles and adults), however, was dependent on patch size and background lateral advection rates. Lateral advection rates were simply the proportion of ambient juveniles and/or adults that were picked up and transported across the seabed at any given time. In nature, this rate might be affected by wave climate, current speed, species size-structure, species position in the sediment column, sediment characteristics, etc. For simplicity we kept the advection rate constant over time. Patch size can affect the influx of immigrants that are being transported horizontally across the seabed because of differential edge-to-surface area ratios. Small patches have much more edge in proportion to their area (edge:interior varies as a function of 2/patch radius), and therefore will fill with immigrants faster than large patches experiencing the same background advection rates. Since the relationship between background rate of advection and patch size is a multiplicative one, the smaller the background rate of advection the more patch size will matter, and vice versa. We explicitly modeled patch size by using edge:interior ratios based on the radius of the patch being simulated.

We modeled dynamics of adult worms arriving to a previously defaunated patch via three alternative routes: (1) larval colonization, subsequent survival through the juvenile stage, and maturation into the adult life stage, (2) juvenile immigration, followed by adult maturation, and (3) adult immigration. The number of adults produced by these three immigration routes were summed at each time step to yield totals for the adult pools of each species (Fig. 1).

The larval pool was dependent upon the density and fecundity of adults in the ambient area, the survivorship of those larvae that were produced, and larval stage duration (number of weeks from egg to juvenile). *Nephtys* was modeled with lower ambient adult densities, pulsed reproductive events (2 pulses yr^{-1}), relatively high larval survivorship, and a 12-wk larval duration (Table 2). *Polydora*, in contrast, had higher ambient adult densities, continuous reproduction, lower larval survivorship, and only a 2-wk larval maturation time. At each time step, a larva either matured into a juvenile, died, or survived but remained a larva. It was assumed that larvae of all ages were equally abundant in the ambient larval pool. Given this assumption, when a simulation was started, larvae competent to settle immediately began to colonize the previously defaunated patch. The number of larvae transitioning into the juvenile pool was inversely proportional to larval maturation time (i.e., after the number of time steps equaled the maturation time, or the average amount of time for an egg produced at time zero to become a juvenile). The next transitions (from the juvenile pool into adulthood and from adulthood to death of old age) were modeled in the same way. Obviously, survivorship and stage duration parameters were modified to reflect the life stage and species being considered.

Juveniles that have just settled from the water column must wait until they have matured before they transition into the adult life stage. For juvenile immigrants, however, a proportion of them could immediately become adults, assuming they were ready to mature at the time of horizontal advection

Table 2. Life-history parameters of *Polydora cornuta* (adapted from Zajac 1991) and *Nephtys incisa* (modified from Zajac & Whitlatch 1989) used in the recolonization model simulations.

Life-history Parameter	Nephtys incisa	Polydora cornuta
Larval survivorship	0.03	0.003
Duration as larvae	12 wk	2 wk
Instantaneous survival (survivorship per time step)	0.75	0.05
Juvenile survivorship	0.6	0.415
Duration as juvenile	80 wk	4 wk
Instantaneous survival (survivorship per time step)	0.99	0.8
Adult survivorship	0.8	0.35
Duration as adult	108 wk	8 wk
Instantaneous survival (survivorship per time step)	1	0.88
Reproduction	2.5 eggs adult^{-1} wk^{-1}	675 eggs adult^{-1} wk^{-1}
Reproductive period	2 pulses yr^{-1}	Continuous

across the seabed. Therefore, when horizontal advection rates for juveniles were set > 0, both juvenile and adult pools were allowed to fill immediately (juvenile pool by immigration and adult pool by maturation). Juveniles of different developmental ages within the ambient patch were assumed to be in equal proportions. Further, we assumed juveniles were advected equally well regardless of their maturation level (i.e., young and old juveniles immigrated in equal proportions), and we did not model potential effects that flow direction or position of the patch within the ambient area may have had on recruitment dynamics. In the field, an empty patch immediately after a defaunation event would have a high net increase in density of immigrants. However, as a disturbed patch begins to be recolonized, some of those immigrants would be advected out of the patch. Therefore, the density flux of immigrants into a defaunated patch was modeled as a decelerating function with respect to time (asymptotically approaching the density of juveniles in the ambient patch). Net immigration of juvenile and adult life stages, therefore, was calculated as:

$$(L_x) \times (A_x) \times (P_x) \times (1-[I_m/L_x])$$

where: L_x = density of a particular life stage in the ambient sediments
   A_x = advection rate of a given life stage
   P_x = size of disturbed patch
   I_m = density of immigrants in the immigrate pool.

Individuals reaching the adult pool by any means had three fates: death, survival to remain in the adult pool, or transition out of the adult pool (e.g., death due to old age). Adult immigration was modeled in the same way as juveniles; older adult immigrants began to die of old age immediately (Fig. 1). This contrasts with adults that matured after juvenile immigration or after larval settlement and survived through the juvenile life stage.

The effect of juvenile and adult immigration on patch population recovery time for the three different patch sizes is shown in Figs. 2 and 3. For both *Polydora* and *Nephtys*, every combination of background juvenile and adult life-stage immigration rates (except when $A_x = 0$) yields faster recovery in small patches than large ones. Also, juvenile immigration is relatively more important to the recovery process than adult immigration into disturbed patches. Small patches also showed the most pronounced changes in recolonization; small changes in background advection of post-settlement life stages translated to large changes in the recovery time of disturbed patches.

Model results also reveal interesting differences between the two contrasting polychaete life histories. The effect of post-settlement immigration was less dramatic on recovery of *Polydora* than *Nephtys*. Only in the small patches did post-settlement immigration strongly reduce the time necessary for recovery of *Polydora*. For this species, the larger larval pool and more rapid maturation time made post-settlement immigration mostly irrelevant except in the relatively small patches. *Nephtys*, in contrast, clearly benefitted from post-settlement immigration even in the largest patches. Post-settlement immigration reduced recovery time most likely because it offset the slow maturation, lower fecundity, and lower ambient densities of reproductively active adults of this late successional stage species.

To more fully understand the role of larval recruitment on patch recovery, we ran simulations where the larval supply for *Nephtys* and *Polydora* was reduced by roughly an order of magnitude at the time of the defaunation event (Fig. 4). When comparing recovery times presented in Fig. 3, *Nephtys* showed slightly slower recovery times (e.g., ~10% longer) only when juvenile and adult immigration levels were fairly low. If post-settlement life-stage immigration levels were high enough, there was little effect of reduced larval supply on recovery time (e.g., compare small and medium size patches in Fig. 3 with similar sized size patches of *Nephtys* in Fig. 4). Reducing larval inputs by roughly an order of magnitude for *Polydora*, compared with results without reduced larval supply, did not lead to pronounced differences in the effects of post-settlement immigration on patch recovery time (Fig. 4) (e.g., compare small and medium size patches in Fig. 2 with similar sized patches of *Polydora* in Fig. 4). This result further indicates that post-settlement immigration is more important to *Nephtys* than *Polydora* in the recovery process.

Effects of Hydrodynamics on Recruitment: Contrasting Larval vs. Post-Settlement Life Stages

While the degree of dispersal ability should vary predictably with organism life stage (e.g., Gunther 1992; Whitlatch et al. 1998), the relative importance of larval versus post-larval recruitment will be dependent upon a number of things; including differences in abundance of the colonist pools, organism size and position in the sediment column, sediment stability, and habitat conditions. We have been studying one of the extrinsic components; namely how the hydrodynamic regime influences the magnitude of recruitment of the two colonist pools.

Field experiments were conducted using an automated sampler, which repeatedly exposes sediments at discrete time intervals and maintains unexposed sediments in a environment that does not result in unnatural mortality of organisms recruiting to the sediments (Whitlatch & Osman 1998). Two 5-d experiments were conducted at a shallow-water site (ca. -1 m at MLW) where tidal flows are 10–15 cm s^{-1} at maximum flood and ebb periods. The recruitment device was positioned with the collection surface flush with the surrounding sediment surface in order to allow access to both larval recruitment and seabed lateral movement of post-settlement life stages. The device does not allow immigration by organisms burrowing through the sediments. Defaunated sediment (medium sand, 0.35% organic content, frozen for 24 h) was thawed, sieved through a 1-mm screen, and placed in the sediment cups of the device. The sediment exposure schedule for the experiments was coincident with tidal state: 1 h before and after high tide (high slack water); 1 h after high tide to 1 h before low tide (ebb tide); 1 h before and after low tide (low slack water); and 1 h after low tide to 1 h before high tide (flood tide). Times for the sediment exposure schedule were taken from a continuously recording tide gauge located ca. 100 m from the deployment site. Upon retrieval, sediment samples were fixed in 3% buffered formalin, transferred to 70% ethanol, sieved through a 0.1-mm mesh screen, and sorted under a dissecting microscope. Of the species recruiting, *Polydora cornuta* comprised 95+% of the individuals found in the samples. These individuals were sized by counting the number of body setigers (e.g., Zajac 1991; Shull 1997).

The abundance of *Polydora* recruiting to the defaunated sediment varied significantly with tidal state; larger numbers were found in substrates exposed during the low, slack water period than during any of the other exposure periods (Fig. 5). There were also differences in the size-frequency distributions of individuals found in the four exposure periods (Fig. 5). *Polydora* recruiting during both high and low slack water periods were generally individuals that averaged 12–15 setigers in length. Since this species generally settles at 15 setigers (Blake 1969), these individuals are considered to be newly settled larvae. In contrast, recruits found in sediment treatments exposed during flood and ebb tide periods were generally larger individuals (e.g., 20+ setigers in length); size-class representative of juvenile and early adult life stages (e.g., Zajac 1991). While these data are admittedly preliminary in nature, they do indicate the presence of tidally-dependent recruitment in *Polydora* and suggest recruitment of the two colonist pools differs according to the hydrodynamic regime. Larval settlement is more common during slack water periods (particularly at low tide) while post-settlement movement occurs during periods of water movement; suggesting that post-settlement life stages are being entrained and dispersed during periods of tidal flow.

Discussion

Benthic recolonization in soft-sediment habitats depends on two sources of colonists: larvae settling from the water column and post-settlement juvenile and adult life stages passively or actively advected across the seabed. While recruitment regulation is considered to be controlled by larval inputs in many marine hard-substrate communities (e.g., Roughgarden et al. 1985), extreme caution should be exercised in generalizing this conclusion to soft-sediment habitats. Post-settlement recruitment rates can rival larval recruitment rates and the relative importance of the two colonist pools is likely dependent upon a number of intrinsic and extrinsic factors.

Results from the recolonization model clearly illustrate the importance of the two colonist pools in affecting recovery rates of disturbed patches by species with different life histories. Simulations indicate that juvenile and adult immigration is particularly important for species possessing "late successional" stage life history traits. We predict these types of species should possess morphological and behavioral traits that promote immigration into disturbed areas. For example, large infaunal species like *Nephtys* often display nocturnal emergence from the sediment and have good swimming abilities (e.g., Dean 1978; Dauer et al. 1982). Other, smaller species are known to exhibit behaviors that facilitate their lateral transport across the seabed (e.g., Cummings et al. 1995; Shull 1997). We also predict post-settlement immigration rates will be positively related with the rate of nocturnal emergence and/or lateral movement of the infauna. Since emergent invertebrates likely originate from surrounding sediments, the rate at which sediment patches recover from disturbance is predicted to

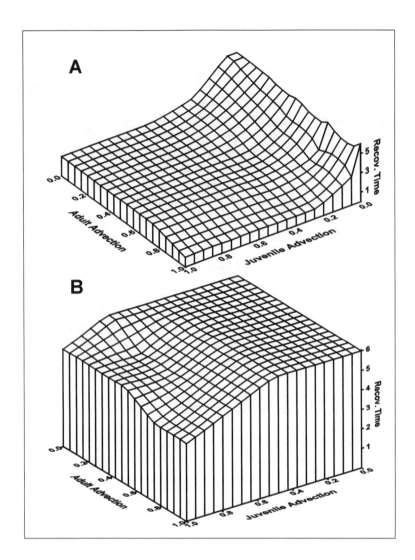

Fig. 2. Model simulations showing the effects of varying the degree of immigration of juvenile and adult life stages of *Polydora cornuta* on recovery time in small (A) and medium (B) size defaunated patches (see text). Recovery time in large patches was 6 wk regardless of the juvenile and adult advection rates. Recov. time = time (in weeks) for the patch to mimic the ambient adult population abundance.

Fig. 3. Model simulations showing the effect of varying the degree of immigration of *Nepthys incisa* juvenile and adult life stages on recovery of (A) small, (B) medium, and (C) large patch sizes (see text). Recov. time = time (in weeks) for the patch to mimic the ambient adult population abundance.

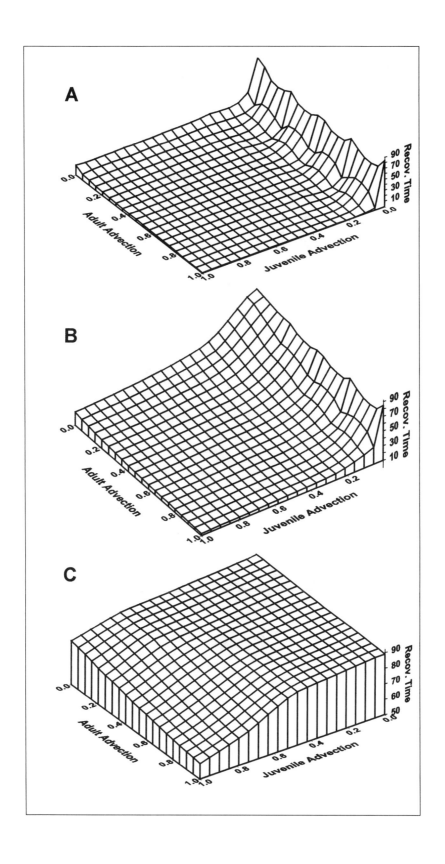

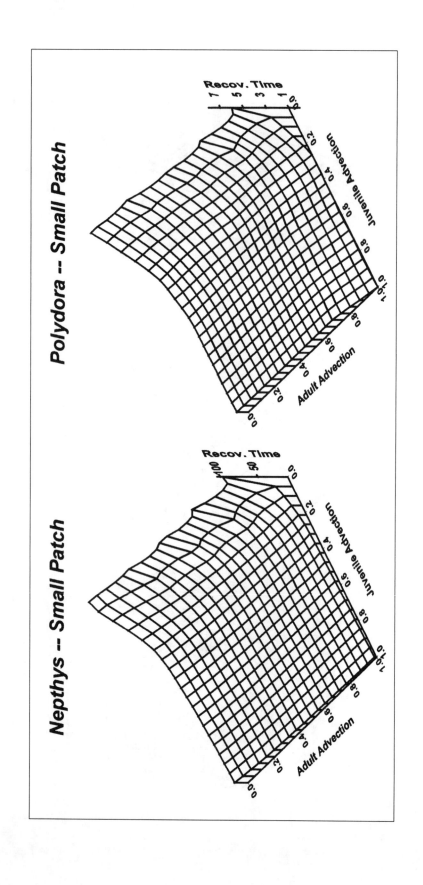

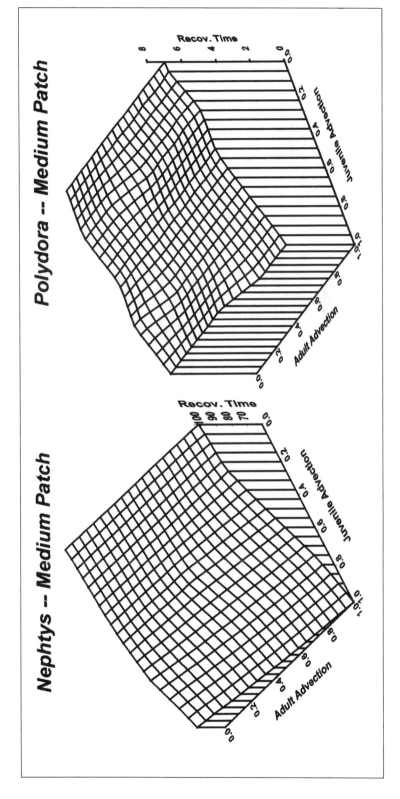

Fig. 4. Model simulations showing the effect reduced larval settlement of *Nephtys incisa* and *Polydora cornuta* on recovery of two different patch sizes. Note differences in recovery times (in weeks) presented in Figs. 1 and 2 for similar patch sizes. Larval production was reduced from 2.5 to 0.25 eggs adult^{-1} wk^{-1} for *Nephtys* and 675 to 67.5 eggs adult^{-1} wk^{-1} for *Polydora*.

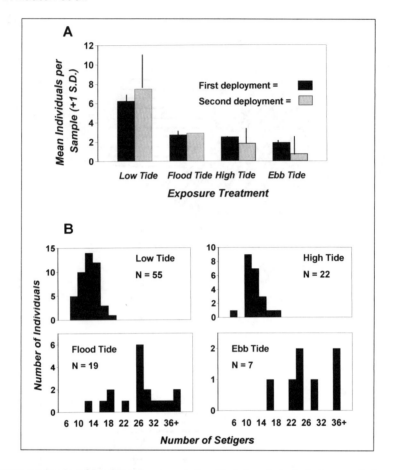

Fig. 5. (A) Average number (± 1 SD) of *Polydora cornuta* found in defaunated substrates repeatedly exposed at four different tidal states for two different 5-d deployment intervals of the benthc recruitment device (see text); (B) Size frequency distributions (number of setigers) of *Polydora cornuta* found in the defaunated sediments repeatedly exposed at four different tidal states (see text). N = total number of individuals found two 5-d deployments of the benthic recruitment device.

be proportional to the adult:larval recruitment rate. Communities in which most of the colonists are post-settlement life stages are predicted to recover faster than communities in which recruits are primarily from the larval pool. Field experiments examining extrinsic factors affecting colonization found that *Polydora* recruitment varied with tidal state. This result was consistent with previously published data (Whitlatch & Osman 1998) and supports the prediction of Gross et al. (1992) that benthic invertebrate larvae should settle at or near periods of slack water. This phenomenon may be very general as other studies have documented the release of algal spores, invertebrate larvae, or gametes during calm water or low tide periods in order to enhance fertilization or settlement success (e.g., Paine 1979; Vadas et al. 1992; Worcester 1994; Sarrao et al. 1996; Pearson et al. 1998).

Data showing strong differences in larval and juvenile-adult recruitment relative to tidal state indicate the need to measure both sources of immigrants simultaneously. Size-frequency distributions of recruits are necessary to identify the life-history stages of immigrants. As with larval recruitment, considerable experimental work is needed to examine under what conditions post-settlement colonists play a role in the recolonization process. Development of sampling devices similar to that described by Whitlatch and Osman (1998) provides one approach for unraveling the problem of assessing the timing and importance of the two colonist pools in structuring the soft-sediment recovery process.

Acknowledgments

Collection of polychaete life-history data was supported by the National Science Foundation and United States Environmental Protection Agency. Model development was supported by the J. B. Cox Charitable Trust and the National Science Foundation (to RBW) and FRST-NZ CO1825 (to SFT). We thank the organizers of the meeting for inviting us and providing a venue for presenting our ideas. Insightful reviews provided by Sara Lindsay, Jon Grant, and Sally Woodin helped to further clarify our thoughts. We dedicate this paper to R. G. Johnson and D. C. Rhoads. The former served as major advisor to Don and the first author of this paper. It would be difficult to find a better pair of role models than Ralph and Don. Their contributions to the field of benthic ecology remain cornerstones of our understanding of soft-sediment disturbance-recovery dynamics.

Literature Cited

Arntz, W. E. & H. Rumohr. 1982. An experimental study of macrobenthic colonization and succession, and the importance of seasonal variation in temperate latitudes. *Journal of Experimental Marine Biology and Ecology* 64:17–45.

Berge, J. A. 1990. Macrofauna recolonization of subtidal sediments. Experimental studies on defaunated sediment contaminated with crude oil in two Norwegian fjords with unequal eutrophication status. I. Community responses. *Marine Ecology Progress Series* 66:103–115.

Blake, J. A. 1969. Reproduction and development of *Polydora* from northern New England (Polychaete: Spionidae). *Ophelia* 7:1–63.

Bonsdorff, E. 1980. Macrobenthic recolonization of a dredged brackish water bay in S.W. Finland. *Ophelia* (suppl.) 1:145–155.

Bonsdorff, E. & C.-S. Osterman. 1985. The establishment, succession, and dynamics of a zoobenthic community - An experimental study, p. 287–297. *In* Gibbs, P. E. (ed.), Proceedings of the Nineteenth European Marine Biology Symposium. Cambridge University Press, Cambridge.

Chesney, E. J. 1985. Succession in soft-bottom benthic environments: Are pioneering species really outcompeted?, p. 277–286. *In* Gibbs, P. E. (ed.), Proceedings of the Nineteenth European Marine Biology Symposium. Cambridge University Press, Cambridge.

Commito, J. A., S. F. Thrush, R. D. Pridmore, J. E. Hewitt, & V.J. Cummings. 1995. Dispersal dynamics in a wind driven benthic system. *Limnology and Oceanography* 40:1513–1518.

Crowe, W. A., A. B. Josefson, & I. Svane. 1987. Influence of adult density on recruitment into soft sediments: A short-term in situ sublittoral experiment. *Marine Ecology Progress Series* 41:61–69.

Cummings, V. J., R. D. Pridmore, S. F. Thrush, & J. E. Hewitt. 1995. Post-settlement movement by intertidal benthic macroinvertebrates: Do common New Zealand species drift in the water column? *New Zealand Journal of Marine and Freshwater Research* 29:59–67.

Dauer, D. M. & J. L. Simon. 1976. Repopulation of the polychaete fauna of an intertidal habitat following natural defaunation: Species equilibrium. *Oecologia* 22:99–117.

Dauer, D. M., R. M. Ewing, J. W. Sourbeer, W. T. Harlan, & T. L. Stokes. 1982. Nocturnal movements of the macrobenthos of the Lafayette River, Virginia. Internationale Revue der Gesamten Hydrobiologie 67:761–775.

Dean, D. 1978. Migration of the sandworm *Nereis virens* during winter nights. *Marine Biology* 45:165–173.

Diaz-Castaneda, V., A. R. Richard, & S. Frontier. 1989. Preliminary results on colonization, recovery and succession in a polluted area of the southern North Sea (Dunkerque's Harbour, France). *Topics in Marine Biology* 53:705–716.

Flemer, D. A., B. F. Ruth, C. M. Bundrick, & G. R. Gaston. 1997. Macrobenthic community colonization and community development in dredged material disposal habitats off coastal Louisiana. *Environmental Pollution* 96:141–154.

Gallagher, E. D., P. A. Jumars, & D. D. Trueblood. 1983. Facilitation of soft-bottom benthic succession by tube builders. *Ecology* 64:1,200–1,216.

Grant, J. 1981. Sediment transport and disturbance on an intertidal sandflat: Infaunal distribution and recolonization. *Marine Ecology Progress Series* 6:249–255.

Gross, T. E., F. E. Werner, & J. E. Eckman. 1992. Numerical modeling of larval settlement in turbulent bottom boundary layers. *Journal of Marine Research* 50:611–642.

Gunther, C. 1992. Dispersal of intertidal invertebrates: A strategy to react to disturbances of different scales? *Netherlands Journal of Sea Research* 30:45–56.

Hall, S. J., D. Raffaelli, & S. F. Thrush. 1994. Patchiness and disturbance in shallow water benthic assemblages, p. 333–375. *In* Hildrew, A. G., P. S. Giller, & D. Raffaelli (eds.), Aquatic Ecology: Scale, Pattern and Processes. Blackwell Scientific, Oxford, England.

Hannon, B. & M. Ruth. 1997. Modeling Dynamic Biological Systems. Springer-Verlag, New York.

Levin, L. A. 1984. Life history and dispersal patterns in a dense infaunal polychaete assemblage: Community structure and response to disturbance. *Ecology* 65:1185–1200.

Lohrer, A. M. & R. B. Whitlatch. In press. Modeling life-stage based dynamics of marine invertebrates in soft-sediment habitats: Some implications for habitat conservation. *In* J. B. Lindholm & M. Ruth (eds.), Dynamic Modeling for Marine Conservation. Springer-Verlag, New York.

McCall, P. L. 1977. Community patterns and adaptive strategies of the infaunal benthos of Long Island Sound. *Journal of Marine Research* 35:221–266.

Neto, S. A. & P. C. Lana. 1994. Effects of sediment disturbance on the structure of benthic fauna in a subtropical tidal creek of southeastern Brazil. *Marine Ecology Progress Series* 106:239–247

Oliver, J. S. & P. N. Slattery. 1985. Destruction and opportunity on the seafloor: Effects of gray whale feeding. *Ecology* 66:1965–1975.

Oliver, J. S., R. G. Kvitek, & R. N. Slattery. 1985. Walrus feeding disturbance: Scavenging habits and recolonization of the Bering Sea benthos. *Journal of Experimental Marine Biology and Ecology* 91:233–246.

Paine, R. T. 1979. Disaster, catastrophe and local persistence of the sea palm, *Postelsia palmaeformis*. *Science* 205:685–687.

Pearson, G. A., E. A. Serrao, & S. H. Brawley. 1998. Control of gamete release in fucoid algae: Sensing hydrodynamic conditions via carbon acquisition. *Ecology* 79: 1725–1739.

Rhoads, D. C., R. C. Aller, & M. Goldhaber. 1977. The influence of colonizing macrobenthos on physical properties and chemical diagenesis of the estuarine seafloor, p. 113–138. *In* Coull, B. C. (ed.), Ecology of Marine Benthos. University of South Carolina Press, Columbia, SC.

Rhoads, D. C., P. L. McCall, & J. Y. Yingst. 1978. Disturbance and production on the estuarine seafloor. *American Scientist* 66:577–586.

Rhoads, D. C. & L. F. Boyer. 1982. The effects of marine benthos on physical properties of sediments: A successional perspective, p. 3–52. *In* P. L. McCall & M. J. S. Tevesz (eds.), Animal-Sediment Relations. Plenum Publishing, New York.

Roughgarden, J., Y. Iwasa, & C. Baxter. 1985. Demographic theory for an open marine population with space-limited recruitment. *Ecology* 6:54–67.

Santos, S. L. & J. L. Simon. 1980. Marine soft-bottom establishment following annual defaunation: Larval or adult recruitment. *Marine Ecology Progress Series* 2:235–241.

Sarrao, E. A, G. A. Pearson, & S. H. Brawley. 1996. Successful external fertilization in turbulent environments. *Proceedings of the National Academy of Sciences* 93:5,286–5,290.

Savage, W. B. & G. L. Taghon. 1988. Passive and active components of colonization following two types of disturbance on an intertidal sandflat. *Journal of Experimental Marine Biology and Ecology* 115:137–155.

Shull, D. H. 1997. Mechanisms of infaunal polychaete dispersal and colonization on an intertidal sandflat. *Journal of Marine Research* 55:153–179.

Smith, C. R. & S. J. Brumsickle. 1989. The effect of patch size and substrate isolation on colonization modes and rate in an intertidal sediment. *Limnology and Oceanography* 34:1263–1277.

Thorson, G. 1950. Reproductive and larval ecology of marine bottom invertebrates. *Biological Reviews* 35:1–45.

Thorson, G. 1966. Some factors influencing the recruitment and establishment of marine benthic communities. *Netherlands Journal of Sea Research* 3:267–293.

Thrush, S. F. 1986. The sublittoral macrobenthic community structure of an Irish sea-lough: Effect of decomposing accumulation of seaweed. *Journal of Experimental Marine Biology and Ecology* 96:199–212.

Thrush, S. F. & D. S. Roper. 1988. Merits of macrofaunal colonization of intertidal mudflats for pollution monitoring: Preliminary study. *Journal of Experimental Marine Biology and Ecology* 116:219–233.

Thrush, S. F., R. D. Pridmore, J. E. Hewitt, & V. J. Cummings. 1991. Impact of ray feeding disturbances on sandflat macrobenthos: Do communities dominated by polychaetes or shellfish respond differently? *Marine Ecology Progress Series* 69:245–252.

Thrush, S. F., R. B. Whitlatch, R. D. Pridmore, J. E. Hewitt, V. J. Cummings, & M. R. Wilkinson. 1996. Scale-dependent recolonization: The role of sediment stability in a dynamic sandflat habitat. *Ecology* 77:2472–2487.

Vadas, R. L., Sr., S. Johnson, & T. A. North. 1992. Recruitment and mortality of early post-settlement stages of benthic algae. *British Phycological Journal* 27:331–351.

VanBlaricom, G. R. 1982. Experimental analyses of structural regulation in a marine sand community exposed to oceanic swell. *Ecological Monographs* 53:283–305

Watzin, M. C. 1986. Larval settlement into marine soft-sediment systems: Interactions with meiofauna. *Journal of Experimental Marine Biology and Ecology* 98:65–113.

Whitlatch, R. B. & R. N. Zajac. 1985. Biotic interactions among estuarine infaunal opportunistic species. *Marine Ecology Progress Series* 21:299–311.

Whitlatch, R. B. & R. W. Osman. 1998. A new device for studying benthic invertebrate recruitment. *Limnology and Oceanography* 43:516–523.

Whitlatch, R. B., A. M. Lohrer, S. F. Thrush, R. D. Pridmore, J. E. Hewitt, V. J. Cummings, & R. N. Zajac. 1998. Scale-dependent recolonization dynamics: Life stage-based dispersal and demographic consequences. *Hydrobiologia*, 375/376:217–226.

Worcester, S. E. 1994. Adult rafting versus larval swimming: Dispersal and recruitment of a botryllid ascidian on eelgrass. *Marine Biology* 121:309–317.

Zajac, R. N. 1991. Population ecology of *Polydora ligni* (Polychaeta: Spionidae). II. Seasonal demographic variation and its potential impact on life history evolution. *Marine Ecology Progress Series* 77:207–220.

Zajac, R. N. & R. B. Whitlatch. 1982a. Responses of estuarine infauna to disturbance. I. Spatial and temporal variations of initial recolonization. *Marine Ecology Progress Series* 10:1–14.

Zajac, R. N. & R. B. Whitlatch. 1982b. Responses of estuarine infauna to disturbance. II. Spatial and temporal variation of succession. *Marine Ecology Progress Series* 10:15–27.

Zajac, R. N. & R. B. Whitlatch 1985. A hierarchical approach to modeling soft-bottom successional dynamics, p. 265–276. *In* Gibbs, P. E. (ed.), Proceedings of the Nineteenth European Marine Biology Symposium. Cambridge University Press, Cambridge.

Zajac, R. N. & R. B. Whitlatch. 1989. Natural and disturbance-induced demographic variation in an infaunal polychaete, *Nepthys incisa*. *Marine Ecology Progress Series* 57:89–102.

Zajac, R. N. & R. B. Whitlatch. 1991. Demographic aspects of marine, soft sediment patch dynamics. *American Zoologist* 31:808–820.

Zajac, R. N., R. B. Whitlatch, & S. F. Thrush. 1998. Recolonization and succession in soft-sediment infaunal communities: The spatial scale and controlling factors. *Hydrobiologia* 375/376:227–240.

Ecological Fidelity of Molluscan Death Assemblages

Susan M. Kidwell

Abstract: *Comparative analysis, still in progress, of marine molluscan faunas and their associated dead shells (so far comprising 80 habitat-level live-dead datasets from 17 study areas) indicates that sedimentary death assemblages are remarkably robust reflections of local community composition. Virtually all live species (mean 89% ± 5) are present in the local death assemblage, dead individuals overwhelmingly belong to species found living in the same habitat (mean 82% ± 10), and the rank abundances of dead species do not diverge significantly from those of live species (80% of datasets tested; p < 0.05). Even small samples of the death assemblage thus capture basic dominance information and habitat preferences of the live fauna, with only slight differences in fidelity among environments (marshes and tidal creeks; intertidal flats; coastal embayments; open marine seafloors). This correspondence is especially striking given the number of post-mortem processes that might act to bias such a record. Because the species richness of a death assemblage is typically 2–3X greater than that of any single census of the local live community, inverse metrics such as "% dead species also present alive" suggest low live-dead agreement. However, the majority of dead-only species are rare and most of the discrepancy (excess dead species richness) is evidently due to undersampling of the live fauna. When limits imposed by sampling are considered, true post-mortem bias from the addition of exotic and relict shells is probably less than 25% of total dead species richness, and would have little effect on abundance-based diversity measures. Molluscan death assemblages thus provide a reliable—plus relatively rapid and inexpensive—means of assessing community composition, both for the purpose of establishing ecological baselines as well as for paleoecological analysis of ancient rocks.*

Introduction

Death assemblages of molluscan shells, sieved from the top few decimeters of sediment in marine habitats, might diverge in composition from the local live community for many reasons. Bias might derive from post-mortem transport of individuals, from differential destruction of species and age-classes (especially shells that are small, fragile, or chemically reactive), and from "time-averaging" of multiple generations and/or community states (because long-term sediment accumulation rates are generally slow compared to population turnover and mixing depths; for reviews, see Powell et al. 1989; Kidwell & Bosence 1991; Kidwell & Flessa 1995). But what are the net effects of these possible biases on the actual composition of death assemblages in modern benthic habitats: that is, to what degree do death assemblages diverge from the local live community? How does live-dead agreement vary among environments, and what are the causes of discrepancies? Quantitative assessment of the ecological fidelity of modern death assemblages is crucial to paleoecological reconstruction, because it indicates the extent to which fossil assemblages can be taken as proxies of original community structure and dynamics. Such tests also indicate the possible utility of death assemblages for environmental impact studies, most particularly their reliability as baselines of pre-impact community state(s). Because dead molluscan shells are typically many times more abundant than live individuals in benthic samples (see data below), they offer an extremely rapid and relatively inexpensive way to establish background conditions in habitats *if* they capture live patterns with sufficiently high fidelity.

Live-dead agreement has been tested in a series of individual studies, mostly by paleontologists during the 1970s and 1980s in response to R. G. Johnson's (1965) seminal study of Tomales Bay, California. The majority of authors reported high live-dead agreement at the habitat- or facies-scale (i.e., after pooling data from multiple samples of a single bottom type), but results were mixed and many were only qualitative (e.g., comparison of cluster analyses). A re-analysis of live-dead studies according to a set of standard metrics by Kidwell & Bosence (1991) indicated that the range in agreement of presence-absence data was largely an artifact of methodological differences among studies, compounded in many instances by undersampling of the live community.

Here I use a larger and more homogeneous collection of datasets to test the fidelity of both relative abundance and presence-absence data at the habitat scale, for environments ranging from vegetated marshes to muddy open shelves. Only half of these datasets are fully electronic (and another dozen are still in earlier stages of vetting and compilation), and so the present analysis is only a progress report (December 1998).

Materials and Methods

Table 1 lists the 17 molluscan datasets used in this analysis. Most studies examined more than a single habitat, and provide either sample-by-sample data on live and dead species or species lists already pooled by habitat. The numbers of samples per habitat range from one to more than 100, and species range from a few to several hundred. In pooling samples, "habitat" is defined operationally as a sedimentary environment that might be recognized as a distinct facies in the rock record. These groupings in almost all cases correspond to clusters based on the composition of the live fauna. A wide variety of bottom-sampling methods were used in the original studies: can cores and trenches in marshes and intertidal flats, and various grab, dredge, and SCUBA-operated suction methods in the subtidal. Samples thus vary widely in volume, depth of penetration (but generally no deeper than 20 cm), bias toward epifauna or deep infauna, and total numbers of specimens (in virtually all studies, live fauna and dead shells come from the same samples). Sieve size also varies (Table 1) along with methods of counting bivalve individuals.

The minimum requirement for a dataset in this re-analysis is unambiguous information on whether species occur alive, dead, or both alive and dead within a habitat. The most common short-coming in candidate datasets, especially older benthic surveys that were not conducted for the purpose of live-dead comparison, is the failure to indicate whether dead material is present when a species occurs alive: authors commonly note when a species is known only from dead material, but do not always stipulate the opposite in published lists. Whenever possible, original authors have been contacted to clarify these and other ambiguities in the composition or mapping of samples. In this same way some published presence-absence datasets have also been amended with relative abundance information and some truncated lists have been expanded to include rare species. Several commonly cited live-dead studies have had to be excluded because insufficient data survive for habitat-level tests (Parker 1963; Johnson 1965), because the original tallies mixed live and dead specimens (Straaten 1960), or because live and dead were retrieved from sieves of different sizes (e.g., central bay habitat of Lingwood 1976a but not other habitats from the same study area). Two other studies are excluded from this analysis (although included in the analysis of Kidwell & Bosence 1991) because the environment is known to have changed significantly through human intervention in the recent past. This includes Mevagissey Bay, England, which is changing from a muddy to a shell-gravel bottom following cessation of China clay waste water input in the 1970s (Knight 1988), and Canso Bay, Nova Scotia, where construction of a causeway in 1954 changed water circulation significantly (Wagner 1975). Live-dead agreement is tested using five different metrics, described with the results below.

Results

RESULTS OF TEST I: PERCENT OF LIVE SPECIES CONTRIBUTING DEAD MATERIAL

Almost all species sampled alive in a habitat are also present in the death assemblage (grand mean 89% ± 5, based on 80 habitat-level measures of live-dead agreement): 91% of live species in vegetated marshes and their tidal creeks, 86% in intertidal flat habitats, 95% in subtidal habitats of coastal embayments, and 85% in open marine sediments (Fig. 1). Restated, this is the percent of known live species richness captured by the local death assemblage. Means do not differ significantly among broad environmental groupings.

Species that are not present in the death assemblage ("live-only" species) are virtually all numerically rare and, in addition, tend to be small-bodied, exceptionally thin-shelled, composed of organic-rich shell types (such as nacreous aragonite), or all of these (e.g., species among small fragile ericynacean bivalves and rissoid and opisthobranch gastropods, and among fragile pinnid, anomiid, and solenid bivalves). These biases match those found by Valentine (1989) in his province-level comparison of living and Pleistocene molluscan faunas from the Californian Province.

RESULTS OF TEST II: FIDELITY OF DEAD INDIVIDUALS TO SPECIES' LIFE HABITATS

Although death assemblages typically contain 2–3X more species than are found alive in the same set of samples, most dead *individuals* in these death assemblages belong to species documented alive in the same habitat (82% ± 10 grand mean of 67 habitat-level measures). Ninety-three percent of dead individuals in marshes and creeks are from species that were also found alive, 76% in intertidal habitats (83% if the Inchon inner flat data point is excluded; live in this habitat consisted of a single specimen), 88% in coastal subtidal habitats, and 72% in open marine habitats (Fig. 2). Restated, this is the percent of the local death assemblage (dead individuals) that could be generated by the local live community based on the current composition of the live community, and shows that death assemblages are preferentially composed of species that are known to live locally.

The remaining dead individuals (from "dead-only" species) may have several explanations. Some are possible exotics, that is, specimens imported from other habitats after death, or possible relics exhumed from older deposits. Intertidal and open marine environments contain the largest proportional numbers of suspect dead individuals (16% and 23%, respectively). A significant

Table 1. Datasets used in analysis of live-dead agreement, indicating data source and number of habitats (bottom types) sampled.

	Vegetated Marsh, Tidal Creek	Intertidal Flats, Channels	Coastal Bay, Ria, Lagoon	Open Marine Shoreface, Shelf	Sieve Size	Limits on Numerical Abundance Data
Washington State to Baja California (MacDonald 1969)	11 paired sets				0.5 mm marsh, 1 mm creeks	
Inchon, Korea (Frey et al. 1988)		3			not specified	
Seto Sea, Japan (Tanabe et al. 1986)		2			5 mm	
Mugu Lagoon, California (Warme 1971 & unpubl. data)		4			3mm	
Choya Bay, Mexico (Fürsich & Flessa 1991)	1	4		1	3 mm	no data for rare dead taxa (≤1%)
Liverpool Bay, United Kingdom (Lingwood 1976a,b)		1		1	5 mm	presence–absence data only
Copano Bay & Laguna Madre, Texas (Staff et al. 1986)			2		0.5 mm	no data for rare taxa (≤1%)
Florida Bay, Florida (Turney & Perkins 1972)			8		1 mm	presence–absence data only
Mannin Bay, Ireland (Bosence 1979 & unpubl. data)			5		< 2mm	

Table 1. Continued.

	Vegetated Marsh, Tidal Creek	Intertidal Flats, Channels	Coastal Bay, Ria, Lagoon	Open Marine Shoreface, Shelf	Sieve Size	Limits on Numerical Abundance Data
Ria de Arosa, Spain (Cadée 1968 & unpubl. data)			4	1	2 mm	no data for rare taxa (≤1%)
Sapelo Island, Georgia (Henderson & Frey 1986)			2	1	1.5 mm	
Yucatan, Mexico (Ekdale 1972)			3	4	3 mm	
Gulf of Mexico, Texas (Staff & Powell 1999)				1	1 mm	no data for rare taxa (≤1%)
Oyster Ground, SE North Sea, Netherlands (Cadée 1984 & unpubl. data)				2	1mm	no data for rare taxa (≤1%)
Helgoland Bight, North Sea, Germany (Reineck et al. 1971)				3	6.3 mm	digitized from spindle diagrams
English Channel, SE United Kingdom (Carthew & Bosence 1986 & unpubl. data)				3 sets of shell gravels	2 mm	
Gulf of Gaeta, Italy (Dörjes 1971; Hertweck 1971)				3	8 mm	digitized from spindle diagrams

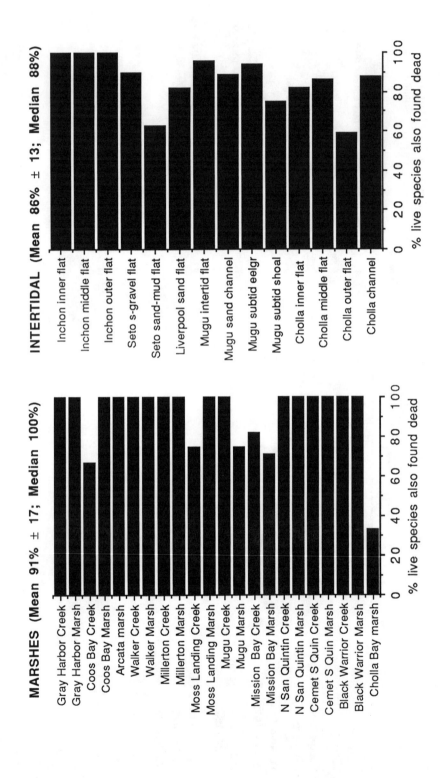

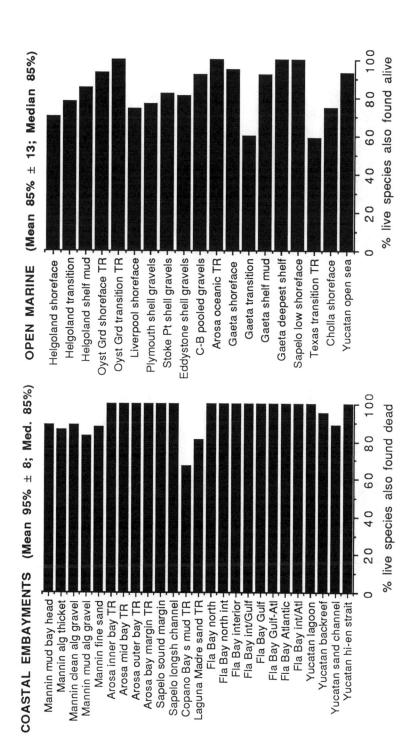

Fig. 1. Live-dead agreement as measured by the percent of live species that were also sampled dead in the same habitat ("% live species also present dead"). Each bar reflects a habitat-level dataset within a study area. Study areas are listed from north (top) to south (bottom), with habitats arrayed onshore to offshore within each study area. TR = truncated dataset lacking data on rare species, usually defined as <1% of individuals. The vast majority of mollusk species leave a record in the local death assemblage (grand mean of all habitat-level datasets is 89% ± 5; n = 80).

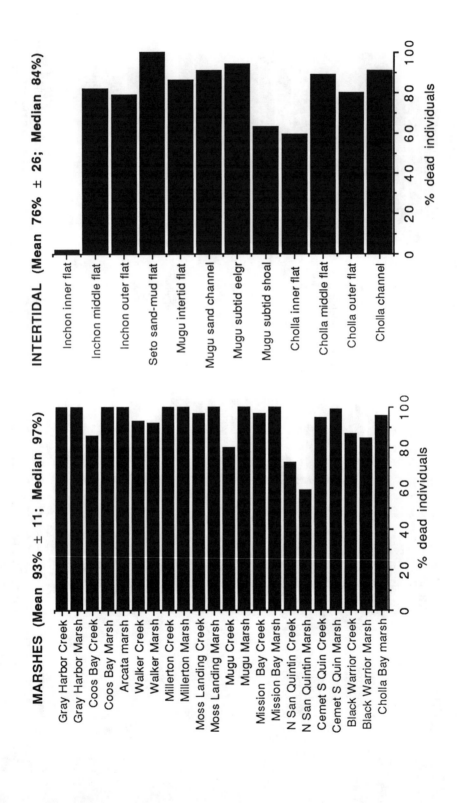

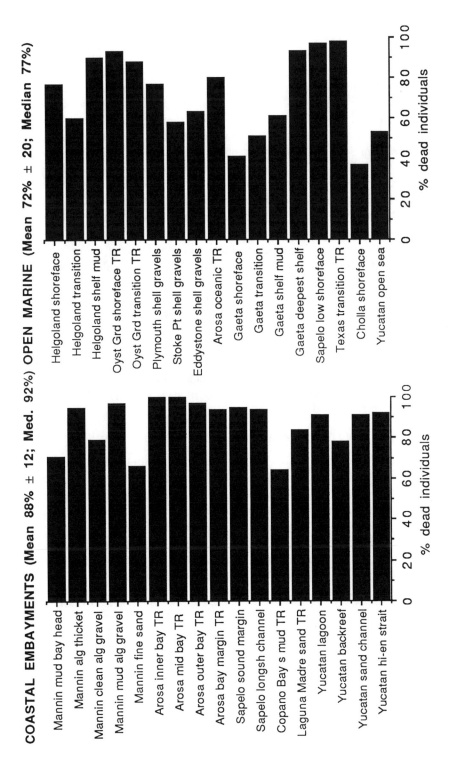

Fig. 2. Live-dead agreement and spatial fidelity as measured by the percent of dead individuals in the death assemblage that are from species also sampled alive in that habitat ("% dead individuals"). The vast majority of dead specimens are from species known to occur in the same habitat (grand mean of all habitat-level datasets is 82% ± 10; n = 67).

number of these might in fact be exotics or relicts, since these environments include the highest energy, most erosive, and most sediment-starved habitats (intertidal flat margins, various tidal channels, foreshore and shoreface sands, starved offshore shell gravels; cf. Cadée 1984). However, dead-only species might also result from undersampling of the live community: most live datasets are based on a single census of the local community and thus are almost certainly incomplete measures of actual live diversity. Single censuses are unlikely to sample all rare live species, especially those that are patchy or seasonal in occurrence within the habitat (cf. Peterson 1976). Opportunistic species, in particular, can depress live-dead agreement by this metric: they can contribute large numbers of dead shells both to local and to exotic habitats, since these shells tend to be small and relatively transportable, and even multiple censuses can fail to encounter live specimens (cf. Levinton 1970). In fact, large numbers of dead opportunistic bivalves flood a few of the habitat-level datasets included in this analysis (e.g., *Lentidium* transported into Gaeta shoreface, transition, and shelf mud facies; Hertweck 1971).

Even taken at face value, these results indicate that habitat preferences are far from obliterated by post-mortem transport. Most death assemblages are overwhelmingly dominated by demonstrably indigenous specimens (40 of 67 habitats measured have $\geq 85\%$ indigenous dead individuals; Fig. 2). In addition to almost always being rare, species not sampled alive tend to be small-bodied, epifaunal or epiphytic (thus raft-able), derived from shallower water habitats (including rocky shores and subtidal outcrops), or all of the above (e.g., seaweed-dwelling rissoid and other small gastropods; freshwater and land gastropods). The shared attributes of dead-only species provide further evidence for selective rather than wholesale post-mortem transport of individuals out of and into habitats. Thus, the most abundant species in a death assemblage are generally meaningful indicators of habitat preferences during life, with potential exceptions being predictable from species and/or habitat characteristics.

RESULTS OF TEST III: FIDELITY OF DEAD RANK ORDER TO THE LOCAL LIVE COMMUNITY

Species rank abundances in death assemblages are strongly correlated with census data from the local live community: in 13 of 16 habitat-level datasets tested so far, the null assumption that live and dead rank orders are mutually independent is rejected at $p < 0.05$ (Spearman rank test; Conover 1980; Fig. 3). Species that dominate the live community (in a single census) usually also dominate the dead; and species that are numerically rare alive are usually also rare dead. The three datasets with poor rank-order agreement have poorly known live faunas based on fewer than 50 live individuals (and pooling of only a few samples from the habitat), in contrast to the other datasets where live rank-order is based on hundreds to thousands of live individuals.

RESULTS OF (INVERSE) TEST IV: PERCENT OF DEAD SPECIES ALSO PRESENT ALIVE

The species richness of a death assemblage is generally 2–3X that of live fauna collected in the same sedimentary volume. Consequently, an impression of relatively poor live-dead agreement results if the metric is "% of dead species also present alive" (grand mean 46% ± 10, 80 habitat-level measures; Fig. 4). Marsh and creek habitats show the best live-dead agreement by this metric (mean 61%), probably because the numbers of species are so low (most datasets have < 10 species). However, the 95% C.I. is so large that this does not differ significantly from the situation in other environments: 42% of dead species are also found alive in intertidal habitats, 38% in coastal subtidal habitats, and 45% in open marine habitats (Fig. 4).

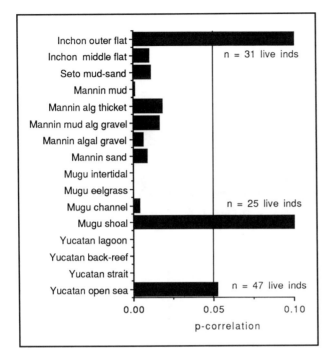

Fig. 3. Live-dead agreement in species' rank order: 13 of the 16 habitat-level datasets tested show no significant difference between live and dead species' rank order, indicating that death assemblages capture dominance information from the live community. The three exceptions have poorly known faunas based on fewer than 50 live individuals.

Across all environments, virtually all dead-only species are numerically rare (represented by only a few specimens and often in only a few samples from the habitat). This is consistent with the results of Test II above (grand mean 83% of dead individuals are from species that are also found alive). It also explains why truncated datasets, from which rare live and dead species have been excluded, usually yield significantly better live-dead agreement than full datasets by this same metric (note bars labeled TR in Fig. 4). For example, in the Ria de Arosa, Spain (Cadée 1968 and his unpublished data), the percentage of dead species also found alive in the open shelf habitat is 88% if rare species are excluded from the comparison but 30% using the full species lists. The same pattern holds for habitats inside the Ria (71% versus 37% in outer bay; 100% versus 25% in middle bay, 53% versus 22% in inner bay, 89% versus ~70% in bay margin; differences are plotted in Fig. 5).

There are many possible explanations for the large number of dead-only species in death assemblages ("excess dead richness"), and more than one may apply to a given dataset, as suggested by most of the original authors of these live-dead studies (references in Table 1). These factors include:

(1) undersampling of the live community. This is almost certainly a factor, since most live datasets in these studies are based on only a single census of the local community, and sediment samples typically yield far fewer live individuals than dead individuals (10–100X; see range in Fig. 6);

(2) enrichment of the death assemblage by exotics. This is almost certainly a factor near rocky substrata, which harbor distinctive fauna that are easily shed into adjacent soft-sediments, and in settings of high surge (e.g., along high-energy foreshores). Because a single exotic specimen adds a species, rare exotic specimens can have a large effect on dead richness but little effect on species relative abundance and rank-order; and

(3) enrichment of the death assemblage with relict shells of species that are now extinct locally. This is almost certainly a source of excess dead species richness in open marine sediments that are palimpsests of Holocene transgression (shelly sand and gravel veneers in shoreface and shelf waters). Relicts may be incorporated by exhumation from older deposits or by exceptionally

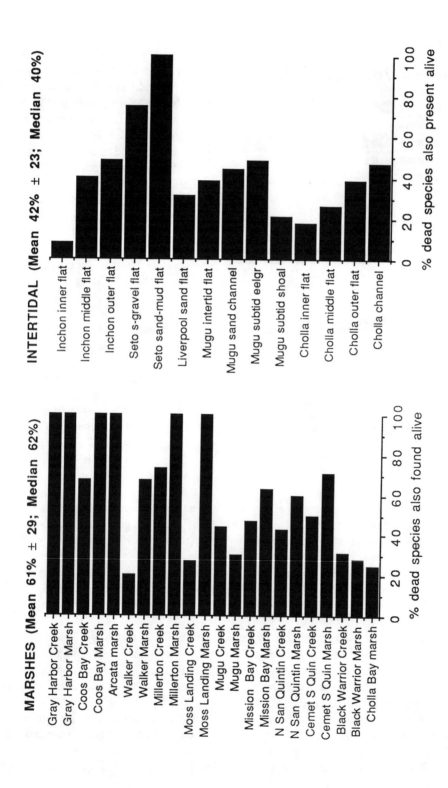

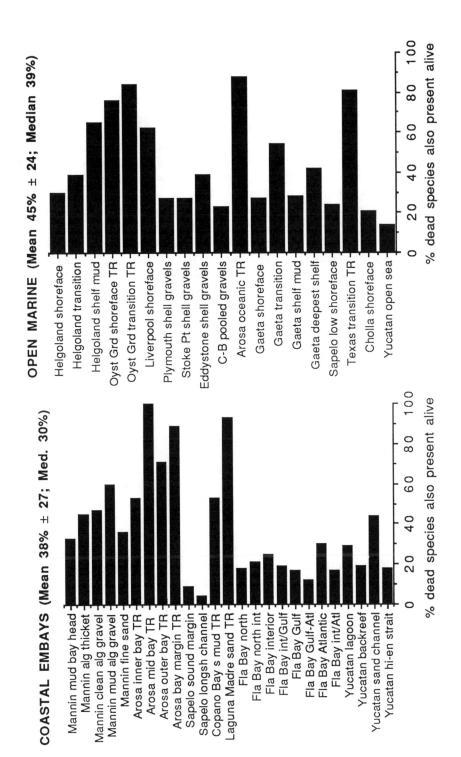

Fig. 4. Live-dead agreement as measured by the percent of dead species that are also present alive in the same habitat ("% dead species also present alive"). Bars as in Fig. 1. Death assemblages typically contain 2–3X the species richness of any single census of live fauna in the habitat, resulting in low live-dead agreement by this metric (grand mean of all habitat-level datasets is 46% ± 10; n = 80).

prolonged time-averaging of shell production, such as over periods of environmental change and local species extinction. Points 2 and 3 are probably part of the explanation why, among all open marine environments, shoreface sands and shell gravels consistently have greater excess dead species (lower % dead species found alive) than do fine-grained habitats (Fig. 5).

Before attributing live-dead discrepancies to taphonomic bias (points 2 and 3 above), the impact of possible undersampling of the live fauna must be evaluated. Pooling of data from replicate censuses of live communities over time, in order to capture indigenous but ephemeral species, is one method of improving species inventories. The few live-dead datasets where single study areas were subjected to replicate sampling programs do show a positive effect on live-dead agreement, supporting a hypothesis of sampling bias (Kidwell & Bosence 1991). In warm-temperate lagoons of southern California, for example, live-dead agreement by this metric (% of dead species also found alive) improved from 20% to 75% as live species richness accrued over 3 yr of successive benthic surveys (Peterson 1976), and a similar but smaller effect was found in open shelf habitats by successive pooling of three censuses over 17 yr (from 37% to 54%; Knight 1988) and of seven censuses over 90 yr (from 23% to 58%; Carthew & Bosence 1986) (calculations by Kidwell & Bosence 1991) . In all three studies, the pooled live data were compared with dead data from a single set of sediment samples. Longer-term sampling programs in more areas are needed to fully test this effect on live-dead agreement.

Increasing the number of samples per habitat also improves inventories of live species, and, thus, one might expect live-dead agreement to be higher among datasets based on dense sampling arrays. Within a habitat, live-dead agreement does increase with pooling of additional samples, up to the maximum number of samples available. However, the results are mixed when different habitat-level datasets are plotted on a single graph, probably because samples vary so much in type among studies and even among habitats in a single study (collecting gear, sieve size, etc.), yielding a large amount of scatter (Fig. 5). For example, marsh, intertidal, and open marine datasets show no correlation between live-dead agreement and sampling density. In coastal embayment habitats, datasets based on larger numbers of samples do have higher live-dead agreements when complete live and dead species lists are compared (truncated species lists consistently generate better live-dead agreement; Fig. 5). In open marine seafloors, shell gravels and shoreface sands generally exhibit lower live-dead agreement by this metric than do fine-grained open shelf substrata. This is consistent with their containing a significant number of relict shells, as mentioned above, but it may also be a consequence in part of shell gravels supporting fewer live individuals than adjacent soft substrata (e.g., Allen 1899), so that the live community is more difficult to inventory.

Live-dead agreement by this metric is correlated positively with the ratio of live to dead individuals, attaining 40–70% levels in datasets where the abundances of live and dead individuals are about equal (truncated datasets excluded from analysis; Fig. 6A). For datasets with larger numbers of live individuals, live-dead agreement rises to 70–100% (Fig. 6A). In contrast, the proportion of live and dead individuals within a dataset has no effect on the percent of live species found dead, which is consistently high (same metric as in Test I; Fig. 6B), showing that the correlation for percent dead species in Fig. 6A is informative. In future work, cumulative species curves will be generated to establish how well these numbers of live and dead individuals capture local or regional richness. But, even the bivariate plot presented here (Fig. 6A) shows the strong negative effect on live-dead agreement caused by undersampling live fauna.

These analyses indicate that the excess dead species richness (large numbers of dead-only species) observed in virtually all datasets are in many instances artifacts of undersampling of the local live community, rather than indicators of severe taphonomic bias in the death assemblage. Correlations between live-dead agreement and various measures of sampling intensity indicate that most dead species richness could accrue through normal time-averaging of spatially and temporally patchy populations indigenous to the habitat, such as might be sampled alive through an ambitious program of replicate censusing.

RESULTS OF (INVERSE) TEST V: OBSERVED LIVE-DEAD AGREEMENT COMPARED TO CALCULATED MAXIMUM POSSIBLE AGREEMENT

Given that so many datasets suffer from undersampling of live faunas, a more suitable method of estimating live-dead agreement might be to compare the observed "% dead species also found alive" (number of dead species also present alive, divided by total number of dead species) against the maximum agreement that is possible in the same dataset (total number of live species, divided by total number of dead species; assumes maximum overlap of the two lists). After all, if a total of only 10 live species are known from the habitat, and dead species richness is 20, then the maximum intersection of dead species with live species (the maximum possible % dead species also found alive) is 50%. A live-dead agreement of only 50% would, in this instance, coincide with the upper mathematical boundary for agreement (100% of maximum agreement possible). If only 40% of dead species are also known alive (that is, some species are live-only), then live-dead agreement is only 80% of the maximum possible, and so on.

Figure 7 plots observed live-dead agreements against their calculated maximum possible agreements; contour lines show how closely individual habitat-level datasets approach maximum possible agreement. In all environmental groupings, virtually all habitat-level datasets plot within 25% of their maximum possible agreement levels (grand mean 88%; graphs include values calculated from both full and truncated species lists where both are available). Live-dead agreement is nearly constant across the 80 habitat-level datasets by this metric, with very few outliers below 75% agreement values.

Conclusions

The tendency for live-dead agreement to rise to ~75% among several different metrics for presence-absence data suggests that 25% is a reasonable liberal estimate of the proportion of dead species richness that might be exotic or relict in origin. This would be a *maximum* estimate of taphonomic bias in death assemblage species richness, because it is possible that some portion of these 25% dead-only species are indeed indigenous but have simply not yet been encountered alive by the sampling program ("stubborn ecological noise"). Only programs of prolonged replicate sampling of the live community, comparable to scales of time-averaging in the death assemblage, will generate live species lists adequate to test this rigorously. However, some portion of this 25% excess dead species richness will comprise species that are truly exotic to the local habitat, and some portion will be relict, especially in habitats with slowly aggrading or erosive seafloors (see Kidwell & Bosence 1991 and Kidwell & Flessa 1995 for discussions of time-scales of faunal mixing in these various circumstances).

It should be realized that various inverse measures of live-dead agreement on species' presence-absence are the most pessimistic assessments of death assemblage fidelity. For example, truncated datasets that exclude rare species (species comprising < 1% of individuals) can yield ~75% or better live-dead agreement levels, even when based on small numbers of samples.

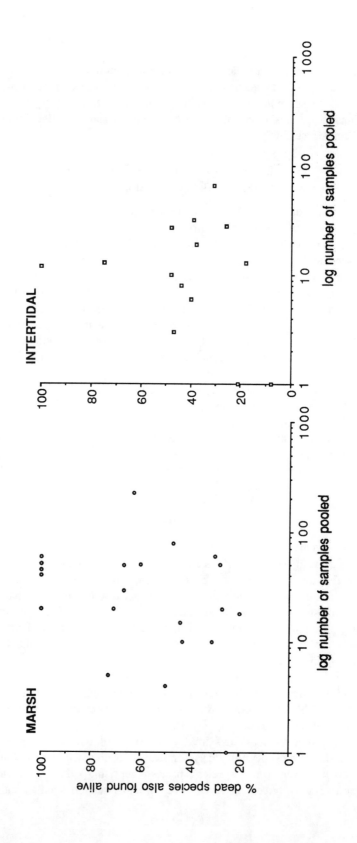

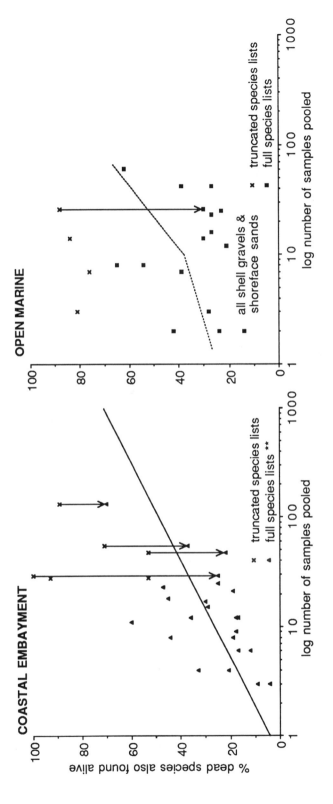

Fig. 5. The percentage of dead species also present alive shows no significant variation with number of samples except for habitats in coastal embayments (** significant at $p < 0.01$). Truncated datasets (x's) consistently have the highest live-dead agreements; arrows point to the lower values that are calculated when full (complete) species lists are used. Among open marine datasets, live-dead agreement is consistently lower in shell-gravels and shoreface sands (points below dashed line) than in the silty sands and muds of transition zone and offshore habitats.

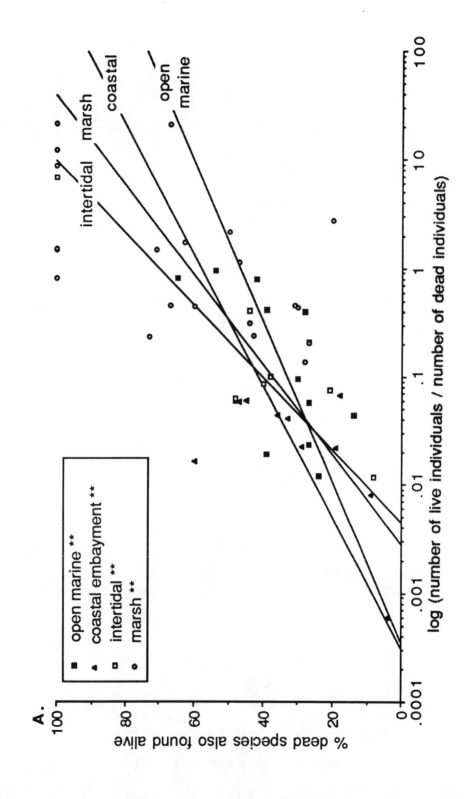

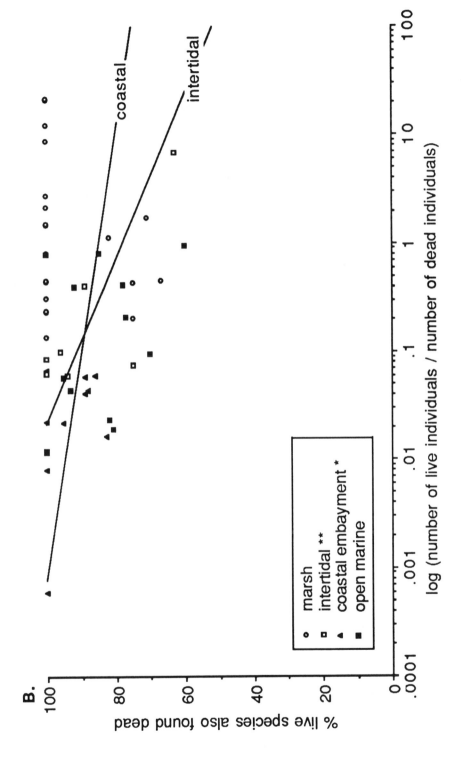

Fig. 6. A) Live-dead agreement as measured by the percent of dead species also found alive is correlated with the ratio of live to dead individuals in the dataset, indicating that much of the excess species richness in death assemblages is an artifact of undersampling of local live fauna. B) In contrast, the percent live species also found dead is unaffected by the proportion of live and dead individuals in the dataset. Truncated datasets not included in analysis. * significant at $p < 0.05$; ** significant at $p < 0.01$.

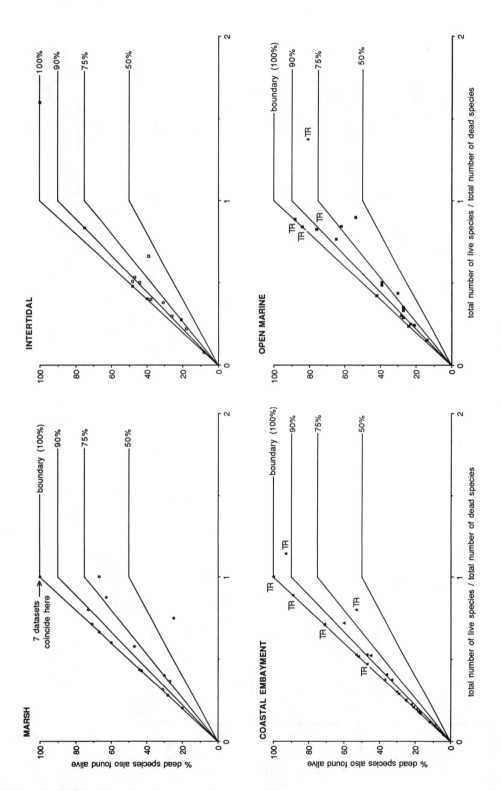

Fig. 7. To evaluate the effect on live-dead agreement of undersampling live fauna, observed values of percent dead species also present alive, divided by total number of dead species (number of dead species also present alive, divided by total number of dead species) are plotted against the maximum possible percent of dead species that can also be present alive (total number of live species, divided by total number of dead species in the same dataset). The majority of datasets have observed live-dead agreements that are within 75% of the calculated maximum agreement possible (grand mean of all habitat-level datasets is 8%; n = 71).

Table 2. Summary statistics by general environmental zone (mean and standard error, median in square brackets, and number of habitat-level datasets).

	marshes	intertidal flats	coastal embay- ments	open marine seafloors	grand means
I. % live species also present dead	91 ± 17 [100] (n = 22)	86 ± 13 [88] (n = 14)	95 ± 8 [100] (n = 25)	85 ± 13 [85] (n = 19)	89 ± 5 (n = 80)
II. % dead individuals from species also present alive	93 ± 11 [97] (n = 21)	76 ± 26 [84] (n = 12)	88 ± 12 [92] (n = 17)	72 ± 20 [77] (n = 17)	82 ± 10 (n = 67)
IV. % dead species also present alive	61 ± 29 [62] (n = 22)	42 ± 23 [40] (n = 14)	38 ± 27 [30] (n = 25)	45 ± 24 [39] (n = 19)	46 ± 10 (n = 80)
V. % dead species scaled to quality of live data	85 (n = 22)	87 (n = 14)	93 (n = 16)	85 (n = 19)	88 (n = 71)

Moreover, forward tests of the contribution of live faunas to death assemblages yield consistently very high live-dead agreement: across all tested habitats, the overwhelming majority of live species (grand mean 89%, Test I) leave identifiable dead material in the life habitat, and the majority of dead individuals (grand mean 82%, Test II) are from species that occur alive in the same habitat (Table 2). Live species thus do leave clear records of their presence in local death assemblages, and post-mortem processes do not homogenize the original habitat distribution of species. Although a more systematic investigation is required, live-only species are apparently not a random draw from the species pool but tend to be rare and exceptionally small or fragile, and thus most absences of species in death assemblages should be ecologically meaningful, especially if morphologically comparable species are present.

Because most dead-only species are rare, excess dead species richness does not compromise the reliability of rank abundance signals in death assemblages: rank orders of species do not differ significantly between live and dead ($p < 0.05$, Test III). Even a few samples from a death assemblage can apparently provide reliable dominance information on the local source live community, with taphonomic uncertainty focused on rare dead-only species. Thus in death assemblages, rank abundance data (and thus presumably the diversity indices based on them) typically show greater live-dead agreement than species richness measures based on presence-absence.

Based on available datasets, molluscan death assemblages thus capture both (1) dominance information, which is not significantly different from a snapshot census of the local community, and (2) information on the time-averaged species richness of the community, including indigenous species that are patchy, ephemeral, and/or sparse in occurrence and thus difficult to inventory in the live community without intense sampling programs. Whether such death assemblage data are deemed adequate depends on the ecological or paleoecologic question at hand—local needs will determine the degree of uncertainty that is acceptable in reconstructing the original biological signal. However, the outlook for practical applications is very good based on these provisional results: molluscan death assemblages show very high quantitative fidelity to the life habitats, taxonomic composition, and dominance structure of local live communities, and the ambiguity in species richness resides among the rare species, as it does even in ecological studies of live communities. Given the great abundance of dead shells compared to live individuals in sediments of most coastal and open marine settings, and the greater ease in processing and archiving such material, death assemblages are extremely promising as a rapid and relatively inexpensive method of benthic community assessment.

Acknowledgments

I am extremely grateful for the encouragement and access to unpublished data provided by the original authors of the live-dead studies. Thanks also to M. M. R. Best for designing the database structure, to M. M. R. Best, D. Jablonski, A. I. Miller, and the Chicago Ev Morph group for valuable feedback and reviews, and to Don Rhoads for his enthusiasm for all things along the sediment-water interface.

Literature Cited

Allen, E. J. 1899. On the fauna and bottom-deposits near the thirty-fathom line from the Eddystone Grounds to Start Point. *Journal of the Marine Biological Association of the United Kingdom, New Series* 4:365–541.

Bosence, D. W. J. 1979. Live and dead faunas from coralline algal gravels, Co. Galway, Eire. *Palaeontology* 22:449–478.

Cadée, G. H. 1968. Molluscan biocoenoses and thanatocoenoses in the Ria de Arosa, Galicia, Spain. *Zoologische Verhandelingen* (Rijksmuseum van Natuurlijke Historie te Leiden) 95:1–121.

Cadée, G. H. 1984. Macrobenthos and macrobenthic remains on the Oyster Ground, North Sea. *Netherlands Journal of Sea Research* 18:160–178.

Carthew, R., & D. W. J. Bosence. 1986. Community preservation in Recent shell-gravels, English Channel. *Palaeontology* 29:243–268.

Conover, W. J. 1980. Practical Nonparametric Statistics, 2nd Ed. Wiley, New York.

Dörjes, J. 1971. Der Golf von Gaeta (Tyrrhensiches Meer). IV. Das Makrobenthos und seine küstenparallele Zonierung. *Senckenbergiana maritima* 3:203–246.

Ekdale, A. A. 1972. Ecology and paleoecology of marine invertebrate communities in calcareous substrates, northeast Quintana Roo, Mexico. M.S. thesis, Rice University, Houston, Texas.

Frey, R. W., J.-S. Hong, & W. B. Hayes. 1988. Physical and biological aspects of shell accumulation on a modern macrotidal flat, Inchon, Korea. *Netherlands Journal of Sea Research* 22:267–278.

Fürsich, F. T. & K. W. Flessa. 1991. Ecology, taphonomy, and paleoecology of Recent and Pleistocene molluscan faunas of Bahia la Choya, northern Gulf of California. *Zitteliana* (Abhandlungen der Bayerischen Staatssammlung für Paläontologie und historische Geologie) 18:1–180.

Henderson, S. W. & R. W. Frey. 1986. Taphonomic redistribution of mollusk shells in a tidal inlet channel, Sapelo Island, Georgia. *Palaios* 1:3–16.

Hertweck, G. 1971. Der Golf von Gaeta (Tyrrhensiches Meer). V. Der Biofaziesbereiche in den Vorstrand- und Schelfsedimenten. *Senckenbergiana maritima* 3:247–276.

Johnson, R. G. 1965. Pelecypod death assemblages in Tomales Bay, California. *Journal of Paleontology* 39:80–85.

Kidwell, S. M. & D. W. J. Bosence. 1991. Taphonomy and time-averaging of marine shelly faunas, p. 115–209. *In* Allison, P. A. & D. E. G. Briggs (eds.), Taphonomy: Releasing the Data Locked in the Fossil Record. Plenum Press, New York.

Kidwell, S. M. & K. W. Flessa. 1995. The quality of the fossil record: Populations, species, and communities. *Annual Review of Ecology and Systematics* 26:269–299.

Knight, A. P. 1988. Ecological and sedimentological studies on China clay waste deposits in Mevagissey Bay, Cornwall. Ph. D. Thesis, University of London, Royal Holloway and New Bedford College, United Kingdom.

Levinton, J. S. 1970. The paleoecological significance of opportunistic species. *Lethaia* 3:69–78.

Lingwood, P. F. 1976a. The marine Mollusca of Liverpool Bay (Irish Sea*). Journal of Conchology* 29:51–56.

Lingwood, P. F. 1976b. The biostratinomy of some inshore marine assemblages. Ph. D. thesis, University of Liverpool, United Kingdom.

MacDonald, K. B. 1969. Quantitative studies of salt marsh mollusc faunas from the North American Pacific Coast. *Ecological Monographs* 39:33–60.

Parker, R. H. 1963. Zoogeography and ecology of some macroinvertebrates, particularly mollusks, in the Gulf of California and the continental slope off Mexico. *Videnskabelige Meddelelser fra Dansk Naturhistorisk Forening* (Kjøbenhavn) 126:1–178.

Peterson, C. H. 1976. Relative abundance of living and dead molluscs in two California lagoons. *Lethaia* 9:137–148.

Powell, E. N., G. M. Staff, D. J. Davies, & W. R. Callender. 1989. Macrobenthic death assemblages in modern marine environments: Formation, interpretation, and application. *CRC Critical Reviews in Aquatic Sciences* 1:555–589.

Reineck, H.-E., J. Dörjes, S. Gadow, & G. Hertweck. 1971. Sedimentologie, Faunenzonierung und Faziesabfolge vor der Ostküste der inneren Deutschen Bucht. *Senckenbergiana lethaea* 49:261–309.

Staff, G. M., R. J. Stanton, Jr., E. N. Powell, & H. Cummins. 1986. Time averaging, taphonomy and their impact on paleocommunity reconstruction: Death assemblages in Texas bays. *Geological Society of America Bulletin* 97:428–443.

Staff, G. M. & E. N. Powell. 1999. Onshore-offshore trends in community structural attributes: Death assemblages from the shallow continental shelf of Texas. *Continental Shelf Research* 19:717–756.

Straaten, L. M. J. U. van. 1960. Marine mollusc shell assemblages of the Rhone delta. *Geologie en Mijnbouw* 39:105–129.

Tanabe, K., T. Fujiki, & T. Katsuta, 1986. Comparative analysis of living and death bivalve assemblages on the Kawarazu shore, Ehime Prefecture, west Japan. *Bulletin of the Japanese Association of Benthology* 30:17–30.

Turney, W. J. & B. F. Perkins. 1972. Molluscan distribution in Florida Bay. *Sedimenta* III. Comparative Sedimentology Laboratory, University of Miami, Miami, Florida.

Valentine, J. W. 1989. How good was the fossil record? Clues from the Californian Pleistocene. *Paleobiology* 15:83–94.

Wagner, F. J. E. 1975. Mollusca of the Strait of Canso area. *Geological Survey of Canada Paper* 75-23:1–22.

Warme, J. E. 1971. Paleoecological aspects of a modern coastal lagoon. *University of California Publications in Geological Sciences* 87:1–110.

Sources of Unpublished Material

Bosence, D. W. J. unpublished data. Supplemental Publication No. SUP 14012 (1979), British Library, Boston Spa, Wetherby, Yorkshire, United Kingdom.

Cadée, G. C., unpublished data. Netherlands Institute of Sea Research, P.O. Box 59, 1790 AB Den Burg, Texel, The Netherlands.

Carthew, R. & D. W. J. Bosence. unpublished data. Department of Geology, Royal Holloway and Bedford New College, University of London, Egham, Surrey, United Kingdom.

Tanabe, K. unpublished data. Geological Institute, Faculty of Science, University of Tokyo, Tokyo, Japan.

Warme, J. E. unpublished data. Department of Geology, Colorado School of Mines, Golden, Colorado.

The Influence of Mangrove Biomass and Production on Biogeochemical Processes in Tropical Macrotidal Coastal Settings

Daniel M. Alongi

Abstract: *The influence of mangrove biomass and production on benthic decomposition processes and sediment chemistry was examined in mangrove forests in different macrotidal coastal settings of the Kimberley region of Western Australia. The forests were (1) a dense Aegiceras corniculatum forest in a ria embayment; (2) a mature Rhizophora stylosa forest in a ria delta; (3) an immature Avicennia marina stand in an open river delta; and (4) a mature Avicennia marina forest in a carbonate-dominated bay. There were no clear differences among forests in rates of O_2 and CO_2 flux measured across exposed sediment surfaces, but total carbon mineralization rates (the sum of the individual decomposition pathways) were generally greater in the larger and mature forests than in the smaller and immature stands. Redox potential and individual decomposition pathways indicated that sediments were dominated by aerobic respiration (56–80% of total carbon oxidation) and, to a lesser extent, by sulfate reduction (13–43% of total carbon oxidation). Manganese and iron reduction were minor processes. No methane was detected. There was some evidence of loss of oxidized carbon, perhaps via lateral advection or groundwater flow facilitated by the large tides. Rates of total carbon oxidation, oxygen consumption, and sulfate reduction correlated positively with canopy production, as did sulfate reduction with dead root biomass, implying a general link between organic matter availability and mineralization. Total carbon mineralization (range: 48–93 mmol C m^{-2} d^{-1}) equaled only 8–10% of net canopy production (range: 642–917 mmol C m^{-2} d^{-1}), suggesting that most mangrove carbon is either refractory and retained within the forest as standing biomass or is lost via tidal export. The physiochemical and biogeochemical properties of these mangrove deposits are attributed to extensive drainage and flooding cycles over a large tidal prism, to intensive bioturbation by crabs, and to mangrove physiological activities via extensive roots and rhizomes.*

Introduction

Mangroves are the dominant organisms modifying the physical and biogeochemical properties of intertidal sediments along many of the world's subtropical and tropical coastlines. Mangroves are trees or shrubs with deep, extensive roots and rhizomes and much greater aboveground biomass that is, on average, more productive than most other intertidal vegetation (Alongi 1998). Despite the importance of mangroves in coastal food chains, in stabilizing shorelines, and as nutrient filters, their impact on sediment biogeochemistry is poorly understood (Wiebe 1989; Sherman et al. 1998). This is reflected in conflicting evidence of the influence of mangrove forest age and type on early diagenesis of sediment organic matter and nutrient recycling (e.g., Middelburg et al. 1996; Alongi et al. 1998).

An understanding of mangrove-sediment relations is crucial because such interactions are likely to be important in regulating tidal water flow, sediment transport, and element cycling within mangrove waterways, and between mangroves and adjacent coastal waters (Alongi 1998). A comparison of functional similarities and differences among mangrove forests in various coastal settings would provide valuable clues as to the importance of the role of these organisms in coastal biogeochemical processes.

Four forests inhabiting different coastal settings (ria, riverine delta, carbonate-dominated bay) in the remote Kimberley region of Western Australia were studied to determine the influence of various ages and types of mangroves on sediment biogeochemistry. The Kimberley coast is a high-energy environment, supporting a mosaic of coastal habitats within a narrow corridor that reflects sharp boundaries in climate, geology, fluvial input, and mangrove vegetation (Semeniuk 1993). The most dominant oceanographic feature is a large semidiurnal tidal range of 6.0–11.5 m. Although this is the largest tidal range in the southern hemisphere, extensive mangrove forests in northwest India, Burma, and northwest Brazil are similarly subjected to large tidal ranges, exceeding 5 m (Spalding et al. 1997). Biogeochemical processes in such mangrove settings are unknown.

Most benthic studies of mangrove biogeochemistry have been conducted in forests inhabiting low- to moderate-energy settings, with tidal ranges usually not exceeding 2.5 m (Robertson & Alongi 1992). It is likely, therefore, that the large semidiurnal flux of tidal water percolating through the Kimberley sediments ameliorates anoxic conditions and decomposition pathways compared with mangroves in lower energy settings where sulfate reduction usually dominates early diagenesis (Nedwell et al. 1994; Kristensen et al. 1995; Middelburg et al. 1996). Moreover, the rates and pathways of early diagenesis and sediment chemistry were examined in relation to aboveground and belowground biomass and net forest productivity. This study was conducted in austral summer (January 1998) when plant and microbial activities are expected to be most intense.

Study Location

Two mangrove forests were situated in different embayments, or rias, of Strickland Bay in the Buccaneer Archipelago of the Kimberley region (Fig.1). The archipelago is a rugged coastline with many rias and islands dominated by sandstone ridges (Semeniuk 1985). These coastal embayments are categorized as rias as they are funnel-shaped estuaries formed by postglacial sea-level transgression, dissecting a rocky coast of moderate to high (> 10 m) relief (Chappell & Woodroffe 1994).

Station SB 1 is a dense thicket of the river mangrove, *Aegiceras corniculatum* (L.) Blanco, located within an unnamed ria of the bay. Station SB 2 is a mature forest of the red mangrove, *Rhizophora stylosa* Griff., within the delta of the Kammargoorh River in an unnamed ria. The third forest, station KS 3, is a immature stand of the grey mangrove, *Avicennia marina* (Forsk.) Vierh., located on the steep banks of Mary Island North, situated off the mouth of the Fitzroy River in King Sound (Fig. 1). This stand sits atop a riverbank composed of a series of alternating steppes of clay and fine quartz sand. This site is fluvially dominated, but river runoff is negligible except in the wet season (Semeniuk 1980). The fourth forest, station BR 4, is a mature *Avicennia marina* stand located at a creek mouth in Roebuck Bay near Broome. The bay borders the Great Sandy Desert and receives

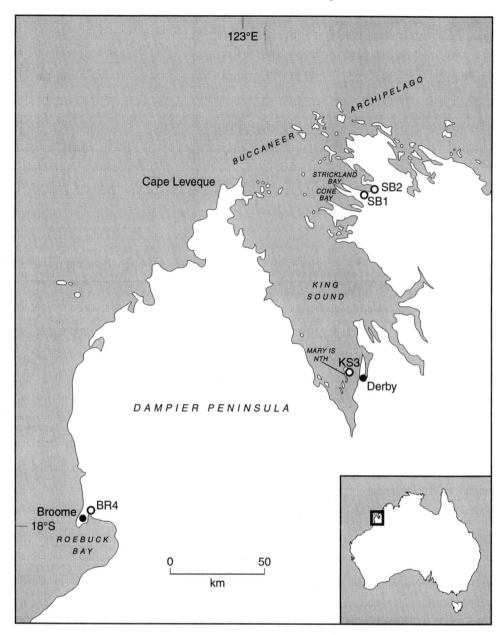

Fig. 1. The location of the four mangrove forests in the Kimberley region of Western Australia.

little, if any, terrestrial input. Sediments of the bay are dominated by carbonate deposits and, in the mangrove-lined creeks, by mixed carbonate mud-quartz sand facies (Semeniuk 1993). Owing to the large tides and the fact that mangroves grow at and above mean sea level, all four forests are situated atop high, stratified embankments (Jennings & Coventry 1973; Semeniuk 1980, 1985).

Wave energy is comparatively low. The four forests reflect a climatic gradient from subhumid (SB 1 and SB 2; 800–900 mm annual rainfall) to semi-arid (KS 3 and BR 4; 500–700 mm annual rainfall). Mean air temperatures range from 31°C to 39°C, with maximum summer temperatures often exceeding 42°C (Beard 1979). During this study, mean water temperature of adjacent tidal waters ranged from 28°C to 31°C and mean salinity ranged from 27.1 to 29.2.

Methods

MANGROVE BIOMASS AND PRODUCTIVITY MEASUREMENTS

A rectangular plot was staked out within each forest. The area occupied by 100 trees delimited the size of each plot: 48.6 m^2 at SB 1, 324.0 m^2 at SB 2, 600.0 m^2 at KS 3, and 399.6 m^2 at BR 4. The number of trees and aboveground biomass were estimated using the procedures and allometric equations in Clough and Scott (1989) and Clough et al. (1997a). Allometric relationships are unknown for *A. corniculatum* but were assumed, based on similar tree morphology, to be identical to those for *A. marina*. Net primary production was estimated using the light interception method as described by Clough (1997). Briefly, measurements of light absorption by the forest canopy (>100 light readings per plot on a sunny day at noon) were used to estimate leaf area index, L (= m^2 leaf area m^{-2} ground area), using the formula

$$L = [\log_e (I)_{mean}] - [\log_e (I_o)_{mean}]/-k$$

where:

$(I)_{mean}$	=	the mean photosynthetically active radiation (PAR) under the canopy
$(I_o)_{mean}$	=	incident PAR
k	=	canopy light extinction coefficient (0.5).

L was corrected to a solar zenith angle (θ) of 20° for the latitude of these forests. The leaf area index (L) was then converted to net canopy photosynthesis (P_N) using the formula

$$P_N = A \times d \times L$$

where:

A	=	average rate of photosynthesis per unit leaf area, assumed here to be 0.216 gC m^{-2} h^{-1} (Clough et al. 1997b; Clough & Sim 1989)
d	=	daylength (h)

Belowground fine-root biomass was estimated by taking three replicate cores (6 cm id) in each plot to a sediment depth of 1 m. Each core was subdivided by 2-cm intervals to 40 cm, then by 5-cm intervals to 1 m. Roots were washed to remove sediment and debris and then frozen until analysis. In the laboratory, roots were thawed, and live and dead roots were separated using the colloidal silica method of Robertson and Dixon (1993).

SEDIMENT CHARACTERISTICS

All samples were taken from the forest floor within each plot using field and laboratory methods detailed in Alongi (1996) and Alongi et al. (1996, 1998, 1999). Briefly, duplicate samples for grain size and water content were taken every 10 cm to a depth of 50 cm using 50-cc syringes with the needle ends cut off. Temperature, redox potential, and pH were measured at 2-cm intervals from duplicate 1-m cores (6 cm id). Samples for interstitial water were taken using the 1-m corer. Cores were often < 1 m due to sediment and root compaction, but most cores were 50–90 cm deep. Porewater samples were obtained by cutting cores under a N_2 atmosphere and squeezing sediment cakes (cut at 2-cm intervals) in a Teflon porewater apparatus. Samples were analyzed for SO_4, Cl, Fe, Mn, NH_4, $NO_2 + NO_3$, PO_4, DOC, and ΣCO_2. Separate samples were analyzed for CH_4 (Aller et al. 1996; Ferdelman et al. 1997). The same squeezed cakes and live and dead roots were dried, ground, and then analyzed for total organic carbon (TOC), total carbon (TC), and total nitrogen (TN). TOC was measured on a Beckman TOC Analyzer, and TC and TN on a Perkin Elmer 2400 CHNS/O Series II Analyzer.

FLUX MEASUREMENTS

Rates of Fe and Mn reduction were estimated using a core incubation method (Aller et al. 1996). Briefly, two sets of duplicate 20-cm-long cores (7 cm id) were subdivided into 4-cm-long sections in a N_2-saturated glove box. Each subsection from each set of cores was placed into a sterile plastic box containing 20 ml of a deoxygenated Na_2MoO_4 (20 mM) solution to inhibit sulfate reduction. Sediments in each box (two boxes per 4-cm interval) were mixed and subsamples were taken for determination of porewater Fe and Mn on days 0, 1, and 3. Total dissolved Fe and Mn were determined on a Varian Liberty 220 ICP-AES.

Rates of sulfate reduction were measured on triplicate 2.7-cm diameter cores taken from each plot using the core injection technique (Fossing & Jorgensen 1989). A two-step distillation procedure was used to determine the fraction of reduced radiolabel in the acid-volatile sulfide (AVS = free sulfide, FeS) and chromium-reducible sulfur (CRS = S^0, FeS_2) pools. Due to compaction, core depths varied among sites as follows: SB 1 = 32 cm, SB 2 = 50 cm, KS 3 = 100 cm, and BR 4 = 38 cm.

Gas (O_2, CO_2, CH_4) and solute (ΣCO_2, DOC, Mn, Fe, HS-, Ca, SiOH, PO_4, DOP, DON, NH_4, $NO_2 + NO_3$) fluxes were measured via the glass chambers placed in the replicate box core (0.027 m^2) samples taken from each plot. The samples were immediately incubated onboard ship in a shaded water bath maintained at ambient seawater temperature (Alongi et al. 1996, 1998, 1999). Water used in all experiments was taken from the mangrove waterways closest to each site.

Gas exchange across the air-sediment interface was measured in clear and opaque chambers to estimate benthic respiration and gross primary production during air-exposed periods (12 h d^{-1}). Benthic respiration from submerged sediments was estimated from the ΣCO_2 flux. An estimate of daily benthic respiration (total carbon oxidation, T_{cox}) at each station, taking into account the effect on sediment of roughly one-half day exposure to air and one-half day submergence, was derived by averaging the CO_2 (exposed condition) and ΣCO_2 (submerged condition) flux rates. Gas measurements were made using a MTI Analytical Instruments P200 gas chromatograph and checked using certified standards. Solutes were measured across the sediment-water interface in separate sets of opaque chambers. ΣCO_2 was determined by the difference in dissolved carbon values pre and post acid digestion (Alongi et al. 1998). All flux measurements were made at 30-min intervals for 3 h.

Differences in parameters among sites were tested using one-way ANOVA and SNK tests. Linear regression analysis and Pearson product-moment correlation were used to explore relationships among sediment and forest variables (Sokal & Rohlf 1995). A significance level of $p = 0.05$ was accepted.

Results

FOREST BIOMASS AND PRODUCTIVITY

The densest vegetation was the *A. corniculatum* stand (SB 1), but the largest stand based on biomass and productivity was the mature *R. stylosa* forest (SB 2), with aboveground biomass estimated at 636 metric tons DW ha^{-1} (Table 1). Both of these forests also had the most extensive belowground roots (Table 1 & Fig. 2). The mature *A. marina* forest (BR 4) had a larger and more productive canopy than the immature *A. marina* stand (KS 3), but the immature forest had proportionally more live than dead roots and greater total root biomass (Table 1). Fine roots of *A. corniculatum* (SB 1) extended to, and peaked at, a depth of 70 cm (Fig. 2). In the mature *A. marina* forest (BR 4), root biomass declined significantly below 30 cm depth; in the immature *A. marina* forest (KS 3), fine roots were found to a depth of 90 cm. Most fine roots of the mature *R. stylosa* trees (SB 2) were found at a depth of 40–100 cm (Fig. 2).

Table 1. Tree density, biomass, and net productivity of mangroves in the Kimberley region. Density = stems ha^{-1}; aboveground biomass (AGB), living belowground biomass (LBGB), and dead belowground biomass (DBGB) = metric tons DW ha^{-1}; net primary production (NPP) = mol C ha^{-2} d^{-1}.

Station	Density	AGB	LBGB	DBGB	NPP
SB 1	20,560	169	13 ± 5	76 ± 21	7,000
SB 2	3,090	636	43 ± 10	99 ± 15	9,167
KS 3	1,333	12	13 ± 4	16 ± 2	6,417
BR 4	2,505	161	5 ± 2	16 ± 2	7,583

SEDIMENT CHEMISTRY

Sediment temperatures were warm when measured at flood tide and increased to 33–35°C during exposure at low tide (Table 2). Water content was low ($\leq 50\%$) at all sites (Table 2), with sediments draining nearly dry (10–17%) during exposure at low tide.

The mature *R. stylosa* (SB 2) and *A. marina* (BR 4) forests inhabited the coarsest deposits, with the latter stand colonizing mixed carbonate-quartz sands (Table 2). Sediments of SB 2 were the most reducing, followed by SB 1, KS 3, and BR 4 (Fig. 3). Redox potential declined significantly with depth at the latter three forests, but variability among cores was high (Fig. 3). Redox potential correlated inversely with dead root biomass (all depths and sites, $r = -0.43$; $p < 0.001$) and positively with live root biomass only at KS 3 ($r = 0.61$) and BR 4 ($r = 0.41$).

At the four sites, the concentrations of nearly all porewater and solid-phase constituents, including pH, Cl$^-$, PO$_4$, TOC, TN, and molar C:N, did not change significantly with increasing sediment depth (Table 2). Free sulfides were not detected at any of the stations. Concentrations of Cl$^-$ were equivalent to overlying water, but SO$_4$ concentrations and the SO$_4$:Cl ratio were in excess relative to overlying water at a few depth intervals at all four sites (data not shown; mean values in Table 2). Most porewater nutrients were highly variable among replicate cores, depths, and sites,

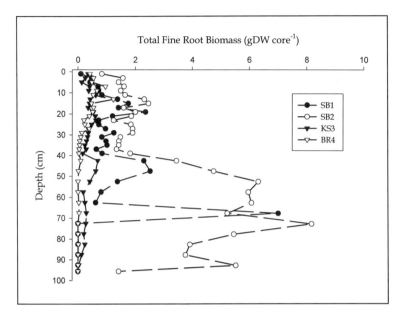

Fig. 2. Vertical distribution of total (live and dead) fine root biomass at each of the four mangrove forests. Values are means of three replicate cores. Core diameter was 6 cm.

precluding any statistical differences. There were, however, some trends: (1) ΣCO_2 concentrations increased with depth at all sites; (2) $NO_2 + NO_3$ were detected in the porewater at most depth intervals at all four sites (Table 2), with the highest concentrations in surface sediments at SB 1 and KS 3; and (3) TOC and TN concentrations and molar C:N ratios were highest at SB 2, followed by SB 1, BR 4, and KS 3 (Table 2).

The total organic carbon content of live roots ranged from 35.5% to 39.3% by DW, and total nitrogen content ranged from 0.34% to 0.46 % by DW, with molar C:N ratios ranging from 90 to 132. Dead roots had greater TN content, ranging from 0.47% to 0.76%. TOC content of dead roots ranged from 33.1% to 42.6%, and C:N ratios varied from 54 to 106. Fine roots made up a significant proportion of sediment TOC and TN. At SB 1, roots accounted for 22.3% and 3.5% of the sediment TOC and TN pools, respectively. At the immature (KS 3) and mature (BR 4) *A. marina* stands, roots accounted for 12.9% and 2.4%, respectively, of the sediment TOC pool and for 8.1% and 1.9% of the sediment TN pool, respectively. At SB 2, roots constituted 22.2% and 8.4% of the sediment TOC and TN pools, respectively.

GAS AND SOLUTE FLUXES

With the exception of ΣCO_2, solute fluxes from submerged sediments were either mostly undetectable or into the sediment (Table 3). There were no measurable fluxes of total dissolved Fe, Ca, $NO_2 + NO_3$, and HS^- at any of the sites (Table 3).

Fluxes of CO_2 and O_2 (Table 4) from exposed sediments were very variable between trials, precluding any clear differences among sites. Methane flux was not detected at any of the four forests. CO_2 fluxes were, on average, slowest at SB 1 (mean = 11.9 mmol m^{-2} d^{-1}) and greatest at BR

Table 2. Sediment chemistry at each of the mangrove forests. Values are mean ± 1 SE (n = 3). Water content and temperatures of submerged sediments are in parentheses. pH and element data are to a depth of 70 cm at stations SB 1, SB 2, and BR 4, and to 90 cm at station KS 3. TOC = total organic carbon; TN = total nitrogen. $CaCO_3$ content = Total Carbon (TOC) × 8.33.

	SB 1	SB 2	KS 3	BR 4
Water content (%)	12 (43)	17 (50)	10 (34)	10 (33)
Temperature (°C)	35 (27)	33 (25)	33 (26)	35 (24)
% Sand	9.6	14.3	3.4	53.1
% Silt + Clay	75.3 + 15.1	81.6 + 4.1	72.5 + 24.1	46.5 + 0.4
pH	7.0 ± 0.1	6.7 ± 0.2	6.8 ± 0.2	7.0 ± 0.2
Cl^- (mM)	579 ± 44	602 ± 12	644 ± 59	621 ± 29
SO_4 (mM)	32.7 ± 1.9	29.8 ± 2.7	31.1 ± 1.1	31.3 ± 2.0
ΣCO_2 (mM)	8.1 ± 1.6	9.7 ± 1.9	8.0 ± 1.9	6.4 ± 1.0
DOC (mM)	3.3 ± 1.8	3.1 ± 1.9	0.8 ± 0.3	0.7 ± 0.2
NH_4 (µM)	13.8 ± 3.4	22.1 ± 15.8	11.3 ± 3.7	58.6 ± 23.1
$NO_2 + NO_3$ (µM)	4.8 ± 1.7	1.2 ± 0.6	5.0 ± 3.4	1.6 ± 0.5
PO_4 (µM)	3.9 ± 2.9	11.5 ± 6.8	20.7 ± 22.5	9.0 ± 2.5
Fe (µM)	2.1 ± 2.4	8.8 ± 7.1	2.4 ± 4.3	5.2 ± 4.1
Mn (µM)	1.5 ± 1.4	0.9 ± 0.7	6.2 ± 6.0	0.2 ± 0.2
TOC (µmol g^{-1})	1,500 ± 166	3,417 ± 416	583 ± 167	750 ± 250
TN (µmol g^{-1})	93 ± 14	107 ± 21	50 ± 21	43 ± 21
C:N (molar)	16	32	12	17
$CaCO_3$ (µmol g^{-1})	1,180	340	1,180	6,320

4 (mean = 41.0 mmol m^{-2} d^{-1}); CO_2 flux rates averaged 17.4 and 18.3 mmol m^{-2} d^{-1} at KS 3 and SB 2, respectively (Table 4). At all sites, O_2 flux rates were not significantly different between clear and dark bottles, indicating no microbenthic autotrophy. Oxygen fluxes in dark bottles were very rapid, averaging 77.5, 119.7, 53.0, and 88.6 mmol m^{-2} d^{-1} at SB 1, SB 2, KS 3, and BR 4, respectively. Rates of CO_2 flux did not correlate with any sediment or forest variables, but total carbon oxidation (Fig. 4A) and oxygen consumption (Fig. 4B) correlated significantly with net mangrove primary production.

IRON, MANGANESE, AND SULFATE REDUCTION

Core incubations for Fe and Mn release resulted in very variable rates of metal reduction among replicate sets of incubations and among depths. Depth-integrated rates of iron reduction (Table 5) ranged from 40 to 3,300 µmol m^{-2} d^{-1}, with site differences as follows (fastest to slowest): BR 4 > SB 1 = KS 3 > SB 2 (SNK test). Rates of manganese reduction were fastest at SB 1 (12,000 µmol m^{-2} d^{-1}), followed in descending order by KS 3, SB 2, and BR 4 (Table 5).

Table 3. Rates of solute flux (μmol m^{-2} d^{-1}) from submerged sediments at the four forests as measured in triplicate chambers. Negative values indicate flux into the sediment. Values are mean ± 95% confidence interval. * = no significant flux.

Solute	SB 1	SB 2	KS 3	BR 4
ΣCO_2	10,700 ± 25,300	29,100 ± 13,600	53,000 ± 20,500	108,200 ± 47,500
DOC	-33,500 ± 22,100	0	0	28,400 ± 9400
Fe	*	0	0	0
Ca	*	0	0	0
Mn	-47 ± 14	0	65 ± 20	0
HS$^-$	*	0	0	0
PO_4	-833 ± 366	-95 ± 20	0	0
DOP	-160 ± 50	*	0	0
Si	*	6,500 ± 2,100	0	0
NH_4	-1,070 ± 480	-420 ± 210	-500 ± 100	-600 ± 300
$NO_2 + NO_3$	*	*	0	0
DON	-1,740 ± 890	*	0	0

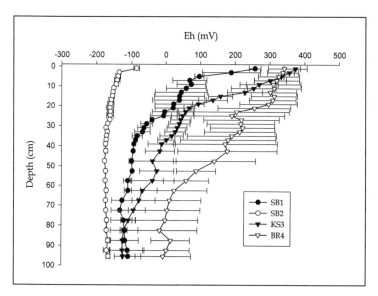

Fig. 3. Vertical profiles of redox potential at each of the four mangrove forests. Values are mean ± 1 SE.

Rates of sulfate reduction (Table 5) were fastest at SB 2, followed in descending order by BR 4, SB 1, and KS 3 (rates at the latter two forests being equivalent). At SB 1, rates of sulfate reduction peaked over the 24–30 cm depth interval (Fig. 5), with little (7%) reduced radiolabel shunted into the acid-volatile sulfides (AVS) fraction. At SB 2, SRR peaked over the 12–18 cm depth interval, with 26% of radiolabel shunted into AVS. The proportion of reduced ^{35}S in the AVS pool was similar at KS 3 and BR 4. At KS 3, peak rates of total sulfate reduction were measured at mid-core (~ 50 cm

Table 4. Rates of benthic metabolism (mmol m^{-2} d^{-1}, mean ± 1 SE) during air exposure at the four forests as measured in triplicate chambers. The two trials at each forest were run on consecutive days.

Station		O$_2$	CO$_2$
SB1	Trial 1	17.3 ± 1.3	13.6 ± 1.1
	Trial 2	137.6 ± 19.0	10.1 ± 1.6
SB2	Trial 1	158.1 ± 14.1	24.6 ± 10.5
	Trial 2	81.3 ± 10.0	12.0 ± 5.2
KS 3	Trial 1	56.6 ± 7.0	21.9 ± 4.6
	Trial 2	49.4 ± 57.5	12.8 ± 2.2
BR4	Trial 1	91.7 ± 6.2	62.5 ± 28.9
	Trial 2	85.4 ± 24.9	19.4 ± 5.7

Table 5. Depth-integrated rates of iron and manganese reduction (mmol m^{-2} d^{-1}) and sulfate reduction (SRR, mmol S m^{-2} d^{-1}) at the four mangrove forests. Values are means ± 1 SE. Values in parentheses are percentage of total sulfate reduction incorporated into the acid-volatile sulfide pathway.

Station	Fe	Mn	SRR
SB 1	800 ± 800	12,000 ± 1,050	4.9 ± 2.6 (7%)
SB 2	40 ± 40	105 ± 20	20.4 ± 2.6 (26%)
KS 3	600 ± 325	7,400 ± 650	3.0 ± 0.3 (26%)
BR 4	3,300 ± 1,500	45 ± 10	14.5 ± 2.0 (32%)

depth), with measurable, but slower, rates to 1 m (Fig. 5). At BR 4, SRR declined with increasing sediment depth (Fig. 5).

Rates of metal reduction did not correlate significantly with any sediment or forest variables, but rates of sulfate reduction correlated significantly with several factors: net mangrove primary production (Fig. 4C), dead root biomass (Fig. 6), sediment TOC (all depths and sites, r = 0.80; $p <$ 0.001), sediment TN (all depths and sites, r = 0.55; $p <$ 0.001), and Eh (all depths and sites, r = - 0.47; $p <$ 0.001). At KS 3, SRR correlated significantly with dead root biomass (r = 0.53; $p <$ 0.01). At BR 4, SRR correlated significantly with dead (r = 0.67; $p <$ 0.01) and live + dead (r = 0.60; $p <$ 0.01) root biomass.

Discussion

Rates of organic matter decomposition in mangrove sediments of the Kimberley appear to be linked to forest productivity rather than to forest type or aboveground biomass, regardless of coastal setting. Fluxes of porewater solutes for submerged sediments indicate that mangrove sediments are a sink for most dissolved nutrients, supporting the notion that they can be efficient nutrient filters in tropical coastal environments (Wiebe 1989; Alongi 1996, 1998). Other sediment processes, however,

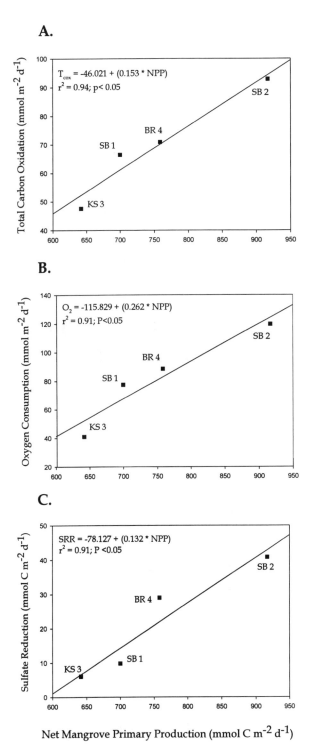

Fig. 4. Relationship between (A) total carbon oxidation (T_{cox} estimated; Table 6), (B) total mean rates of oxygen consumption (Table 4), and (C) total mean rates of sulfate reduction (Table 5) with net primary production (Table 1) of the four mangrove forests. Lines are linear regression fits to the data.

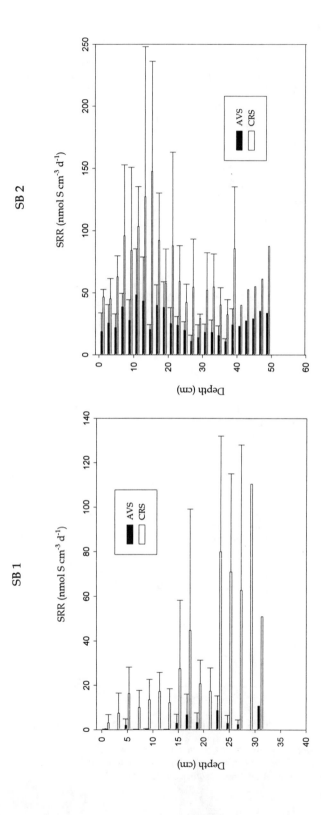

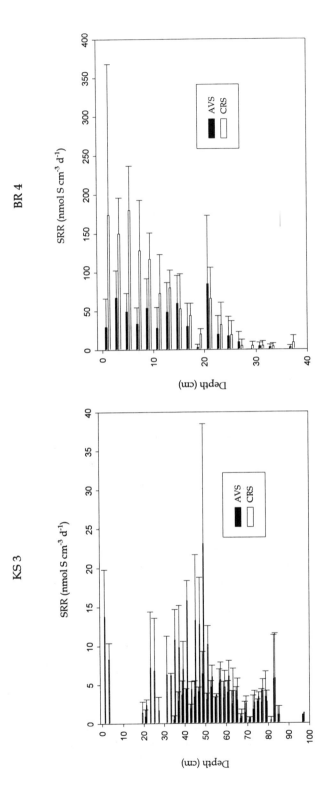

Fig. 5. Vertical profiles of mean rates of sulfate reduction and the proportion of reduced ^{35}S recovered in the acid-volatile sulfide (AVS) and chromium-reducible sulfur (CRS) pools at each of the four mangrove forests. Note differences in depth and SRR scales among forests. Values are means + 1 SE.

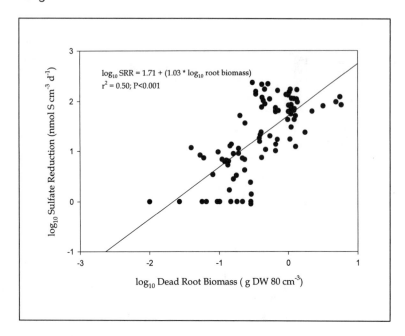

Fig. 6. Relationship between total mean rates of sulfate reduction and dead root biomass (all forests and depths). Data are $\log_{10}$-transformed. Line is linear regression fit to the data.

such as redox chemistry, do appear to be species-specific and linked to belowground roots and their activities.

There was a strong relationship between forest net primary production and rates of organic matter mineralization (Fig. 4). The most productive forests (stations SB 2 and BR 4), which have the most disparate coastal settings and sediment types, exhibited the fastest rates of early diagenesis. The functional relationship, however, is unclear, as mangrove productivity was unrelated to canopy biomass and there was no demonstrable relationship of live and dead roots with porewater nutrient concentrations. Some insights into sediment-plant-microbe links are illustrated by (1) the relationship between dead root biomass and rates of sulfate reduction (Fig. 6), (2) the correlation between sulfate reduction and redox potential, and (3) correlations of dead and live roots with redox potential. These correlations suggest that dead, not living, roots fuel sulfate-reducers. More productive, and presumably more mature, forests with greater belowground biomass can support greater rates of subsurface mineralization to sediment depths of ~ 1 m (Fig. 5).

Living roots exert important changes in redox chemistry in mangrove sediments; both *Avicennia* and *Rhizophora* spp. are known to oxidize surrounding sediments by translocating oxygen to the roots (McKee 1993). Redox status, in turn, appears to influence the distribution and composition of the porewater solute pools. Free sulfides were not detected in any of the sediments and there were consistently detectable concentrations of oxidized solutes (e.g., SO_4, NO_2, NO_3, PO_4). Roots also contributed to the solid-phase element pool, accounting for 8–22 % of sediment TOC and 2–8 % of sediment TN. Both dead root biomass and particulate nutrients correlated with rates of sulfate reduction, implying that root-derived nutrients and the total nutrient pools are fueling sulfate-reducers. A summary of the individual metabolic pathways (Table 6) shows that total organic carbon mineralization is dominated by aerobic respiration at all four forests—at least in the summer wet season, with variable contributions via sulfate reduction and, to a much lesser extent, by iron and manganese reduction. The rates of metal reduction are subject to some uncertainty because the core incubation method does not account for possible adsorption and chelation reactions, which would cause reduction rates to be underestimated, or for possible stimulation effects caused by sediment mixing.

Denitrification was not measured in this study, but denitrification rates are low in other Western Australian mangroves (Alongi 1998, p.72), suggesting that it has a minor role in organic carbon decomposition. The estimates of aerobic respiration are very crude; however, even if all reduced solutes resulting from the microbial (or chemical) reduction of iron, manganese, and sulfate were oxidized (see footnotes, Table 6), most of the total oxygen consumption measured at the sediment-air interface can be accounted for by respiration of aerobic organisms. This finding is supported by the redox potential and oxidized solute profiles, indicating that these deposits were aerobic to suboxic. These sediments were also devoid of free sulfides and methane, with low concentrations of reduced metals and NH_4^+ (Table 2). Further, comparatively little solute flux was detected across the sediment-

Table 6. Summary of the various C oxidation pathways (mmol C m^{-2} d^{-1}) at each of the four mangrove forests. Total carbon oxidation (T_{cox} measured) = average of air-exposed CO_2 flux + submerged ΣCO_2 fluxes, assuming sediments are exposed and submerged daily for roughly equal time periods. Methanogenesis (as CH_4 flux) was not detected and denitrification was presumed to be a minor C oxidation process (Canfield 1993; Alongi 1998). All conversions to carbon were made using the diagenetic equations in Fig. 1.1 in Alongi (1998). Values in parentheses are the percentage of the estimated total carbon oxidation accounted for by each individual pathway.

	SB 1	SB 2	KS 3	BR 4
A. T_{cox} (measured)	63.4	11.5	21.9	29.2
1. O_2[a]	51 (77%)	52 (56%)	38 (80%)	41 (58%)
2. Mn	5.5 (8%)	0.05 (<< 1%)	3.4 (7%)	0.02 (<< 1%)
3. Fe	0.2 (<< 1%)	0.01 (<< 1%)	0.15(<< 1%)	0.83 (1%)
4. SRR	9.8 (15%)	40.8 (43%)	6.0 (13%)	29.0 (41%)
B. T_{cox} (estimated)[b]	66.5	92.9	47.6	70.9
Δ A – B[c]	-55	-71	-18	-8

[a] = aerobic respiration estimated by subtracting oxidation of all reduced solutes (NH_4, HS-, Mn, FeS_2) from the mean total oxygen consumption rates (Table 4) using the following equations: (1) $NH_3 + 2O_2 \rightarrow HNO_3 + H_2O$ (NH_3 production assumed = NH_4 uptake rates in Table 3); (2) $H_2S + 2O_2 \rightarrow H_2SO_4$ (H_2S production assumed from percentage AVS fraction of SRR rates in Table 5); (3) $Mn^{2+} + \frac{1}{2}O_2 + H_2O \rightarrow MnO_2 + 2H^+$ (Mn^{2+} production assumed = rates of Mn reduction in Table 5); (4) $FeS_2 + 3.75 O_2 + 2.5 H_2O \rightarrow FeOOH + 2 H_2SO_4$ (FeS_2 production assumed from percentage CRS fraction of SRR rates in Table 5). As SRR > Fe reduction rates, Fe^+ oxidation is presumed accounted for in the sulfide oxidation calculations above. Oxidation equations are from Langmuir (1997).
[b] = total carbon oxidation as the sum of 1+2+3+4.
[c] = discrepancy between T_{cox} (measured) across sediment surface and T_{cox} (estimated) as the sum of the individual metabolic pathways = (A)T_{cox} measured – (B)T_{cox} estimated.

water interface, suggesting that reduced by-products were immobilized via oxidative processes or that the trees were sequestering the solutes, or both (Table 3). It is unlikely that solutes were assimilated at the sediment surface because the oxygen experiments showed no significant microalgal production when sediments were exposed to air.

The discrepancy between the average $CO_2 + \Sigma CO_2$ fluxes measured across the sediment-air interface (nominally considered here as 'total carbon oxidation') and the sum of the individual metabolic pathways (Table 6) suggests: (1) that some of the individual pathways are overestimates; (2) total carbon oxidation, especially the CO_2 measurements, are underestimates; and/or (3) that some oxidized carbon was lost via some pathway not measured. Other possibilities include systematic error, variability of the many individual measurements, and variability in the use of cores of different sizes. The values in Table 6 also do not reflect the high coefficient of variation for most of the individual measurements..It is difficult to estimate the total carbon oxidation in intertidal sediments accurately, particularly in macrotidal estuaries where the sediments are rapidly flooded at high tide and drained nearly dry during low tide. Given the large changes in sediment temperature and water content between tides, it is likely that measurements of total carbon oxidation based solely on ΣCO_2 fluxes from submerged sediments would not accurately reflect in situ conditions, particularly the roughly one-half day when the sediments are exposed. It is likely that microbial activity slows when sediments are exposed and drain dry, as shown for rates of sediment bacterial productivity and specific growth in mangroves of north Queensland (Alongi 1988). Conversely, CO_2 measurements from air-exposed sediments would normally underestimate total carbon oxidation because any oxidized carbon produced as a result of organic decomposition would be retained as dissolved CO_2 in the interstitial water, in equilibrium with the carbonate system. Our gaseous CO_2 fluxes, however, more realistically mimic exposed conditions, and may not significantly underestimate total carbon mineralization because of the lack of interstitial water (10–17%, Table 2) at low tide.

Despite these problems, it is unlikely that any measurements made across the sediment surface would accurately estimate total carbon oxidation in intertidal sediments subject to large tides. Such an estimate would assume steady-state diagenesis, which is unlikely where interstitial water movements caused by large tides are probably greater than rates of either molecular or ionic diffusion or biological advection. Some CO_2 is probably lost via burrows, fractures, and cracks at the surface of the forest floor (surface topography was highly irregular in these forests). Given the lack of evidence for methanogenesis and surface microalgae, sequestration of CO_2 by mangrove biota is unlikely. The most likely losses of carbon other than across the sediment surface would be via lateral tidal drainage, advection, or groundwater transport. This is a reasonable explanation considering the huge tidal prism and the large number of burrows and other biogenic structures that may serve as conduits for subsurface transport of solutes and gases. The enormous tides and subsequent tidal flows may serve to explain why these sediments are oxic-suboxic, even to considerable depths. The floors of these forests are essentially sponges, networked with extensive roots and rhizomes, burrows, fractures, and fissures. With tidal movement, there must be a large lateral flux of water and associated materials between the interstices of sediment and the adjacent tidal waterways. This flux may be magnified in summer when monsoonal rains drain into the sediment and assist in oxidizing reduced metabolites.

Similar losses of carbon and inorganic nutrients via tidal drainage and groundwater flow have been found to occur in other mangroves (Alongi et al. 1998) and in salt marshes (Howes & Goehringer 1994). Recent studies in mangrove estuaries with smaller (1–2 m) tidal ranges (Ridd 1996; Kitheka 1998) have confirmed the importance of groundwater flow. Ridd (1996) conservatively estimated that total water flow through crab burrows in a 1 km^2 area of forest ranged from 1,000 to 10,000 m^3 per tidal cycle. It is reasonable to assume that the volume of water flow through these macrotidal mangrove settings would be considerably greater.

The rates and pathways of organic matter decomposition, and the distribution and composition of porewater solutes and solid-phase nutrients, represent a unique environment compared with most other benthic ecosystems, including salt marshes and seagrass meadows. Like other rooted halophytes, mangroves fuel microbial decomposition in sediments directly via oxidation by roots and indirectly by providing microbes with a large stable source of organic matter vested in extensive roots and rhizomes, and in litter and wood falling from the forest. Unlike other vegetated aquatic ecosystems, however, these mangrove sediments foster slower rates of sulfate reduction and no detectable methanogenesis (Alongi 1998).

The oxic-suboxic characteristics of these mangrove sediments are also likely maintained by the poor quality of particulate and dissolved organic matter (C:N ratios were high and dissolved organic nutrient pools were small). Assuming carbon losses via lateral tidal exchange, the sum of the individual metabolic pathways would be the most accurate estimate of total carbon oxidation, instead of estimates derived from CO_2 and ΣCO_2 fluxes across the sediment surface. Rates of total carbon oxidation ranged from 48 to 93 mmol C m^{-2} d^{-1} (= 576 to 1,116 mg C m^{-2} d^{-1}) among the four forests, equivalent to those measured in other mangroves (Nedwell et al. 1994; Kristensen et al. 1995) but slower than in other coastal ecosystems (Canfield 1993). As a comparison, rates of total carbon oxidation in salt marshes range from 199 mmol C m^{-2} d^{-1} in the marshes of Sapelo Island, Georgia, to 592 mmol C m^{-2} d^{-1} in Great Sippewissett Marsh, Massachusetts (Howarth & Hobbie 1982; Howarth 1993). On average, aerobic respiration and sulfate reduction account for 2–21% and 67–69% of total carbon oxidation in both marshes, respectively, equating to 48–85% of net marsh primary production (Howarth 1993; Alongi 1998). In contrast, the rates of sediment carbon decomposition in the Kimberley mangroves—forests that appear to be twice as productive as the American salt marshes—equate to only 8–10% of net primary forest production. This proportion is probably smaller as mangrove root production is unknown but likely to be equivalent to canopy production.

Differences between mangroves and salt marshes in the proportion of fixed carbon decomposed by sediment microbes are likely due to many factors, such as differences in tidal export and in the decomposability of mangrove versus salt marsh detritus. In *Rhizophora* forests, stem, branch, and prop root wood ordinarily account for more than 80% of total forest dry weight (Robertson & Alongi 1992). In contrast, a much greater proportion of marsh grasses and seagrasses—being nonwoody plants—are more decomposable than mangrove wood, with a larger fraction of nutrients stored in large sedimentary pools than in plant biomass. Mature mangrove forests may be more similar to tropical rainforests (Archibold 1995) in that roughly one half or more of the organic matter pool is retained in living biomass (average of 58% of the total organic carbon pool in the mature Kimberley mangroves) than is stored in the sediment carbon pool. Storage of a large fraction of organic matter in standing biomass may be a mechanism widespread in tropical forests to retain and conserve essential nutrients.

Acknowledgments

A large number of individuals assisted in this study and I dedicate this paper to them: P. Christoffersen, O. Dalhaus, P. Dixon, F. Tirendi, B. McCarthy, T. Sheppard, C. Payn, S. Boyle, J. Wu Won, T. McKenna, and the crew of the R/V *Lady Basten*. John Bunt, Jeffrey Chanton, and an anonymous reviewer improved an earlier draft of this paper. Contribution No. 944 from the Australian Institute of Marine Science.

Literature Cited

Aller, R. C., N. E Blair, Q. Xia, & P. D. Rude. 1996. Remineralization rates, recycling, and storage of carbon in Amazon shelf sediments. *Continental Shelf Research* 16:753–786.

Alongi, D. M. 1988. Bacterial productivity and microbial biomass in tropical mangrove sediments. *Microbial Ecology* 15:59–79.

Alongi, D. M. 1996. The dynamics of benthic nutrient pools and fluxes in tropical mangrove forests. *Journal of Marine Research* 54:123–148.

Alongi, D. M. 1998. Coastal Ecosystem Processes. CRC Press, Boca Raton, Florida.

Alongi, D. M., A. Sasekumar, F. Tirendi, & P. Dixon. 1998. The influence of stand age on benthic decomposition and recycling of organic matter in managed mangrove forests of Malaysia. *Journal of Experimental Marine Biology and Ecology* 225:197–218.

Alongi, D. M., F. Tirendi, P. Dixon, L. A. Trott, & G. J. Brunskill. 1999. Mineralization of organic matter in intertidal sediments of a tropical semi-enclosed delta. *Estuarine, Coastal and Shelf Science* 48:451–467.

Alongi, D. M., F. Tirendi, & A. Goldrick. 1996. Organic matter oxidation and sediment chemistry in mixed terrigenous-carbonate sands of Ningaloo Reef, Western Australia. *Marine Chemistry* 54:203–219.

Archibold, O. W. 1995. Ecology of World Vegetation. Chapman and Hall, London.

Beard, J. S. 1979. Vegetation Survey of Western Australia: Kimberley. University of Western Australia Press, Perth, Western Australia.

Canfield, D. E. 1993. Organic matter oxidation in marine sediments, p. 333–363. *In* Wollast, R., F. T. MacKenzie, & L. Chou (eds.), Interactions of C, N, P and S in Biogeochemical Cycles and Global Change. Springer, Berlin.

Chappell, J. & C. D. Woodruffe. 1994. Macrotidal estuaries, p. 187–218. *In* Carter, R. W. G. & C. D. Woodruffe (eds.), Coastal Evolution—Late Quaternary Shoreline Morphodynamics. Cambridge University Press, Cambridge.

Clough, B. F. 1997. Mangrove ecosystems, p. 119–196. *In* English, S., C. Wilkinson, & V. Baker (eds.), Survey Manual for Tropical Marine Resources, 2nd ed. Australian Institute of Marine Science, Townsville, Australia.

Clough, B. F., P. Dixon, & O. Dalhaus. 1997a. Allometric relationships for estimating biomass in multi-stemmed mangrove trees. *Australian Journal of Botany* 45:1023–1031.

Clough, B. F., J. E. Ong, & W. K. Gong. 1997b. Estimating leaf area index and photosynthetic production in canopies of the mangrove *Rhizophora apiculata*. *Marine Ecology Progress Series* 159:285–292.

Clough, B. F. & K. Scott. 1989. Allometric relationships for estimating above-ground biomass in six mangrove species. *Forest Ecology and Management* 27:117–127.

Clough, B. F. & R. G. Sim. 1989. Changes in gas exchange characteristics and water use efficiency of mangroves in response to salinity and vapour pressure deficit. *Oecologia* 79:38–44.

Ferdelman, T. G., C. Lee, S. Pantoia, J. Harder, B.M. Bebout, & H. Fossing. 1997. Sulfate reduction and methanogenesis in a *Thioploca*-dominated sediment off the coast of Chile. *Geochimica et Cosmochimica Acta* 61:3065–3079.

Fossing, H. & B. B. Jorgensen. 1989. Measurement of bacterial sulfate reduction in sediments: Evaluation of a single-step chromium reduction method. *Biogeochemistry* 8:205–222.

Howarth, R. W. 1993. Microbial processes in salt-marsh sediments, p. 239–259. *In* Ford, T. E. (ed.), Aquatic Microbiology: An Ecological Approach. Blackwell, Boston.

Howarth, R. W. & J. E. Hobbie. 1982. The regulation of decomposition and heterotrophic microbial activity in salt marsh soils: A review, p. 183–207. *In* Kennedy, V. S. (ed.), Estuarine Comparisons. Academic Press, New York.

Howes, B. L. & D. D. Goehringer. 1994. Porewater drainage and dissolved organic carbon and nutrient losses through the intertidal creeks of a New England salt marsh. *Marine Ecology Progress Series* 114:289–301.

Jennings, J. N. & R. J. Coventry. 1973. Structure and texture of a gravelly barrier island in the Fitzroy estuary, Western Australia, and the role of mangroves in the shore dynamics. *Marine Geology* 15:145–167.

Kitheka, J. U. 1998. Groundwater outflow and its linkage to coastal circulation in a mangrove-fringed creek in Kenya. *Estuarine, Coastal and Shelf Science* 47:63–75.

Kristensen, E., M. Holmer, G. T. Banta, M. H. Jensen, & K. Hansen. 1995. Carbon, nitrogen and sulfur cycling in sediments of the Ao Nam Bor mangrove forest, Phuket, Thailand: A review. *Phuket Marine Biological Center Research Bulletin* 60:37–64.

Langmuir, D. 1997. Aqueous Environmental Geochemistry. Prentice Hall, New Jersey.

McKee, K. L. 1993. Soil physicochemical patterns and mangrove species distribution–Reciprocal effects? *Journal of Ecology* 81:477–487.

Middelburg, J. J., J. Nieuwenhuize, F. J. Slim, & B. Ohowa. 1996. Sediment biogeochemistry in an East African mangrove forest (Gazi Bay, Kenya). *Biogeochemistry* 34:133–155.

Nedwell, D. B., T. H. Blackburn, & W. J. Wiebe. 1994. Dynamic nature of the turnover of organic carbon, nitrogen and sulfur in the sediments of a Jamaican mangrove forest. *Marine Ecology Progress Series* 110:223–231.

Ridd, P. V. 1996. Flow through animal burrows in mangrove creeks. *Estuarine, Coastal and Shelf Science* 43:617–625.

Robertson, A. I. & D. M. Alongi (eds.). 1992. Tropical Mangrove Ecosystems, Coastal and Estuarine Studies No. 41. American Geophysical Union, Washington, D.C.

Robertson, A. I. & P. Dixon. 1993. Separating live and dead fine roots using colloidal silica: An example from mangrove forests. *Plant and Soil* 157:151–154.

Semeniuk, V. 1980. Mangrove zonation along an eroding coastline in King Sound, North-western Australia. *Journal of Ecology* 68:789–812.

Semeniuk, V. 1985. Development of mangrove habitats along ria shorelines in north and northwestern tropical Australia. *Vegetatio* 60:3–23.

Semeniuk, V. 1993. The mangrove systems of Western Australia: 1993 Presidential Address. *Journal of the Royal Society of Western Australia* 76:99–122.

Sherman, R. E., T. J. Fahey, & R.W. Howarth. 1998. Soil-plant interactions in a neotropical mangrove forest: iron, phosphorus and sulfur dynamics. *Oecologia* 115:553–563.

Sokal, R. R. & F. J. Rohlf. 1995. Biometry, 3rd ed. Freeman and Company, New York.

Spalding, M. D., F. Blasco, & C. D. Field (eds.). 1997. World Mangrove Atlas. The International Society for Mangrove Ecosystems, Okinawa, Japan.

Wiebe, W .J. 1989. Phosphorus, sulfur, and nitrogen cycles in mangrove forests, p. 312–317. *In* Hattori, T., Y. Ishida, Y. Maruyama, R.Y. Morita, & A. Uchida (eds.), Recent Advances in Microbial Ecology. Japan Scientific Societies Press, Tokyo.

The Biogeochemistry of Carbon in Continental Slope Sediments: The North Carolina Margin

Neal Blair*, Lisa Levin, David DeMaster, Gayle Plaia, Chris Martin, William Fornes, Carrie Thomas, and Robin Pope

Abstract: *The responses of the continental slope benthos to organic detritus deposition were studied with a multiple tracer approach. Study sites offshore (850 m water depth) of Cape Fear (I) and Cape Hatteras (III), North Carolina, were characterized by different organic C deposition rates (III > I), macrofaunal densities (III > I), and taxa. Natural abundances of ^{13}C and ^{12}C in particulate organic carbon (POC), dissolved inorganic carbon (DIC), and macrofauna indicate that the reactive organic detritus is marine in origin. Natural abundance levels of ^{14}C in benthic animals reveal they incorporate a relatively young component of carbon into their biomass. ^{13}C-labeled diatoms (Thalassiosira pseudonana Hasle et Heimdal) tagged with ^{210}Pb, slope sediment tagged with ^{113}Sn, and ^{228}Th-labeled glass beads were emplaced in plots on the seafloor at both locations, and the plots were sampled after 30 min, 1–1.5 d, and 14 mo. At site I, the tracer-labeled diatom was intercepted at the surface primarily by protozoans and surface-feeding annelids. Little of the diatom C penetrated below 2 cm even after 14 mo. Oxidation of organic carbon appeared to be largely aerobic. At site III, annelids were primarily responsible for the initial uptake of tracer. On the time scale of days, diatom C was transported to a depth of 12 cm and was found in animals collected between 5 cm and 10 cm depth. Oxidation of the diatom organic carbon was evident to at least 10 cm depth. At both sites, 10% or less of the tracers remained in the plots after 14 mo. Horizontal transport, which was probably biologically mediated, may be responsible for much of this loss. The horizontal transport may represent an important but nearly untraceable diagenetic process if it is associated with the ingestion and egestion of organic matter.*

Introduction

Deposition of organic carbon on the continental slope is an important component of the global and marine biogeochemical carbon cycles. Nearly 90% of the sedimentary inventory of organic carbon deposited during the Holocene resides on the slope and rise (Premuzic et al. 1982; Romankevich 1984; Walsh et al. 1985; Hedges & Keil 1995). While the detailed sources and fates of organic matter on the slope are significant to understanding the roles of continental margins in historical and future changes in the global carbon cycle, they are poorly understood.

Considerable spatial heterogeneity exists in terms of sediment deposition and carbon cycling in this transitional environment between shallow and deep waters. Bathymetry ranges from gently sloping, nearly featureless bottoms to steep slopes that are cut with canyons, sometimes over small

243

areas, such as on the North Carolina margin (Blair et al. 1994; DeMaster et al. 1994). Non-steady-state deposition and mass wasting may characterize the steep and incised environments (Thomas 1998). Organic carbon depositional fluxes and benthic oxidation rates can vary 1–2 orders of magnitude and 2–3X respectively over 100-km scales (Anderson et al. 1994; Blair et al. 1994; DeMaster et al. 1994; Thomas 1998). Benthic organism abundances and species compositions reflect those variations (Schaff et al. 1992). Accordingly, the pathways through which the carbon is processed are expected to vary as well.

The objective of this study is to elucidate the short-term fate of organic matter deposited on the continental slope off North Carolina. We examined the source and age of naturally occurring organic carbon in sediments and fauna, and with tracer experiments, examined particle mixing mechanisms and selectivity, carbon residence time, and the roles of benthic organisms in these processes. Comparisons are drawn between two continental slope locations with contrasting carbon depositional regimes and faunal assemblages. This paper is a synthesis of work previously published by the authors and new data.

The study sites, I and III, are offshore of North Carolina in approximately 850 m of water (Fig. 1). Site I (32°52'N, 76°27'W) is 170 km southeast of Cape Fear, North Carolina, and has a gently sloping seafloor. Site III (35°24'N, 74°50'W) is 61 km northeast of Cape Hatteras, North Carolina. The region in the vicinity of site III is extensively carved by small canyons and gullies (Mellor & Paull 1994). The convergence of the Gulf Steam and Virginia Current offshore of Cape Hatteras focuses settling particulates into the site III region, producing an organic depositional flux that is 20–200X greater than at site I (DeMaster et al. 1994; Thomas 1998). Macrofauna ($\geq$ 300 µm) are 2–6X more abundant at site III than at site I as a result of the elevated food supply (Schaff et al. 1992; Levin et al. 1999).

A multi-tracer approach was used in this study. Natural abundance levels of ^{12}C, ^{13}C, and ^{14}C provided information about the sources and fates of the organic matter in the two environments. In addition, diatoms artificially enriched in ^{13}C (95%) along with radiotracer-tagged sediment and glass beads were used in situ to identify particulate and solute transport processes in the seabed, determine organic carbon oxidation rates, and identify the organisms responsible for the uptake and mixing of recently deposited labile material. The time scales of the in situ experiments ranged from 1 d to 14 mo. The combined multi-tracer approach provided a novel view of the benthic carbon cycle on the continental slope.

Methods

A particle mixture containing ^{13}C- and ^{210}Pb-labeled diatoms (freeze-dried onto kaolin), ^{113}Sn-labeled local slope sediment (10–60 µm in size), and ^{228}Th-labeled glass beads (105–149 µm diameter) was spread in 40 x 40 cm plots on the seafloor via submersible in August 1994 and October 1995. Details concerning the multi-tracer methods used in these and previous experiments are provided in Blair et al. (1996), Levin et al. (1997), and Fornes et al. (1999). Tracer particles were selected to represent a variety of natural particle types and sizes. The pelagic diatom *Thalassiosira pseudonana* was used because diatoms in this genus are found in surface sediments on the North Carolina slope and beneath coastal upwelling regions (Cahoon et al. 1994). Subsurface sediment (> 100 yr in burial age) collected near site III was used to represent local sediment low in labile organic

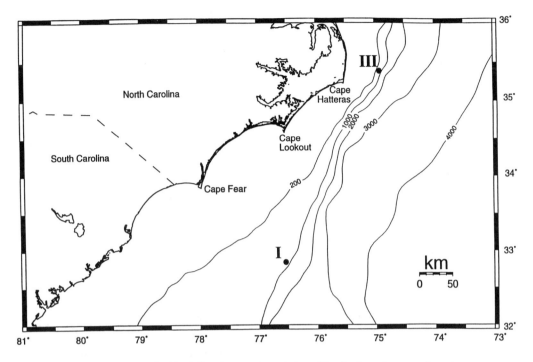

Fig. 1. Study site locations on the North Carolina continental slope. The depth contours are in meters.

matter. Sand-sized glass beads also were chosen because approximately 32% of the naturally occurring sediment at the study site is this size (Kelchner 1992).

Algae, slope sediment, and glass beads were tagged with γ-emitting tracers to permit rapid detection following core recovery from the seabed. In the case of the tagged diatom, the ^{210}Pb also served as a conservative tracer to detect particle removal from the plots for the longer 14-mo experiments. The seafloor plots were sampled once using Ekman box cores after either 30 min of tracer spreading (t = 0 plots), ~1–1.5 d, or 14 mo. Each Ekman box core (15 x 15 x 15 cm) contained four subcores (a–d). Sediments of the subcores were analyzed for (a) δ^{13}C of dissolved inorganic carbon (DIC) and particulate organic carbon (POC), (b) δ^{13}C of infauna sorted onboard ship, then frozen, (c) radiotracers and glass beads, and (d) infaunal abundance and composition. Cores of background sediment were collected within 10 m of the plots for comparative purposes. All cores were stored onboard ship at in situ temperature (4–6°C) until they could be subsampled, usually within several hours.

Sediment porewater was separated from the solid phase via centrifugation, and the two fractions were stored frozen until analysis. After thawing, 1-ml porewater samples were acidified. The resulting CO_2 was stripped with helium, dried via passage through magnesium perchlorate and Nafion tubing (Perma-Pure), and then trapped cryogenically for isotopic analysis. Carefully homogenized subsamples of the centrifuged sediment were used for the POC analysis. CO_2 for isotopic analysis was produced from the POC as described in Blair and Carter (1992). The ^{13}C/^{12}C content of the CO_2 from both the DIC and POC was measured via isotope ratio mass spectrometry (Blair et al. 1996).

Analyses of background cores were used to correct for the natural occurrence of ^{13}C (Blair et al. 1996). Concentrations are reported in terms of total C ($^{13}C + {}^{12}C$) derived from the 95% ^{13}C-labeled diatom.

Faunal samples for ^{13}C analysis were sorted from sediments retained on a 0.3-mm mesh screen and then frozen at -20°C. Subcore sediments targeted for infaunal counts were sieved through a 0.3-mm screen, and the animals retained were preserved in 8% buffered formalin and seawater (Levin et al. 1997).

In the laboratory, frozen animals were acidified with 4N HCl in silver boats and dried in vacuo. The samples were then treated in an identical fashion as the POC for $^{13}C/^{12}C$ analyses. Samples used for the determination of ^{14}C-age were prepared similarly except that the CO_2 produced by combustion was analyzed by accelerator mass spectrometer at the Lawrence Livermore Laboratory. Preserved specimens were sorted from sediments at 12X magnification and identified to species using a compound microscope with higher magnification when necessary.

Gamma activities from the tracers were measured on wet sediment with an intrinsic germanium γ-detector (DeMaster et al. 1994; Fornes et al. 1999). The ^{210}Pb and ^{228}Th activities in the experimental plots were corrected for naturally occurring ^{210}Pb and ^{228}Th, based on profiles from background sediment cores.

The experimental design avoided excessive organic carbon enrichment of sediments; labeled diatom carbon accounted for < 1.1% of existing POC levels (1.2–1.7% C) in the upper 1 cm of sediment. The depositional flux of tracer when integrated over the 20-min particle settling time in the experiments was 200–1,000X greater than the mean background depositional flux of metabolizable organic C, ~ 1 μmol C cm^{-2} d^{-1} (Blair et al. 1994; Thomas 1998). Despite the apparently large difference, the contribution of the tracer to the total DIC benthic flux by in situ oxidation of labeled diatom was < 1% over a 24-h period in a site I closed-chamber experiment. An enhancement of < 3% was observed in previous laboratory 1.5-d experiments using site III sediments (Blair et al. 1996).

Results and Discussion

Sources of Organic Matter

The organic carbon concentrations of the surface sediments (0–15 cm depth) average 1.2% and 1.7% on a dry weight basis for sites I and III respectively (DeMaster et al. 1994). The organic matter at both locations is largely marine in origin as indicated by the respective $\delta^{13}C$ values of -18.7 ± 0.1‰ and -21.2 ± 0.1‰ (Blair et al. 1994). These values are characteristic of those found along the shelf and slope of the eastern United States (Hunt 1970; Tanaka et al. 1991). The different $^{13}C/^{12}C$ compositions may be attributed to the mixing of marine and terrestrial sources with approximate end member $\delta^{13}C$ values of ~ -19 ± 1‰ and -26 ± 2‰ (Hedges & Parker 1976; Fry & Sherr 1984).

The proximity of site III to sources of terrigenous material, such as the Chesapeake Bay and the North Carolina sounds, is consistent with the difference in sedimentary $^{13}C/^{12}C$ compositions, as well as reported elemental C/N ratios (site I = 7.5, site III = 10; Blair et al. 1994). Lipid indicators of terrigenous organic matter have been identified in sediments from the Cape Hatteras slope near the site III location (Harvey 1994). However, even though the evidence points toward a gradient in marine-terrestrial sources from site I to site III, we cannot rule out the possibility that the isotopic difference originates, at least in part, from multiple marine sources with either varying species composition or diagenetic history (Gearing et al. 1984; Fry & Sherr 1984).

The sedimentary marine organic pool has multiple sources. Microscopic analyses reveal planktonic and benthic diatoms, foraminifera, and seagrass fragments (Kelchner 1992; Cahoon et al. 1994; L. Levin unpublished data). Lipid distributions of Cape Hatteras slope sediments suggest inputs from dinoflagellates and marine heterotrophs (Harvey 1994). The abundances of microbial and zooplankter lipids indicate extensive heterotrophic recycling of the organic mixture.

The seaward deflection and convergence of the Gulf Stream and Virginia Current offshore of Cape Hatteras focus shelf-derived debris onto the slope (Csanady & Hamilton 1988; Blair et al. 1994; Walsh 1994; Thomas 1998). This material likely originates from a combination of terrestrial, estuarine, and marine production and has been heterotrophically reworked during storage and/or transport on the shelf and upper slope. Delivery to the site III seafloor appears to occur primarily via the slow settling of a dense suspension of flocculant snow punctuated by episodic lateral transport events (Hecker 1990; Blair et al. 1996). The complexity of the physics and geology in the Cape Hatteras area virtually guarantees that the organics are a mixture of partially reworked young and old materials.

The mechanisms of organic matter delivery to site I are less well characterized. Downslope mass wasting should not be an important factor due to the gentle slope (2–3°; Schaff et al. 1992). In addition to local productivity, an additional source of material may be transport from the south via the Gulf Stream. Gulf Stream meanders may entrain nearshore production from the west and transport it seaward to the site as well.

PARTICLE TRANSPORT WITHIN THE SEABED

Biological particle mixing rates are dramatically different at the two sites. The biodiffusion coefficients, D_b, estimated using naturally occurring excess [234]Th profiles in the seabed, were 0.3 ± 0.1 and 13 ± 7 cm^2 yr^{-1} for sites I and III, respectively (Fornes 1996; Fornes et al. 1999). Infaunal activities are controls on the D_b values. Metazoan macrofaunal densities were nearly 3X greater at site III during the study period (Table 1), which is comparable to past observations (Schaff et al. 1992; Blake & Grassle 1994; Fornes et al. 1999; Levin et al. 1999). Macrofauna were found distributed deeper at site III than at site I. Of the approximately 25,000 individuals m^{-2} present in 1994–1995 at site III, about 15% were found below 5 cm. In contrast, 8% of 9,000 individuals m^{-2} were collected from below 5 cm at site I (Fornes et al.1999; Levin et al. 1999). Polychaetes were the dominant metazoan macrofaunal taxon at both sites, although the major polychaete taxa differed between sites I and III (Table 1; Schaff et al. 1992; Levin et al. 1999). Agglutinated protozoa (foraminifera and xenophyophores) were common at both sites yet were slightly more abundant and much more diverse at site I (Hughes 1996; Gooday unpublished data). Larger infauna (megafauna) were not quantified by our Ekman cores but may be important bioturbators.

An artificial multi-tracer mixture containing ^{13}C-labeled *Thalassiosira pseudonana*, tagged sediment, and glass beads was emplaced in plots at both sites to determine the general nature of particle mixing processes. At site I, little ^{13}C-tracer was detected below 1 cm depth in the seabed in the t = 0 controls (4 ± 5%, n = 5 for 1994–1995) and the t = 1 (1.5-d) plots (4 ± 5%, n = 6; Fig. 2). The tagged sediment and glass beads exhibited parallel distributions (Fornes 1996; Fornes et al. 1999). The similarities between the t = 0 and t = 1 (1.5-d) plots indicate that passive settling of tracer down burrows and tubes and/or artifacts associated with the experimental procedure cannot be discounted as important modes of delivery of the tracers below 1 cm in these experiments. Evidence for biologically mediated transport was observed in a portion of a 1.5-d plot at site I that was subsampled for the γ-emitter measurements. In that subcore, subsurface peaks of [210]Pb (tagged

Table 1. Densities and taxonomic composition of North Carolina slope benthic macrofauna (> 0.3 mm). Data are from August 1994 background cores and experimental plots.

	Site I	Site II
No. of box cores	9	8
Individuals m^{-2}	10,249	29,362
Standard error	1,145	3,734
Percentage of major taxa		
Polychaetes	36.1	32.0
Oligochaetes	14.6	3.9
Crustaceans	14.2	23.3
Molluscs	16.1	21.1
Cnidarians	9.3	11.6
Others	9.7	8.1

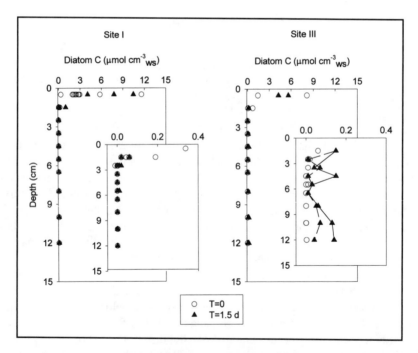

Fig. 2. The concentration of tracer carbon in the POC fraction as a function of depth in the seabed at sites I and III. Open circles represent t = 0 controls. The closed triangles are concentrations after 1.5 d on the seafloor. Graphical inserts have expanded concentration scales so that the subsurface peaks can be seen. The subsurface peaks near 10 cm at site III are associated with maldanid feeding chambers.

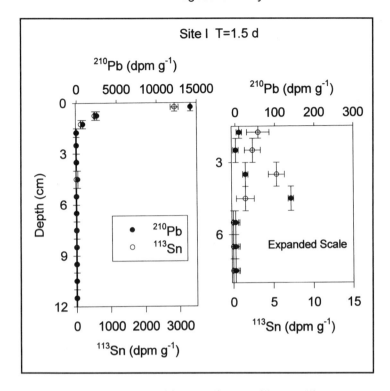

Fig. 3. Depth distribution in the seabed of tracer ^{210}Pb and [113]Sn. The ^{210}Pb and [113]Sn were sorbed to the labeled diatom and old slope sediment, respectively. At site I in the 1.5-d experiments the two tracers typically tracked each other, indicating that transport was not particle selective. However, in the situation shown, diatoms were selectively transported to 4–5 cm relative to the tagged sediment, suggesting biological mediation. The plot on the right has an expanded activity scale to show the subsurface peak.

diatom) and [113]Sn (sediment) were observed at 3–5 cm. The peaks accounted for 2% and 0.4% of the total respective inventories (Fig. 3). This particular transport event appears to have been selective for the diatom as indicated by the difference between the ^{210}Pb/[113]Sn ratios for the peak (24 ± 8) and the total core (5 ± 0.2; Fornes et al. 1999).

Losses of tracer over a year time scale were estimated by comparing 14-mo (n = 3) and 1.5-d (n = 3) plots that were established in 1994 at site I. The t = 0 controls were not used for the comparison because of observed tracer loss caused by the bow wave of the Ekman corer. The tracer mixture was better incorporated into the sediment fabric after 1.5 d, and, as a result, ^{13}C and ^{210}Pb inventories were less variable and greater than in the t = 0 controls (Fornes 1996).

Spatial heterogeneity within and between plots is an important concern when comparing tracer inventories from different treatments. Variability in the inventories resulted from heterogeneous tracer deposition and natural processes. Placement of the Ekman box corer was standardized throughout the study to minimize the variability associated with spatial heterogeneities. The standard deviation about the mean inventory (1 σ) for a treatment (t = 1.5 d or 14 mo) in the site I experiments was propagated through the calculation of the amount of tracer lost during the experiments. Duplicate t = 1.5-d and

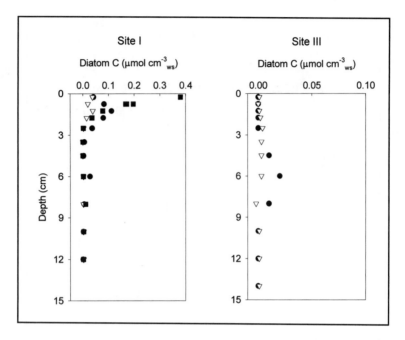

Fig. 4. The concentration of tracer carbon in the POC fraction after 14 mo. Note the difference in concentration scales between this figure and Fig. 2. Most of the loss from the plots was due to horizontal transport rather than in situ oxidation as indicated by tracer ^{210}Pb inventories. The different symbols denote data from two separate experimental plots.

14-mo plots were compared at site III, and the range about the mean inventory of each treatment was used as the uncertainty that was propagated through the calculation.

After 14 mo, 4.1 ± 3.6% of the ^{13}C-tracer remained in the plots at site I, with the bulk sequestered in the upper 2 cm (Fig. 4). Transport below 2 cm was minimal with the exception of the creation of a subsurface peak centered at 6 cm in one subcore. Bioturbation over time scales of days to 14 mo thus seems to be best characterized by slow diffusive mixing coupled with occasional nonlocal transfer of material to depths of 3–6 cm (Fornes et al. 1999).

Tracer loss from the plots occurred by both in situ oxidation and horizontal transport. The relative importance of those two processes was resolved semiquantitatively using the inventory of tracer ^{210}Pb. The portion of tracer ^{210}Pb remaining after 14 mo, 10 ± 8.6%, indicates that approximately 90% of the diatom may have been lost via horizontal transport. This is based on the assumption that the ^{13}C and ^{210}Pb physically tracked each other reasonably well prior to the oxidation of the ^{13}C–labeled organic matter, and that the time dependencies of the oxidation and transport rates were comparable. If oxidation significantly outpaced horizontal transport over the first weeks to months, the 90% estimate is an upper limit. The mode of horizontal removal could have been either biological or physical. We currently do not have a means to distinguish between the two at this site.

Mixing occurred more rapidly and deeper at site III. In the t = 0 control plots (n = 2), ≤ 2% of the ^{13}C-labeled diatom was found below 2 cm (Fig. 2). In contrast, 7–17% of the diatom tracer was found below 2 cm in the t = 1.5-d plots (n = 2). Multiple subsurface peaks were observed with the most prominent occurring at 10–12 cm. The tagged glass beads and sediment exhibited similar distributions (Levin et al. 1997; Fornes et al. 1999).

The whitish tracer mixture was observed at the base of a maldanid tube (*Praxillella* sp. Verrill) at 13 cm depth in one t = 1.5 d subcore. The individual in the tube had not ingested the tracer as indicated by its $\delta^{13}C$ value, which suggests that the tracer was hoed to depth by this head-down deposit-feeder (Levin et al. 1997). Shallow-water members of the same subfamily (Euclyminae) appear to drag fresh surface material down their tubes into the feeding cavities as well (Mangum 1964; Kudenov 1978; Dobbs & Whitlach 1982). The ratios of tagged diatom, sediment, and glass beads indicate that this transport process was nonselective for the tracer particle types (Fornes 1996; Levin et al. 1997; Fornes et al. 1999).

Shallower subsurface peaks in tracer concentration were observed in these (Fig. 2) and previous in situ experiments that used ^{13}C-*Chlorella* at this site (Blair et al. 1996). The transport processes that created the peaks were not significantly selective for any of the particle types in the tracer mixture (Fornes et al. 1999). Candidates for the nonlocal transport, based on ^{13}C-incorporation (see below), faunal vertical distributions within the sediment, and the construction of distinct tube and burrow systems that might serve as repositories of tracer, include the polychaetes *Scalibregma inflatum* (Rathke)and *Aricidea* spp., several species of maldanids, and the nereid *Ceratocephale loveni* (Malmgrem) (Blair et al. 1996; Levin et al. 1999).

The ^{13}C-diatom organic matter and the γ-emitting tracers were largely lost from the plots at site III over 14 mo. The portions of ^{13}C and ^{210}Pb remaining were $0.9 \pm 0.5\%$ and $1.7 \pm 0.9\%$ respectively and were distributed to at least 8 cm depth (Fig. 4). Biological removal was suspected because of the extensive tracks and trails on the sediment surface. In addition, flags and stakes marking some of the 14-mo plots were pushed out radially from the center, suggesting disturbance from within the plot as opposed to a unidirectional movement across plots as might be expected by a physical process. Certain demersal fish species, such as the eelpout, *Lycenchelys verrilli* (Goode and Bean), have population densities nearly an order of magnitude higher on the middle slope of Cape Hatteras than adjacent slope regions, possibly the result of abundant infaunal prey (Hecker 1994). Eelpout were observed wallowing in plots shortly after tracer emplacement. These same predators may be responsible for rapid horizontal particulate transport.

Vertical particle mixing at site III was primarily nonlocal for days after tracer emplacement. Tracer profiles attained a diffusive appearance over time scales of weeks to months based on model simulations (Blair et al. 1996) and tracer profiles from 3-mo in situ experiments (not shown; Fornes et al. 1999). Over the same period, large surface-dwelling animals moved the tracer laterally. The rapid vertical transport provided a means for subsurface organisms to deal with the competitive pressures on the surface for the same food resources (Jumars et al. 1990) and thus may have been critical for the support of the subsurface ecosystem (Levin et al. 1997).

ORGANIC CARBON UPTAKE, ASSIMILATION, AND OXIDATION IN THE SEABED

The natural abundances of ^{12}C, ^{13}C, and ^{14}C provide information concerning the source of the organic carbon that is assimilated into benthic biomass. The $\delta^{13}C$ of annelids (-16.5 ± 0.3‰, n = 16), non-annelid metazoans (-18.2 ± 0.6‰, n = 7), and agglutinating protozoans (-18.7 ± 0.2‰, n = 25) at site I were the same as or more positive than that of the sedimentary organic matter in background sediments (Levin et al. 1999). The isotopic composition of the protozoans likely reflects, in part, the carbon associated with their tests of sediment particles. The ^{13}C-enrichment exhibited in the annelids relative to the bulk sediment may have been the result of either the selective incorporation of an organic fraction from the sedimentary pool or a metabolic isotope effect by the animal itself (Hentschel 1998; Levin et al. 1999). All of the $\delta^{13}C$ values indicated a diet dependent on marine organic sources (Fry & Sherr 1984).

Table 2. Natural abundance ^{14}C contents (ages) of fauna and sediment from the Cape Hatteras continental slope region. Negative ^{14}C ages reflect the addition of ^{14}C produced by the atmospheric testing of nuclear weapons. Positive ^{14}C ages are in years before present (1950). All polychaetes were collected from the upper 15 cm of the sediment column.

Fauna/Sample	Number of samples	^{14}C age (years)
Lumbrineridae[a]	2	-592 ± 63[c]
Onuphidae[a]	2	-643 ± 24
Maldanidae[a]	3	-429 ± 40
Trichobranchidae[a]	2	-364 ± 16
Goniadidae[a]	1	-388
Cirratulidae[a]	1	-160
Lysenchelys verrilli[b]	1	-589
Surface sediments upper 10 cm	14	1,316 ± 616

[a] polychaete

[b] demersal fish

[c] The uncertainty for n = 2 is the range around the mean, for n $\geq$ 3 it is the standard deviation (1σ).

Similar patterns existed at site III. Annelid, non-annelid, and agglutinating protozoan δ^{13}C values were -19.0 ± 0.2‰ (n = 30), -19.8 ± 0.8‰ (n = 8), and -20.0 ± 0.1‰ (n = 5), respectively. These values were comparable with that of the metabolizable organic fraction, ~ -18‰ to 20‰, estimated from DIC production and fluxes (Thomas 1998).

Natural ^{14}C abundances of animals from the Cape Hatteras slope indicated clearly that young organic matter was preferentially incorporated into biomass (Table 2). All animals, ranging from the demersal eelpout to subsurface-dwelling annelids, had modern ^{14}C ages despite a $\geq$ 700-yr age of the surface organic matter. Particle discrimination, possibly by feeding on young surficial floc, and the selective incorporation of biochemically recognizable fractions may have contributed to the young ages.

The macrofauna that accessed the fresh phytodetritus first were identified by their elevated ^{13}C-content following the 1–2 d exposure to the tracer diatoms (Table 3). Approximately 60% of the agglutinating protozoan and annelid specimens collected in the 1–1.5 d experiments had ^{13}C contents in large excess of that found in the t = 0 controls (Fig. 5; Levin et al. 1999). Non-annelid metazoans (molluscs, crustaceans, and other taxa) incorporated far less tracer on average. Surface-deposit feeding annelids were important initial consumers of the tracer, whereas subsurface-deposit feeders and carnivores were not (Levin et al. 1999). Both annelids and protozoans retained significant levels of diatom carbon 14 mo after tracer emplacement, with annelids apparently retaining more (Fig. 5). After 14 mo, the tracer C was distributed through surface- and subsurface-deposit-feeders as well as carnivorous annelids.

Annelids were the primary consumers of diatom tracer in the 1.5-d treatments at site III (Fig. 5; Levin et al. 1999). Over 60% of the individuals collected for isotope analysis from the 1.5-d plots

Table 3. Examples of heavily ^{13}C-labeled macrofauna from 1–1.5 d experiments. Background (nonlabeled) δ^{13}C values of annelids and agglutinating protozoans at site I were -16.5 ± 0.3‰ and -18.7 ± 0.2‰, respectively. Background annelid, non-annelid, and agglutinating protozoan δ^{13}C values were -19.0 ± 0.2‰, -19.8 ± 0.8‰, and -20.0 ± 0.1‰ at site III.

Site I	δ^{13}C	Site III	δ^{13}C
Annelids		Annelids	
Melinna sp. (Malgrem)	+166	*Melinna* sp.	+1160
Cirratulid	+47	Maldanid	+1670
Ceratocephale sp.	+317	*Aricidea* sp. (Webster)	+7830
		Oligochaete	+67
Protozoans			
Mudball (foram)	+305	Other	
Astrorhiza-like	+380	Bivalve	+50
Bathysiphon rufum (de Folin)	+930		

were ^{13}C-enriched. *Aricidea* spp. and maldanids consumed the most diatom tracer (Table 3). The ^{13}C-tracer was retained in some species (e.g., *Ceratocephale loveni*, *Levinsenia* sp. (Mesnil), and *Cossura* sp. (Webster and Benedict)) after 14 mo, though levels were considerably lower than at site I. The between-site difference in retention reflects the greater loss of tracer from the site III plots.

Oxidation of the ^{13}C-labeled diatom led to the accumulation of ^{13}C-enriched dissolved inorganic carbon (DIC) in the sediment porewater at both sites (Fig. 6). At site I, oxidation occurred principally at the surface and the tracer DIC appeared to diffuse into the seabed. The results from a reaction-transport model that simulates tracer DIC production after tracer addition indicates solute transport was dominated by molecular diffusion (see Appendix). In other words, bioirrigation was not a significant factor at site I.

The oxidation of organic carbon is typically treated as a pseudo first-order process (Berner 1980). The rate coefficient, k, is dependent on a variety of factors, including molecular structure (Henrichs & Doyle 1986; de Leeuw & Largeau 1993), sample matrix (e.g., diatom frustules, clay aggregates; Burdige & Martens 1988; Hedges & Keil 1995; Cowie & Hedges 1996), age of organic carbon (Middelburg 1989), and environment (Cowie & Hedges 1991; Sun et al. 1993; Canfield 1994). In the in situ experiments, microbial colonization of the newly emplaced tracer should have influenced k as well. The fit of the model output to the observed tracer DIC profiles was improved by the inclusion of a time-dependent k that increased from zero or near-zero to realistic values (-10 to 20 yr^{-1}; Westrich & Berner 1984) in less than 2 d. A similar microbial response to the input of fresh detritus to deep-sea sediments has been observed (Turley & Lochte 1990). The projected half-life of the tracer based on the modeled k was < 1 mo, yet the presence of excess ^{13}C in the 14-mo plots indicated that a portion of the diatomaceous organic matter was longer lived. The material that persisted may have been protected within the diatom frustule (Burdige & Martens 1988; Cowie & Hedges 1996) or sequestered as a diagenetic product, which could have included benthic biomass.

Tracer DIC penetration into the seabed was deeper at site III (Fig. 6). Coincident subsurface peaks in tracer POC and DIC, centered near 10 cm, indicated that the deep ^{13}C-enriched DIC was the

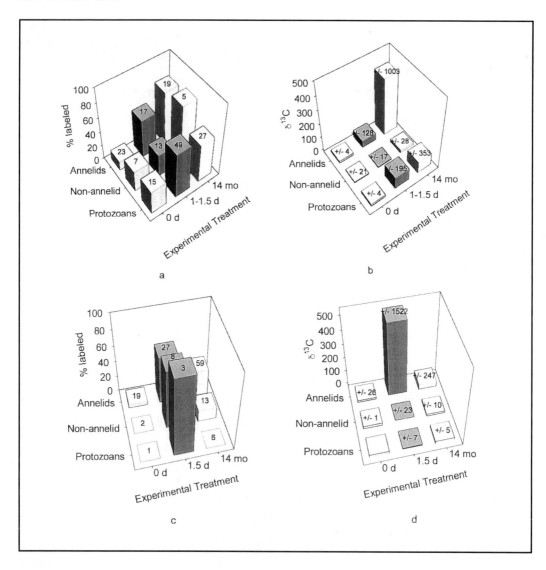

Fig. 5. Incorporation of tracer carbon by infauna. Plots 5a and c illustrate the percent of analyzed individuals that were labeled as a function of major taxon (annelids, non-annelid metazoans, and protozoans) and time of exposure to tracer at sites I and III. The $\delta^{13}C$ values of -12‰ and -15‰ were used as thresholds to determine label uptake at sites I and III, respectively. The numbers shown on the bars are the number of individuals analyzed. Plots 5b and d show the mean $\delta^{13}C$ values of the infauna. Values on the bars are the standard deviations about the means.

organic result of the oxidation of subducted diatom tracer (Levin et al. 1997). The subsurface oxidation of carbon at this site is dominated by anaerobic processes as suggested by the natural abundances of $^{13}C/^{12}C$ in the background DIC pool (Blair et al. 1994). The $\delta^{13}C$ of the DIC added to the background porewater pool was ~ -18‰ to -20‰, reflecting the oxidation of marine organic matter with no corresponding dissolution of carbonate (~ 0‰). Carbonate dissolution would be

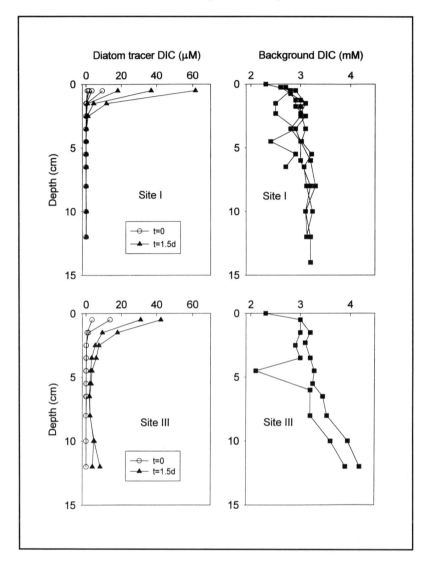

Fig. 6. The concentration of tracer diatom carbon ($^{13}C + ^{12}C$) in the DIC fraction after 1.5 d. Oxidation of emplaced diatoms was largely localized at the surface at site I. DIC transport into the seabed was controlled by molecular diffusion. At site III, oxidation occurred at depth. Note the subsurface peaks near 10 cm, which correspond with those in the POC fraction. Also shown are the background DIC concentrations for comparative purposes. The irregular, non-ideal shapes of the background DIC profiles from site III suggest that the DIC is lost from the seabed via processes other than diffusion. This could include bioirrigation.

expected if oxidation occurred aerobically. If the fate of the diatom tracer was characteristic of what happens to natural detrital falls, then the subduction activities of macrofauna play a significant role in determining the aerobic-anaerobic exposure history of reactive organic matter (Graf 1989; Levin et al. 1997; Gerino et al. 1998). This in turn will influence the pathways and rates of organic diagenesis and the resulting geochemical record.

The results from the reaction-transport model could not be reconciled with observed $DI^{13}C$ profiles at site III when molecular diffusion was assumed to be the dominant mode of solute transport. The shape of background DIC porewater profiles (Fig. 6) along with observations of numerous burrows and tubes, the larger of which often had rust-colored linings, indicate that irrigation was an important solute transport process. The model output also suggested that either irrigation rates were higher at the surface than in the subsurface, or k was larger for subducted material. The former scenario is plausible because infaunal densities decrease with depth (Levin et al. 1999). Aller (1988) discussed the relationship between burrow density and irrigation intensity. Subduction of newly deposited organic matter may have increased k by mixing the material with sediment and thus providing an inoculum for microbial degradation. The relative importance of the two scenarios has not been determined.

Summary and Conclusions

The organic detritus deposited on the North Carolina continental slope is derived predominantly from marine sources. Much of it has been recycled heterotrophically. Components of the organic mixture are modern to greater than 700 yr in age. The youngest material is preferentially incorporated into the benthic biomass presumably because it is biochemically recognizable by organisms for anabolic processes and the older material is less so.

Benthic response to the deposition of organic detritus appears to be linked to the average rate of deposition. Organic matter deposited at site I, an environment that normally experiences a low particulate flux, tends to be consumed aerobically on or near the sediment surface. Agglutinating protozoans and surface-feeding annelids are the primary macrofaunal consumers of phytodetritus shortly after its deposition. Nonlocal transport of material to the subsurface occurs sporadically. Microbial colonization of the deposited material occurs over the time scale of days.

Organic matter falling on a seabed that is characterized by high particle deposition rates (site III) is rapidly mixed into the seabed and much of it is decomposed anaerobically. Offshore of Cape Hatteras, nonlocal transport of material down to > 10 cm by annelids is a frequent phenomenon and may be an important source of food for subsurface-feeding animals and the anaerobic ecosystem.

Particles are effectively moved horizontally across the seafloor surface at these sites. If the horizontal transport is largely physical in nature or associated with faunal locomotion, the most obvious impact on the carbon biogeochemistry will be the homogenization of the surface sediments. On the other hand, if transport is the consequence of ingestion and egestion, it could represent a significant, yet virtually untraceable, diagenetic process. The importance of horizontal transport on the organic geochemical record merits further investigation.

The interrelationship between sedimentary organic carbon and the benthos is complex. The character and depositional flux of organic carbon have a profound influence on faunal composition and density. At the same time, the attributes of the benthic organisms control the fate of the organic matter after deposition. The resulting molecular organic record and the organic carbon flux are linked intimately. Delineation of that linkage awaits future studies of organism-sediment interactions.

Acknowledgments

The skilled contributions of the crews, pilots, and captain of the R/V *Edwin Link* and *Johnson SeaLink* were essential for this project. We thank the many participants of the SLOPEX cruises, including D. Beauregard, S. Boehme, E. Clesceri, R. Collman, J. Crooks, D. Daley, C. DiBacco, D. Dinsmore, A. McCray, T. Pease, C. Reimers, D. Savidge, W. Savidge, and D. Talley, for their technical assistance. D. Albert, R. C. Aller, L.

Benninger, M. Alperin, R. Jahnke, and C. S. Martens were of great assistance during the DOE Ocean Margins Program portion of this research. C. E. Reimers, W. R. Martin, and R. C. Aller are thanked for their reviews of this manuscript. This research was supported by grants from the NOAA National Undersea Research Program at Wilmington, North Carolina, the National Science Foundation (OCE 93-11711 and 93-01793), and the Department of Energy (DE-FG05-95ER62082).

Literature Cited

Aller, R. C. 1988. Benthic fauna and biogeochemical processes in marine sediments: The role of burrow structures, p. 310–338. *In* Blackburn, T. H. & J. Sorensen (eds.), Nitrogen Cycling in Coastal Marine Environments. J. Wiley & Sons, New York.

Anderson R F., G. T. Rowe, P. Kemp, S. Trumbore, & P. E. Biscaye. 1994. Carbon budget for the mid-slope depocenter of the Mid-Atlantic Bight. *Deep-Sea Research II* 41:669–703.

Berner, R. A. 1980. Early Diagenesis: A Theoretical Approach. Princeton University Press, Princeton, New Jersey.

Blair, N. E. & W. D. Carter, Jr. 1992. The carbon isotope biogeochemistry of acetate from a methanogenic marine sediment. *Geochimica et Cosmochimica Acta* 56:1247–1258.

Blair, N. E., L. A. Levin, D. J. DeMaster, & G. Plaia. 1996. The short-term fate of fresh algal carbon in continental slope sediments. *Limnology and Oceanography* 41:1208–1219.

Blair, N. E., G. R. Plaia, S. E. Boehme, D. J. DeMaster, & L. A. Levin. 1994. The remineralization of organic carbon on the North Carolina continental slope. *Deep-Sea Research II* 41:755–766.

Blake, J. A. & J. F. Grassle. 1994. Benthic community structure on the U.S. South Atlantic slope off the Carolinas: Spatial heterogeneity in a current dominated system. *Deep-Sea Research* 41:834–874.

Burdige, D. J. & C. S. Martens. 1988. Biogeochemical cycling in an organic-rich coastal marine basin: 10. The role of amino acids in sedimentary carbon and nitrogen cycling. *Geochimica et Cosmochimica Acta* 52:1571–1584.

Cahoon, L. B., R. A. Laws, & C. J. Thomas. 1994. Viable diatoms and chlorophyll *a* in continental slope sediments off Cape Hatteras, North Carolina. *Deep-Sea Research II* 41:767–782.

Canfield, D. E. 1994. Factors influencing organic carbon preservation in marine sediments. *Chemical Geology* 114:315–329.

Cowie, G. L. & J. I. Hedges. 1991. The role of anoxia in organic matter preservation in coastal sediments: Relative stabilities of the major biochemicals under oxic and anoxic depositional conditions. *Advances in Organic Geochemistry* 19:229–234.

Cowie, G. L. & J. I. Hedges 1996. Digestion and alteration of the biochemical constituents of a diatom (*Thalassiosira weissflogii*) ingested by an herbivorous zooplankton (*Calanus pacificus*). *Limnology and Oceanography* 41:581–594.

Csanady, G. T. & P. Hamilton. 1988. Circulation of slopewater. *Continental Shelf Research* 8:565–624.

de Leeuw, J.W. & C. Largeau. 1993. A review of macromolecular organic compounds that comprise living organisms and their role in kerogen, coal and petroleum formation, p. 23–72. *In* Engel, M. H. & S.A. Macko (eds.), Organic Geochemistry - Principles and Applications. Plenum Press, New York.

DeMaster, D. J., R. H. Pope, L. A. Levin, & N. E. Blair. 1994. Biological mixing intensity and rates of organic carbon accumulation in North Carolina slope sediments. *Deep-Sea Research II* 41:735–753.

Dobbs, F. C. & R. B. Whitlach. 1982. Aspects of deposit-feeding by the polychaete *Clymenella torquata*. *Ophelia* 21:159–166.

Fornes, W. L. 1996. Bioturbation and particle transport in Carolina slope sediments: A radiochemical approach. M. S. thesis. North Carolina State University, Raleigh, North Carolina.

Fornes, W. L., D. J. DeMaster, L. A. Levin, & N.E. Blair. 1999. Bioturbation and particle transport in Carolina slope sediments: A radiochemical approach. *Journal of Marine Research* 57:335–355.

Fry, B. & E. B. Sherr. 1984. $\delta^{13}C$ measurements as indicators of carbon flow in marine and freshwater ecosystems. *Contributions in Marine Science* 27:13–47.

Gearing, J. N., P. J. Gearing, D. T. Rudnick, A. G. Requejo, & M. J. Hutchins.1984. Isotopic variability of organic carbon in a phytoplankton-based temperate estuary. *Geochimica et Cosmochimica Acta* 48:1089–1098.

Gerino, M., R. C. Aller, C. Lee, J. K. Cochran, J. Y. Aller, M. A. Green, & D. Hirschberg. 1998. Comparison of different tracers and methods used to quantify bioturbation during a spring bloom: 234-Thorium, luminophores and chlorophyll *a*. *Estuarine, Coastal and Shelf Science* 46:531–547.

Graf, G. 1989. Benthic-pelagic coupling in a deep-sea benthic community. *Nature* 341:437–439.

Harvey, H. R. 1994. Fatty acids and sterols as source markers of organic matter in sediments of the North Carolina continental slope. *Deep-Sea Research* 41:783–796.

Hecker, B. 1990. Photographic evidence for the rapid flux of particles to the sea floor and their transport down the continental slope. *Deep-Sea Research* 37:1773–1782.

Hedges, J. I. & R. G. Keil 1995. Sedimentary organic matter preservation: An assessment and speculative synthesis. *Marine Chemistry* 49:81–115.

Hedges, J. I. & P. L. Parker. 1976. Land-derived organic matter in surface sediments from the Gulf of Mexico. *Geochimica et Cosmochimica Acta* 40:1019–1029.

Hentschel, B. 1998. Intraspecific variations in $\delta^{13}C$ indicate ontogenetic diet changes in deposit-feeding polychaetes. *Ecology* 79:1357–1370.

Henrichs, S. M. & A. P. Doyle. 1986. Decomposition of ^{14}C-labeled organic substances in marine sediments. *Limnology and Oceanography* 31:765–778.

Hughes, J. A. 1996. Benthic foraminiferal assemblages on the North Carolina continental slope. M. S. thesis, University of Southhampton, United Kingdom.

Hunt, J.M. 1970. The significance of carbon isotope variations in marine sediments, p. 27–35. *In* Hobson, G. D. & G. C. Speers (eds.), Advances in Organic Geochemistry, 1966, Pergamon Press, Oxford.

Jumars, P. A., L. M. Mayer, J. W. Deming, J. A. Baross, & R. A. Wheatcroft. 1990. Deep-sea deposit-feeding strategies suggested by environmental and feeding constraints. *Philosophical Transactions of the Royal Society of London Series A* 331:85–101.

Kelchner, C. 1992. Organic-rich depositional regimes, North Carolina continental slope: Evaluating source sites and sediment transport pathways. M. S. thesis, North Carolina State University, Raleigh, North Carolina.

Kudenov, J. D. 1978. The feeding ecology of *Axiothella rubrocincta* (Johnson) (Polychaeta: Maldanidae). *Journal of Experimental Marine Biology and Ecology* 31:209–221.

Levin, L., N. Blair, D. DeMaster, G. Plaia, W. Fornes, C. Martin, & C. Thomas. 1997. Rapid subduction of organic matter by maldanid polychaetes on the North Carolina slope. *Journal of Marine Research* 55:595–611.

Levin, L., N. Blair, C. Martin, D. DeMaster, G. Plaia, & C. Thomas. 1999. Macrofaunal processing of phytodetritus at two sites on the Carolina margin: In situ experiments using ^{13}C-labeled diatoms. *Marine Ecology Progress Series* 182:37–54.

Mangum, C. 1964. Activity pattern in the metabolism and ecology of polychaetes. *Comparative Biochemistry and Physiology* 11:239–256.

Mellor, C. A. & C. K. Paull. 1994. Sea beam bathymetry of the Manteo 467 lease block off Cape Hatteras, North Carolina. *Deep-Sea Research* 41:711–718.

Middelburg, J. J. 1989. A simple rate model for organic matter decomposition. *Geochimica et Cosmochimica Acta* 53:1577–1581.

Premuzic, E. T., C. M. Benkovitz, J. S. Gaffney, & J. J. Walsh. 1982. The nature and distribution of organic matter in the surface sediments of world oceans and seas. *Organic Geochemistry* 4:63–77.

Romankevich, E. A. 1984. Geochemistry of Organic Matter in the Ocean. Springer-Verlag, Berlin.

Schaff, T., L. Levin, N. Blair, D. DeMaster, R. Pope, & S. Boehme. 1992. Spatial heterogeneity of benthos on the North Carolina continental slope: Large (100 km)-scale variation. *Marine Ecology Progress Series* 88:143–160.

Sun, M.-Y., C. Lee, & R. C. Aller. 1993. Anoxic and oxic degradation of ^{14}C-labeled chloropigments and a ^{14}C-labeled diatom in Long Island Sound sediments. *Limnology and Oceanography* 38:1438–1451.

Tanaka N., K. K. Turekian, & D. M. Rye. 1991. The radiocarbon, $\delta^{13}C$, ^{210}Pb, and ^{137}Cs record in boxcores from the continental margin of the Middle Atlantic Bight. *American Journal of Science* 291:90–105.

Thomas, C. J. 1998. Individual deposit-feeder, community and ecosystem level controls on organic matter diagenesis in marine benthic environments. Ph. D. thesis, North Carolina State University, Raleigh, North Carolina.

Turley, C. M. & K. Lochte. 1990. Microbial response to the input of fresh detritus to the deep-sea bed. *Palaeogeography, Palaeoclimatology, Palaeoecology (Global and Planetary Change Section)* 89:3–23.

Ullman, W. J. & R. C. Aller. 1982. Diffusion coefficients in near-shore marine sediments. *Limnology and Oceanography* 27:552–556.

Walsh, J. J. 1994. Particle export at Cape Hatteras. *Deep-Sea Research II* 41:603–628.

Walsh, J. J., E. T. Premuzic, J. S. Gaffney, G. T. Rowe, G. Harbottle, R. W. Stoenner, W. L. Balsam, P. R. Betzer, & S. A. Macko. 1985. Organic storage of CO_2 on the continental slope off the mid-Atlantic bight, the southeastern Bering Sea, and the Peru coast. *Deep-Sea Research* 32:853–883.

Westrich, J. T. & R. A. Berner. 1984. The role of sedimentary organic matter in bacterial sulfate reduction. *Limnology and Oceanography* 29:236–249.

Unpublished Material

Andrew Gooday. unpublished results. Southampton Oceanography Centre, European Way, Southampton SO14 3ZH.

Appendix

A reaction-transport model that employs nonlocal as well as biodiffusive mixing was used to estimate the in situ oxidation rate of the ^{13}C-labeled diatom. The model was also used to determine if transport of the DIC derived from the oxidation of the ^{13}C-labeled diatom was dominated by molecular diffusion or was heavily influenced by an advective process such as irrigation. Portions of this model have been used to reproduce subsurface profiles of ^{13}C-labeled *Chlorella* emplaced in situ on North Carolina slope sediments (Blair et al. 1996), and ^{13}C-labeled diatoms and DIC in laboratory experiments using slope and shelf sediments (Thomas 1998). The model describes the distribution of tracer in two pools: the instantaneously deposited surface layer and the subsurface distribution. The concentration of the particulate tracer in the deposited tracer layer, C_o ($0 \leq x \leq \varepsilon$), was described by

$$(1) \qquad \frac{\partial C_o}{\partial t} = \frac{\partial}{\partial x}\left(D_b \frac{\partial C_o}{\partial x}\right) - [k + \underline{R}] \, C_o$$

where:

C = concentration of particulate organic tracer C from diatom (μmoles C cm^{-3})
t = time (yr)
x = depth in sediment (cm)
D_b = biological mixing coefficient (cm^2 yr^{-1})
k = first-order decay coefficient for the diatom (yr^{-1})
$\underline{R}$ = nonlocal transport removal term (yr^{-1})
ε = thickness of the tracer layer (1 mm)

Nonlocal transport of tracer from the surface layer was assumed to be first-order with respect to surface concentration and to be unidirectional. It was also assumed that the 1-mm thick tracer layer remained well mixed and that deposit feeding was depth independent within the layer. The concentration profile of tracer POC below the tracer layer ($\varepsilon \leq x \leq L$) was simulated using

$$(2) \qquad \frac{\partial C}{\partial t} = \frac{\partial}{\partial x}\left(D_b \frac{\partial C}{\partial x}\right) - kC + s_x \overline{C_o}$$

where:

L = depth of the mixed layer (~15 cm)
s_x = nonlocal transport supply term (yr^{-1})

$$(3) \qquad \overline{C_o} = \frac{1}{\varepsilon} \int_0^\varepsilon C_o \, dx$$

When there were multiple subsurface peaks, the nonlocal removal term, $\underline{R}$, was given by

$$(4) \qquad \underline{R} = \sum_{x=1}^{n} r_x$$

where n was the number of peaks. For any given peak, the nonlocal supply term, s_x, was related to r_x by

$$(5) \qquad \int_0^\varepsilon r_x \, dx = \int_{x_i}^{x_j} s_x \, dx$$

Equations 1 and 2 were solved numerically with an explicit finite difference method. The two equations were coupled by the nonlocal transport term, $s_x \overline{C}_0$ and the boundary condition described in Eq. 6b, which establishes conservation of tracer flux across the interface at $x = \varepsilon$. Boundary conditions for the solutions were

$$(6a) \qquad \partial C_0 / \partial x = 0 \qquad \qquad x = 0$$

$$(6b) \qquad \partial C_0 / \partial x = \partial C / \partial x \qquad x = \varepsilon$$

$$(6c) \qquad \partial C / \partial x = 0 \qquad \qquad x = L$$

The reaction portion of the model simulated DIC production from the tracer diatom in the surface tracer layer and subsurface porewater pools. The concentration of tracer DIC ($^{12}C + {}^{13}C$) oxidatively produced was given by

$$(7) \qquad \frac{\partial [DIC]}{\partial t} = \frac{\partial \Phi \, D_s}{\partial x} \left(\frac{\partial [DIC]}{\partial x} \right) + \frac{kC}{\Phi}$$

$$(8) \qquad D_s = D_o \times \Phi^2 \quad \text{(Ullman \& Aller 1982)}$$

where:

D_o = temperature/viscosity corrected free solution diffusivity of HCO_3^-
Φ = sediment porosity

At time zero, the concentration of tracer DIC in both the sediment and overlying water was set at zero. The overlying water was assumed to be well mixed, and the only source of tracer DIC to the overlying water was the flux out of the sediment. All model profiles were integrated over 1-cm intervals to facilitate comparison with data. Time steps of 0.1 h were adequate to maintain stability.

Initially, the tracer POC profiles from each 1–1.5 d plot were reproduced with the model using the total tracer POC inventory of each plot as C_0 at t = 0, and D_b's generated from profiles of naturally occurring ^{234}Th. The nonlocal transport term, s_x, and in some cases D_b, were adjusted iteratively to provide the best fit of the model output to observed tracer POC profiles.

After fitting the POC data, k was adjusted to reproduce the tracer-derived DIC distributions. At site I, solute transport characterized by D_s was adequate to fit model results to field measurements. A time-dependent k in the general functional form of At or Be^{Ct}, improved the fit. At site III, measured and modeled results could not be reconciled if molecular diffusion was assumed to be the sole means of solute transport. In addition, the ingrowth of tracer DIC appeared more rapid at depth than at the surface in the field experiments when compared to model predictions. In other words, the model underpredicted subsurface DIC production when its output was forced to fit the surface portion of the DIC profile. This may be because solute transport is more rapid at the surface as a result of irrigation, which was not included in the model, and/or the oxidation of the tracer POC is more rapid at depth. Those two explanations cannot be resolved with the available data.

Mechanisms of Age-Dependent Bioturbation on the Bathyal California Margin: The Young and the Restless

Craig R. Smith*, David J. DeMaster, and William L. Fornes

Abstract: Bioturbation in deep-sea habitats frequently is tracer dependent, with short-lived tracers (in particular ^{234}Th) apparently mixed at higher rates than longer lived counterparts (e.g., ^{210}Pb). We postulate three plausible mechanisms that could yield such tracer-dependent mixing: (1) selective pickup and ingestion of young, food-rich particles throughout the mixed layer by deposit feeders, (2) rapid bioadvection of young, food-rich particles into the sediment through the feeding and caching activities of benthos, and (3) passive cascade of depositing particles into open burrows. The first two mechanisms are variations of age-dependent mixing, wherein deposit feeders differentially consume and mix recently deposited, food-rich materials. These three mechanisms yield contrasting predictions concerning excess ^{234}Th activity in the gut sediments of deposit feeders, and the relative fates of young (nutrient-rich) and old (nutrient-poor) particles experimentally introduced to the seafloor. We tested the predictions in three bathyal basins (Santa Catalina, San Nicolas, and San Clemente) off southern California by conducting gut-sediment measurements and particle-introduction experiments at the seafloor. As predicted by the selective-pickup and rapid-bioadvection models, gut sediments of deposit feeders were dramatically enriched (frequently 6–74 fold) in excess ^{234}Th activity compared with their general feeding zones. Our particle-introduction experiments, conducted in Santa Catalina Basin with ^{139}Ce-labeled diatoms (Ditylum), ^{88}Y-labeled surface sediments, and ^{113}Sn-labeled subsurface sediments, indicate that all particles types penetrated to depths of 2 cm within 1 h, apparently cascading down open burrows. On 74-d time scales, diatom detritus was mixed or subducted into the sediment 1.5–3 times faster than less food-rich sediments, as predicted by the selective-pickup model. However, the strongest pattern in our experiments was a 10-fold reduction in mixing rates of all particle types over 4 d to 74 d and over 74 d to ~500 d, as expected under the rapid-bioadvection model. We conclude that tracer-dependent mixing in Santa Catalina Basin (and other deep-sea sites) results largely from differentially rapid, initial subduction of young, food-rich particles. In other words, the nature of bioturbation varies dramatically with particle age. Our findings indicate a strong need to match the time scales of bioturbation tracers and modeled reactants in studies of diagenesis at the deep-sea floor.

Introduction

Nearly all surficial marine sediments are "bioturbated" by benthos. Field studies and scaling arguments suggest that solid-phase bioturbation is controlled by larger deposit-feeding animals, in particular the macrofauna and megafauna (Aller 1982; Thayer 1983; Wheatcroft et al. 1990). Deposit

263

feeder control of mixing seems especially likely in the deep sea, where low sedimentary organic content should yield high weight-specific deposit-feeding rates (Cammen 1980; Smith 1992).

Knowledge of the mechanisms and rates of bioturbation is critical to understanding a variety of processes at the ocean floor. For example, the rate constants calculated for organic-matter degradation, pollutant breakdown, silica dissolution, and manganese reduction are influenced by bioturbation intensity (e.g., Schink et al. 1975; Berner 1980; Aller 1990; Hammond et al. 1996). In addition, the burial rate of organic matter in sediments, and the utility of carbon accumulation as a paleo-productivity indicator, may depend on sediment mixing (Emerson 1985; Rabouille & Gaillard 1991). Finally, reconstructions of pollution histories and paleoclimates from sediment cores must account for the smearing of input signals by animal activities (e.g., Wheatcroft 1990; Kramer et al. 1991). Clearly, if we wish to understand the role of marine sediments as biogeochemical sources and/or sinks, we must elucidate the nature of bioturbation.

Models of sediment diagenesis generally assume that animals bioturbate all particles uniformly, that is, sediment grains differing in size, organic content, density, etc., are all mixed similarly (Smith et al. 1993). Based on this assumption, one need only measure a single, particle-associated tracer to characterize the intensity and depth of bioturbation in a particular environment. Thus, naturally occurring, particle-reactive radiotracers such as ^{210}Pb (half-life = 22.3 yr) or ^{234}Th (half-life = 24.1 d) are often used in isolation to characterize sediment mixing regimes (see Aller 1982 and Smith et al. 1993 for reviews). Both isotopes are scavenged from the water column by sinking particles, yielding an excess activity that is used to trace particle behavior for about five half-lives after arrival at the seafloor (e.g., Aller & DeMaster 1984; Pope et al. 1996).

However, this assumption of uniform mixing of sediment particles may be seriously in error. For example, size-dependent mixing may be common in marine sediments (e.g., Wheatcroft 1992; but see Smith & Kukert 1996 for a counter example). In addition, we have shown that bioturbation in deep-sea sediments can be highly tracer dependent, with short-lived radiotracers such as ^{234}Th mixed 10–100 times faster than their longer lived counterparts such as ^{210}Pb (Smith et al. 1993, 1997). We hypothesized that this negative relationship between mixing rate and tracer half-life results from *age-dependent mixing*, wherein recently sedimented, food-rich particles are ingested and mixed at higher rates by deposit feeders than are older, food-poor particles. Because excess ^{234}Th ($^{234}Th_{xs}$) is restricted to freshly deposited particles (i.e., those < 100 d old) and the bulk of $^{210}Pb_{xs}$ remains on particles for decades after deposition, $^{234}Th_{xs}$ differentially records the higher mixing rates associated with young sediment grains (Smith et al. 1993).

We can envision three specific mechanisms likely to yield enhanced mixing rates of $^{234}Th_{xs}$, relative to $^{210}Pb_{xs}$, in deep-sea sediments (Smith et al. 1993) (Fig. 1); the first two are variations on the theme of age-dependent mixing. (1) *Continuous selective pickup and ingestion of young, food-rich particles by deposit feeders throughout the mixing layer*. In the simplest form, this mechanism would result from selective feeding of both surface and subsurface deposit feeders, and would enhance diffusive bioturbation of young particles throughout the mixing layer (Smith et al. 1993). (2) *Rapid bioadvection of recently deposited, food-rich particles into the sediment, followed by relatively slow diffusive subsurface mixing*. Such rapid bioadvection could result from subduction of surficial grains shortly after deposition events (cf. Graf 1989; Levin et al. 1997) or in patches where young particles are differentially accumulating (e.g., in pits; Yager et al. 1993). Rapid bioadvection could be driven by direct ingestion or burrow caching of food-rich particles by reverse-conveyor-belt feeders such as echiurans, sipunculans, and some maldanid polychaetes (Smith et al. 1986, 1993, 1996; Graf 1989; Wheatcroft et al. 1990; Levin et al. 1977). (3) *Passive cascade of depositing material into open*

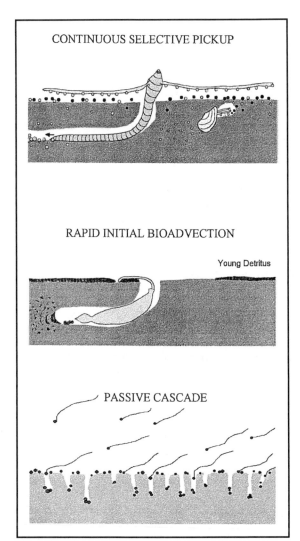

CONTINUOUS SELECTIVE PICKUP

RAPID INITIAL BIOADVECTION

Young Detritus

PASSIVE CASCADE

Fig. 1. The three mechanisms potentially causing tracer-dependent mixing in deep-sea sediments. Top: Selective pickup and ingestion of young, food-rich particles by deposit feeders throughout the mixing layer. Here, the younger particles are lighter in color and are selectively ingested and displaced throughout the mixed layer by both surface and subsurface deposit-feeders (e.g., spionid polychaetes and proto-branch bivalves, respectively). Middle: Rapid initial bioadvection of recently deposited, food-rich particles (darker layer), followed by slow, nonselective, diffusive mixing. In this example, an echiuran is ingesting and subducting a recently deposited layer of phytodetritus. Bottom: Passive cascade of depositing particles (darker shading) into open animal burrows. Young particles penetrate the sediment at the time of deposition as they fall into tubes and burrows.

animal burrows, followed by slow diffusive subsurface mixing. Such a process would entrain depositing particles into the sediment on very short time scales and independently of food quality.

These three mechanisms yield contrasting predictions concerning (a) $^{234}Th_{xs}$ activity in the gut sediments of deposit feeders, and (b) the fate of particles of differing ages (and food quality) experimentally introduced to the seafloor (Table 1). Specifically, under mechanism number 1 (*continuous selective pickup*), $^{234}Th_{xs}$ activity in the guts of surface feeders and subsurface deposit feeders (particularly those controlling mixing) should be enriched relative to their general feeding zones. Under mechanism number 2 (*rapid bioadvection*), the surface feeders responsible for rapid subduction should have gut sediments enriched in $^{234}Th_{xs}$ activity, whereas subsurface deposit feeders need not. Finally, the *passive-cascade* model (mechanism number 3) predicts no enrichment of $^{234}Th_{xs}$ activity in deposit-feeder gut sediments (Table 1).

Predictions regarding introduced particles of varying ages are as follows. Consider an experiment (as outlined in Smith et al. 1993) in which similarly sized particles of the following types are introduced as a pulse to the seafloor: (a) fresh diatoms (= young particles), (b) particles from the top 0.5 cm of the sediment column (= middle-aged particles), and (c) particles from 30-cm deep in the sediment column (= old particles). The *continuous selective-pickup* model (mechanism number 1) predicts that the fresh diatoms will be "bioturbated" throughout the mixing layer at higher intensities than middle-aged or old particles for at least the first ~100 d of the experiment (i.e., over the characteristic time scale of $^{234}Th_{xs}$ activity; see modeling results of Smith et al. 1993). The *rapid initial bioadvection* model (mechanism number 2) predicts that introduced particles will penetrate the sediment on time scales of days (cf. Graf 1989; Levin et al. 1997), and then be mixed nonselectively within the sediment. The *passive-cascade* model (mechanism no. 3) predicts that introduced particles of all ages will penetrate the sediments similarly on very short time scales (< 1 h) as they settle to the seafloor, and then be nonselectively mixed within the sediments. Note that each model predicts a different time scale of rapid penetration and/or mixing of young particles into the sediment; passive cascade predicts < 1 h, rapid bioadvection predicts days, and general selective feeding predicts tens of days. These predictions are summarized in Table 1.

In this paper, we present an overview of tests of these predictions at bathyal depths off southern California. Specifically, we compared the $^{234}Th_{xs}$ activity in deposit-feeder guts and surface sediments from three California basins. In addition, we conducted in situ experiments to trace the behavior of fresh diatoms, surface sediments, and subsurface sediments introduced as a pulse to the Santa Catalina Basin floor. We find strong evidence that age-dependent mixing contributes substantially to the rapid bioturbation of $^{234}Th_{xs}$ in deep-sea sediments.

Materials and Methods

GUT MEASUREMENTS OF $^{234}TH_{XS}$ ACTIVITY

Megafauna were collected with an 8-m semiballoon otter trawl as in Miller et al. (2000) from the floors of three basins off southern California: 1,240 m deep in Santa Catalina Basin (~33° 10.7' N, 118° 27.2' W), 1,780 m deep in San Nicolas Basin (~33°4.0'N, 119° 4.0' W), and 1,900 m deep in San Clemente Basin (~32°31.8'N, 118° 8.0' W) (see Emery 1960 for a description of these basins). Trawl sampling in Santa Catalina Basin occurred in December 1995, January 1996, and August 1997, in San Nicolas Basin in June 1997, and in San Clemente Basin in April 1998. Large deposit-feeders were dissected within 2 h of recovery, and their sediment-laden guts removed and frozen. For larger holothurians (i.e., *Pannychia moseleyi* Théel from Santa Catalina Basin, holothurian species 1 from San Nicolas Basin, and *Mesothuria lactea* (Théel) from San Clemente Basin) only sediments from the foregut were assayed for $^{234}Th_{xs}$ activity. For smaller species, whole guts from a number of individuals were combined as follows to provide adequate sediment mass for analysis: 3–5 individuals of *Bathybembix bairdii* (Dall); 10 individuals of sipunculan species 1 from San Nicolas; 10 individuals of *Nuculana* sp.; 12 individuals of *Malletia* sp.; 3–5 individuals of *Eremicaster pacificus* (Ludwig); and 2–4 individuals of *Molpadia musculus* (Risso) from San Clemente Basin. In the laboratory, $^{234}Th_{xs}$ activity in gut sediments was measured as in Pope (1992) and Miller et al. (2000). Briefly, ^{234}Th activity was measured on a low-level, gas-flow, anticoincidence beta counter, and ^{238}U activity was measured with alpha spectroscopy. $^{234}Th_{xs}$ was then calculated by subtracting ^{238}U activity from total ^{234}Th activity and correcting for decay since sample collection. Activities are expressed in dpm g^{-1} dry sediment corrected for salt content.

MEASUREMENTS OF SURFACE-SEDIMENT $^{234}TH_{XS}$ ACTIVITY

During the same time intervals as the trawls, seafloor surface sediments were collected using a multiple core that collected 10-cm diameter tube cores (as in Pope et al. 1996). Cores were carefully extruded and the top 0.5 cm of sediment removed and assayed for $^{234}Th_{xs}$ activity using the methods of Pope et al. (1996). In San Nicolas Basin, $^{234}Th_{xs}$ activity was also measured for surface floc carefully siphoned off the tops of four multiple core tubes, using the methods of Smith et al. (1997).

Table 1. Predicted answers to questions posed in this study for each of the mechanisms potentially yielding tracer-dependent mixing in deep-sea sediments. The third and fourth rows of the table describe the predicted penetration of particles of different ages experimentally introduced as a pulse to the seafloor. Y = fresh diatoms (i.e., young sediment), MA = sediments from 0 to 0.5 cm sediment depths (i.e., middle-aged sediment), and O = sediments from 30-cm deep in the sediment column (i.e., old sediment).

Question	Tracer-Dependent Mixing Mechanism		
	(1) Selective Particle Pickup Throughout Mixed Layer	(2) Rapid Initial Bioadvection	(3) Passive Cascade into Burrows at Deposition
Surface-deposit-feeder guts enriched in $^{234}Th_{xs}$?	yes	yes	no
Subsurface-deposit-feeder guts enriched in $^{234}Th_{xs}$?	yes	no	no
Relative penetration rates of particles of different ages?	Y > MA ≥ O	Y = MA = O	Y = MA = O
Time scale of most rapid particle penetration or mixing?	10–100 days	~ 10 days	< 1 hour

All $^{234}Th_{xs}$ activities from core sediments are expressed in dpm g^{-1} dry sediment (corrected for salt content).

PARTICLE INTRODUCTION EXPERIMENTS

Particle Introduction Experiments (PIEs) were carried out in Santa Catalina Basin to explore the mixing of particles of different "ages" into the sediment. Four similarly sized (roughly 20–90 m diameter) particle types were used: (1) diatoms, *Ditylum brightwelli* (West), freshly grown in the laboratory, (2) the top 0.5 cm of Catalina Basin surface sediments collected by multiple corer, (3) Catalina Basin sediments from 30–30.5 cm deep in multiple cores, and (4) glass beads. *Ditylum* (mean cell size ~ 60–90 m diameter) was grown in batch culture in the laboratory, labeled with the highly particle reactive radionuclide ^{139}Ce (half-life = 138 d) and then freeze dried or refrigerated for 1–2 wk until use. Sediments for PIEs were collected in Santa Catalina Basin by multiple corer approximately 1 mo prior to experiment emplacement. Immediately after multiple-core recovery, sediments were extruded and the 0–0.5 cm and 30–30.5 cm intervals removed and stored separately under refrigeration. The $^{234}Th_{xs}$ activity in the 0–0.5 cm layer of Santa Catalina sediments typically falls between 7 and 60 dpm g^{-1}, while $^{234}Th_{xs}$ activity in the 30–30.5 cm layer is zero (Smith et al. 1993; Fornes 1999). Because Santa Catalina Basin has a sedimentation rate of ~ 13 cm ky^{-1} (Fornes

1999), the 30–30.5 cm sediments are, on average, several thousand years old. After collection, the PIE sediments were shipped to the laboratory and allowed to settle in a refrigerated seawater bath; particles settling at velocities characteristic of Stokes diameters of > 20 μm were retained. The sediments from 0 to 0.5 cm were labeled with the highly particle-reactive isotope [88]Y (half-life = 107 d) and those from 30 to 30.5 cm labeled with the highly particle-reactive isotope [113]Sn (half-life = 115 d). Sediments were kept refrigerated throughout storage and handling. Glass beads (Glas Shot manufactured by Cataphote) were passed through a 63-μm sieve and collected on a 42-μm sieve and washed in distilled water. On shipboard prior to experiment emplacement, 6 g of *Ditylum*, 8 g of surface sediments, 8 g of subsurface sediments, 30 g of glass beads, and 10 g of kaolin were mixed and placed in shaker apparatuses. Shakers were then taken to the seafloor using the research submersible *Alvin* and one shakerfull, each, of particles was dispersed inside a plexiglass canopy over randomly located, replicate 0.16-m[2] experimental plots using methods similar to Fornes et al. (1999). Particles were allowed > 50 min to settle to the seafloor within canopies (a vertical distance < 40 cm), and canopies were then gently removed. Sites were marked for relocation by *Alvin* with plastic stakes and triplane acoustic reflectors. Implanted particle treatments contained ~ 0.25 g of *Ditylum* organic carbon, an amount roughly equivalent to 90 d of natural sedimentation to 0.16 m[2] of Catalina Basin floor. Experimental methods for PIEs are detailed in Fornes et al. (1999).

After emplacement, replicate PIE plots were sampled destructively from the *Alvin* using 400-cm[2] Ekman cores that were internally subdivided into four 100-cm[2] subcores (Kukert & Smith 1992). Single Ekman cores were collected on replicate plots at the following times: immediately after canopy removal (t = 0), t = 4 d, t = 74 d, t = 520 d, and t = 594 d. One 100-cm[2] subcore from each plot was sectioned into 0.5-cm vertical intervals to a depth of 2 cm, and into 1-cm intervals from 2 cm to 10 cm depths. The peripheral "rind" of each section was removed to preclude smearing artifacts from core sampling, leaving 30–60 ml of sediment per sample for radiotracer analyses. The activities of [139]Ce, [88]Y, and [113]Sn in each section were then counted on a gamma detector as detailed in Fornes et al. (1999).

Diffusive mixing intensity, or D_b, for each particle type within each profile was estimated using a numerical, impulse-tracer model as described in Fornes (1999) and Fornes et al. (in press). This model assumes constant diffusive mixing, as well as constant porosity, to a sediment depth greater than 10 cm (the depth of our core samples). Maximum subduction velocities for each particle type within each profile were estimated by dividing the maximum penetration depth of tracer (either in a continuous front or discontinuous peak) by the time interval. Differences among particles types in D_b or subduction velocity at a given time interval were evaluated with the Mood Median Test (Minitab version 12.1) because it is not sensitive to outliers. An $\alpha = 0.05$ was used as the criterion for statistical significance.

Results

GUT MEASUREMENTS OF [234]TH$_{xs}$ ACTIVITY

The gut sediments of surface deposit feeders from all three basins were heavily enriched in [234]Th$_{xs}$ activity compared to surface sediments (Fig. 2). In Santa Catalina and San Clemente basins, enrichment factors (i.e., gut [234]Th$_{xs}$ activity/surface sediment [234]Th$_{xs}$ activity) for surface deposit feeders ranged from 10 to 74. In San Nicolas Basin, enrichment factors were somewhat lower, ranging from 6 to 13. The lowest value (6) was associated with very high variance; of the three holothurians, *Pseudostichopus mollis* Théel, collected from San Nicolas Basin, one had an enrichment

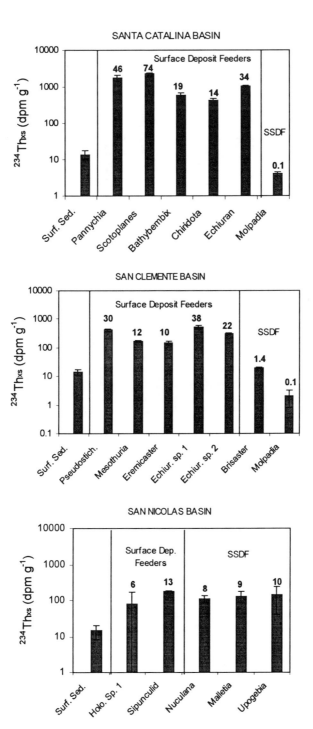

Fig. 2. Excess ^{234}Th activities in surface sediments (0–0.5 cm) and the gut sediments of deposit feeders from Santa Catalina, San Clemente, and San Nicolas basins. SSDF = subsurface deposit-feeders. Means ± 1 se are given. Numbers above columns are enrichment factors of gut sediments relative to surface sediments from the same basin.

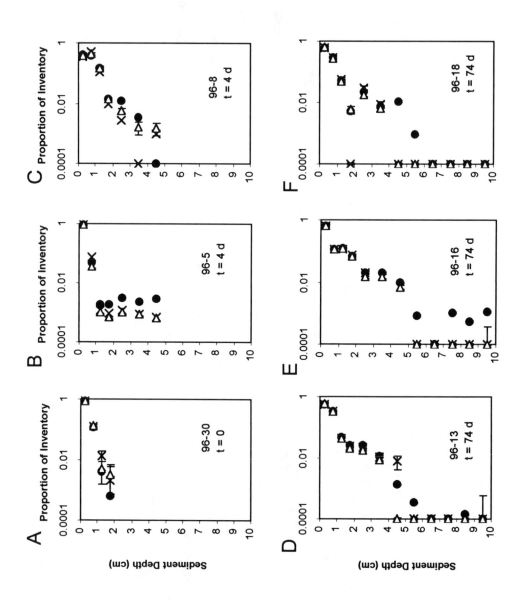

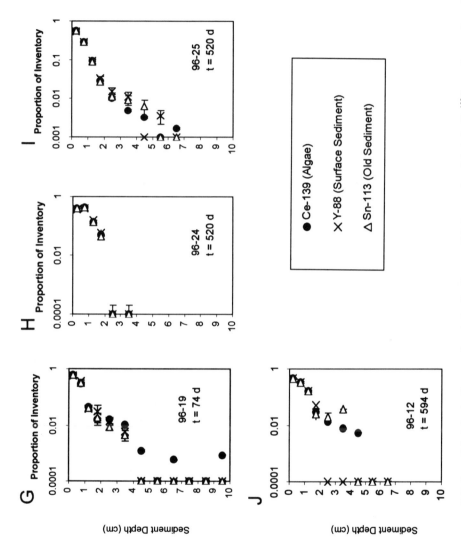

Fig. 3. Depth profiles of gamma activity in particle-introduction experiments for the three particle types: algae labeled with ^{139}Ce, surface sediment (0–0.5 cm level) labeled with ^{88}Y, and old sediment (30–30.5 cm level) labeled with ^{113}Sn. Time (t) is the number of days since canopy removal. Means of the proportion of the tracer inventory in the core found in each sampled interval are given. Error bars are ± 1 standard counting error and are not shown if they are no wider than data symbols. Numbers beginning with "96" are treatment numbers. Points on the y-axis do not differ significantly from zero. In many cases (e.g., all levels in treatment 96-24) points are overlapping and thus difficult to resolve individually.

factor of 17, while the other two had no ^{234}Th$_{xs}$ enrichment. Thus, this species appears to be highly capable of selective feeding on ^{234}Th$_{xs}$-rich particles but does not do so invariably. In San Nicolas Basin, the average gut sediments of surface deposit feeders were 3–6 times richer in ^{234}Th$_{xs}$ activity than floculent material pipetted off the tops of multiple cores (mean = 32 dpm g^{-1}, se = 15, n = 4 for flocculent material). We conclude that the gut sediments of all 12 species of surface deposit feeders sampled in the three basins are heavily enriched in ^{234}Th$_{xs}$ activity compared with the top 0.5 cm of sediment. In San Nicolas Basin (the only site where surface floc data are available), the gut sediments of surface deposit feeders are also enriched in ^{234}Th$_{xs}$ activity when compared with flocculent material at the sediment-water interface.

Interestingly, the gut sediments of all six putative subsurface deposit feeders also have non-zero ^{234}Th$_{xs}$ activities (Fig. 2). For *Molpadia* sp. from Santa Catalina Basin, the presumed feeding depth is 7–10 cm; ^{234}Th$_{xs}$ profiles from Santa Catalina Basin reveal no ^{234}Th$_{xs}$ activity in average sediments from 7 cm to 10 cm depth (Smith et al. 1993; Fornes 1999; Miller et al. 2000). Thus, the gut sediments of *Molpadia* sp. appear to be substantially enriched in ^{234}Th$_{xs}$ activity relative to average sediments at the depth of its feeding zone. In San Clemente Basin, five of six sediment profiles had no detectable ^{234}Th$_{xs}$ activity below 1 cm (Fornes 1999); nonetheless, the gut sediments of one putative subsurface deposit feeder from this basin, the irregular urchin *Brisaster* sp., are equivalent in ^{234}Th$_{xs}$ activity to surface sediments (Fig. 2B). This indicates enrichment of *Brisaster* guts in ^{234}Th$_{xs}$ activity relative to average sediments at its feeding depth. In San Nicolas Basin, all three putative subsurface deposit feeders have gut sediments with an excess (8–10 fold) of ^{234}Th$_{xs}$ as compared with the top 0.5 cm of sediments (Fig. 2C). Their gut ^{234}Th$_{xs}$ activities are also 4–5 fold higher than that in surface floc. Thus, we conclude that the gut sediments of these species are enriched relative to their general feeding zones in ^{234}Th$_{xs}$ activity, whether they are classified as subsurface or surface deposit feeders.

PARTICLE INTRODUCTION EXPERIMENTS

At t = 0 (i.e., after introduced particles had settled for ~ 1 h), gamma-emitting tracers indicated that all three particle types had penetrated significantly into the sediment, reaching a maximal depth of 2.0 cm (Fig. 3A). There was very little difference among particle types in the shapes of profiles at this time, indicating that the penetration process did not select among young, middle-aged, or old particles. We interpret this rapid penetration of particles to 2.0 cm to result from passive cascade of particles down open burrows, although it may also reflect, in part, artifacts caused by jostling of cores during recovery. For subsequent sampling times, we interpreted penetration of particles below the 2.0-cm horizon to result from bioturbation.

At t = 4 d, tracers for all three types of particles had penetrated substantially below 2 cm (Fig. 3B,C), reaching depths of 4 cm to ≥ 5 cm. In one replicate treatment (96-5, Fig. 3B), a larger fractional inventory of ^{139}Ce than either ^{88}Y or ^{113}Sn was found at all depths below 1.5 cm, suggesting that the algal detritus had been transported downward selectively. The second replicate treatment (96-8, Fig. 3C) showed a similar, though less marked, pattern suggestive of selective algal penetration.

At t = 74 d (Fig. 3D-G), ^{139}Ce-labeled particles (i.e., algae) had penetrated to sediment depths of 6 cm to ≥ 10 cm, with some variability among treatments. In contrast, the depth penetration of middle-aged and old particles was 5–6 cm (i.e., a depth penetration similar to that at the 4-d sampling interval). As a consequence, in all four replicates, the tracer for algal detritus (^{139}Ce) had greater proportional inventory deeper in the sediment than ^{88}Y or ^{113}Sn. Greater penetration of the algal tracer is evident both in the form of continuous fronts extending downward from the 2-cm horizon, and in the form of isolated tracer peaks deeper in the cores.

At t = 520–594 d (Fig. 3H-J), tracers had not penetrated substantially deeper than at t = 74 d (i.e., 6–10 cm) and the general shapes of profiles (including the top 2 cm) resembled those from 74 d. Two of the three replicate treatments exhibited somewhat deeper penetration of ^{139}Ce than the other tracers, suggesting selective penetration of algal detritus.

Estimated D_b and maximum subduction velocities exhibited variability as a function of particle type. At t = 74 d, the estimated D_b for ^{139}Ce (i.e., algal detritus) was 1.5 times that of ^{88}Y and ^{113}Sn (tracers for surface and old sediment, respectively), and this difference was statistically significant at p < 0.02 with the Mood Median test (Fig. 4). Similarly, the maximum subduction velocity below the 2-cm level for the algal detritus was nearly three times that for the middle-aged and old sediments at t = 74 d (p < 0.02, Mood Median test; Fig. 4). At the other sampling times (t = 0, 4 d, and 520–594 d), there were no significant differences among particles types in D_b or maximum subduction velocity.

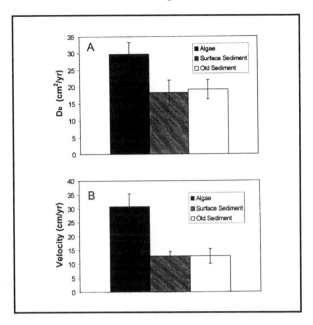

Fig. 4. Mean (± 1 se) for D_b (panel A) and maximum subduction velocity V (panel B) for the three particles types in particle-introduction experiments at 74 d. V is calculated for the portions of profiles below 2 cm (i.e., below the depth of passive cascade).

Estimated D_b and maximum subduction velocities for all three particle types declined dramatically with time from particle introduction (Fig. 5). Mean D_b and maximum subduction velocities dropped roughly 10-fold from 4 to 74 d, and again from 74 to 520–594 d. This time-dependent change in mixing and/or subduction rates was substantially larger than the differences among particle types observed within the 74-d sampling interval (Fig. 5).

Results from the particle-introduction experiments can be summarized as follows. All particle types penetrated sediments to depths of roughly 2 cm within a time scale of 1 h. Over the next 4 d, particles penetrated an additional 2 cm to reach depths ≥ 4 cm, with some evidence of differential penetration of algal (i.e., young) particles. Over the next 70 d, algal particles penetrated an additional 1 cm to reach depths ≥ 5 cm, while profiles of middle-aged and older particles changed little, suggesting selective bioturbation of algal detritus. Between sampling intervals of 74 d and ~ 500 d, there was relatively little change in the depth profiles of all three particle types, implying relatively low rates of particle displacement within this time interval. Intensities of particle displacement, whether expressed as D_b or a subduction velocity, declined dramatically with time from particle introduction.

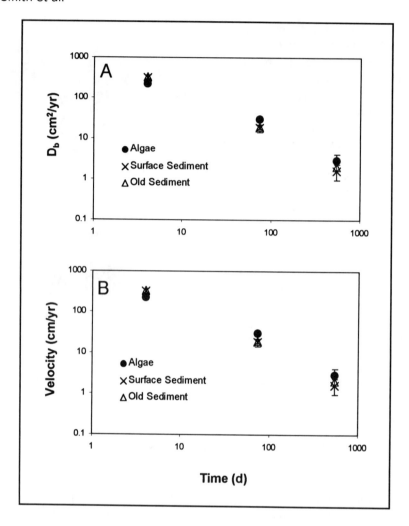

Fig. 5. Mean (± 1 se) for D_b (panel A) and maximum subduction velocity V (panel B) versus incubation time for the particle types in particle-introduction experiments. Error bars are not shown if they are no wider than data symbols.

Discussion

Our results provide evidence that all three hypothesized mechanisms (i.e., passive cascade, selective pickup, and rapid initial bioadvection) contribute to tracer-dependent mixing in the bathyal basins off southern California. The nonselective penetration of tracers within 1 h of particle introduction is consistent with the predictions of the *passive-cascade* model (Table 1). Our experiments suggest that passive cascade is restricted to the top 2 cm of Catalina Basin sediments, but this conclusion must be tempered by the single replicate treatment we obtained at t = 0. Nonetheless, our experimental design is likely to have overestimated the importance of passive cascade. Our introduced particles were allowed to settle in still water inside a canopy, likely yielding essentially vertical particle trajectories to the seafloor. In contrast, naturally depositing particles will generally approach the seafloor at shallow angles due to horizontal displacement by water flow, including low-energy deep-sea currents (e.g., Yager et al. 1993). Thus, our experimental particles are more likely than naturally depositing particles to have fallen down open burrows during the deposition process.

As predicted by the *selective-pickup* model, the gut sediments of both surface and subsurface deposit feeders were substantially enriched in $^{234}Th_{xs}$ activity compared with their general feeding zones. In addition, our particle-introduction experiments demonstrated differential penetration of young, food-rich particles (i.e., algal detritus) on ~ 100-d time scales. This differential penetration of algae is most parsimoniously explained by selective pickup and subsequent egestion of young particles by deposit feeders. However, this effect is relatively modest, yielding only a 1.5–3 fold increase in D_b or subduction velocities for young, as compared with older, particles. It thus appears that selective particle pickup can only partially explain the 10–100 fold differences in D_b previously documented for $^{234}Th_{xs}$ compared with $^{210}Pb_{xs}$ in Santa Catalina Basin (Smith et al. 1993).

Our results are most consistent with a *rapid initial bioadvection* mechanism driving tracer-dependent (and age-dependent) mixing in Santa Catalina Basin sediments. During our introduction experiments, all particles types penetrated the sediment to 5-cm depths over time scales of days. This rapid particle penetration, as well as the $^{234}Th_{xs}$ enrichment in the guts of surface deposit feeders, matches the predictions of the bioadvection model (Table 1). Initial rapid bioadvection over time scales of days, followed by much less efficient diffusive mixing, can easily explain our large time-dependent changes in D_b and/or subduction velocities (Fig. 5). It is interesting to note that the order of magnitude difference in D_b between the 74-d and ~ 500-d time scales for our introduced particles approaches the differences in D_b observed between ^{234}Th and ^{210}Pb in Catalina Basin (Smith et al. 1993). This suggests that *rapid initial subduction* may explain a substantial proportion of the tracer-dependent bioturbation in Catalina Basin.

Such rapid bioadvection of recently deposited particles is clearly a form of age-dependent mixing in that young, food-rich particles (e.g., algal detritus) are bioturbated by more efficient transport mechanisms than are older particles (e.g., the bulk sediment) (Smith et al. 1993). In other words, there is an age dependency in the mechanism and efficiency of the transport process. Although we cannot as yet identify a particular animal responsible for the rapid bioadvection in Catalina Basin, a number of potentially responsible taxa are common in the basin, including maldanid, paraonid, and cirratulid polychaetes, chiridotid holothurians, and echiuran worms (Smith et al. 1986; Kukert & Smith 1992; Miller et al. 2000). Rapid bioadvection of young, food-rich particles has been documented in a number of other bathyal and abyssal sites (e.g., Graf 1989; Smith et al. 1996; Pope et al. 1996; Levin et al. 1997; Fornes et al. 1999), suggesting that this mechanism is likely to yield age-dependent mixing in many deep-sea areas.

Our results have major ramifications for studies of organic carbon diagenesis in marine sediments. In Santa Catalina Basin, and in many deep-sea settings, recently deposited, labile organic matter is likely to be mixed at different intensities and by different mechanisms than are refractory carbon and bulk sediments. Thus, in order to evaluate organic-carbon kinetics accurately, the time scales of mixing tracers and modeled reactants should be matched. In particular, in quiescent deep-sea settings, $^{234}Th_{xs}$ is likely to be a good tracer for mixing of labile organic carbon, while $^{210}Pb_{xs}$ should be more suitable for evaluating the behavior of bulk sediments or particles of little nutritive value to deposit feeders (e.g., tests of planktonic foraminiferans). Such tracer-reactant pairing is likely to substantially improve estimates of reaction-rate constants for labile chemical species in deep-sea sediments (cf. Hammond et al. 1996; Soetert et al. 1998).

Acknowledgments

We thank R. Miller, B. Glaser, A. Baco, A Jones, S. Doan, R. Pope, C. Thomas,D. Daley, D. Stokes and many volunteers for help at sea. The *Alvin* group provided their usual outstanding support. This work was supported by National Science Foundation grant OCE-9521116 to C. R. S. and D. J. D. This is contribution no. 5288 from the School of Ocean and Earth Science and Technology, University of Hawaii at Manoa.

Literature Cited

Aller, R. C. 1990. Bioturbation and manganese cycling in hemipelagic sediments, p. 51–68. *In* Charnock, H., H. Charnock, J. Edmond, I. McCave, A. Rice, & T. Wilson (eds.), The Deep Sea Bed: Its Physics, Chemistry, and Biology. Cambridge University Press, Cambridge.

Aller, R. C. 1982. The effects of macrobenthos on chemical properties of marine sediment and overlying water, p. 53–102. *In* McCall, P. L. & M. J. S. Tevesz (eds.), Animal Sediment Relations, Plenum, New York.

Aller, R. C. & D. J. DeMaster. 1984. Estimates of particle flux and reworking at the deep-sea floor using Th-234/U-238 disequilibrium. *Earth and Planetary Science Letters* 67:308–318.

Berner, R. A. 1980. Early Diagenesis: A Theoretical Approach. Princeton University Press, Princeton, New Jersey.

Cammen, L. M. 1980. Ingestion rate: An empirical model for aquatic deposit feeders and detritivores. *Oecologia* 44:303–310.

Emerson, S. 1985. Organic carbon preservation in marine sediments, p. 78–87. *In* Sundquist, E. & W. Broecker (eds.), The Carbon Cycle and Atmospheric CO_2: Natural Variations Archean to Present. Geophysical Monograph 32. American Geophysical Union, Washington, D.C.

Emery, K. O. 1960. The Sea Off Southern California. John Wiley and Sons Inc., New York.

Fornes, W. L. 1999. Mechanisms and Rates of Bioturbation and Sedimentation in California Borderland Sediments. Ph.D. Dissertation, North Carolina State University, Raleigh, North Carolina.

Fornes, W. L., D. J. DeMaster, L. A. Levin, & N. E. Blair. 1999. Bioturbation and particle transport in Carolina Slope sediments: A radiochemical tracer approach. *Journal of Marine Research* 57:335–355.

Fornes, W. L., D. J. DeMaster, & C. R. Smith. In press. A particle introduction experiment in Santa Catalina Basin sediments: Testing the age-dependent mixing hypothesis. *Journal of Marine Research*.

Graf, G. 1989. Benthic-pelagic coupling in a deep-sea benthic community. *Nature* 341:437–439.

Hammond, D. E., J. McManus, W. M. Berelson, T. E. Kilgore, & R. Pope. 1996. Early diagenesis of organic material in Equatorial Pacific sediments: Stoichiometry and kinetics. *Deep-Sea Research II* 43:1365–1412.

Kramer, K., R. Misdorp, G. Berger, & R. Duijts.1991. Maximum pollution concentrations at the wrong depth: A misleading pollution history in a sediment core. *Marine Chemistry* 36:183–198.

Kukert, H. & C. R. Smith. 1992. Disturbance, colonization and succession in a deep-sea sediment community: Artificial-mound experiments. *Deep-Sea Research* 39:1349–1371.

Levin, L., N. Blair, D. DeMaster, G. Plaia, W. Fornes, C. Martin, & C. Thomas. 1997. Rapid subduction of organic matter by maldanid polychaetes on the North Carolina Slope. *Journal of Marine Research* 55:595–611.

Miller, R. J., C. R. Smith, D. J. DeMaster, & W. L. Fornes. 2000. Feeding selectivity and rapid particle processing by deep-sea megafaunal deposit feeders: A [234]Th tracer approach. *Journal of Marine Research* 58:653–673.

Pope, R. H. 1992. Particle mixing at two continental margin sites: A multi-tracer radiochemical approach. M.S. Thesis, North Carolina State University, Raleigh, North Carolina.

Pope, R. H., D. J. DeMaster, C. R. Smith, & H. Seltmann, Jr. 1996. Rapid bioturbation in Equatorial Pacific sediments: Evidence from excess Th-234 measurements. *Deep-Sea Research II* 43:1339–1364.

Rabouille, C. & J. F. Gaillard. 1991. A coupled model representing the deep-sea organic carbon mineralization and oxygen consumption in surficial sediments. *Journal of Geophysical Research* 96:2761–2776.

Schink D. R., N. L. Guinasso, & K. A. Fanning. 1975. Processes affecting the concentration of silica at the sediment-water interface of the Atlantic Ocean. *Journal of Geophysical Research* 80:3013–3031.

Smith, C. R. 1992. Factors controlling bioturbation in deep-sea sediments and their relation to models of carbon diagenesis, p. 375–393. *In* Rowe, G. T. & V. Pariente (eds.), Deep-Sea Food Chains and the Global Carbon Cycle. Kluwer, Dordrecht, The Netherlands.

Smith, C. R., W. Berelson, D. J. DeMaster, F. C. Dobbs, D. Hammond, D. J. Hoover, R. H. Pope, & M. Stephens. 1997. Latitudinal variations in benthic processes in the abyssal equatorial Pacific: Controls by biogenic particle flux. *Deep-Sea Research II* 44:2295–2317.

Smith, C. R., D. J. Hoover, S. E. Doan, R. H. Pope, D. J. DeMaster, F. C. Dobbs, & M. A. Altabet.1996. Phytodetritus at the abyssal seafloor across 10° of latitude in the central equatorial Pacific. *Deep-Sea Research II* 43:1309–1338.

Smith, C. R., P. A. Jumars, & D. J. DeMaster. 1986. In situ studies of megafaunal mounds indicate rapid sediment turnover and community response at the deep-sea floor. *Nature* 323:251–253.

Smith, C. R. & H. Kukert. 1996. Macrobenthic community structure, secondary production and rates of bioturbation and sedimentation at the Kaneohe-Lagoon Floor. *Pacific Science* 50:211–229.

Smith, C. R., R. H. Pope, D. J. DeMaster, & L. Magaard. 1993. Age-dependent mixing of deep-sea sediments. *Geochimica et Cosmochimica Acta* 57:1473–1488.

Soetart, K., P. M. J. Herman, J. J. Middelburg, & C. Heip. 1998. Assessing organic matter mineralization, degradability, and mixing rate in an ocean margin sediment (Northeast Atlantic) by diagenetic modeling. *Journal of Marine Research* 56:519–534.

Thayer, C. W. 1983. Sediment-mediated biological disturbance and the evolution of marine benthos, p. 475–625. *In* Tevesz, M. J. S & P. L. McCall (eds.), Biotic Interactions in Recent and Fossil Communities. Plenum Press, New York.

Wheatcroft, R. A.1990. Preservation potential of sedimentary event layers. *Geology* 18: 843–845.

Wheatcroft, R. A. 1992 Experimental tests for particle size-dependant bioturbation in the deep ocean. *Limnology and Oceanography* 37:90–104.

Wheatcroft, R.A., P. A. Jumars, C. R. Smith, & A. R. M. Nowell. 1990. A mechanistic view of the particulate biodiffusion coefficient: Step lengths, rest periods and transport directions. *Journal of Marine Research* 48:177–207.

Yager, P. L., A. R. Nowell, & P. A. Jumars. 1993. Enhanced deposition in pits: A local food source for benthos. *Journal of Marine Research* 51:209–236.

In situ Effects of Organisms on Porewater Geochemistry in Great Lakes Sediments

Frederick M. Soster[*], Gerald Matisoff, Peter L. McCall, and
John A. Robbins

Abstract: *Soft-bottom habitats in the Great Lakes are numerically dominated by deposit-feeding oligochaete annelids, tube-building filter-feeding chironomid insect larvae, and deposit-feeding amphipods, this last group being relatively abundant only in deeper oligotrophic bottoms. The abundance of macrobenthos colonizing uninhabited sediment trays and residing in natural bottom sediments at two sites in Lake Erie were compared with geochemical profiles obtained from porewater peepers. Differences in geochemical profiles were due more to differences in abundances of chironomid larvae (Chironomus plumosus) than to differences in tubificid oligochaete abundance. Porewater concentrations of soluble reactive silicate (SRS), ammonium, carbonate alkalinity, and soluble reactive phosphate (SRP) were lower in western Lake Erie tray sediments, where C. plumosus were more abundant, than in natural bottom sediments where C. plumosus were relatively are. The opposite abundance pattern was observed in the central basin, where C. plumosus were more abundant in natural sediments and porewater concentrations of SRS, ammonium, and ferrous iron were lower than in tray sediments from the same site. In similar sediments from Lower Saginaw Bay, Lake Huron, geochemical fluxes were determined from changes in concentrations of water overlying sediment cores incubated shipboard at in situ temperatures. There were significant correlations between ammonia flux, SRS flux, and chironomid abundance, between SRP flux and mature tubificid abundance, and between nitrate flux (into the sediment) and immature tubificids. A faunal succession from a community with abundant Chironomus larvae to one with abundant tubificid worms might cause a geochemical succession. Early colonizing Chironomus larvae decrease porewater concentrations of SRS, ammonium, and bicarbonate alkalinity and increase the flux out of the sediment (and for ammonium, also may indirectly increase the rate of nitrification). Concentrations of ferrous iron and SRP are also depressed, but there is no enhanced flux because iron is precipitated near burrow walls after contact with oxygen and adsorbs phosphate, so iron and phosphate do not accumulate in the burrow pore waters. Porewater concentrations of SRS, ammonium, bicarbonate alkalinity, ferrous iron, and SRP are high in the presence of slower colonizing tubificid worms, but the effect that worms have on the flux of most of these species is more complicated.*

Introduction

Freshwater soft-bottom habitats are dominated by oligochaete annelids, chironomid insect larvae, pisidiid and unionid bivalves, and amphipods (Brinkhurst 1974). These macrobenthos bioturbate the sediments as a result of their feeding, burrowing, and water-pumping activities, and consequently, have a profound impact on both the physical and chemical properties of the sediments. Physically, bioturbation alters sediment porosity, texture, compactness, and shear velocity necessary to resuspend particles (McCall & Tevesz 1982). Chemically, infaunal benthos affect sediment porewater geochemical profiles and the flux of materials across the sediment-water interface (SWI) (Fisher 1982).

Although laboratory studies have demonstrated the abilities of freshwater benthos to affect the movement of materials across the SWI (e.g., Robbins et al. 1979; Fisher 1982; Karickhoff & Morris 1985; Matisoff et al. 1985; Fukuhara & Sakamoto 1987; Soster et al. 1992), we are aware of no in situ studies, similar to those performed in marine systems (e.g., Aller et al. 1983; Kristensen & Blackburn 1987; Martin & Banta 1992), that actually demonstrate that such effects do occur in freshwater lakes under natural conditions. We present results here from two studies designed to assess the effect of Great Lakes infaunal benthos on sediment porewater geochemistry and the flux of materials across the SWI under natural conditions.

Materials and Methods

LAKE ERIE

Tray colonization experiments were conducted simultaneously at two stations in Lake Erie from May through September in 1981. Station WB was located in the southeastern portion of the western basin west of Green Island (41°38'N, 82°57'W, 9 m deep) and station CB was located in the central basin northwest of Cleveland, Ohio (41°32'N, 81°47'W, 16 m deep). Areas of physical bottom disturbances were simulated by placing 1-m^2 wooden trays on the lake floor at each site. Each tray held 100 plastic freezer containers (100 cm^2, 12 cm deep) filled with defaunated sediment . The sediment was collected from each site and defaunated by forcing it by hand through a 250-μm sieve. Despite the small size of the trays, they were meant to simulate a relatively "large" disturbance, because the container walls and tray bottoms prohibited infauna from burrowing into the containers from the margins of the disturbed "patch" or from deep sediment refuges below the "patch." Porewater "peepers" (see description below) were placed into the sediment inside 12 of the freezer containers. Trays were covered with a plastic sheet and lowered to the lake floor. Each experiment began when a diver removed the plastic sheet covering the trays. We had no way of knowing a priori which species would colonize the tray sediments, nor what temporal abundance patterns the colonizing species would exhibit; however, we did expect large contrasts in species temporal abundance patterns between tray and natural bottom sediments because Lake Erie is a good freshwater analog to temperate, coastal, marine subtidal environments where infaunal successional sequences have been well documented (e.g., Grassle & Grassle 1974; Dauer & Simon 1976; McCall 1977; Rhoads et al. 1978).

Replicate samples (2–4) for faunal analysis were collected by divers who carefully removed plastic freezer containers from the trays and immediately fastened watertight plastic lids to the top of each container. Samples also were collected from undisturbed areas of natural bottom surrounding the trays by divers using either 9.8-cm or 4.7-cm diameter core tubes and rubber stoppers. Natural bottom and tray samples were collected 0, 11, 39, 73, and 115 d after the start of the western basin experiment, and 0, 11, 33, 75, and 118 d after the start of the central basin experiment. All tray and natural bottom samples were returned to the laboratory, gently washed through a 250-μm sieve, and preserved in 4% formalin. Organisms were sorted from sieve residue with the aid of a binocular dissecting microscope (22.5X magnification) and identified.

Sediment porewaters in natural bottom and tray freezer containers were sampled using peepers (Hesslein 1976). Individual peepers were mounted into 0.6 cm x 16.5 cm PVC panels in which 12 holes had been drilled in a staggered pattern along the length of the panel to accommodate the peepers. Similar assemblies were constructed to sample the natural bottom porewaters, except that the PVC panels were longer (45 cm) and accommodated more (17) individual peepers. An individual peeper was constructed from a 2 cm x 1.7 cm OD x 1.3 cm ID piece of Teflon tubing. Submerged in a container of deionized water, a peeper was inserted through a hole in the panel and a piece of Nucleopore® 0.2-µm polycarbonate membrane filter was applied to each end of the tubing. The peeper assemblies remained submerged in deionized water while they were transported to the field. Just prior to deployment of a colonization tray, peeper assemblies were inserted vertically into 12 of the tray freezer containers. The top four peepers on each assembly were exposed to water overlying the freezer containers and the bottom eight peepers sampled porewater beneath the SWI inside the containers. Although the containers had a bottom that prevented the diffusion of nutrients into or out of the sediments below the trays, the 10 cm of sediment depth inside containers was deep enough so that there would be no evidence of bottom boundary effects over the length of the experiments. All peepers remained in the sediment at least 1 mo to equilibrate; consequently, any oxygen introduced into the sediment by the peepers would be consumed rapidly prior to retrieval. An assumption inherent in this study design is that faunal abundances in the vicinity of natural bottom and tray peepers were similar to the population estimates derived from the replicate faunal samples. Peepers were retrieved from both tray and natural bottom sediments 39, 73, and 115 d after the start of the western basin tray experiment, and 33, 75, and 118 d after the start of the central basin tray experiment. Freezer containers with mud and peeper assemblies were removed from the trays by divers and placed inside 16-cm diameter metallic cylinders (i.e., three-pound coffee cans) that were then capped with plastic lids. Peepers that had been emplaced in the natural bottom sediments near the tray on the previous sampling date were recovered by divers, who used 7-cm diameter polycarbonate tubes to obtain an intact sediment core with a contained peeper. Both ends of the core tube were then capped with plastic lids. Tray and natural bottom peepers were immediately refrigerated and transported to the laboratory where each peeper assembly was removed from the container, placed in a nitrogen-filled glove bag, and rinsed with deionized water. The peeper waters were collected by removing the filter at one end and pouring the water into a clean, labeled LPE vial. Samples for metal analysis were acidified with Ultrex nitric acid to pH = 2.

Chemical analytical methods were modified from standard, published methods to permit use of ~ 2 ml sample volumes. Concentrations of soluble reactive silicate (Strickland & Parsons 1972), ammonium (Strickland & Parsons 1972), total carbonate alkalinity (Standard Methods 1975), soluble reactive phosphate (EPA Method 365.1), and ferrous iron (Eaton 1979) were determined for most samples.

LOWER SAGINAW BAY, LAKE HURON

During the period from April to November in 1978, sediment cores were collected with a 3-inch diameter gravity corer from several stations near the center of Lower Saginaw Bay (43°51'N, 83°40'W, 10 m deep). Cores were hydraulically extruded into short plastic liners approximately 25 cm long. Care was taken to preserve approximately 10 cm of overlying water and to disturb the core minimally during the transfer process. The short sections of core with overlying water were placed in a water-bath incubator, which was adjusted to match the in situ sediment temperature. The overlying water in the cores was continuously aerated and mixed by introduction of filtered air from Tygon tubing extending a few cm into the water overlying each core. Fluxes were inferred from measurement of the concentration of nutrients in overlying water as a function of time. Approximately every 24 h for 1 wk, 20 ml of water were withdrawn from each core (about 5–7% of the total volume of overlying water), filtered immediately through a phosphate-free, prerinsed, 0.45-μm (Millipore®) filter, and frozen for subsequent analysis. Concentrations of nutrients were determined by colorimetric methods as described by Strickland and Parsons (1972).

Results and Discussion

LAKE ERIE

Tray sediments placed on the lake floor at each station were colonized by organisms present on the surrounding natural bottom. Of the macrobenthos present at the study sites, only tubificid worms and *Chironomus plumosus* larvae possessed the right combination of functional adaptations, size relative to other organisms, and population densities at the study sites that might cause detectable effects on porewater geochemistry at the size scale of our investigation. Most Lake Erie tubificids are generally less than 3.0 cm in contracted preserved length and average less than 1.0 mg dry weight. However, their high population densities ($10,000–50,000$ m^{-2}) and their head-down, "conveyor-belt" style of deposit-feeding, which mixes sediments to a depth of 6–9 cm (Fisher et al. 1980) and produces pelletized surface layers up to 1 cm thick (McCall 1979), can influence porewater geochemistry. *C. plumosus* larvae were not as abundant as the tubificid worms, but their large size, which averages 3.5 mg for fourth-instar larvae (Soster 1984), and their burrowing and water pumping activities likely also cause detectable alterations in porewater geochemical profiles. *C. plumosus* larvae are filter-feeders who usually construct U-shaped tubes that may extend 8 cm below the sediment-water interface and who irrigate their tubes with overlying water when feeding (Walshe 1947).

The porewater geochemical effects of the other macrofauna present at our Lake Erie study sites were likely small relative to the effects of tubificid worms and *C. plumosus* larvae. Because the other macrofauna do not have the sizes, abundances, or abilities to mix sediments or irrigate burrows to significant depths, we think that any effects on porewater geochemistry seen in our tray experiments can be interpreted as being caused by tubificid oligochaetes and *C. plumosus* larvae.

Abundance patterns of *C. plumosus* larvae and tubificid worms in natural bottom and tray sediments are compared in Fig. 1(A & B) and Fig. 2(A & B). In the western basin, *C. plumosus* larvae colonized tray sediments in disproportionately high numbers and tray abundances remained higher than natural bottom abundances for the duration of the experiment (Fig. 1A). However, in the central basin, *C. plumosus* larvae colonized the tray sediments slowly and abundances remained lower than or similar to natural bottom abundances (Fig. 2A). The tubificid worms colonized tray sediments slowly at both study sites, and tray abundances were considerably lower than natural bottom abundances during most of the experiment (Figs. 1B and 2B).

Several laboratory studies have shown that *C. plumosus* larvae increase the flux of silica from lake sediments (Tessenow 1964; Granéli 1979; Matisoff et al. 1985). Matisoff et al. (1985) reported that silica concentrations in the top 6 cm of sediment were significantly lower when *C. plumosus* larvae were present. They attributed this to the water pumping activities of the larvae and the

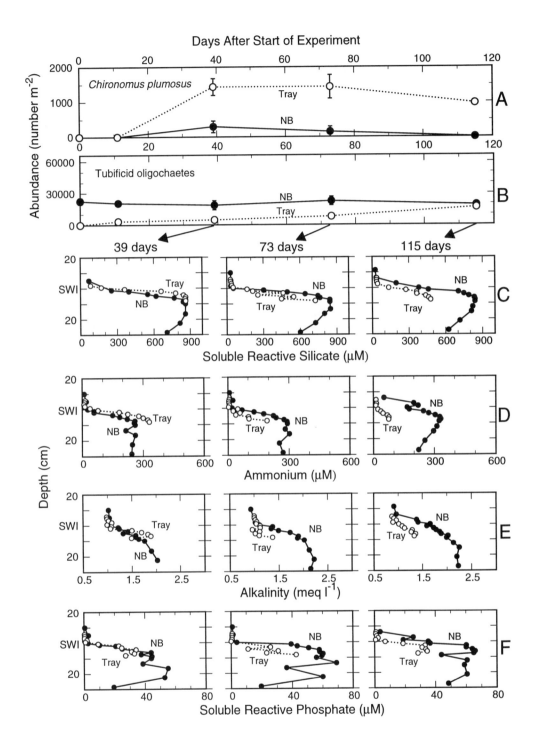

Fig. 1. Abundances of *Chironomus plumosus* larvae (A) and tubificid oligochaetes (B) in western basin natural bottom and tray sediments. Vertical bars indicate the standard error of the mean. Standard errors less than the width of the symbol are not shown. Natural bottom and tray sediment porewater profiles of soluble reactive silicate (C), ammonium (D), alkalinity (E), and soluble reactive phosphate (F) 39 d, 73 d, and 115 d after the start of the experiment. NB refers to natural bottom and SWI refers to sediment-water interface.

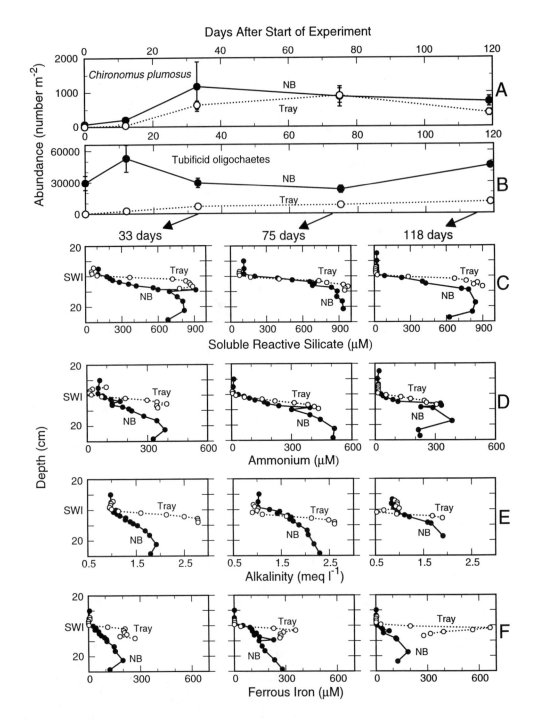

Fig. 2. Abundances of *Chironomus plumosus* larvae (A) and tubificid oligochaetes (B) in central basin natural bottom and tray sediments. Vertical bars indicate the standard error of the mean. Standard errors less than the width of the symbol are not shown. Natural bottom and tray sediment porewater profiles of soluble reactive silicate (C), ammonium (D), alkalinity (E), and ferrous iron (F) 33 d, 75 d, and 118 d after the start of the experiment. NB refers to natural bottom and SWI refers to sediment-water interface.

associated flushing of their burrows. Subsequently, Matisoff (1995) showed that bioirrigation was sufficient to explain the enhanced silica flux. The impact of tubificid oligochaetes on porewater silica concentrations is less pronounced. In a time series laboratory study, Matisoff et al. (1985) found only slightly higher concentrations of silica in water overlying sediments populated by worms than in water overlying sediments with no worms. Direct measurements of porewater silica concentrations were nearly identical in microcosms with and without tubificids, confirming that tubificids had little impact on silica concentrations. Based on the results of these laboratory studies, we would expect that porewater silica concentrations should be relatively high when the abundance of *C. plumosus* larvae is low and relatively low when larval abundance is high. Tubificids should have little effect on silica concentrations.

The prediction that silica concentrations are inversely related to *C. plumosus* abundances is generally confirmed. In the central basin (Fig. 2C), silica concentrations are highest in sediments with lowest larval densities, in this case the tray sediments. Central basin natural bottom sediments have relatively lower silica concentrations and higher larval densities. In the western basin (Fig. 1C), larval abundances are higher in tray sediments than in natural bottom sediments and, on two of the three sampling dates, silica concentrations are lower in tray sediments than in natural bottom sediments. Only on day 39 did the pattern not hold: larval abundances and silica concentrations were higher in tray sediments than natural bottom sediments. Figure 1A shows that *C. plumosus* larvae colonized the western basin tray sediments rapidly between the day 11 and the day 39 sampling dates. Mud-water systems generally require 10–14 d to reach biological and chemical equilibrium (Hargrave 1975). It may be that larval abundances in the tray sediments were not high enough for a sufficient length of time prior to the first geochemical sampling date (day 39) to depress silica porewater concentrations.

The inverse relationship between chironomid density and silica concentration also holds for between-station comparisons. Higher chironomid abundances in the western basin tray resulted in lower silica concentrations in the pore water compared with central basin tray sediments. For example, on the last sampling date at a depth of 5 cm, silica concentrations were about 400 μM in the western basin tray and about 800 μM in the central basin tray (Figs. 1C & 2C). Just the opposite pattern is found in the natural bottom sediments where chironomid densities were higher in the central basin. On this same sampling date and depth, silica concentrations were only about 400 μM in the central basin natural bottom and were about 800 μM in the western basin.

Ammonium concentrations in sediment pore water are altered by both tubificid oligochaetes and chironomid larvae. Tubificids apparently increase ammonium concentrations in sediment pore water (Kikuchi & Kurihara 1977; Matisoff et al. 1985; Fukuhara & Sakamoto 1987) because they excrete it directly (Gardner et al. 1983; Gardner et al. 1987) and because they enhance its formation by increasing microbial activity in the sediment (Matisoff et al. 1985). Infaunal, filter-feeding chironomid larvae decrease ammonium concentrations because of bioirrigation (enhanced diffusional flux) (Matisoff et al. 1985; Fukuhara & Sakamoto 1987) and nitrification. We therefore expect ammonium concentrations in pore waters to be highest in sediments that contain high densities of tubificid oligochaetes and/or low densities of chironomid larvae. This expectation is at least partially confirmed by the results from the western basin: natural bottom sediments contained higher abundances of worms and lower abundances of larvae than the tray sediments. On two of the three sampling dates, porewater concentrations of ammonium were lower in the tray sediments than in the natural bottom sediments (Fig. 1D). As in the case for silica, we think that low larval abundances in the tray sediments early in the experiment likely provided insufficient time for ammonium concentrations to be affected. In the central basin, both worm and chironomid densities were higher in natural bottom sediments than in tray sediments. Porewater ammonium concentrations were always lower in natural bottom sediments (Fig. 2D). These data suggest that chironomids may have a more profound impact on ammonium concentrations than the worms. The tubificid density difference between natural bottom and tray sediments was much larger than the chironomid density difference. Because tubificids increase ammonium concentrations in pore waters whereas chironomids decrease ammonium concentrations, we would expect ammonium concentrations to be higher in the natural bottom sediments where worm abundance is high. However, ammonium concentrations are lower in the natural bottom sediments, implying that the differences in the ammonium profiles between the tray and natural bottom sediments are primarily due to the differences in the chironomid densities and not the tubificid densities.

There are two possibilities for the fate of the porewater ammonium in the central basin natural bottom sediments. One possibility is that mixed populations of tubificid worms and chironomid larvae may be extremely efficient at increasing the flux of ammonium from natural lake sediments. Tubificid worms increase the concentration of ammonium in sediment pore water and their bioturbation of the near-surface layers of sediment increases the rate of diffusion of this ammonium into the overlying water (Matisoff et al. 1985; Fukuhara & Sakamoto 1987). Chironomids also excrete ammonium (Gardner et al. 1983) and they increase the rate of ammonium flux into overlying water because they maintain and irrigate burrows connected to the overlying water (Matisoff et al. 1985; Fukuhara & Sakamoto 1987). Porewater concentrations decrease as ammonium is flushed out of the sediment by the water pumping activities of the chironomids. A second possibility is that the chironomid larvae are creating a cascading trophic effect that increases nitrification of the porewater ammonium, similar to the effects described by Lavrentyev et al. (1997), who demonstrated that grazing of nitrifying bacteria by protozoa could potentially control nitrification rates in aquatic systems. Using experimental manipulations, they found that when the protozoan grazers were removed, the nitrifying bacteria flourished and ammonium was rapidly nitrified. In our study, removal of protozoan bacterial grazers by the filter-feeding chironomids would allow increased bacterial populations and increased nitrification of the ammonium released by both the tubificids and the chironomids. This would explain why the chironomids have a more pronounced effect on porewater ammonium concentrations, even when tubificids are abundant. Furthermore, there apparently are some experimental data to support this scenario. Lavrentyev et al. (1997) indicate that they have found an increase in nitrification rates and a decrease in the ratio of protistan to bacterial carbon in an experimental sediment system populated with the filter-feeding bivalve *Dreissena polymorpha*. Whatever the mechanism, our field studies show that mixed populations of tubificids and chironomids result in lower porewater ammonium concentrations compared with sediments having lower abundances of these benthos, and that chironomid larvae are lowering porewater ammonium concentrations either directly via enhanced flux or indirectly by causing increased nitrification rates.

The effects of tubificid worms and chironomid larvae on sediment porewater alkalinity have not been well studied, and the results from the few studies that have been performed are ambiguous. Granéli (1979) measured alkalinity in water overlying lake sediments before and after the addition of *C. plumosus* larvae and found no significant changes. However, Matisoff et al. (1985) measured higher alkalinity in water overlying sediments with *C. plumosus* larvae than in water overlying sediment without larvae. In the same study, they found that tubificid worms also increased the alkalinity in water overlying sediments with worms compared with water overlying sediments without worms. Direct measurements of porewater alkalinity in separate laboratory experiments showed that worms caused a slight increase in porewater alkalinity at sediment depths below about 3 cm and that *C. plumosus* larvae decreased porewater alkalinity at sediment depths below 5 cm.

Results from the western basin tray and natural bottom sediments are in general agreement with the Matisoff et al. (1985) laboratory study (Fig. 1E). Porewater alkalinity was lower in tray sediments, which contained higher densities of larvae, than in natural bottom sediments on two of the three sampling dates. Alkalinity was higher in tray sediment pore water on day 39, but we think that there was insufficient time between establishment of the high larval densities and the sampling date for the larvae to affect the porewater alkalinity. Results from the central basin are harder to interpret (Fig. 2E). Porewater alkalinity was higher in tray sediments than in natural bottom sediments on day 33. On days 75 and 118, however, alkalinity was lower in the top 3 cm of tray sediment than natural bottom sediment, yet higher in the deeper levels of tray sediment. This pattern is not easily related to faunal abundance patterns because both worm and larval abundances were higher in the natural bottom sediments throughout the experiment, because worms and larvae have opposite effects on porewater alkalinity concentrations, and because there may have been some confounding effect by high abundances of Pisidiidae clams on the central basin natural bottom sediments. Pisidiids were only abundant in central basin natural bottom sediments and exceeded 10,000 ind m^{-2} on day 75, when alkalinity in natural bottom sediments was highest. Unfortunately, little is known about pisidiid effects on porewater geochemistry and we are unable to evaluate this possibility at this time.

The activity of ferrous iron in sediment pore water is controlled largely by redox reactions in the oxidized layer near the sediment-water interface (Sundby et al. 1986) and in oxidized sediment that surrounds the irrigated burrows of infaunal organisms (Matisoff et al. 1985). Ferrous iron concentrations are low in oxidized sediment because it is precipitated as ferric iron in the presence of oxygen. *C. plumosus* larvae are particularly efficient at decreasing the concentration of ferrous iron in sediment pore waters because burrow irrigation produces a large volume of oxidized sediment around the burrows that contains no dissolved iron (Matisoff et. al. 1985). Granéli (1979) found that the addition of *C. plumosus* larvae to aerobic sediment cores from two Swedish lakes increased the flux of iron into overlying water, but concentrations returned to pre-addition concentrations within 60 d. He attributed this to burrow penetration of the oxidized surface layer, where upward diffusing iron is normally precipitated. The effect of tubificid worms on the distribution of ferrous iron in sediments is less clear. Kikuchi and Kurihara (1977) showed that the activities of two tubificid worms, *Limnodrilus socialis* and *Branchiura sowerbyi*, destroyed the oxidized layer in submerged ricefield soils, which allowed the activity of ferrous iron in the upper soil layers to remain high relative to soil with no worms. Destruction of the oxidized layer by tubificids increased the diffusion rate of ferrous iron into the overlying water, whereas in the absence of worms, an oxidized layer developed where ferric iron was precipitated. However, Matisoff et al. (1985) did not find significant differences in porewater ferrous iron profiles in laboratory microcosms with and without worms. If, based on the results of the few studies described above, *C. plumosus* exert a significant influence on porewater ferrous iron profiles and the effects of tubificids are of only second-order importance, then we would expect higher concentrations in sediments with low abundances of chironomids and worms, and this expectation is confirmed by data from the central basin experiment (Fig. 2F). Ferrous iron concentrations are almost always higher in tray sediments, where both larvae and worm abundances are generally lower, and two of the three profiles from the tray sediments exhibit a well-defined subsurface iron concentration maximum similar to those described by Matisoff et al. (1985) for laboratory control microcosms. Both Granéli (1979) and Matisoff et al. (1985) believed the lower ferrous iron concentrations in the presence of larvae were the direct result of the burrowing and water pumping activities of the larvae, which not only flushes reduced compounds out of the sediment but prevents the buildup of reduced compounds in the oxidized sediment surrounding the burrows. Moreover, the subsurface iron concentration maximum in the tray sediments may be the result of upward diffusion of ferrous iron to the redox potential discontinuity, where it is then precipitated as ferric iron (Froelich et al. 1979; Gaillard et al. 1989).

The effects of chironomid larvae and tubificid worms on phosphorus flux from lake sediments has received considerable attention. *Chironomus* larvae increase phosphorus flux from sediments into overlying water (Gallepp et al. 1978; Gallepp 1979; Granéli 1979) and the effect is density- and size-dependent: large populations and large larvae produce greater release rates (Gallepp et al. 1978; Gallepp 1979). However, Gallepp (1979) found that the measured phosphorous flux from sediments with *Chironomus* larvae was considerably less than that expected from theoretical considerations and suggested that precipitation of phosphorous in oxidized burrow linings, among other possibilities, could account for this difference. Matisoff et al. (1985) found much lower phosphate concentrations in sediment pore waters when *C. plumosus* larvae were present compared with sediments without larvae. But, in a related laboratory experiment, they found no increase in phosphate concentrations in water overlying chironomid-inhabited sediments. They attributed this to irrigation of the chironomid burrows, which produced oxidized sediments around the burrow walls where phosphorus was adsorbed onto ferric oxyhydroxides. Thus, sediment porewater phosphorus concentrations were reduced, yet there was no increase in phosphorus flux into the overlying water.

The effect of tubificid worms on phosphorus flux is less clear. Davis et al. (1975) did not detect any movement of ^{32}P injected 2 cm below the sediment-water interface in laboratory sediments inhabited by tubificids, but they did note that worms reduced the proportion of PO_4-phosphorus in the uppermost sediment. Gallepp et al. (1978) concluded that tubificids were of minor importance in enhancing phosphate release from sediments, a conclusion that appears to be confirmed by the laboratory experiment of Matisoff et al. (1985), who reported no significant differences in porewater soluble reactive phosphate (SRP) concentration profiles in sediments with and without worms. However, McCall and Fisher (1980) found that, after the onset of anoxia, tubificid worms inhibited the release of phosphate from laboratory sediments compared with sediments with no worms. They reasoned that in sediments with no worms, a phosphorus-rich iron hydroxide layer formed near the SWI during oxic conditions, and then dissolved after the onset of anoxia, releasing phosphorus to the overlying water. The sediment mixing activity of the worms prior to the onset of anoxia prevented the formation of such a phosphorus-rich layer.

Results from the western basin tray experiment are in good agreement with the majority of these studies (Fig. 1F). We would expect porewater SRP concentrations to be lower in sediments with high abundances of *C. plumosus* larvae, in this case, the tray sediments. SRP profiles in tray and natural bottom sediment pore waters are similar on day 39, but SRP concentrations are markedly lower in tray sediment pore waters on days 73 and 115. It is not possible to draw any conclusions from these data with regard to the effect of the tubificid worms on SRP flux because they were considerably more abundant on the natural bottom than in the tray sediments for most of the experiment. However, tray and natural bottom abundances were similar by the last sampling date, yet SRP concentrations in tray sediments were still lower than natural bottom sediments. This suggests that if tubificids do affect SRP flux, the effect is minor and only of second-order importance when *C. plumosus* larvae are present, which agrees with the work of Davis et al. (1975), Gallepp et al. (1978), and Matisoff et al. (1985).

LOWER SAGINAW BAY, LAKE HURON

Average abundances of Saginaw Bay macrofauna were determined from nine sediment cores collected in October and November 1978. The Saginaw Bay community was dominated by mature tubificid worms (913 m^{-2}, se ± 262), immature tubificid worms (24,822 m^{-2} ± 6,783), naidid worms (4,046 m^{-2} ± 1,710), and chironomids (438 m^{-2} ± 142), and thus was broadly similar to the Lake Erie communities.

Average nutrient fluxes from sediment cores collected at several stations in October and November 1978 are shown in Table 1 and correlations between nutrient fluxes and benthos densities are shown in Table 2. Because of the limited number of observations, most correlations in Table 2 are not significant. However, there was a correlation between ammonia flux and chironomid abundance, and a high correlation between silica flux and chironomid abundance. These results are consistent with the Lake Erie data, where we found depressed porewater concentrations of ammonium and silica in sediments with high densities of chironomid larvae, implying increased fluxes of these nutrients from the sediments. Thus, enhanced flux of ammonium into the overlying water appears to be an important mechanism by which filter-feeding chironomid larvae depress porewater ammonium concentrations. The relative importance of increased rates of nitrification in chironomid-inhabited sediments is not known. Although there is a correlation between phosphate flux and mature tubificid worm abundance in Saginaw Bay, there is not a strong or significant correlation between phosphate flux and immature worm abundance. The immature worms compose the majority of the worm population at both of the Lake Erie stations and so the Saginaw Bay data are in good agreement with the western basin Lake Erie data, which suggest little if any effect on phosphate flux by worms. Finally, there was a correlation between nitrate flux and immature tubificid worm abundance, but comparable data from Lake Erie with which to make comparisons do not exist. Because nitrate concentrations are higher in the overlying water than in the pore water, it seems likely that this apparent increase in the flux of nitrate from sediments in the presence of worms is a consequence of nitrification of ammonia that has accumulated in the overlying water.

Table 1. Average nutrient fluxes (± standard error of the mean) in sediment cores collected from Lower Saginaw Bay. A negative value indicates a loss from overlying water. N refers to the number of cores.

Date 1978	Soluble Reactive Silicate		Ammonia Nitrogen		Soluble Reactive Phosphate		Nitrate Nitrogen	
	μg cm^{-2} h^{-1}	n	μg cm^{-2} h^{-1}	n	μg cm^{-2} h^{-1}	n	μg cm^{-2} h^{-1}	n
October	0.160 (± 0.030)	10	0.013 (± 0.003)	10	-0.040 (± 0.040)	9	-0.010 (± 0.003)	8
November	0.190 (± 0.030)	14	0.007 (± 0.002)	14	-0.100 (± 0.050)	14	-0.003 (± 0.002)	14

Table 2. Correlations between nutrient fluxes and benthic organism densities in cores from Lower Saginaw Bay, Lake Huron, in October and November 1978. Correlations that are significant on both sampling dates are underlined.

Organism	Silicon	Ammonia	SRP	Nitrate	Sulfate
October 1978					
Tubifcidae					
Mature	-0.07	-0.09	_0.93_	-0.17	-0.75
Immature	-0.36	-0.82	0.07	_0.69_	-0.54
Naididae	-0.29	-0.49	-0.19	0.13	-0.41
Chironomidae	_0.97_	_0.74_	0.11	0.04	0.04
Total	-0.26	-0.69	0.06	_0.55_	-0.56
November 1978					
Tubifcidae					
Mature	0.88	0.92	_0.41_	0.23	0.97
Immature	0.49	0.14	0.23	_0.93_	0.22
Naididae	0.13	0.63	-0.30	0.78	-0.50
Chironomidae	_0.99_	_0.63_	0.09	0.25	0.76
Total	0.62	-0.05	-0.90	_0.94_	0.01

The relationship between soluble reactive silicate flux and chironomid density is shown in Fig. 3. The results for silica indicate an enhanced flux of about 2,000 µg Si cm^{-2} yr^{-1} at a density of about 1,000 ind m^{-2}. These results are in agreement with experimental studies that report enhanced silica fluxes in the presence of chironomids (Tessenow 1964; Granéli 1979). The results shown in Fig. 3 suggest that in shallow waters of the Great Lakes where chironomid-inhabited, fine-grained sediments are found, these larvae may play a major role in the regeneration of silica from sediments. Our Saginaw Bay and Lake Erie data indicate that burrow irrigation by chironomids increases silica fluxes but reduces porewater silica gradients.

GEOCHEMICAL SUCCESSION

C. plumosus larvae and tubificid worms possess contrasting functional adaptations (filter feeding versus deposit feeding) and they also respond differently to the provision of open space in the form of defaunated sediments (Soster & McCall 1990). *C. plumosus* larvae actively migrate into unoccupied sediment early in their life cycle and may reach population densities twice as high as in surrounding, occupied sediments in a few weeks time. Tubificid worms migrate into unoccupied sediments slowly, and may take several months to over a year to reach population densities similar to those in surrounding areas. Furthermore, it appears that the spatial and temporal distributions of many of the macrobenthos in western Lake Erie are correlated with disturbance levels (McCall & Soster 1990). We hypothesize that disturbance-initiated faunal successions of adaptive and functional types may also cause geochemical successions. Our results suggest that porewater concentrations and

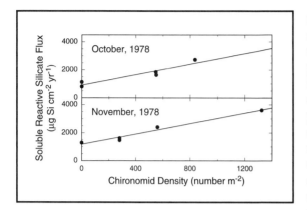

Fig. 3. Relation between the soluble reactive silicate flux and the density of chironomid larvae in Saginaw Bay during October and November 1978.

fluxes of nutrients should change as abundances of colonizing infauna change (Table 3). The rapid build-up of a large population of *C. plumosus* larvae responding to an unoccupied area of lake bottom, such as one that might be produced by physical disturbance by storm waves or currents, would result in high fluxes of silica, ammonium, and bicarbonate into the overlying water and a corresponding decrease in porewater concentrations of these chemical species. *C. plumosus* larvae also would increase the flux of nitrate into the sediment and increase the porewater concentrations of nitrate. As the faunal composition of the disturbed area gradually changed toward a tubificid-dominated assemblage, porewater concentrations and fluxes of nutrients would also change. Porewater concentrations, low because of the water pumping activities of the chironomids, would begin to increase, but fluxes out of the sediments would initially be low until a concentration gradient from the sediment to the overlying water was re-established. Silica fluxes would slow, primarily because of the absence of the water pumping activity of the chironomids, but also because the worms have little effect on silica. Consequently, silica porewater concentrations would increase. Porewater ammonium concentrations would increase and the flux of ammonium would remain high because the worms excrete it and because they stimulate bacterial activity, which enhances ammonium formation (Matisoff et al. 1985). Likewise, both porewater concentrations and flux of bicarbonate would be high because the worms increase bicarbonate concentrations in pore waters. Nitrate concentrations would be low because enhanced denitrification in the presence of worms will consume any additional nitrate supplied by an increase in the flux across the sediment-water interface.

The geochemical succession would be different for ferrous iron and soluble reactive phosphate. Although their concentrations are depressed in pore waters in the upper portion of the sediment column during the early successional stage, they would not exhibit an enhanced flux. The reason for this is that the water pumping activity of the chironomids increases the flux of dissolved oxygen into the sediment. The downward and radially inward diffusing oxygen oxidizes the ferrous iron before it can accumulate in the burrows, because the oxidation rate of ferrous iron is fast relative to its rate of diffusive transport. Therefore the iron is oxidized and precipitated in a zone around the burrows and does not accumulate in the burrows. This is visually apparent in laboratory microcosms where an orange rind develops around chironomid burrows (Matisoff et al. 1985). Dissolved phosphate is sorbed onto the freshly precipitated iron hydroxide, so its porewater concentration is also depressed and it, too, does not accumulate in the burrows, which results in a lower flux of phosphate in the presence of chironomids. Thus, in the early successional stage, both ferrous iron and phosphate show depressed porewater concentration profiles in the irrigated zone that are similar to those of silicate, ammonium, and alkalinity, but iron and phosphate do not show the enhanced flux that these other species do. Instead, the irrigated (and oxygenated) layer is a zone of accumulation of iron and phosphorus in the solid phase. It is not clear from our results or from the few laboratory studies that

Table 3. Generalized geochemical succession following a mortality-causing disturbance of a lake community that is numerically dominated by filter-feeding chironomid larvae and deposit-feeding tubificid oligochaetes. All fluxes are out of the sediment with the exception of nitrate, which is into the sediment. Faunal succession from Soster and McCall (1990). PWC is porewater concentration, SRS is soluble reactive silicate, and SRP is soluble reactive phosphate.

Successional Stage:		Early	Late
Faunal Assemblage:		High chironomid abundances Low oligochaete abundances	Low chironomid abundances High oligochaete abundances
SRS	PWC	Low	High
	Flux	High	Intermediate
Ammonium	PWC	Low	High
	Flux	High	High
Alkalinity	PWC	Low	High
	Flux	High	High
Ferrous Iron	PWC	Low	High
	Flux	Low	Low
SRP	PWC	Low	High
	Flux	Low	Intermediate
Nitrate	PWC	High	Low
	Flux*	High	High

*Nitrate flux is <u>into</u> the sediment.

have been done what effect the later colonizing worms would have on ferrous iron and phosphate concentrations or fluxes. If the worms destroy the burrowed layer, then oxygen exchange would decrease and porewater ferrous iron and phosphate concentrations would increase in a manner similar to that described by Kikuchi and Kurihara (1977). This might explain why iron concentrations in pore waters are sometimes low and sometimes high. Iron flux would remain low, because it would be precipitated when it came into contact with oxidized sediment near the sediment-water interface. Phosphate flux would likely increase because the worms excrete it (Gardner et al. 1981; Fukuhara & Yasuda 1985). If, however, the effect of the worms is to thicken the oxidized surface layer, rather than destroy it (see, for example, Soster et al. 1992), and if upward diffusing ferrous iron is precipitated as ferric iron at the base of this zone, then ferrous iron porewater concentrations and fluxes would remain low.

Summary

In situ studies conducted at two sites in Lake Erie and one site in Saginaw Bay, Lake Huron demonstrate that tube-building chironomid larvae (*Chironomus* sp.) decrease porewater concentrations of soluble reactive silicate, ammonium, and bicarbonate alkalinity and increase the flux out of the sediment. Concentrations of ferrous iron and soluble reactive phosphate also are depressed, but there is no enhanced flux because these redox sensitive species are precipitated near burrow walls and do not accumulate in the burrow pore waters. Porewater concentrations of these chemical species are high in the presence of tubificid worms when chironomids are rare, but fluxes remain low until a porewater concentration gradient to the overlying water is established. Our results are in good agreement with laboratory studies of systems. A faunal succession from a *Chironomus*-dominated early assemblage to a tubificid-dominated late assemblage might also cause a geochemical succession.

Acknowledgments

We thank David S. White, Z. Batac-Catalan, Bert Fisher, and Brian Soster for help with the field and laboratory work, and Wayne S. Gardner and an anonymous reviewer for thorough reviews that significantly improved the manuscript. Funds were provided by grants from the National Science Foundation (OCE-77-24613 and OCE-80-05103), National Oceanic and Atmospheric Administration (NA-80-RED-00036), US Environmental Protection Agency (R804686), and the DePauw University Fisher Fund and Faculty Development Fund.

Literature Cited

Aller, R. C., J. Y. Yingst, & W. J. Ullman. 1983. Comparative biogeochemistry of water in intertidal *Onuphis* (polychaeta) and *Upogebia* (crustacea) burrows: Temporal patterns and causes. *Journal of Marine Research* 41:571–604.

American Public Health Association.1975. Standard Methods for the Examination of Water and Waste Waters (14th ed.). American Public Health Association, Washington D. C.

Brinkhurst, R. O. 1974. The Benthos of Lakes. St. Martin's Press, New York.

Chatarpaul, L., J. B. Robinson, & N. K. Kaushik. 1980. Effects of tubificid worms on denitrification and nitrification in stream sediment. *Canadian Journal of Fisheries and Aquatic Sciences* 37:656–663.

Dauer, D. M. & J. L. Simon. 1976. Habitat expansion among polychaetous annelids repopulating a defaunated marine habitat. *Marine Biology* 37:169–177.

Davis, R. B., D. L. Thurlow, & F. E. Brewster. 1975. Effects of burrowing tubificid worms on the exchange of phosphorus between lake sediment and overlying water. *Verhandlungen der Internationalen Vereinigung fuer Theoretische und Angewandte Limnologie* 19:382–394.

Eaton, A. 1979. Removal of 'soluble' iron in the Potomac River estuary. *Estuarine and Coastal Marine Science* 9:41–49.

Fisher, J. B. 1982. Effects of macrobenthos on the chemical diagenesis of freshwater sediments, p. 177–218. *In* McCall, P. L. & M. J. S. Tevesz (eds.), Animal-Sediment Relations: The Biogenic Alteration of Sediments. Plenum Press, New York.

Fisher, J. B., W. J. Lick, P. L. McCall, & J. A. Robbins. 1980. Vertical mixing of lake sediments by tubificid oligochaetes. *Journal of Geophysical Research* 85:3997–4006.

Froelich, P. N., G. P. Klinkhammer, M. L. Bender, N. A. Luedtke, G. R. Heath, D. Cullen, P. Dauphin, D. Hammond, B. Hartman, & V. Maynard. 1979. Early oxidation of organic matter in pelagic sediments of the eastern equatorial Atlantic: Suboxic diagenesis. *Geochimica et Cosmochimica Acta* 43:1075–1090.

Fukuhara, H. & M. Sakamoto. 1987. Enhancement of inorganic nitrogen and phosphate release from lake sediment by tubificid worms and chironomid larvae. *Oikos* 48:312–320.

Fukuhara, H. & K. Yasuda. 1985. Phosphorus excretion by some zoobenthos in a eutrophic freshwater lake and its temperature dependency. *Japanese Journal of Limnology* 46:287–296.

Gaillard, J-F., H. Pauwels, & G. Michard. 1989. Chemical diagenesis in coastal marine sediments. *Oceanologica Acta* 12:175–187.

Gallepp, G. W. 1979. Chironomid influence on phosphorus release in sediment-water microcosms. *Ecology* 60:547–556.

Gallepp, G. W., J. F. Kitchell, & S. M. Bartell. 1978. Phosphorus release from lake sediments as affected by chironomids. *Verhandlungen der Internationalen Vereinigung fuer Theoretische und Angewandte Limnologie* 20:458–465.

Gardner, W. S., T. F. Nalepa, M. A. Quigley, & J. M. Malczyk. 1981. Release of phosphorus by certain benthic invertebrates. *Canadian Journal of Fisheries and Aquatic Sciences* 38:978–981.

Gardner, W. S., T. F. Nalepa, D. R. Slavens, & G. A. Laird. 1983. Patterns and rates of nitrogen release by benthic Chironomidae and Oligochaeta. *Canadian Journal of Fisheries and Aquatic Sciences* 40:259–266.

Gardner, W. S., T. F. Nalepa, & J. M. Malczyk. 1987. Nitrogen mineralization and denitrification in Lake Michigan sediments. *Limnology and Oceanography* 32:1226–1238.

Granéli, W. 1979. The influence of *Chironomus plumosus* larvae on the exchange of dissolved substances between sediment and water. *Hydrobiologia* 66:149–159.

Grassle, J. R. & J. P. Grassle. 1974. Opportunistic life histories and genetic systems in marine benthic polychaetes. *Journal of Marine Research* 32:253–384.

Hargrave, B. T. 1975. Stability in structure and function of the mud-water interface. *Verhandlungen der Internationalen Vereinigung fuer Theoretische und Angewandte Limnologie* 19:1073–1079.

Hesslein, R. H. 1976. An in situ sampler for close interval pore water studies. *Limnology and Oceanography* 21:912–914.

Karickhoff, S. W. & K. R. Morris. 1985. Impact of tubificid oligochaetes on pollutant transport in bottom sediments. *Environmental Science and Technology* 19:51–56.

Kikuchi, E. & Y. Kurihara. 1977. In vitro studies on the effects of tubificids on the biological, chemical, and physical characteristics of submerged ricefield soil and overlying water. *Oikos* 29:348–356.

Kristensen, E. & T. H. Blackburn. 1987. The fate of organic carbon and nitrogen in experimental marine sediment systems: Influence of bioturbation and anoxia. *Journal of Marine Research* 45:231–257.

Lavrentyev, P. J., W. S. Gardner, & J. R. Johnson. 1997. Cascading trophic effects on aquatic nitrification: Experimental evidence and potential implications. *Aquatic Microbial Ecology* 13:161–175.

Martin, W. R. & G. T. Banta. 1992. The measurement of sediment irrigation rates: A comparison of the Br⁻ tracer and $^{222}Rn/^{226}Ra$ disequilibrium techniques. *Journal of Marine Research* 50:125–154.

Matisoff, G. 1995. Effects of bioturbation on solute and particle transport in sediments, p. 201–272. *In* Allen, H. E. (ed.), Metal Contaminated Aquatic Sediments. Ann Arbor Press, Chelsea, Michigan.

Matisoff, G., J. B. Fisher, & S. Matis. 1985. Effects of benthic macroinvertebrates on the exchange of solutes between sediments and freshwater. *Hydrobiologia* 122:19–33.

McCall, P. L. 1977. Community patterns and adaptive strategies of the infaunal benthos of Long Island Sound. *Journal of Marine Research* 35:221–266.

McCall, P. L. 1979. The effects of deposit feeding oligochaetes on particle size and settling velocity of Lake Erie sediments. *Journal of Sedimentary Petrology* 49:813–818.

McCall, P. L. & J. B. Fisher. 1980. Effects of tubificid oligochaetes on physical and chemical properties of Lake Erie sediments, p. 253–317. *In* Brinkhurst, R. O. & D. G. Cook (eds.), Aquatic Oligochaete Biology. Plenum Press, New York.

McCall, P. L. & M. J. S. Tevesz. 1982. The effects of benthos on physical properties of freshwater sediments, p. 105–176. *In* McCall, P. L. & M. J. S. Tevesz (eds.), Animal-Sediment Relations: The Biogenic Alteration of Sediments. Plenum Press, New York.

McCall, P. L. & F. M. Soster. 1990. Benthos response to disturbance in western Lake Erie: Regional faunal surveys. *Canadian Journal of Fisheries and Aquatic Sciences* 47:1996–2009.

Rhoads, D. C., P. L. McCall, & J. Y. Yingst. 1978. Disturbance and production on the estuarine seafloor. *American Scientist* 66:577–586.

Robbins, J. A., P. L. McCall, J. B. Fisher, & J. R. Krezoski. 1979. Effect of deposit feeders on migration of ^{137}Cs in lake sediments. *Earth and Planetary Science Letters* 42:277–287.

Soster, F. M. 1984. Colonization by benthic invertebrates in western Lake Erie following sediment disturbance. Ph.D. thesis, Case Western Reserve University, Cleveland, Ohio.

Soster, F. M. & P. L. McCall. 1990. Benthos response to disturbance in western lake Erie: Field experiments. *Canadian Journal of Fisheries and Aquatic Sciences* 47:1970–1985.

Soster, F. M., D. T. Harvey, M. R. Troksa, & T. Grooms. 1992. The effects of tubificid oligochaetes on the uptake of zinc by Lake Erie sediments. *Hydrobiologia* 248:249–258.

Strickland, J. D. H. & T. R. Parsons. 1972. A Practical Handbook of Seawater Analysis. *Bulletin of the Fisheries Research Board of Canada* 167 (2nd ed.). Ottawa, Canada.

Sundby, B., L. Anderson, P. O. J. Hall, A. Iverfeldt, M. M. R. van der Loeff, & S. F. G. Westerlund. 1986. The effect of oxygen on release and uptake of cobalt, manganese, iron, and phosphate at the sediment-water interface. *Geochimica et Cosmochimica Acta* 50:1281–1288.

Tessenow, U. 1964. Experimental investigations concerning the recovery of silica from lake mud by Chironomid larvae (Plumosus group). *Archive für Hydrobiologie* 60:497–504.

Walshe, B. M. 1947. Feeding mechanism of *Chironomus* larvae. *Nature* 160:474–476.

The Effect of Burial on Shell Preservation and Epibiont Cover in Gulf of Mexico and Bahamas Shelf and Slope Environments After Two Years: An Experimental Approach

Karla M. Parsons-Hubbard, Eric N. Powell*, George M. Staff, W. Russell Callender, Carlton E. Brett, and Sally E. Walker

Abstract: *The introduction of biologic material into the fossil record occurs at various rates and via different taphonomic pathways, depending on the environment of deposition (EOD) and bathymetric depth. A comprehensive study of taphonomic rates and modes in a wide variety of depositional environments was undertaken by the Shelf and Slope Experimental Taphonomy Initiative (SSETI). Shells of the mussel* Mytilus edulis *and the surf clam* Arctica islandica *were deployed from 15 to 570 m depth in the Gulf of Mexico and the Bahamas and retrieved 2 yr later. Taphonomic alteration and biont coverage were influenced by a complex interaction between degree of burial, sediment texture, depth, and geographic region. Degree of burial was the single most important factor overall. Exposed shells showed significantly higher loss of color and dissolution than buried shells, but dissolution varied little among the sediment types. Color loss, however, was highest on hard grounds in the Bahamas and in carbonate sands in the Gulf of Mexico. Total biont coverage declined with burial, but burial influenced the accumulation of skeletonized over nonskeletonized forms of bionts. While many soft-bodied epibionts such as tunicates, fleshy algae, and sponges do not survive long periods of burial, the skeletons of epibionts such as encrusting foraminiferans, bryozoans, and tube-forming polychaetes withstand shallow burial and become proportionally more dominant after periods of burial and exhumation (although the bionts may be killed by burial). Where exhumation cycles do not occur (generally in deeper water), burial severely limits biont coverage. Shallow burial may not limit the total surface coverage by preservable (skeletonized and boring) bionts but may affect physical taphonomic patterns. Burial rate is an important factor in shell preservation; however, the characteristics of the enclosing sediment appear to be less influential.*

Introduction

Reconstructing paleocommunities from fossil assemblages requires a clear understanding of the variety of possible effects that post-mortem processes have on skeletal remains within their depositional environment. Because of taphonomic alteration, the original community is rarely if ever fully preserved. In order to understand the bias imposed on the fossil record, we must know the rates or types of taphonomic alteration in different depositional environments (comparative taphonomy, Brett & Baird 1986). Although rates are difficult to measure in fossil assemblages, we can establish

297

modern rates of destruction and extrapolate these to ancient depositional environments. Laboratory studies examining the rates of destruction of organic remains have provided important information on the mechanisms of taphonomic decay (Flessa & Brown 1983; Plotnick 1986; Kidwell & Baumiller 1990). While some studies have looked at taphonomic rates in shallow nearshore environments (Cummins et al. 1986; Fürsich & Flessa 1987; Flessa et al. 1993; Cutler 1994; Walker & Carlton 1995; Nebelsick et al. 1997), only one study has addressed rates of taphonomic processes in an offshore setting (Callender et al. 1994; at a hydrocarbon seep site). This study reports results of a much broader experimental analysis of taphonomic rates and processes over a 2-yr period on both the Gulf of Mexico continental shelf and slope and along a depth gradient from 15 to 260 m in the Bahamas. The study focuses on the relationships that shell preservation and biont coverage have with bottom conditions (sediment type and burial rate) over a wide range of depths, in different geographic regions, and in a suite of areally extensive and areally restricted environments of deposition (EODs).

Experimental Design

The Shelf and Slope Experimental Taphonomy Initiative (SSETI) project is a controlled series of experiments designed to measure and compare rates of taphonomic processes over a wide range of continental shelf and slope environments of deposition. SSETI was designed to compare the range of common (areally extensive) EODs typical of the shelves and slopes of the Gulf of Mexico and Caribbean Sea, as well as a series of unique (areally restricted) EODs of potential importance in the fossil record, such as brine seeps, collapsed basins, and carbonate-capped topographic highs. Because of the suspected slow rates of some important taphonomic processes in some EODs, each of the experiments was designed as a long-term deployment, some destined for a minimum of 10 yr on the seafloor.

The experiments consisted of shell material deployed in mesh bags (1.3 cm diamond mesh) attached to PVC rods, which allowed easy deployment and retrieval by submersible. A polyethylene float was attached 1.5 m above the array by polypropylene line and a 2.3-kg weight was added to counter its buoyancy. This float was easily seen by submersible and aided in relocating the sites, particularly after burial of the rods. Each bag contained five shells of each of five species of mollusc. Bivalves included *Mytilus edulis* (Linné), *Arctica islandica* (Linné), *Codakia orbicularis* (Linné), and, at some locations, *Mercenaria mercenaria* (Linné), *Glycymeris undata* (Linné), and/or *Argopecten irradians* (Lamarck). Two species of gastropod were also deployed. This analysis will address *M. edulis* (size range of ~6 to 9 cm) and *A. islandica* (range of 9 to 12 cm in longest dimension) because they were consistently deployed at the most sites and offer contrasting shell microstructures and thicknesses. With a few exceptions, four deployment sites were established in each EOD. In some cases, EODs were relatively uniform in topography and sediment texture. However, some EODs offered an array of habitats so that each single site represented some point in a complex gradient of topographies and sediment textures. Four bag arrays plus accompanying shells were deployed loose on the seafloor at each of the four sites in each EOD (Parsons et al. 1997; Parsons-Hubbard et al. 1999).

All sites were surveyed by video immediately after deployment and after 2 yr. Upon collection, all shells were photodocumented and then underwent a taphonomic analysis, including evidence of dissolution and color loss (methods described in Davies et al. 1990), and a biont analysis, including percent cover for each distinguishable taxon. Each bivalve shell was evaluated over eight standard shell areas (e.g., umbo, outer margin) on the inner and outer surface. We focus on three taphonomic indices in this analysis: average dissolution, maximum dissolution, and shell color. Average dissolution was obtained by grading each shell area on the internal and external surfaces semiquantitatively on a 4-point scale: 0 = no dissolution, 1 = chalky or slightly pitted surface, 2 = major dissolution present (e.g., 25%–90% pitting, surface soft, sculpture enhanced), 3 = extreme dissolution present (surface sculpture gone, dissolution extending into inner shell layers). An average dissolution level was then calculated as the surface area-weighted average (because some shell areas were larger than others) of the 16 evaluations (eight areas each on the inner and outer surfaces) on each shell. Maximum dissolution used the same 4-point scale (see above), but was determined by the shell area(s) with the most extreme alteration. Color loss was graded as 0 = original color, 1 = somewhat faded, 2 = faded to white, and 3 = discolored. Discolored shells were generally grey or black due to iron sulfide staining (Pilkey et al. 1969), or orange-staining due to oil impregnation (at petroleum seeps).

Degree of infestation by encrusting and boring bionts was determined by a visual estimate of percent cover by each taxon over each of the eight standard shell areas on the inner and outer surfaces (16 determinations per shell). Bionts were identified to species where possible and morphospecies were used for the remainder. Consequently, calculation of taxon richness provides a minimum value. Taxa were counted regardless of whether they were live or dead. For this report, bionts were lumped into broad categories (e.g., all nonskeletonized forms, all skeletonized forms including borings). To do so, total coverage of each shell area was calculated as the arithmetic sum of the coverage of each morphospecies in that shell area. Then, total shell coverage was calculated as the surface-area-weighted sum (because some shell areas were larger than others) of each of the 16 shell areas. This provides the total percent cover for the shell for a given biont category (not the mean of the 16 shell areas).Therefore, n always refers to the number of shells in each analysis because data are presented for the whole shell in all cases.

Statistical analysis used an ANCOVA with water depth as the covariate. Main effects included geographic location (Bahamas, Gulf of Mexico), sediment type and degree of burial (Parsons-Hubbard et al. 1999; Table 1). Significant differences between treatments were identified within significant ANCOVAs using Tukey's Studentized Range Test at $\alpha = 0.05$.

SSETI Deployment Sites

Bag arrays were deployed by submersible in 1993 and 1994, and were collected in 1995 and 1996. The *Johnson Sea Link* was used in the Gulf of Mexico and the *Nekton Gamma* and *Clelia* in the Bahamas. Parsons et al. (1997) and Parsons-Hubbard et al. (1999) provide detailed site descriptions. Experimental arrays were deployed in the Gulf of Mexico at 21 locations that encompassed a variety of terrigenous and carbonate settings on the Texas-Louisiana continental shelf and slope (Fig. 1a). The sites ranged in depth from 75 m to 190 m on the continental shelf and to 363 m on the continental slope. Sample arrays were also set out in less areally extensive EODs such as brine seeps, petroleum seeps, hardgrounds, a collapsed carbonate bank, and deep-water reefs. The deepest sites, at petroleum seeps, exceeded 550 m.

Bag arrays were deployed in four different depositional environments along the Atlantic side of Lee Stocking Island, Bahamas (Fig. 1b,c). These were the shallow coral reef and inter-reef (15–33 m), near-vertical wall (70–88 m), talus slope (183–226 m), and deep-water relict dune field (256–267 m). The reef sites included both sand and hard bottom areas of a larger reef and patch-reef complex. The wall drops off from a depth of 33 m at a slope exceeding 60°, to depths greater than 200 m. Wall sites are characterized by hard bottom and consolidated sediment surfaces affected by moderate current velocities sufficient to move bag arrays short distances and are populated by a variety of attached biota. Bag arrays were deployed along narrow ledges projecting from the wall. Below the wall, the slope begins to moderate at about 220 m. Carbonate sands with occasional outcroppings of carbonate rock dominate the bottom in this area (Fig. 1c). Further down-slope, the deep-water dune field consists of carbonate mud with a sparse macrofaunal assemblage.

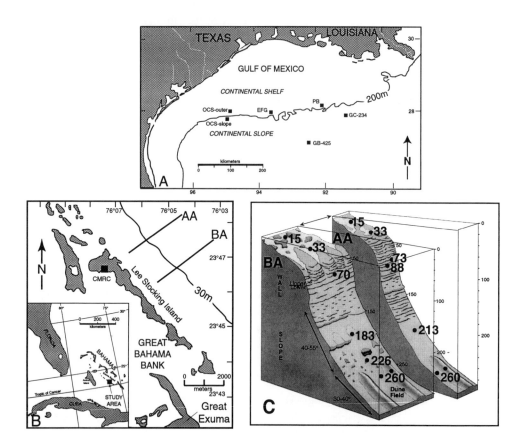

Fig. 1. Location map for experimental sites. A) Gulf of Mexico deployments were made at 183 m on the outer continental shelf (OCS-outer) and on the continental slope at 363 m (OCS-slope) off Galveston, Texas. The East Flower Garden (EFG) and Parker Bank (PB) are surface manifestations of underlying salt diapirism (Rezak & Bright 1981; Powell et al. 1986). Typically these salt domes have carbonate caprock. EFG sites consisted of a shallow carbonate reef hardground, a deeper carbonate sand locality down-slope from the reef, and several sites located within a brine pool and brine canyon caused by dissolution of the salt dome. Parker Bank is a collapsed carbonate structure. Sites included the central basin and the carbonate hardground surrounding the basin. Green Canyon (GC) 234 and Garden Banks (GB) 425 are petroleum seep sites located on the continental slope (Kennicutt et al. 1988; Callender & Powell 1992). B) Bahamian sites were located on two transects in Exuma Sound (McNeil & Grammer 1993; Colin 1995) offshore of Lee Stocking Island and the Caribbean Marine Research Center (CMRC). Transect AA is a higher energy transect located directly off a break in the Exuma Island chain. C) Profile of the shelf and slope along transect lines AA and BA at Lee Stocking Island, Bahamas. The black dots shown along the profiles are placed at the approximate locations of experiment deployment.

Table 1. Site designations and descriptions.

Locale[a]	Site Name	Ocean Basin[b]	Depth (m)	n[c] M.e.	n[c] A.i.	Bottom Type	Exposure
EFG	EFLGDN1	GOM	67	10	5	HG	Exposed
EFG	EFLGDN2	GOM	69	10	5	HG	Exposed
EFG	EFLGDN4	GOM	60	10	5	BP	Buried
EFG	EFLGDN5	GOM	78	10	5	CS	Exposed
EFG	EFLGDN6	GOM	71	10	5	HG	Exposed
EFG	EFLGDN7	GOM	80	10	5	CS	Buried
GB425	GB425S1	GOM	570	10	10	TM	Dusted
GB425	GB425S4	GOM	570	10	10	TM	Buried
GC234	GC234S2	GOM	570	10	5	TM	Buried
OCS-outer	OCSG2P1	GOM	183	10	10	TM	Buried
OCS-outer	OCSG1P2	GOM	183	7	10	TM	Buried
OCS-outer	OCSG1P1	GOM	183	10	10	TM	Buried
OCS-slope	SLOPES1	GOM	362	10	5	TM	Buried
OCS-slope	SLOPES2	GOM	359	10	5	TM	Buried
OCS-slope	SLOPES4	GOM	358	10	5	TM	Buried
PB	PARKR21	GOM	122	10	5	CM	Buried
PB	PARKR22	GOM	122	10	5	CM	Buried
PB	PARKR31	GOM	97	10	5	HG	Dusted
PB	PARKR41	GOM	91	10	5	CS	Dusted
AA	50RID	BAH	15	10	10	CS	Exposed
BA	50MUD	BAH	15	10	0	VH	Buried
BA	50SND	BAH	15	10	0	CS	Buried
BA	100NO	BAH	33	10	5	CS	Buried
BA	100SO	BAH	33	10	10	CS	Buried
AA	100S2	BAH	33	10	10	CS	Buried
AA	240LE	BAH	73	10	10	HG	Exposed
AA	290WA	BAH	88	10	10	VH	Exposed
BA	230WA	BAH	70	10	0	VH	Exposed
BA	230NO	BAH	70	10	0	VH	Exposed
BA	600CL	BAH	183	10	0	CS	Dusted
AA	700CR	BAH	213	10	10	VH	Dusted
BA	740BO	BAH	226	10	0	CS	Dusted
AA	875TR	BAH	267	10	10	CS	Exposed
AA	865CR	BAH	264	10	10	CS	Dusted
BA	850TN	BAH	259	10	0	VH	Dusted
BA	830CU	BAH	253	10	0	CS	Buried
BA	860DP	BAH	262	10	0	CS	Buried

[a] Details on site descriptions and exposure in Parsons-Hubbard et al. (1999). EFG, East Flower Garden; GB425, Garden Banks lease block 425; GC234, Green Canyon lease block 234; OCS-outer, outer continental shelf, Galveston, TX; OCS-slope, continental slope, Galveston, TX; PB, Parker Bank; AA, Transect "AA" Lee Stocking Island, Bahamas; BA Transect "BA" Lee Stocking Island, Bahamas.

[b] Ocean basins are designated GOM (Gulf of Mexico) and BAH (Bahamas).

[c] Sample size for *Mytilus edulis* (*M.e.*) and *Arctica islandica* (*A.i.*).

Table 2. ANCOVA results for variables determining taphonomic signature for *Arctica islandica* and *Mytilus edulis*[1].

	Mytilus edulis			*Arctica islandica*		
	Average Dissolution	Maximum Dissolution	Shell Color	Average Dissolution	Maximum Dissolution	Shell Color
Ocean[b]	0.02	0.0001	0.0001	-	-	0.03
Sediment type	0.007	0.0001	0.0001	-	-	0.0001
Exposure	0.0001	0.0003	0.0001	0.0002	0.0001	0.03
Depth	-	0.003	0.0001	-	-	-
Depth x Ocean	0.05	0.0001	0.0001	-	-	0.03
Depth x Sed. type	0.04	0.0001	0.0001	-	-	0.0001
Depth x Exposure	0.0003	0.0001	0.0002	-	-	-

[1] Values given are significant *p*-values ≤ 0.05; *p*-values listed as 0.0001 are ≤ 0.0001.
[A] Ocean refers to Gulf of Mexico or Bahamas.

Results

DEGREE OF BURIAL

The degree of burial of shells should have a significant impact on taphonomic condition (Fürsich & Flessa 1991). Each experimental site was designated as either a buried site, an exposed site, or a site having only a dusting of sediment on the shells (Parsons-Hubbard et al. 1999). The degree of burial and/or exposure had a significant impact on taphonomic signature in *M. edulis* and *A. islandica* (Table 2). In *M. edulis*, the interaction between the degree of burial and water depth was also significant for each of the three taphonomic attributes: average dissolution, maximum dissolution and shell color. The interaction terms with depth were not significant for *A. islandica*.

The ANCOVA revealed a significant difference in taphonomic signature due to exposure for *M. edulis* and *A. islandica* (Table 2 & Fig. 2a). With the exception of shell color, the source of the variation could not be identified in *M. edulis* using a Tukey's Studentized Range Test because other main effects and interaction terms with other EOD variables, such as sediment type and geographic location, were significant. Therefore, *M. edulis* showed no significant difference in dissolution due to burial. Exposed *A. islandica* were significantly more dissolved than buried shells, while dusted shells fell between. Color loss in both shell species was significantly greater when shells were exposed (Fig. 2a).

Exposed shells had significantly greater biont coverage than either dusted or buried shells (Table 3 & Fig. 2b). This held true for nonskeletonized and skeletonized bionts when analyzed separately. Interestingly, for skeletonized bionts, the buried shells, while still having less than half the biont coverage of exposed shells, nevertheless had significantly greater biont coverage than the shells dusted with sediment. Biont taxon richness was also greatest on exposed shells.

CARIBBEAN VERSUS THE GULF OF MEXICO

Gulf of Mexico sites differed significantly from Caribbean sites in many respects. All taphonomic attributes of *M. edulis* differed significantly between these two regions, as did total cover by bionts, skeletonized bionts, and biont species richness (Fig. 3). Interaction terms with depth were routinely significant for taphonomic attributes, indicating a strong differential effect of depth on taphonomic status (Table 3). Depth was much less important for biont coverage. Interestingly, *A. islandica* showed an opposite pattern: depth was less important for taphonomic signature but more important to biont cover.

Table 3. Results of ANCOVA for variables describing biont coverage for *Arctica islandica* and *Mytilus edulis* [1]. Nonsignificance is indicated by a dash (-).

	Total Biont Cover	Skeletonized Biont Cover	Nonskeletonized Biont Cover	Taxon Richness	Eveness
Mytilus edulis					
Ocean [a]	-	0.004	-	0.003	-
Sediment type	-	-	-	0.0001	-
Exposure	0.0001	0.0001	0.0001	-	-
Depth	-	-	-	-	-
Depth x Ocean	-	-	-	0.03	-
Depth x Sed. type	-	-	-	0.0001	-
Depth x Exposure	0.0003	0.002	0.007	-	0.03
Arctica islandica					
Ocean [a]	0.0001	-	0.0001	0.0008	-
Sediment type	0.0001	0.006	0.0001	0.0001	-
Exposure	0.0001	0.006	0.0001	0.0002	-
Depth	0.0001	0.04	0.0001	0.0001	-
Depth x Ocean	0.0001	-	0.0001	0.002	-
Depth x Sed. type	0.0001	0.04	0.0001	0.0001	-
Depth x Exposure	0.0003	-	0.0001	0.01	-

[1] Values given are significant p-values ≤ 0.05; p-values listed as 0.0001 are ≤ 0.0001.

[a] Ocean refers to Gulf of Mexico or Bahamas.

Looking at taphonomic condition in more detail, and recognizing that other significant main effects and interaction terms restrict the value of simple comparisons, a Tukey's Studentized Range Test detected pronounced geographic effects in *M. edulis* (Fig. 3a). Shells were more severely dissolved (i.e., average dissolution was greater) in the Gulf of Mexico; the maximum degree of dissolution was higher, and color loss was more severe. No significant effect of geography on taphonomic condition was noted in *A. islandica* (Fig. 3a). Significantly more biont encrustation occurred on both shell types in Bahamas locations than in Gulf of Mexico locations (Fig. 3b). However, the difference was due to a much larger coverage of skeletonized bionts in the Bahamas. Non-skeletonized biont coverage was about the same in both regions. Biont taxon richness was significantly greater in the Bahamas in both species.

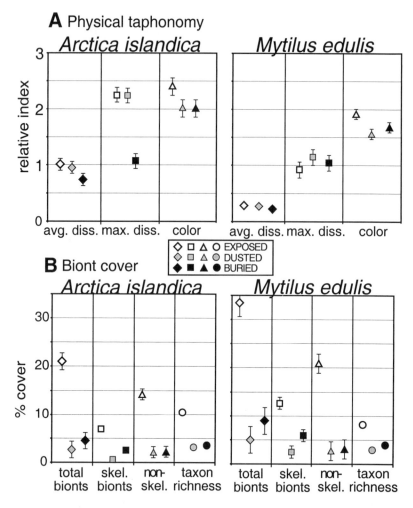

Fig. 2. The effect of exposure state on taphonomic condition and percent cover by bionts. A) Physical taphonomy. Average dissolution measured on a scale of 0 (no dissolution) to 3 (dissolution extending into the inner shell layers). Maximum dissolution records the most severe dissolution observed anywhere on the shell. Color refers to loss of color on a scale of 0 (no color loss) to 3 (complete loss of shell color or discolored). After 2 yr, average dissolution was low in both species. Exposed *Arctica islandica* were significantly more dissolved than buried (black symbols) for both average and maximum dissolution. Error bars are calculated from Tukey's Studentized Range Test at α = 0.05. Error bars that do not overlap indicate a significant difference. B) Biont cover. For both shell species, exposed shells had significantly higher cover than either dusted or buried shells. Taxon richness was also greater on exposed shells.

Effect of Sediment Type

The differences in taphonomic alteration may be linked to the type of sediment present at each location. The bottom-type categories chosen for this study were 1) carbonate mud, 2) carbonate sand, 3) veneered hardground (a hard-ground with a thin covering of sediment), 4) hardground, 5) terrigenous mud, and 6) brine (see Fig. 4 for sample sizes in each category). Although 'brine' is not a category of

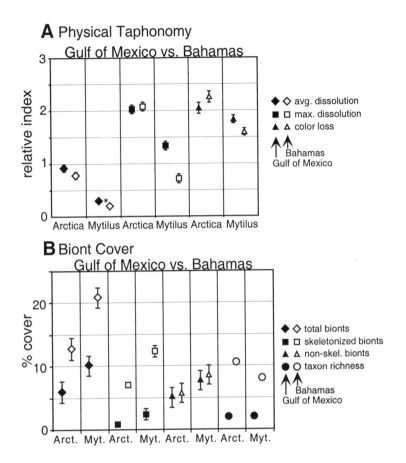

Fig. 3. Comparison of means for taphonomic criteria and biont cover between the Gulf of Mexico (black symbols) and the Bahamas (open symbols). A) Physical taphonomy. Average dissolution was generally low after 2 yr on the seafloor. Maximum dissolution was higher in *Arctica islandica* and statistically similar between regions. *Mytilus edulis* showed greater maximum dissolution in the Gulf of Mexico (*). Color loss was high in both regions for both shell types. In *A. islandica*, color loss was similar among the two regions; in *M. edulis* color loss was greater in the Gulf of Mexico. B) Biont cover. Total biont cover and skeletonized biont cover were significantly higher on Bahamas shells than on Gulf of Mexico shells. However, nonskeletonized forms showed no regional differences. Taxon richness was statistically higher in the Bahamas as well. Error bars are calculated using Tukey's Studentized Range Test at α= 0.05. Significant differences between means for each regional pair exist when error bars do not overlap. *Symbols too large to show error bars, but the means are significantly different.

sediment, the effects of the brine on the shells far outweigh the sediment type; therefore these sites were placed in their own sediment-type category.

Sediment type had a limited effect on taphonomic condition in *A. islandica*, except for loss of shell color. In contrast, sediment type significantly influenced the degree of dissolution in *M. edulis*, as well as its shell color (Table 2). The degree of dissolution was influenced by burial but not by sediment type. Certain bottom types precluded burial, of course; hardgrounds being a prime example. However, considering the suite of EODs, taphonomic condition was much more directly related to degree of burial than to sediment type. Assuming that sediment type is a good surrogate for a suite of edaphic factors, EOD-dependent factors other than degree of burial were much more important in *M. edulis* than in *A. islandica*.

Only carbonate sands and hardgrounds were common to both the Gulf of Mexico and the Bahamas. Consequently, and because of the number of significant effects of region in the ANCOVA (Tables 2 & 3), further analysis of the effect of sediment type was conducted for each region separately (Fig. 4, open symbols vs. black symbols). No individual sediment type affected the degree of dissolution on either shell type more than any other in the Gulf of Mexico or the Bahamas (Fig. 4a). In the Bahamas, color loss was greatest on hardgrounds (and veneered hardgrounds for *A.*

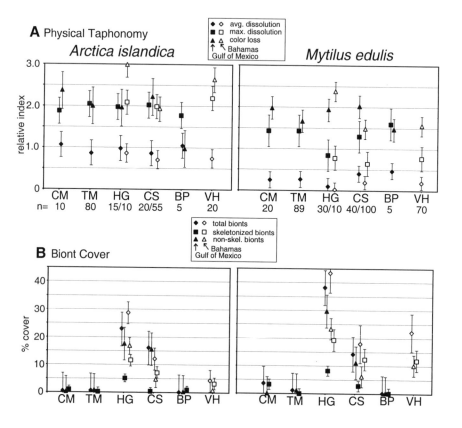

Fig. 4. Taphonomic condition and biont cover for each sediment type for the Gulf of Mexico (black symbols) and the Bahamas (open symbols). A) Taphonomic condition including dissolution and color loss. Note that the only significant differences in condition among sediment types occur in shell color (triangles). Neither average nor maximum dissolution varied with sediment type for either shell species. B) Biont cover for shells from each sediment type. Biont cover of shells was greatest on hardgrounds, with carbonate sands having abundant biont cover. On veneered hardgrounds, high biont cover was found for *Mytilus edulis* but not *Arctica islandica*. Values are means for all study sites having the same sediment type, regardless of burial state, and for all depths combined. Error bars are the calculated using Tukey's Studentized Range Test at α = 0.05. Any two means with overlapping error bars were not significantly different. CM - carbonate mud, TM - terrigenous mud, HG - hardground, CS - carbonate sand, BP - brine pool, VH - veneered hardground.

islandica). Loss of color appears to be decoupled from dissolution in both geographic regions and therefore should not be used as an indicator of chemical dissolution on pigmented shells.

Sediment type had only a limited effect on biont coverage in *M. edulis* (Table 3). Only species richness showed a significant main effect. Not so for *A. islandica*, where sediment type had a significant effect on all biont categories except evenness and where the interaction terms with depth were consistently significant. Again, examining sediment types by region (Fig 4b), biont infestation was significantly greater on shells in the hardground sites in both the Bahamas and Gulf of Mexico. This is likely related to their lack of sediment cover. Shells in carbonate sands showed high cover in both regions, and *M. edulis* had high cover on veneered hardgrounds.

Biont taxon richness reached a mean of six morphospecies for both *A. islandica* and *M. edulis* deployed on Gulf of Mexico hardgrounds, significantly more than other sediment types. In the Bahamas, taxon richness was also highest on the hardgrounds (a mean of 23 morphospecies for *A. islandica* and 14 for *M. edulis*) and dropped off considerably in other sediment types.

EFFECT OF EXPOSURE WITHIN SEDIMENT TYPE

We examined the influence of burial on taphonomy and biont coverage within sediment types. Shells at hardground sites were always exposed, those at carbonate mud and brine sites were always buried. Therefore comparisons of exposure condition within these bottom types could not be made.

Exposed shells of *M. edulis* and *A. islandica* (open symbols, Fig. 5) deployed on carbonate sands in the Gulf of Mexico experienced more dissolution than shells either lightly dusted with sediment or completely buried and these differences were often significant (particularly for *M. edulis*). No differences existed between dusted and buried shells in the terrigenous mud localities for either shell species. In the Bahamas, shells showed few effects of dissolution on average (avg. dissolution ≤ 1), but the maximum severity of dissolution reached on any shell area was higher for exposed shells on carbonate sands and for dusted shells on veneered hardgrounds.

Loss of color was high for *M. edulis* in both the Gulf of Mexico and the Bahamas on carbonate sand and terrigenous mud (Fig. 5a). Burial condition did not affect color loss on veneered hardgrounds, however. *A. islandica* shells were significantly faded on all bottom types in both geographic regions (Fig. 5b). The only significant differences based on exposure condition occurred in the Gulf of Mexico on carbonate sands where buried shells had greater loss of color than dusted shells (n.b., this is based on only one site for each exposure condition but at comparable depths). In the Bahamas, exposed shells had significantly greater color loss than buried shells (again, this is not depth related, one exposed site was at 15 m and the other at 267 m).

Buried *M. edulis* tended to have significantly less biont cover than either dusted or fully geographic regions (Fig. 5b). The only significant differences based on exposure condition occurred in the Gulf of Mexico on carbonate sands where buried shells had greater loss of color than dusted shells (n.b., this is based on only one site for each exposure condition but at comparable depths). In the Bahamas, exposed shells had significantly greater color loss than buried shells (again, this is not depth related, one exposed site was at 15 m and the other at 267 m). exposed shells, except on veneered hardgrounds in the Bahamas (Fig. 6). In the Gulf of Mexico, dusted shells had greater biont coverage than buried shells in all bottom types. Nonskeletonized biont coverage tracked total coverage, whereas skeletonized biont coverage was markedly different. Terrigenous muds in the Gulf of Mexico had characteristically low biont cover while carbonate sands in both regions were characteristically biont-rich for both *M. edulis* and *A. islandica*. Exposed shells on carbonate sands

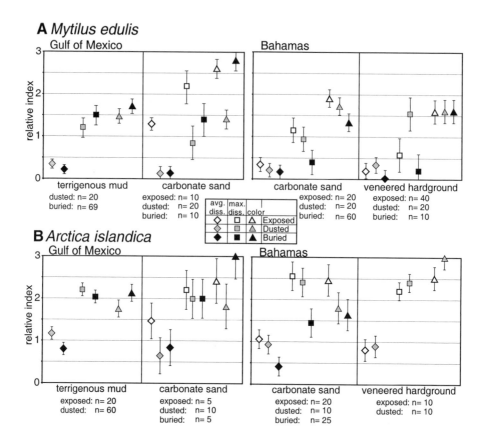

Fig. 5. Effects of sediment type and burial on physical taphonomy of *Mytilus edulis* and *Arctica islandica* deployed for 2 yr in the Gulf of Mexico and Bahamas. Scale as in Fig. 4. A) Results for *M. edulis* (sample size is listed below each graph). Average dissolution was low in both the Gulf of Mexico and Bahamas for all sediment types and exposure conditions, except for shells exposed on carbonate sand in the Gulf of Mexico where shells displayed minor pitting and etching (avg. = 1.3). Color loss was generally higher on exposed shells on carbonate sands whereas exposure condition had no effect on veneered hardgrounds. B) Results for *Arctica islandica*. Average dissolution had more effect on *A. islandica* than on *M. edulis* shells, with exposed shells having somewhat higher average and maximum dissolution values. Color loss showed little difference based on exposure condition with the exception of buried shells having more color loss than dusted shells on carbonate sands in the Gulf of Mexico. Error bars are calculated using Tukey's Studentized Range Test for comparison of means after ANOVA at α = 0.05.

had significantly higher biont cover than dusted or buried shells. *M. edulis* shows an interesting trend on Bahamas veneered hardgrounds, however. Buried shells had significantly more biont cover than exposed shells in terms of total cover, skeletonized cover, and taxon richness (nonskeletonized cover was higher, but not significantly so). Therefore burial seems to have enhanced biont infestation on shells in this setting.

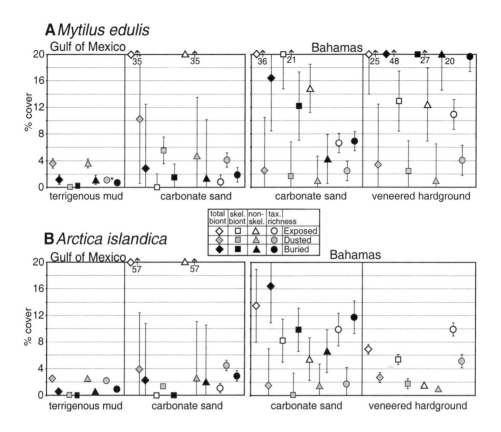

Fig. 6. Effects of burial and sediment type on biont cover for *Mytilus edulis* and *Arctica islandica* deployed for 2 yr in the Gulf of Mexico and Bahamas. A) Results for *M. edulis*. Shells on carbonate sands were highly infested by bionts in both regions, with exposed shells having the greatest cover. Interestingly, buried shells on veneered hardgrounds had higher cover than exposed shells. B) *A. islandica* also shows the most biont cover on carbonate sands and very low cover on terrigenous muds. Sample sizes are given in Fig. 5. Error bars calculated using Tukey's Studentized Range Test for comparison of means after ANOVA at α = 0.05. *Symbols too large to show error bars, but the means are significantly different.

Biont taxon richness was highest on buried, and lowest on dusted *M. edulis* in carbonate sands and veneered hardgrounds in the Bahamas. Dusted shells consistently had the highest taxon richness in the Gulf of Mexico for both shell species. Interestingly, only on one shell species (*A. islandica*) in one environment (veneered hardground) did exposed shells rank highest in taxon richness. Some amount of sediment cover seemed to enhance the diversity of epibiont taxa on these shells.

Discussion

In the ANCOVA analyses, most main effects (sediment type, geographic region, and burial state) were significant for most taphonomic or biont attributes in both species. The covariate, water depth, was also significant in most cases. Significant results were common enough to suggest that shell condition is overwhelmingly controlled by a combination of EOD-specific factors, including water depth, sediment type, and burial rate. General trends would seem to be the exception rather than the rule, and use of mean values for describing taphonomic alteration or biont coverage must be evaluated cautiously. Clearly, interpretation of the preservational process in these EODs requires the simultaneous consideration of a number of important variables.

Nevertheless, some general trends can be discerned. The degree of exposure of shells at the sediment-water interface exerted a strong control on the taphonomic condition of shells after 2 yr on the seafloor. The degree of burial was far and away the most significant factor controlling the preservational process in these EODs. Exposed shells were generally more encrusted with biont overgrowths and more dissolved than buried shells. Exposed shells generally had lost more color than buried shells. However, after 2 yr, the degree of taphonomic alteration was not extreme in either species. The average degree of dissolution rarely exceeded 1.0 on a 0–3 scale. Average shell color indices exceeding 2.0 on a 0–3 scale occurred over a wide range of EODs in both geographic regions. Shell color was rapidly lost in many of these EODs, probably because even the smallest degree of dissolution significantly affects the look of the shell. One basic dogma of taphonomy is that shell preservation is enhanced with rapid burial (Driscoll 1970; Powell 1992). Although some interesting exceptions exist in this data set, overall, this series of experiments supports the importance of rapid burial in preservation.

The two species did not differ much in the overall pattern of biont coverage and in the effect of taphonomic processes. Rates were clearly different in a number of cases. In the Bahamas, for example, M. edulis had much higher biont coverage than A. islandica. This was not the case in the Gulf of Mexico. Most of the differences observed, however, were not species-dependent; they were EOD-dependent. Geographic region also exerted a significant influence. Some of these differences were due to the mix of EODs in the two regions. Terrigenous muds, for example, were common in the Gulf of Mexico, but not present in the Bahamas. However, within common sediment types, bionts were more abundant in the Bahamas. But here again, other factors may come into play. The Bahamas had more sites within the photic zone (15–150 m). Voight and Walker (1995) also observed a significant influence of geography on biont coverage in the Caribbean and off South America. Dissolution rates were higher in the Gulf of Mexico. Thus, the large-scale physical environment exerted a significant influence on these EODs, albeit not as large an influence as the more local processes of burial.

Sediment type exerted a remarkably limited influence on taphonomic condition. Loss of shell color was significantly more rapid on hardgrounds in the Bahamas and shells in brine retained their original coloration better than at other sites.Overall, dissolution and color loss appear to be controlled by exposure of the shells more so than by sediment type. Shells deployed on hardgrounds had consistently chalky shell surfaces and shells on carbonate sands showed significantly less dissolution for buried shells than for exposed shells, for example. Similar degrees of dissolution and loss of shell color occurred on shells deployed on carbonate and terrigenous mud bottoms. Even for biont coverage, the effect of sediment type was not very influential. Biont coverage was higher on shells deployed on hardgrounds and higher overall at the Bahamian sites. No bionts were present on shells deployed in the East Flower Garden brine pool. However, biont coverage did not differ significantly between carbonate sands, carbonate muds, and terrigenous muds in the Gulf of Mexico or between carbonate sands and veneered hardgrounds in the Bahamas. Therefore, sediment type did not exert an overwhelming influence on either the taphonomic process or the occupation of shell substrate by bionts. Only biont taxon richness, of all the shell attributes measured, was consistently affected significantly by changes in sediment type. Of course, 2 yr is a relatively short time in the taphonomic life of a shell, so the present results may simply be due to the insufficient passage of time necessary for sediment type to impose a fingerprint on taphonomic signature. However, after 2 yr, a shell may be buried deeply enough to be removed from the taphonomically active zone. Thus, effects on shells deployed for 2 yr are important, equally as important as longer term exposure. These experiments continue to run and will be retrieved at 6 yr and beyond to compare shorter term effects to long-term effects.

Biont coverage was highest on shells deployed on hardgrounds, followed by those on carbonate sands. Many of these sites were in relatively shallow water, within the photic zone, in comparison with the muddy sites, so that the influence of depth cannot be discounted as a confounding variable (Parsons et al. 1997). Even so, the coverage of bionts on shells deployed in muddy sites is extremely low. Rapid burial at these sites is the most likely explanation.

Even a dusting of sediment tended to reduce taphonomic decay. Powell (1992) emphasized the importance of surficial sediments in the taphonomic process. In theory, oxidation of reduced substances should provide an acidic environment that would enhance the rate of dissolution at or near the sediment surface. It is unlikely that any of our shells were buried deeper than the taphonomically-active zone in any of these EODs. Nevertheless, in comparison with previous treatments (e.g., Powell 1992), the relatively dramatic reduction in the rate of taphonomic alteration with even a dusting of sediment is an important finding. Very likely, the most significant step in preservation is the movement of a shell even millimeters below the sediment surface. This would seem to be true even though the oxic zone varies greatly in thickness among these EODs (Lin & Morse 1991). Our results emphasize the need to focus on the influence of water column exposure on the taphonomic process.

Biont taxon richness was strongly influenced in both species by geographic region and sediment type, as well as by depth and degree of burial in *A. islandica*. For both shell species, taxon richness was unique for nearly all sediment types.

Biont coverage was lowest on dusted shells and highest on exposed and buried shells in the Bahamas. However, dusted shells occurred only in deep water (180–264 m), so low cover is likely influenced by depth in this case. On the veneered hardgrounds, buried *M. edulis* had the highest biont coverages for that sediment type. A tendency for skeletonized bionts to be distributed differently with respect to burial state than nonskeletonized forms was present in both species and in the Bahamas and the Gulf of Mexico. In both species of shell at exposed sites, skeletonized bionts covered only half as much surface area as nonskeletonized forms. We hypothesize that nonskeletonized bionts out-compete skeletonized forms, on the average. Exposed shells develop a rich biont fauna and flora and the majority of these will be nonpreservable forms. However, whether the biont lives or dies, skeletons are much more likely to withstand burial. Accordingly, while total biont coverage is reduced by burial, the ratio of skeletonized (live and dead) to nonskeletonized forms is increased.

Despite the general trend toward a proportional increase in coverage by skeletonized bionts on buried shells, nonskeletonized bionts were unusually common on buried shells on veneered hardgrounds. Nonskeletonized bionts have low percent coverage on buried shells in general; however, on these veneered hardgrounds, one site at 15 m reached nearly 20% cover on *M. edulis* while cover on exposed shells reached only 12%. Veneered hardgrounds were sampled only in the Bahamas and occurred at 15 m on shallow hummocks stabilized by gorgonians and *Halimeda* (one site, buried), on ledges jutting out from the near-vertical shelf-edge wall (three sites, all exposed; 70–88 m), and in the troughs between the large dunes on the deepest part of the Bahamas transects (one dusted site at 250 m). These sites may be characterized by shifting loose sediments on stabilized bottoms. Thus the burial condition may be only temporary and although the shells were buried when they were collected, they may have experienced periods of intermittent burial and exposure over the 2 yr of the deployment. One cannot discount the possibility that rapidly shifting sediments permit biont coverage by nonskeletonized taxa to remain high by reducing exposure to predators while limiting mortality associated with longer term burial. This could account for the significantly higher nonskeletonized biont coverage at these sites.

In the Bahamas, the dusted sites had fewer bionts. A tempting hypothesis is that shells at these sites do not experience burial-exhumation cycles. Under this hypothesis, the dusting is a permanent cover and these shells, like most of the buried shells in the Gulf of Mexico, have only a limited biont fauna. It is striking that, under appropriate conditions, in both the Bahamas and Gulf of Mexico and on a variety of bottom types, just a little dusting of sediment dramatically reduces biont infestation as part of the preservational process.

One of the curious exceptions to this general trend occurs on carbonate sands in the Gulf of Mexico. Here, all of the biont cover on exposed shells was nonskeletonized, whereas skeletonized biont cover dominated the dusted and buried shells and tended to be highest on dusted shells. Probably dusted shells, being very near the surface, offered substrate for aerobic skeletonized forms while limiting their exposure to predators.

Shells that reside on exposed substrates are likely to experience biont infestation. Both preservable bionts and nonpreservable bionts normally occupy exposed shells, but nonskeletonized bionts are proportionately more abundant. The skeletonized forms become increasingly represented if certain burial processes occur. It is tempting to consider the use of biont coverage to identify quiet water settings in the fossil record where shells lay exposed on the seafloor for some time. Our data identify a danger in too easily accepting this view. Exhumation and burial would seem to provide a mechanism to increase skeletonized biont coverage and preferentially preserve them. Thus, the dramatic difference between biont coverage on exposed and buried shells, noted for total bionts, may be much less clear once the shell is preserved. The results of this set of experiments suggest that long-term exposure and rapid burial-exhumation cycles may not be differentiated easily in the fossil record using biont coverage alone.

Conclusions

Based on the results of the first 2 yr of a long-term experimental deployment of bivalves on continental shelf and slope environments in the Gulf of Mexico and Bahamas, burial plays a very important role in rate of shell degradation. Burial rate is often touted as the most important factor in shell preservation, but this hypothesis has previously been tested only over the short term in shallow near-shore environments. The results of this study support this position. The characteristics of the enclosing sediment (such as carbonate vs. terrigenous mud) were not as important. Biont cover was also affected by shell burial. Overall, biont cover was reduced on buried shells. However, when preservable and nonpreservable (soft-bodied) bionts were analyzed separately, preservable bionts with hard skeletons (including borings) showed an increase in surface area cover with burial, particularly at shallow sites where shells were periodically exposed and re-buried. Soft-bodied epibionts were easily killed and removed by the burial process and therefore were more prevalent on exposed shells.

The degree of burial exerted the greatest influence on shell condition. However, the general geographic region also affected the taphonomy (e.g., dissolution was stronger in the Gulf of Mexico while biont coverage was higher in the Bahamas). Depth also strongly affected certain factors such as the extent of biont infestation, but even on shells in shallow, exposed settings, burial was the most important controlling factor.

Acknowledgments

The submersible work required for the deployment and recovery of experiments was made possible through a series of grants from the National Oceanic and Atmospheric Administration's NURP programs at UNC-Wilmington (Gulf of Mexico sampling) and at the Caribbean Marine Research Center (Bahamas sampling). We would like to thank these NURP programs for the consistent funding of the four major field efforts that permitted deployment and recovery over this wide regional area. We would like to thank M. Ellis for field and laboratory logistical support and the support crews of the *Johnson Sea Link*, *Clelia*, and *Nekton Gamma* submersibles. We appreciate the efforts of the CMRC staff on Lee Stocking Island and the NURP personnel from UNCW and CMRC that took part in these field programs. We would also like to thank R. Aller, H. Lescinsky, and K. Flessa for greatly improving this manuscript.

Literature Cited

Brett, C. E. & G. C. Baird. 1986. Comparative taphonomy: A key to paleoenvironmental interpretation based on fossil preservation. *Palaios* 1:207–227.

Callender, W. R. & E. N. Powell. 1992. Taphonomic signature of petroleum seep assemblages on the Louisiana upper continental slope: Recognition of autochthonous shell beds in the fossil record. *Palaios* 7:388–408.

Callender, W. R., E. N. Powell, & G. M. Staff. 1994. Taphonomic rates of molluscan shells placed in autochthonous assemblages on the Louisiana continental slope. *Palaios* 9:60–73.

Colin, P. L. 1995. Surface currents in Exuma Sound, Bahamas, and adjacent areas with reference to potential larval transport. *Bulletin of Marine Science* 56:48–57.

Cummins, H., E. N. Powell, R. J. Stanton Jr., & G. Staff. 1986. The rate of taphonomic loss in modern benthic habitats: How much of the potentially preservable community is preserved? *Palaeogeography, Palaeoclimatology, Palaeoecology* 52:291–320.

Cutler, A. H. 1987. Surface textures of shells as taphonomic indicators, p. 164–176. *In* Flessa, K. W. (ed.), Paleoecology and Taphonomy of Recent to Pleistocene Intertidal Deposits: Gulf of California. Special publication no. 2. The Paleological Society.

Davies, D. J., G. M. Staff, W. R. Callender, & E. N. Powell. 1990. Description of a quantitative approach to taphonomy and taphofacies analysis, p. 328–350. *In* Miller, W., III (ed.), Paleocommunity Temporal Dynamics: The Long-term Development of Multispecies Assemblies. Special publication no. 5. The Paleological Society.

Driscoll, E. G. 1970. Selective bivalve shell destruction in marine environments, a field study. *Journal of Sedimentary Petrology* 40:898–905.

Flessa, K. W. & T. J. Brown. 1983. Selective solution of macroinvertebrate calcareous hard parts: A laboratory study. *Lethaia* 16:193–205.

Flessa, K. W., A. H. Cutler, & K. H. Meldahl. 1993. Time and taphonomy: Quantitative estimates of time-averaging and stratigraphic disorder in a shallow marine habitat. *Paleobiology* 19:266–286.

Fürsich, F. T. & K. W. Flessa. 1987. Taphonomy of tidal flat molluscs in the Northern Gulf of California: Paleoenvironmental analysis despite the perils of preservation. *Palaios* 2:543–559.

Fürsich, F. T. & K. W. Flessa. 1991. The origin and interpretation of Bahia la Choya (northern Gulf of California) taphocoenoses: Implications for paleoenvironmental analysis. *Zitteliana* 18:165–169.

Kennicutt, M. C., II, J. M. Brooks, R. R. Bidigare, & G. J. Denoux. 1988. Gulf of Mexico hydrocarbon seep communities. I. Regional distribution of hydrocarbon seepage and associated fauna. *Deep-Sea Research* 35:1639–1651.

Kidwell, S. M. & T. M. Baumiller. 1990. Experimental disintegration of regular echinoids: Roles of temperature, oxygen and decay thresholds. *Paleobiology* 16:247–271.

Lin, S. & J. W. Morse. 1991. Sulfate reduction and iron sulfide mineral formation in Gulf of Mexico anoxic sediments. *American Journal of Science* 291:55–81.

McNeil, D. F. & G. M. Grammer. 1993. A reconnaissance characterization of NOAA/NURP dive transects Lee Stocking Island, Exuma Sound. Caribbean Marine Research Center Report. Perry Institute, Tequesta, Florida.

Nebelsick, J. H., B. Schmid, & M. Stachowitsch. 1997. The encrustation of fossil and recent sea-urchin tests: Ecological and taphonomic significance. *Lethaia* 30:271–284.

Parsons, K. M., E. N. Powell, C. E. Brett, S. E. Walker, & W. R. Callender. 1997. Shelf and Slope Experimental Taphonomy Initiative (SSETI): Bahamas and Gulf of Mexico. *Proceedings of the 8th International Coral Reef Symposium* 2:1807–1812.

Parsons-Hubbard, K. M., W. R. Callender, E. N. Powell, C. E. Brett, S. E. Walker, A. L. Raymond, & G. M. Staff. 1999. Rates of burial and disturbance of experimentally-deployed molluscs: Implications for preservation potential. *Palaios* 14:337–351.

Pilkey, O. H., B. W. Blackwelder, L. J. Doyle, & E. L. Estes. 1969. Environmental significance of the physical attributes of calcareous sedimentary particles. *Transactions Gulf Coast Association of Geological Societies* 19:113–114.

Plotnick, R. E. 1986. Taphonomy of a modern shrimp: Implications for the arthropod fossil record. *Palaios* 1:286–293.

Powell, E. N. 1992. A model for death assemblage formation. Can sediment shelliness be explained? *Journal of Marine Research* 50:229–265.

Powell, E. N., T. J. Bright, & J. M. Brooks. 1986. The effect of sulfide and an increased food supply on the meiofauna and macrofauna at the East Flower Garden brine seep. *Helgolander Meeresuntersuchungen* 40:57–82.

Rezak, R. & T. J. Bright. 1981. Seafloor instability at East Flower Garden Bank, northwest Gulf of Mexico. *Geo-Marine Letters* 1:97–103.

Walker, S. E. & J. T. Carlton. 1995. Taphonomic losses become taphonomic gains: An experimental approach using the rocky shore gastropod *Tegula funebralis*. *Palaeogeography, Palaeoclimatology, Palaeoecology* 114:197–217.

Voight, J. R. & S. E. Walker. 1995. Geographic variation of shell bionts in the deep-sea snail *Gaza*. *Deep-Sea Research* 42:1261–1271.

Effects of Particle and Solute Transport on Rates and Extent of Remineralization in Bioturbated Sediments

R. C. Aller, J. Y. Aller, and P. F. Kemp

Abstract: *Diagenetic reactions in surface sediments are dramatically affected by macrofaunal and meiofaunal activities. Of the many factors simultaneously influenced by infauna, alterations of the diagenetic transport regime during particle reworking, burrow formation, and irrigation play particularly important roles in controlling net degradation of organic material and nutrient recycling. Particle and fluid transport in the bioturbated zone influences the overall rates of reaction, reaction distributions, degradation pathways, synthetic reactions, coupled redox reaction balances, and the eventual extent of organic matter degradation (preservation). Macrofaunal and meiofaunal effects on transport are highly density, size, and species specific. In general, overall suboxic decomposition pathways and microbial activities are greatly enhanced by both particle and solute transport. Reaction zonation and coupling become more complex and highly dynamic in the presence of infauna. Experimental studies of both bulk sedimentary organic matter and decay of specific compounds, such as lipids and photopigments, demonstrate that macrobenthic activity results in more rapid and complete decomposition relative to deposits without bioturbation. Net rate increases (typically ~ 2–10X) are due both to alteration of the effective reaction rate constants governing net decomposition (k) of organic material, and to enhancement of the effective pool size of material available to react ($\hat{C}_r$). Effects on pool size are directly related to reaction pathway, synthetic reactions, and increased total decomposition under steady or oscillating oxic-suboxic conditions. Reactivity, parameterized as a rate constant, k, is a conditional function of solute transport, and at least for the refractory organic pool, also of redox reaction path. Enhanced solute transport resulting from density-dependent irrigation clearly promotes net rates of remineralization independent of oxidant availability. Microbial populations are more abundant and individual cells larger and more active as solute exchange increases, even under strictly anoxic conditions. The biogeochemical functioning of time-dependent, three-dimensional transport-reaction regimes, such as characterize the bioturbated sediment zone, remains a rich area of study critical to understanding elemental cycling and its geological evolution in surficial environments.*

Introduction

Early diagenetic reactions and elemental cycling in surface sediments are extensively affected by benthic macrofauna and meiofauna. The decomposition of sedimentary organic matter and nutrient remineralization are important sets of diagenetic reactions subject to both direct and indirect alteration by animal activities. Many of the multiple factors that govern decomposition processes are influenced simultaneously by benthos (Table 1). Of these factors, biogenic changes in the diagenetic transport regime during particle reworking, burrow formation, and irrigation are of primary importance in altering remineralization processes and the interaction of sediments with other surficial reservoirs. Bottom deposits are transformed during feeding, irrigation, and burrow construction activity, from upwardly accreting bodies dominated by one-dimensional vertical transport, into temporally dynamic and fully three-dimensional biogeochemical mosaics of sediment and interspersed overlying water. Particle and fluid transport processes within the bioturbated zone influence the absolute rates of reactions, the relative rates and thus balances between reactions, degradation pathways, spatial and temporal distributions of reactions, the extent of reaction, and the eventual sedimentary storage of reaction products (Demaison & Moore 1980; Andersen & Kristensen 1991; Aller & Aller 1998; Table 1 references). Whereas many of these effects are obvious and often readily demonstrated, others are more subtle, poorly understood, or potentially subject to conceptual confusion. In the present contribution we briefly review and examine, within the context of the most commonly utilized organic matter decomposition models, a few selected effects of particle reworking and solute exchange on organic matter reactivity, reaction rates, and overall remineralization processes in the bioturbated zone, and point out several areas where fundamental understanding remains minimal or confused.

Models of Organic Matter Decomposition

The time-dependent decomposition of sedimentary organic matter can be represented by reactions and kinetic functions of the form (modified after Paul & Van Veen 1978; Van Cappellen et al. 1993; Van Cappellen & Gaillard 1996; Boudreau 1997):

$$\hat{C}_r + Ox_i + \mathbf{B} \rightarrow CO_2 + Red_i + y_i\mathbf{B} \tag{1a}$$

$$\frac{d\hat{C}_r}{dt} = -k_i(t)\hat{C}_r \frac{(Ox)_i}{(K_{ox,i} + (Ox)_i)} In_{ij} \tag{1b}$$

where:

$\hat{C}_r$	=	reactive carbon substrate pool or specific compound concentration
Ox_i	=	concentration of oxidant i
$\mathbf{B}$	=	active bacterial and faunal biomass mediating reaction pathway i
y_i	=	biomass yield from pathway i (can be time dependent)
Red_i	=	reduced reaction products from oxidant i
t	=	time
$k_i(t)$	=	reaction rate coefficient for oxidant i
$K_{ox,i}$	=	reaction half saturation constant for oxidant i
In_{ij}	=	inhibition of oxidant i by all other oxidants j.

Table 1. Macrofaunal effects on organic C decomposition, remineralization[1].

Macrofaunal Activity	Effect	Decomposition Rate and/or Extent[a]
Particle manipulation	Substrate exposure, fragmentation surface area increase	+
Grazing, digestion	Microbe consumption, Bacterial growth stimulation Surfactant, enzyme bath Redox oscillation	+
Excretion, secretion	Mucus substrate, nutrient release, bacterial growth stimulation	+
Construction, secretion	Synthesis of refractory or inhibitory structural products (tube linings, halophenols, body structural products)	−
Irrigation	Soluble reactants supplied, metabolite build-up lowered, increased reoxidation redox oscillation	+
Particle transport	Transfer between major redox zones, increased reoxidation, redox oscillation, substrate priming	+

[1] after: King 1986; Andersen & Kristensen 1991; Kristensen et al. 1991; Alkemade et al. 1992; Graf 1992; Woodin et al. 1993; Aller 1994; Mayer et al. 1997; Aller & Aller 1998

[a] stimulation = + sign , inhibition = − sign

In this kinetic formulation (Eq. 1b), reactive substrate ($\hat{C}_r$) is an overall limiting reactant, resulting in a first-order dependence on reductant. A hyperbolic dependence (Michaelis-Menten or Monod kinetics) on the concentration of a specific oxidant, i, is presumed. Inhibition of secondary oxidants (e.g., $SO_4^=$) by the availability of higher order oxidants (e.g., O_2, NO_3^-, Mn(III, IV)) is accounted for in the function In_{ij}, ensuring successive utilization of oxidants in the order of progressively decreasing free-energy yield. In some situations, however, multiple oxidants may be used simultaneously. The effective reaction rate coefficient, $k(t)$, is time-dependent, reflecting both differential utilization of multiple substrate pools within $\hat{C}_r$, and/or the progressive evolution of $\hat{C}_r$ during reaction. The rate coefficient also incorporates the functional role of the bacterial community and the intrinsic stability of the reactant pool, including physical attributes such as exposed surface area.

In general, environmentally available oxidants are sufficiently abundant and the inhibition relations between oxidants are such that decomposition rates of reductant are essentially maximized at all stages and thus approximated closely by (Middelburg 1989; Middelburg et al. 1993; Van Cappellen & Gaillard 1996):

$$\frac{d\hat{C}_r}{dt} = -k(t)\hat{C}_r \tag{2a}$$

or by (Berner 1980; Westrich & Berner 1984; Boudreau & Ruddick 1991):

$$\frac{d\hat{C}_r}{dt} = -\sum_n k_n \hat{C}_{r,n} \tag{2b}$$

The formulation of Eq. (2a) emphasizes the time-dependent evolution of the overall reactive reductant pool, while Eq. (2b) emphasizes the spectrum of reactive subpools present at any given time. Equation (2a) can also be expressed in terms of the total measurable carbon pool, $\sum \hat{C}$, as opposed to an operationally observed reactive portion. Both model equations can reproduce the observed patterns of net decomposition in sediments. Neither model explicitly accounts for synthetic processes, such as biomass formation ($y_i \mathbf{B}$), during decomposition.

Over relatively short intervals of time or restricted depth intervals in sedimentary deposits, net decomposition relations further simplify to a simple pseudo first-order kinetic relationship:

$$\frac{d\hat{C}_r}{dt} = -k\hat{C}_r \tag{3}$$

Equation (3) represents the most commonly utilized quantitative conceptual model for sedimentary organic matter decomposition (Berner 1980; Westrich & Berner 1984; Boudreau 1997). In this deceptively simple idealized formulation, net remineralization depends in a relatively straightforward way on both a reactive pool size, $\hat{C}_r$, and a characteristic reactivity, k, of that pool. The reactive pool size is operationally defined as the difference between an apparently stable residue remaining after an extended period of decomposition, $\hat{C}_\infty$, and the total measurable quantity of organic carbon or specific compound, $\sum \hat{C}$. Such first-order relations seem reasonable a priori, in that, when environmental oxidants are in excess, remineralization rates should depend largely on the intrinsic stability and form of the particular organic compounds or general substrate types present (reactivity k), as well as on their quantity (pool $\hat{C}_r$). There are, however, no generally accepted a priori ways of accurately determining $\hat{C}_r$ relative to an observed value of $\sum \hat{C}$ in natural sediments.

In a typical practical application of Eq. (3) or Eqs. (1b)–(2) to sedimentary deposits, the reactive pool represents a fraction of an initially present total sedimentary carbon pool that decays away approximately exponentially over time and depth with an apparent attenuation $1/k$. A refractory residual portion, $\hat{C}_\infty$, of the total sedimentary pool, does not react and is presumably preserved (Fig. 1). The net decay of the reactive carbon pool drives the stoichiometric recycling of nutrients, uptake of oxidants, and diagenetic generation of reduced products of reaction such as S and Fe^{++}. The residual pool represents the quantity of organic material removed from recycling on ecologically significant time scales and, when integrated over sedimentary environments globally, is a critical control on longer term elemental cycling (Berner 1989).

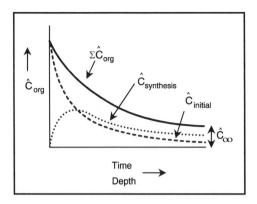

Fig. 1. Conceptual model for time-dependent decomposition of a complex substrate or single compound (after Paul & Van Veen 1978). The initially present substrate, $\hat{C}_{initial}$, decays during simultaneous synthesis of biomass and geopolymers, $\hat{C}_{synthesis}$, resulting in a net decomposition pattern given by $\sum \hat{C}$ (= $\hat{C}_{initial}$ + $\hat{C}_{synthesis}$). The residual organic material, $\hat{C}_{\infty}$, is composed of both residual initial substrate and compounds synthesized during decomposition. Most sedimentary decomposition models emphasize the net pattern of degradation given by the overall reactive pool, $\hat{C}_r = (\sum \hat{C} - \hat{C}_{\infty})$, and corresponding conceptual interpretations often neglect synthetic processes.

The simplicity of Eq. (3), or Eqs. (1b)–(2), belies an extraordinarily complicated process subject to a range of *conditional*, in addition to *intrinsic*, controls. These conditional factors, which affect both the reactivity and the reactive pool size, include biological community properties such as decomposer community biomass (**B**), population turnover, and bacterial species composition, and a range of physical-chemical properties such as inhibitory metabolite concentrations, oxidant availability, and fluctuating redox conditions. For example, synthetic reactions generate new biomass and suites of more or less reactive new compounds simultaneously with the net degradation of total substrate (Paul & van Veen 1978; Blackburn 1979; Jenkinson & Ladd 1981; Rice & Hanson 1989; Canuel & Martens 1993; Gong & Hollander 1996). Viable biomass, **B**, itself is one factor determining apparent reactivity, k, (e.g., $k \sim k^*\mathbf{B}$), while at the same time representing potentially reactive substrate, $\hat{C}_r$, following death. The decomposition reaction pathway (for example, the specific oxidant involved) affects the yield of bacterial biomass during degradation, suites of possible abiogenic reactions, and also the potential pool size of reactive reductants. Net decomposition therefore represents a dynamic balance between complex sets of degradation and synthetic reactions, each of which is potentially subject to control and reaction evolution by a range of conditional environmental factors such as oxidant availability and metabolite product buildup (⇒ transport regime). The composition of the total decomposing reactant pool, $\hat{C}_r$, is not fixed but changes continuously, and the fate of an initially present compound group can differ substantially at any given time from that of the total organic pool (Fig. 1). In natural open systems, regular progressive decomposition patterns of an initially present substrate can also be further complicated by periodic introduction of new reactive material from external sources (e.g., priming), perturbing microbial populations from a local steady state, and enhancing the potential for co-oxidation or metabolism of otherwise refractory compounds (Paul & Van Veen 1978; Graf 1992; Canfield 1994).

Despite the underlying complexity and potential for substantial conceptual confusion (e.g., changes in actual metabolized substrate, $\hat{C}_{initial}$, versus measures of net remineralization, d$\hat{C}_r$/dt), model Eq. (3), or moderate modifications thereof, has proved to be a robust practical basis for quantification of remineralization and comparative analysis of dominant processes in different depositional environments (Fig. 2). Within the context of this first-order decomposition model, the effects of fauna on sedimentary decomposition can be interpreted for comparative purposes in terms of possible conditional alterations of the overall reactivity, k, the available reactant pool, $\hat{C}_r$, and the refractory residual, $\hat{C}_{\infty}$. In terms of the corresponding biogeochemical consequences, effects on the overall reactivity alone change only the time-dependence of elemental recycling and not the extent.

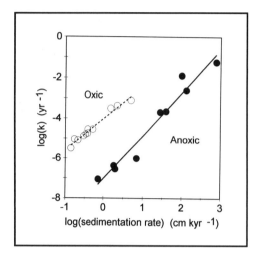

Fig. 2. Application of the single reactive pool decomposition model allows for general environmental patterns in basic decomposition processes to be discerned. For example, although there are many factors confounding simple interpretations, the average reactivity, k, of surficial sedimentary organic material appears to vary regularly as a function of both sediment accumulation rate and oxygenation of overlying water (after Middelburg et al. 1993). Macrofaunal bioturbation is significant only under oxygenated conditions and contributes to the differences observed.

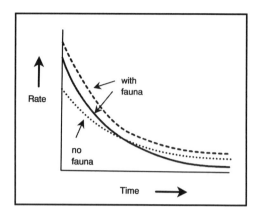

Fig. 3. A variety of studies indicate that bioturbation activity by benthic fauna results in an increase in the rate of sedimentary organic matter decomposition relative to deposits lacking bioturbation. The preponderance of experimental and environmental distribution data indicates that fauna alter both the overall net reactivity, k, and the available pool of decomposing material, $\hat{C}_r$, thus producing a pattern where integrated decomposition with time is greater in the presence than in the absence of fauna (area under rate curve [dashed line] is greater than the no fauna case [dotted line]). The case where initial reactivity is altered by faunal activities, but decomposing pool sizes are not, is indicated by the solid line (long-term net decomposition identical with no fauna case). Most experimental studies of bioturbation cannot differentiate between the two 'with fauna' cases shown.

Alteration of $\hat{C}_r$ relative to $\hat{C}_\infty$, however, changes both the time-dependence of remineralization and the absolute mass of recycled components (Fig. 3). It can be shown that modifications of the sedimentary transport regime by benthos alone, including changes in oxidant supply and metabolite exchange, result in alteration of *both* the effective overall reactivity and the effective reactive pool size of decomposing organic debris.

Oxidant Type and Reaction Path

Particle reworking and burrow irrigation result in the enhanced penetration of O_2, continuous re-exposure and re-oxidation of reduced material such as S and Fe^{++}, and the repetitive oscillation of redox conditions in surface sediments over multiple time and space scales (Rhoads 1974; Chanton et al. 1987; Kristensen 1988; Aller 1994). How do these transport-related effects on oxidant supply and distribution per se alter net remineralization relationships of sedimentary organic C as embodied in Eq. (3)? The initial net decomposition rates of fresh planktonic debris and many small organic molecules are apparently relatively independent of oxidant type and the corresponding heterotrophic respiratory pathway (Westrich & Berner 1984; Henrichs & Reeburg 1987; Lee 1992). Degradation of specific compounds, however, need not be independent of oxidant, and progressive decomposition

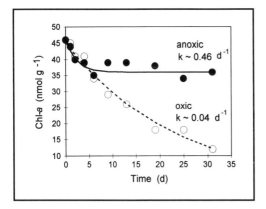

Fig. 4. Degradation pathway can dramatically alter decomposition patterns of specific compounds. For example, degradation of Chl a in diffusively open sediment differs under oxic and anoxic conditions. Although initial decomposition rates are comparable under oxic or anoxic conditions, reactive pool size differs substantially, corresponding to an apparently greater reactivity (k) under anoxic conditions but a much larger reactive pool, $\hat{C}_r$, and greater net decomposition under oxic conditions (after Sun et al. 1993).

through particular pathways can form new compounds or leave residua more or less resistant to oxidants weaker than O_2 (Stout et al. 1981; Hedges & Keil 1995). Oscillations between redox conditions also typify the bioturbated zone, and result in repetitive microbial successions and enhanced net remineralization (Schink 1988; Aller 1994).

One example of a common compound whose natural decomposition behavior is sensitive to oxidation pathway is the photopigment chlorophyll a (Sun et al. 1993). Chlorophyll a (Chl a) is not subject to resynthesis during heterotrophic decomposition, and in that sense its behavior is simple, but it is present in different forms or associations more or less available to degradation by particular pathways (Sun et al. 1993). Initial total rates of natural sedimentary Chl a decomposition in the absence of macrofauna are similar under oxic and anoxic conditions; however, reactive and residual pool sizes in each case differ substantially (Fig. 4).

Periodic brief exposures to oxic conditions interspersed with periods of anoxia result in overall decomposition near that of continuously oxic conditions. Thus, in terms of model Eq. (3), the reactivity of Chl a can be higher under anoxic than oxic conditions (that is, $k_{an} > k_{ox}$), but decomposition is clearly more sustained and complete under continuously oxic or fluctuating oxic conditions ($\hat{C}_{r,ox} >> \hat{C}_{r,an}$; $k_{ox}\hat{C}_{r,ox}(t > t^*) >> k_{an}\hat{C}_{r,an}(t > t^*)$; where t* represents a time scale characteristic of initial behavior). At least a part of these differences may be due to properties of the entire decomposer community, including protozoa and meiofauna, rather than strictly a result of biogenic or abiogenic reactions (Lee 1992).

Microcosm experiments with the deposit-feeding bivalve *Yoldia limatula* are used here to demonstrate potential influences of macrofaunal activities on redox conditions and Chl a decomposition. *Yoldia* is representative of the protobranch bivalves, an infaunal group that is typically highly mobile and that oxidize the upper few centimeters of surface sediment (visually apparent) (Rhoads 1968). Two sets of experimental microcosms were utilized. One set initially contained homogenized anoxic sediment with or without *Yoldia*. In a second, otherwise similar set, a layer of fresh algae was added to the sediment surface to simulate deposition of an algal bloom and a range in abundance of *Yoldia* was added. Overlying water was continuously oxygenated. Chl a shows depth-dependent decomposition rates below the surficial sediment-water interface (Fig. 5; Ingalls et al. 2000). Loss of Chl a, whether present as an enriched labile layer or not, is relatively enhanced in nonbioturbated surface sediment, decreasing with depth. Introduction of *Yoldia* greatly increases the overall decomposition rates of Chl a relative to controls, and also substantially accentuates the observed depth-dependence of decomposition in proportion to *Yoldia* abundance (Figs. 5 & 6A).

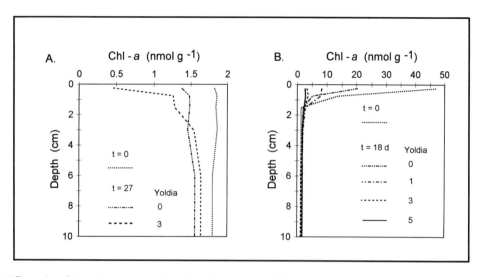

Fig. 5. Examples of time-dependent depth profiles of sedimentary Chl *a* in experimental microcosms with and without the protobranch bivalve *Yoldia limatula* (after Ingalls et al. 2000). A) Vertical profiles of Chl *a* in unamended homogenized surface sediment (~ 0–2 cm) from central Long Island Sound at time zero (t = 0) and after 27 d in the presence and absence of *Yoldia* (3 per ~ 50 cm^2 microcosm [~ 600 m^{-2}]). B) Vertical profiles of Chl *a* in sedimentary microcosms initially enriched with an algal surface layer (~ 0.5 cm) at t = 0 and subsequently exposed to different abundances of *Yoldia* (0, 1, 3, or 5 per ~ 50 cm^2). Sedimentary Chl *a* decomposes more rapidly and extensively in oxic surface sediments, and its degradation is greatly enhanced by the presence of *Yoldia*.

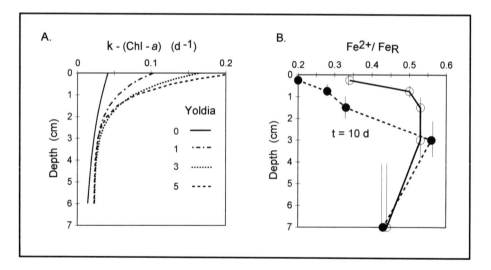

Fig. 6. A) The effects of *Yoldia* on Chl *a* degradation can be modeled in terms of an increase in net reactivity, *k*, of the reactive Chl *a* pool (assumed identical in all cases). *Yoldia* increases both the overall apparent reactivity, *k*, and the depth dependence of decomposition of Chl *a* (experiments as in Fig. 5B, after Ingalls et al. 2000). Reactivity (*k*) depth profiles were derived from a nonsteady state transport reaction model that assumed diffusive mixing of particles and first-order decomposition of Chl-a. B) *Yoldia* greatly enhances suboxic conditions in surface sediment as illustrated by the oxidation state distribution of reactive Fe, Fe$_R$, in the microcosm experiments (example t = 10 d; symbols as in A). Fe^{+2} released with Fe$_R$, where Fe$_R$ = total HCl leachable reactive Fe (15 min, T = 22°C), and corresponds approximately to ferrihydrite, lepidochrosite, mackinawite-greigite, and siderite pools.

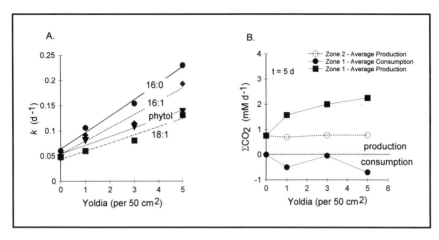

Fig. 7. The activities of *Yoldia* enhance organic matter decomposition generally in surface sediment and in proportion to intensity of bioturbation, as illustrated by A) the regular increase in the net decomposition reactivity of representative lipid compounds (fatty acids 16:1(w7), 16:0, 18:1(w9) and phytol), and B) the total release of ΣCO_2 into pore water as a function of *Yoldia* abundance (experiments as in Figs. 5 & 6; after Sun et al. 1999). Zone 1 indicates the upper 3 cm of deposit, zone 2 represents 3–10 cm. Consumption represents loss of alkalinity or chemoautotropic uptake of CO_2 during oxidation reactions. Biogenic reworking clearly increases the complexity of reaction balances and reaction coupling in addition to overall net degradation rates.

In the experiments illustrated, the measured dissolved O_2 penetration depth was only modestly increased by the activities of *Yoldia* (e.g., ~ 3.5 mm versus ~ 4 mm); however, measurement of solid phase Fe oxidation state distributions demonstrated that suboxic, nonsulfidic conditions are substantially extended by reworking and are correlative with Chl *a* loss (Fig. 6B). Particle reworking of surface sediment thus produces decomposition patterns of Chl *a* consistent with biogenic modification of redox conditions and the oxidant-dependent behavior such as observed in incubation experiments lacking macrofauna (Fig. 4). Direct digestion by *Yoldia* likely also plays a role (Bianchi et al. 1988) and is included in the decomposition parameters. If the residual pool ($\hat{C}_\infty$) at all depths in the sediment is assumed to be that eventually attained in the surfacemost layer, then the depth-dependent rate of Chl *a* loss in each treatment case can be parameterized in terms of a regular depth variation in the effective value of k. The average magnitude of decomposition and, therefore, k, correlated directly with the number of *Yoldia* present and the intensity of bioturbation (Fig. 6; Ingalls et al. 2000). Alternatively, reaction patterns could be modeled as depth-dependent change in the reactive pool size, or a combination of changes in both k and $\hat{C}_r$ of Chl *a*. These types of results are not restricted to Chl *a* but are also found for a range of representative lipid compounds in the same experiments (Fig 7A; Sun et al. 1999). Porewater ΣCO_2 production rates in the *Yoldia* microcosm experiment further demonstrate that enhancement effects, which may depend on more than redox processes alone, extend to the complete oxidation of the bulk organic matter in the reworked zone (Fig. 7B). The increased production of porewater ΣCO_2 and the enhanced loss of Chl *a* from surficial sites, as opposed to dominant subsurface consumption, argue against direct digestion by *Yoldia* as an exclusive governing factor. Therefore, as illustrated by Chl *a*, selected lipids, and ΣCO_2 production, the extension of oxic and suboxic conditions by *Yoldia* and other infauna (see, e.g., Andersen & Kristensen 1991) generally enhance both the net rate ($k\hat{C}_r$) and the extent ($\hat{C}_r /(\hat{C}_r + \hat{C}_\infty)$) of specific organic compound degradation over time scales of at least months (seasonal or experimental timescales).

It is possible, as noted previously, that given enough time, the extent of remineralization under anoxic conditions could become comparable with that generated by either constant oxic or periodic re-exposure to oxic or suboxic conditions. That is, $\hat{C}_\infty$ could be independent of redox reaction pathway as time becomes long (Henrichs & Reeburg 1987). The deduction that, for all practical purposes, $\hat{C}_\infty$ is oxidant dependent, however, is supported by a variety of direct and circumstantial evidence, including environmental distribution patterns of specific compound preservation, patterns of bulk carbon preservation, theoretical considerations, and by additional types of manipulative experiments (Emerson & Hedges 1988; Hedges & Keil 1995; Hulthe et al. 1998). High concentrations of Chl a, for example, are typically preserved only in organic-rich sediments underlying anoxic waters and lacking significant bioturbation (Furlong & Carpenter 1988; Sun et al. 1994). Oxidation of otherwise refractory sedimentary carbon within individual deep-sea turbidites occurs only in surficial portions of the turbidite layer during progressive re-exposure to O_2 or metal oxides (suboxic) (Hedges & Keil 1995). When once deeply buried nearshore sediments are exhumed and re-exposed to environmental oxidants in laboratory experiments mimicking biogenic mixing, elevated ΣCO_2 production is observed only in the presence of O_2 and not in its absence (Hulthe et al. 1998). Whereas fresh algal debris degrades at rates largely independent of oxidant, relatively refractory substrates are observed in experiments to decompose more rapidly under oxic compared to anoxic conditions (Kristensen et al. 1995). Finally, theoretical considerations indicate that oxic or suboxic conditions are required for certain specific oxidation reactions. These types of observations and correlations of $\hat{C}_\infty$ with sedimentary environments (Demaison & Moore 1980; Emerson 1985; Canfield 1993) imply that the relative enhancement of oxic-suboxic remineralization by re-exposure and re-oxidation during reworking and irrigation in the bioturbated zone likely translates into permanent decreases in carbon preservation ($\hat{C}_\infty$) compared to conditions of unidirectional burial or uniformly anoxic conditions.

Redox Reaction Balances and Burrow Distributions

The high mobility and deposit-feeding activity of fauna such as *Yoldia* produce a relatively uniform extension of the oxic-suboxic zone. In general, however, the geometry and time-dependence of oxic-suboxic conditions in the bioturbated zone are much more complex than a simple uniform expansion of the upper oxic layer of a two-layer system. The relative distribution of oxic-suboxic and sulfidic conditions in the bioturbated zone depends strongly on the life habit, mobility, burrow geometry, and spacing between individual inhabitants, the latter closely related to overall population abundance. Irrigation patterns within individual burrows can create periodic or irregular fluctuations of redox conditions and reactions within entire or individual sectors of burrows over a variety of time scales (Kristensen 1988; Forster 1991; Mayer et al. 1995). Regardless of the time-dependence of redox conditions within individual burrows, geometrical spacing alone of even the most simple burrow structures creates complex redox reaction patterns, variations in the coupling between oxidation and reduction reactions having soluble reactants or products, and provision of different forms of remineralized constituents (N, P, DOC) for re-incorporation into biomass. A given number of irrigated burrows per area of seafloor may be distributed in ways that result in entirely different balances between oxic and anoxic reactions (Fig. 8A). For example, particular distributions and spacings of burrows within sediment having otherwise identical overall rates of organic matter remineralization can produce widely different relative rates of coupled redox reactions such as

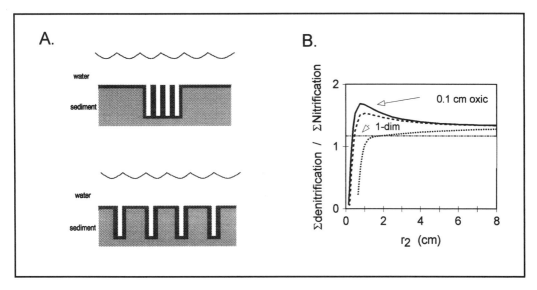

Fig. 8. A) Different geometrical distributions of oxygenated burrow walls embedded in otherwise anoxic sediment can produce widely varying balances in diffusively coupled redox reactions such as nitrification-denitrification. For example, oxygenated zones of closely spaced burrows can meld together, lowering the overall interaction between oxic and immediately surrounding anoxic regions compared with more widely spaced oxic microenvironments. B) Transport-reaction model calculations demonstrating the changing balances between denitrification and nitrification rates expected in sediment containing burrows of different sizes ($r_1 = 0.05$ cm, solid; 0.1 cm, dashed; 0.5 cm, dotted) and average spacing ($2r_2$) but otherwise having the same burrow wall oxic zone thicknesses (0.1 cm) and first-order denitrification rate constant (after Aller 1988, 2000). The reaction rate ratio from a strictly planar two-layer model (surface 0.1 cm oxidized overlying anoxic) having the same oxic zone nitrification rate and denitrification rate constant is also shown.

nitrification-denitrification because the proximity and coupled masses of oxidized and reduced regions depend strongly on burrow distribution patterns (Fig. 8B; Aller 1988, 1999).

The dependence of the oxic zone thickness on average population abundance or small-scale spacing of individual burrow sectors, and associated re-oxidation of soluble anaerobic metabolites, can be readily demonstrated experimentally by simply measuring the penetration depth of O_2 from overlying water into sediment layers of identical composition but variable thickness. In such a case, variable thicknesses of sediment simulate roughly the effects of variable spacing between burrows. An impermeable base below each layer ensures the zero flux condition (gradient) that must approximately exist halfway between any two burrow structures. If re-oxidation of diffusing metabolites did not occur, then the depth of the oxic zone would be independent of sediment thickness. Oxic zone penetration, however, is a function of total sediment scaling, directly reflecting the diffusive transport and oxidation of reduced metabolites in the oxic zone (Fig. 9). As the ratio of the anoxic zone to oxic zone thickness decreases, the relative supply of reduced metabolites decreases and the oxic zone expands.

The interaction of oxic and anoxic zone processes through diffusive transport coupling is clearly closely related to both burrow distribution, transport properties of burrow or tube boundaries, and the detailed geometry of burrow sectors (Forster & Graf 1992; Mayer et al. 1995; Aller 2000). The dependence of such reaction balances on scaling patterns alone can complicate elucidation of species-

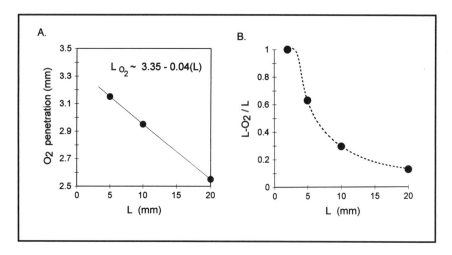

Fig. 9. A) Oxygen penetration depths in sediment slices of varying overall thickness but otherwise identical composition and overlying water oxygen concentration (sediment from Long Island Sound, 0–2 cm). Diffusion of reduced metabolites into the oxygenated zone lowers the penetration depth as the total sediment thickness increases (comparable to inter-burrow spacing) (data after Aller et al. unpublished). B) Fraction of sediment that is completely oxic (L-O_2 in example A) as a function of sediment thickness (L) or, by analogy, burrow sector or burrow spacing.

specific effects on remineralization, as opposed to the effects of size and abundance patterns per se. Exactly how the specifics of redox reaction coupling within particular diffusion geometries created by burrow structures, as opposed to periodic and fluctuating re-oxidation during feeding or irrigation, interact with the overall net decomposition of organic matter in surface sediment is not well known. For instance, chemoautotrophic synthesis of biomass must differ substantially as a function of burrow scaling and packing patterns in the bioturbated zone, thus altering the apparent overall net k and $\hat{C}_r$ of bulk sedimentary organic matter due to synthetic reactions.

Solute Exchange, Bacteria, and Decomposition Rates

Whereas enhanced exchange of solutes between sediments and overlying water clearly alters redox reaction dominance and reaction balances because of the transport and provision of specific oxidants in complex patterns, an additional intriguing observation is that regardless of redox condition, diffusively open sedimentary systems behave differently with respect to decomposition than do relatively closed systems. Subtle indications of such effects come from changes in remineralization rates in sediment regions adjacent to but not actually within the bioturbated zone. For example, measurements of net remineralization rates in the *Yoldia* microcosm experiment described earlier and in previous comparable experiments (Aller 1978; Aller & Yingst 1985; Sun et al. 1999), demonstrated that ammonification rates are modestly enhanced *below* as well as within the surficial reworked zone. Such remote or far-field effects implicate solute transport as a governing factor and also raise the question as to whether a portion of the elevated remineralization typically observed near the sediment-water interface of bottom deposits is due to general solute exchange rather than specific oxidant or labile substrate supply.

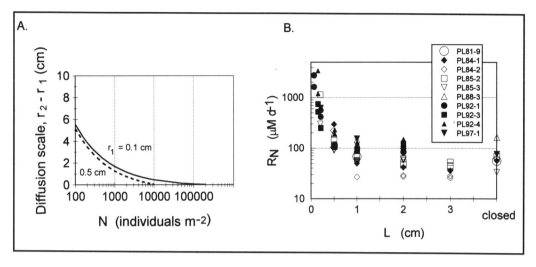

Fig. 10. A) The average diffusion scale for all solutes, not just oxygen, changes dramatically with the abundance or spacing of burrow structures in sediment. A typical diffusion length-scale is given roughly by the difference between average half-distance between burrows (r_2) and irrigated burrow radius (r_1). For example, the expected diffusion scale throughout the bioturbated zone at population abundance of ~ 500–1,000 sedentary individuals m^{-2} is ~ 1–2 cm. Faunal mobility further decreases characteristic diffusion scales. B) Net ammonification under anoxic conditions in sediment plugs increases with decreasing scale of diffusion in organic-rich nearshore sediments, as shown by a variety of experimental comparisons of net ammonification in diffusively open and closed anoxic treatments (after Aller & Aller 1998; PL97-1 from Hulth et al. 1999). (A part, but not all, of the increase at small scales in these examples is an artifact resulting from ignoring the effect of diffusive boundary layers in rate calculations).

The conditional effects of variable diffusive exchange on decomposition can be examined experimentally in a straightforward way by comparing decomposition processes in sediment layers of variable thickness but otherwise identical composition and dominant oxidant (Aller & Aller 1998). The thinner the sediment layer, the more efficient the diffusive exchange with overlying water. Scaling of such layers corresponds roughly to transport scaling distances generated between individual irrigated burrows or burrow sectors in the bioturbated zone (Fig. 10A). Results from experiments where layers of sediments of different thicknesses are exposed to completely anoxic overlying waters commonly demonstrate an increase in net remineralization with decreasing diffusive transport scale (Fig. 10B). Specific oxidants and redox coupling do not affect response in these completely anoxic conditions (in contrast to case of Fig. 8). Such results indicate that net remineralization rates in sedimentary deposits, that is, $k\hat{C}_r$, are affected by conditional transport factors other than specific oxidant supply.

It is important to note that O_2 is seldom available in sediments without an associated increase in diffusive solute exchange by proximity to a larger reservoir. In natural systems, therefore, O_2 supply is closely coupled to efficient solute exchange. However, in the Chl *a* decomposition experiments described earlier, both redox treatments had equivalent transport conditions, thus separating the two classes of effects (Fig. 4).

Enhanced net remineralization in diffusively open systems apparently results from several factors acting simultaneously, including decreased efficiency of back reactions into biomass or other synthetic pools, loss of inhibiting metabolites, and desorption of substrate. That is, both gross and net

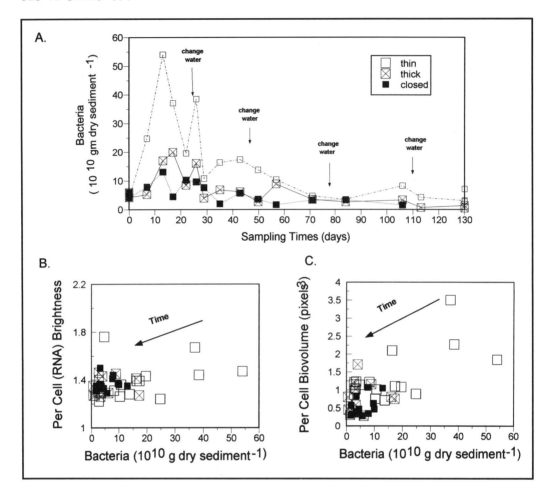

Fig. 11. Sedimentary bacterial abundances and activity are enhanced by the diffusive openness resulting from burrow irrigation. A) Bacterial abundance as a function of time in otherwise identical sediment incubated under diffusively closed or open anoxic conditions ('thin' = 0.17 mm thick sediment layer; 'thick' = 1.4 cm thick sediment layer; after Aller & Aller 1998). B) In the same experiments, diffusively open sediment apparently has enhanced per cell RNA content relative to conditions of lowered solute exchange during initial stages of incubation, implying greater per cell activity (RNA measured using methods of Lee et al. 1993; Kemp 1995). C) In addition to being more abundant and apparently more active, bacterial cells are larger under diffusively open compared with progressively closed conditions.

rates of decomposition are subject to conditional transport controls. Evidence that the absolute or gross rates of decomposition are enhanced come in part from changes in the abundance, size, and activity of sedimentary bacteria, all of which increase as diffusive transport increases (Fig. 11). These responses imply that diffusive exchange per se is an important factor controlling bacterial communities regardless of the dominant oxidant present, and that net remineralization increases, such as NH_4^+ release, are not only due to decreased efficiency of back reaction of remineralized solutes into biomass (e.g., N re-incorporation) but are also due to absolute increases in overall rate. Clearly,

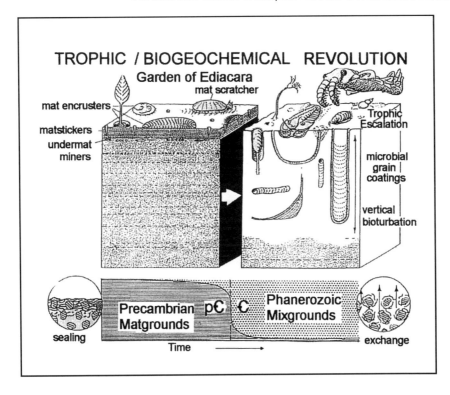

Fig. 12. As illustrated by the few examples of bioturbated zone processes given here, the transition during geologic time from a seafloor dominated in the Precambrian by bacterial and algal planar mat systems to present day fully four-dimensional bioturbated deposits must have produced a dramatic change in benthic material fluxes and decomposition processes, associated coupling between sedimentary redox reactions, and the storage pattern of reduced compounds (after Seilacher & Pflüger 1994; courtesy A. Seilacher).

the geometric structure of individual burrow sectors and the integrated geometry of the bioturbated zone as a whole play major roles in controlling diffusive solute exchange, bacterial community properties, sedimentary decomposition processes, and redox reaction coupling between oxic and anoxic zones. The details of these interactions and their mechanistic basis remain relatively unknown.

Conclusions

Alteration of the sedimentary transport regime by benthic organisms increases both the rate and extent of sedimentary organic matter decomposition. The flux of remineralized constituents between biogeochemical reservoirs is thereby enhanced and the existing balance between the composition and mass of reservoirs sustained. Increased net decomposition in the presence of macrofauna results from complex interactions between reaction pathways (e.g., oxidant availability, redox oscillation patterns), synthetic reactions (e.g., biomass synthesis-consumption), coupling between reaction zones (e.g., geometry of reaction distributions), and reactant-product concentrations (e.g., soluble metabolite exchange). While these phenomena have been demonstrated directly or deduced from consistent correlations, the mechanistic bases of the multiple effects are not very well understood and decomposition processes cannot be modeled deterministically to any significant degree. Much of the attention of the marine geochemical community has centered on net decomposition patterns and their relation to factors such as the intrinsic stability of organic material, organic matter associations with sedimentary debris, and oxidant type. Future models of sedimentary decomposition processes will presumably better incorporate the role of competing biosynthetic-remineralization reactions and the corresponding conditional governing factors that are determined in large part by the benthic community. A more complete understanding of such conditional controls, which clearly vary among present day environments, will provide substantial insight into development of global biogeochemical cycles as the biosphere evolved over geologic time (Fig. 12).

Acknowledgments

Numerous colleagues and previous coauthors have interacted with us in carrying out the research summarized here. We thank Liti Haramaty (Brookhaven National Laboratory) and Christina Heilbrun for aid in the laboratory. Neal Blair and Erik Kristensen provided constructive critical reviews. This effort was supported largely by National Science Foundation grant OCE9730933.

Literature Cited

Alkemade, R., A. Wielemaker, S. A. de Jong, & A. J. J. Sandee. 1992. Experimental evidence for the role of bioturbation by the marine nematode *Diplolaimella dievengatensis* in stimulating the mineralization of *Spartina anglica* detritus. *Marine Ecology Progress Series* 90:149–155.

Aller, R. C. 1978. Experimental studies of changes produced by deposit feeders on pore water, sediment, and overlying water chemistry. *American Journal of Science* 278:1,185–1,234.

Aller, R. C. 1988. Benthic fauna and biogeochemical processes in marine sediments: The role of burrow structures, p. 301–338. *In* Blackburn, T. H. & J. Sorensen (eds.), Nitrogen Cycling in Coastal Marine Environments. John Wiley & Sons Ltd., New York.

Aller, R. C. 1994. Bioturbation and remineralization of sedimentary organic matter: Effects of redox oscillation. *Chemical Geology* 114:331–345.

Aller, R. C. 2000. Solute transport and biogeochemical reactions in the bioturbated zone: Structure and biogeochemical function, p. 269–301. *In* Boudreau, B.B. & B.B. Jørgensen (eds.), The Benthic Boundary Layer. Transport Processes and Biogeochemistry. Oxford Press, Oxford.

Aller, R. C. & J. Y. Yingst. 1985. Effects of the marine deposit-feeders *Heteromastus filiformis* (Polychaeta), *Macoma balthica* (Bivalvia), and *Tellina texana* (Bivalvia) on averaged sedimentary solute transport, reaction rates, and microbial distributions. *Journal of Marine Research* 43:615–645.

Aller, R. C. & J. Y. Aller. 1998. The effect of biogenic irrigation intensity and solute exchange on diagenetic reaction rates in marine sediments. *Journal of Marine Research* 56:905–936.

Andersen, F. Ø. & E. Kristensen. 1991. Effects of burrowing macrofauna on organic matter decomposition in coastal marine sediments. *Symposia of the Zoological Society of London.* 63:69–88.

Bianchi, T. S., R. Dawson, & P. Sqwangwon. 1988. The effects of macrobenthic deposit-feeding on the degradation of chloropigments in sandy sediments. *Journal of Experimental Marine Biology and Ecology* 122:243–255.

Blackburn, T. H. 1979. Method for measuring rates of NH_4^+ turnover in anoxic marine sediments, using a 15N-NH_4^+ dilution technique. *Applied Environmental Microbiology* 37:760–765.

Berner, R. A. 1980. Early Diagenesis: A Theoretical Approach. Princeton University Press, Princeton, New Jersey.

Berner, R. A. 1989. Biogeochemical cycles of carbon and sulfur and their effect on atmospheric oxygen over Phanerozoic time. *Paleogeogr. Paleoclim. Paleoecol.* 75:97–122.

Boudreau, B. P. 1997. Diagenetic Models and Their Implementation. Modelling Transport and Reactions in Aquatic Sediments. Springer-Verlag, New York.

Boudreau, B. P. & B. R. Ruddick 1991. On a reactive continuum representation of organic matter diagenesis. *American Journal of Science* 291:507–538.

Canfield, D. E. 1993. Organic matter oxidation in marine sediments, p. 333–363. *In* Wollast, R., F. Mackenzie, & L. Chou (eds.). Interactions of C, N, P and S Biogeochemical Cycles. NATO-ASI Ser. v. I 4, Springer, Berlin.

Canfield, D. E. 1994. Factors influencing organic carbon preservation in marine sediments. *Chemical Geology* 114:315–329.

Canuel, E. A. & C. S. Martens. 1993. Seasonal variations in the sources and alteration of organic matter associated with recently-deposited sediments. *Organic Geochemistry* 20:563–577.

Chanton, J. P., C. S. Martens, & M. B. Goldhaber. 1987. Biogeochemical cycling in an organic-rich coastal marine basin, 7. Sulfur mass balance, oxygen uptake and sulfide retention. *Geochimica et Cosmochimica Acta* 51:1,187–1,199.

DeMaison, G. L. & G. T. Moore. 1980. Anoxic marine environments and oil source bed genesis. *AAPG Bulletin* 64:1,179–1,209.

Emerson, S. 1985. Organic carbon preservation in marine sediments, p. 78–87. *In* Sundquist, E. T. & W. S. Broecker (eds.), The Carbon Cycle and Atmospheric CO_2: Natural Variations Archean to Present. American Geophysical Union, Washington, D.C.

Emerson, S. & J. I. Hedges. 1988. Processes controlling the organic carbon content of open ocean sediments. *Paleoceanography* 3:621–634.

Forster, S. 1991. Die Bedeutung biogener Strukturen für den Sauerstofffluß ins sediment. Ph.D. dissertation, Universität Kiel, Kiel.

Forster, S. & G. Graf. 1992. Continuously measured changes in redox potential influenced by oxygen penetrating from burrows of *Callianassa subterranea*. *Hydrobiologia* 235/236:527–532.

Furlong, E. T. & R. Carpenter. 1988. Pigment preservation and remineralization in oxic coastal marine sediments. *Geochimica et Cosmochimica Acta* 52:78–99.

Gong, C. & D. J. Hollander. 1996. Differential contribution of bacteria to sedimentary organic matter in oxic and anoxic environments, Santa Monica Basin. *Organic Geochemistry* 26:5,435–5,463.

Graf, G. 1992. Benthic-pelagic coupling: A benthic view. *Oceanography and Marine Biology Annual Review* 30:149–190.

Hedges, J. I & R. G. Keil. 1995. Sedimentary organic matter preservation: An assessment and speculative synthesis. *Marine Chemistry* 49:81–115.

Henrichs, S. M. & W. S. Reeburg 1987. Anaerobic mineralization of marine sediment organic matter: Rates and the role of anaerobic processes in the oceanic carbon economy. *Geomicrobiology Journal* 5:191–237.

Hulth, S., R. C. Aller, & F. Gilbert. 1999. Coupled anoxic nitrification/manganese reduction in marine sediments. *Geochimica et Cosmochimica Acta* 63(1):49–66.

Hulthe, G., S. Hulth, & P. O. J. Hall. 1998. Effect of oxygen on degradation rate of refractory and labile organic matter in continental margin sediments. *Geochimica et Cosmochimica Acta* 62(8):1319–1328.

Ingalls, A. E., R. C. Aller, C. Lee, and M-Y. Sun 1999. The influence of macrofauna on chlorophyll-*a* degradation in coastal marine sediments. *Journal of Marine Research.* 58:631–651.

Jenkinson, D. S. & J. N. Ladd 1981. Microbial biomass in soil: Measurement and turnover, p. 415–471. *In* Paul, E. A. & J. N. Ladd (eds.), Soil Biochemistry, vol. 5. M. Dekker, New York.

Kemp, P. F. 1995. Can we estimate bacterial growth rates from ribosomal RNA content? *In* Joint, I. (ed.), Molecular Ecology of Aquatic Microbes. Springer-Verlag. *NATO ASI Series* 38:279–302.

King, G. M. 1986. Inhibition of microbial activity in marine sediments by a bromophenol from a hemichordate. *Nature (London)* 323:257–259.

Kristensen, E. 1988. Benthic fauna and biogeochemical processes in marine sediments: Microbial activities and fluxes, p. 275–299. *In* Blackburn, T. H. & J. Sørensen (eds.), Nitrogen Cycling in Coastal Marine Environments. John Wiley & Sons, Chichester, England.

Kristensen, E., S. I. Ahmed, & A. H. Devol. 1995. Aerobic and anaerobic decomposition of organic matter in marine sediment: Which is fastest? *Limnology and Oceanography* 40:1,430–1,437.

Kristensen, E., R. C. Aller, & J. Y. Aller. 1991. Oxic and anoxic decomposition of tubes from the burrowing sea anemone *Ceriantheopsis americanus*: Implications for bulk sediment carbon and nitrogen balance. *Journal of Marine Research* 49:589–617.

Lee, C. 1992. Controls on organic carbon preservation: The use of stratified water bodies to compare intrinsic rates of decomposition in oxic and anoxic systems. *Geochimica et Cosmochimica Acta* 56:3,323–3,338.

Lee, S. H., C. Malone, & P. F. Kemp. 1993. Use of multiple 16S rRNA-targeted fluorescent probes to increase signal strength and measure cellular RNA from natural planktonic bacteria. *Marine Ecology Progress Series* 101:193–201.

Mayer, L. M., L. L. Schick, R. F. L.Self, P. A. Jumars, R. H. Findlay, Z. Chen, & S. Sampson. 1997. Digestive environments of benthic macroinvertebrate guts: Enzymes, surfactants and dissolved organic matter. *Journal of Marine Research* 55:785–812.

Mayer, M. S., L. Schaffner, & W. M. Kemp 1995. Nitrification potentials of benthic macrofaunal tubes and burrow walls: Effects of sediment NH^+_4 and animal irrigation behavior. *Marine Ecology Progress Series* 121:157–169.

Middelburg, J. J. 1989. A simple rate model for organic matter decomposition in marine sediments. *Geochimica et Cosmochimica Acta* 53:1577–1581.

Middelburg, J. J., T. Vlug, & F. J. W. A. van der Nat. 1993. Organic matter mineralization in marine systems. *Global Planetary Change* 8:47–58.

Paul, E. A. & J. A. van Veen. 1978. The use of tracers to determine the dynamic nature of organic matter. *Transactions of the 11th International Congress of Soil Science* 3:61–89.

Rhoads, D. C. 1963. Rates of sediment reworking by *Yoldia limatula* in Buzzards Bay, Massachusetts, and Long Island Sound. *Journal of Sedimentary Petrology* 33:723–727.

Rhoads, D. C. 1974. Organism-sediment relations on the muddy sea floor. *Oceanography and Marine Biology Annual Review* 12:263–300.

Rice, D. L. & R. B. Hanson. 1989. A kinetic model for detritus nitrogen: Role of the associated bacteria in nitrogen accumulation. *Bulletin of Marine Science* 35:326–340.

Seilacher, A. & F. Pflüger 1994. From biomats to benthic agriculture: A biohistoric revolution, p. 97–105. *In* Krumbein, W. E., D. M. Paterson, & L. J. Sal (eds.), Biostabilization of Sediments. Oldenburg, Bibliotheks; und Informationssystem der Universität Oldenburg, Oldenburg.

Schink, B. 1988. Principles and limits of anaerobic degradation: Environmental and technological aspects, p. 771–846. *In* Zehnder, A. J. B. (ed.), Biology of Anaerobic Microorganisms. Wiley, New York.

Stout, J. D., K. M. Goh, & T. A. Rafter 1981. Chemistry and turnover of naturally occurring resistant organic compounds in soil, p. 1–73. *In* Paul, E. A. & J. N. Ladd (eds.), Soil Biochemistry, vol. 5. M. Dekker, New York.

Sun, M-Y., C. Lee, & R. C. Aller. 1993. Anoxic and oxic degradation of ^{14}C-labeled chloropigments and a ^{14}C-labeled diatom in Long Island Sound sediments. *Limnology and Oceanography* 38:1,438–1,451.

Sun, M-Y., R. C. Aller, & C. Lee. 1994. Spatial and temporal distributions of sedimentary chloropigments as indicators of benthic processes in Long Island Sound. *Journal of Marine Research* 52:149–176.

Sun, M-Y., R. C. Aller, C. Lee, & S. G. Wakeham. 1999. Enhanced degradation of algal lipids by benthic macrofaunal activity: Effect of *Yoldia limatula*. *Journal of Marine Research* 57:775–804.

Van Cappellen, P., J.-F. Gaillard, & C. Rabouille. 1993. Biogeochemical transformations in sediments: Kinetic models of early diagenesis, p. 401–445. *In* Wollast, R., F. Mackenzie, & L. Chou (eds.), Interactions of C, N, P and S Biogeochemical Cycles. NATO-ASI Ser. vol. 14, Springer, Berlin.

Van Cappellen, P. & J.-F. Gaillard. 1996. Biogeochemical dynamics in aquatic sediments. *In* Litchner, P. C., C. I. Steefel, & E. H. Oelkers (eds.), Reactive Transport in Porous Media. Mineralogical Society of America. *Reviews in Minerology* 34:335–376.

Westrich, J.T. & R.A. Berner 1984. The role of sedimentary organic matter in bacterial sulfate reduction: The G-model tested. *Limnology and Oceanography* 29:236–249.

Woodin, S. A., R. L. Marinelli, & D. E. Lincoln. 1993. Biogenic brominated aromatic compounds and recruitment of infauna. *Journal of Chemical Ecology* 19:517–530.

Two Roads to Sparagmos:
Extracellular Digestion of Sedimentary Food by
Bacterial Inoculation Versus Deposit-feeding

Lawrence M. Mayer*, Peter A. Jumars, Michael J. Bock, Yves-Alain Vetter, and Jill L. Schmidt

Abstract: Extracellular digestion is a frequent mode of dissolution of dispersed organic detritus in marine sediments. Two geometric modes of deploying this digestive attack are extracorporeal: prokaryotic unicells, and intracorporeal: metazoans with flow-through guts. Enclosed digestive geometry gives metazoans greater ability both to retain digestive agents and to obtain digestive products, allowing them to create digestive conditions more intense than are generally found with communities of sedimentary bacteria. Metazoans hence can dissolve food substrate more quickly. Extracellular hydrolytic enzymes are sorptive for bacteria and dissolved for animals, enhancing both speed and net digestive gain for their respective geometries. Experiments showed little evidence for mass loss of digestive agents to sediments transiting deposit feeders. Standing stocks of sedimentary bacteria—ubiquitously on the order of 10^9 bacteria (cc pore water)$^{-1}$—set a cap on digestive rates by bacterial inoculations of sediment, and provide metazoans a further kinetic advantage at food concentrations greater than the bacterial biomass. Metazoans therefore have digestive advantage on substrates amenable to concentrated, quick dissolution, suited to their role as high wattage consumers of sedimentary food. Deposit feeding shows strong analogy to laundry technology.

The Opportunity

Particulate organic matter settles as planktonic detritus from the water column. It may be converted into biomass and energy, if successfully rended, dissolved, and hydrolyzed to oligomer size (*sparagmos*; "nature grinding down and dissolving matter to energy," Paglia 1990) and then absorbed into cells. This nutritional detritus is mixed, by physical and biological means, into the top several centimeters of sediment, becoming increasingly diluted (if averaged over size scales of > 1 mm) by the indigestible sedimentary matrix.

In chemical terms, a dilute fuel, or reductant, is available for combination with an oxidant. Metabolizable organic matter in the presence of oxidants forms a galvanic cell, or battery waiting to discharge. Heterotrophic prokaryotes and eukaryotes are catalysts capable of physically and then chemically combining the oxidants and fuel—discharging the battery. A conceptual chemical reaction for this process is

$$\rightarrow \text{Biomass} \rightarrow$$
$$\text{Organic matter} + \text{oxidant} \rightarrow \text{Remineralization products} + \text{biomass} \qquad (1)$$

The battery discharges with a voltage determined by the relative redox states of the oxidant and reductant, and with an amperage governed by the amount of reactants. This reaction is partially autocatalytic, in that its progress yields additional catalyst that can accelerate the reaction, as evidenced by an induction of biomass after deposition of bloom material (Graf et al. 1982; Lehtonen & Andersin 1998). Modeling of organic matter decay must, therefore, explicitly take into account the concentration or activity of biotic catalyst, especially if this catalyst term varies among sites or times being considered.

The heterotrophic opportunity requires a sequence of processes including encounter between organism and food particle, digestion of the food, digestate transport to and absorption by the consuming cells, and subsequent cellular metabolism. Two organismal plans adapted for responding to this opportunity are dispersed microbes (unicells) and multicellular animals. Each plan has various advantages with respect to the several tasks involved in using detrital substrate. Here we emphasize the digestive step, comparing two strategies designed to digest the largely polymeric food substrates available in sediments.

Digestive Response Strategies

FRAME OF REFERENCE

Digestive optimization can result from enhancing dissolution of the polymeric substrate and restricting costs of doing so. Digestion rate will be some positive function of the activity of digestive agents and substrate concentration and lability (ease of dissolution). One schematic possibility that relates substrate mass (M), time (T) and volume (V), in Michaelis-Menten form, is

$$
Digestion\ Rate = \sum \left[\{DA\} \frac{k_Q S_Q}{\left(K_Q + S_Q\right)} \right] \tag{2}
$$

where,
k_Q is the lability (T^{-1}) of a polymer substrate of quality Q in the food mixture,
S_Q is substrate concentration ($M\ V^{-1}$),
K_Q is a half-saturation constant ($M\ V^{-1}$), and
DA is the amount of the digestive agent(s) deployed against the substrate.

Digestion Rate and DA can be normalized to either biomass or system volume (if normalized to system volume, then the units would be $M\ V^{-1}\ T^{-1}$ and $M\ V^{-1}$, respectively). The measurement and parameterization of DA are problematic and depend on interactions among various enzymes and other digestive agents such as surfactants. The lability parameter (k_Q) is a rate constant that results from an interaction between the digestive agents and substrate. It could be dependent on factors such as susceptibility of substrate bonds to hydrolysis by DA, or physical access such as the amount of surface area of food substrate available to DA. In the latter case, k_Q can be enhanced by grinding, which may be significant in animals with muscular digestive chambers or jaws.

This formulation can be thought of as a normal Michaelis-Menten expression in which the V_{max} term has been decomposed into a rate constant (k_Q) times the total concentration of digestive agents ($\{DA\}$), the latter being dependent on biomass or digestive agent secretion. This equation also makes explicit the substrate compartmentalization of the well-known multi-G models (Berner 1980). Thus

Eq. (2) accounts for organismal participation in organic matter decay, as did the multi-B model of Smith et al. (1992), albeit in different form. Digestive optimization, then, requires increasing the $\Sigma^Q k_Q S_Q/(K_Q + S_Q)$ term and decreasing the cost of the {DA} terms.

This paper examines dispersed unicellular prokaryotes and metazoan deposit feeders that utilize primarily extracellular digestion. However, the former digest extracorporeally and the latter digest intracorporeally. There are other digestive approaches (primarily intracellular, e.g., protozoans, mollusks), but this paper will focus on comparative aspects of the extracellular types. An important reason for this focus is that the terms in Eq. (2) have received more attention for extracellular digesters, and quantitative examination of these terms for intracellular digesters has not begun.

In terms of biomass distribution, bacterial and animal populations differ in the degree of patchiness of cells. They can have similar total biomass in sediments (Schwinghamer 1983); however, bacterial cells are highly dispersed and animal cells are aggregated into larger units that are in turn dispersed. To access food substrate, bacteria spread into (inoculate) the sedimentary food matrix, whereas animals gather and concentrate food matrix inside their multicellular bodies. Each may secrete cell-free, extracellular enzymes and perhaps other digestive agents into the food matrix, and absorb the resulting dissolved hydrolyzate. Additionally, each probably performs some hydrolysis with external but membrane-bound enzymes, to complete the more distant hydrolysis performed by the cell-free enzymes (Ugolev 1972). In energetic terms, there is advantage to the organism in restricting the loss of digestive agents (DA) to the environment and enhancing the capture of hydrolyzate that they create. There are strong contrasts between abilities of animals and bacteria to achieve each of these goals.

BACTERIA

Assuming cell-free enzymes, bacteria will rely on diffusion to transport both secreted digestive agents and the resultant hydrolyzate. Bacteria's small cell sizes and high surface area:volume ratios set them up for transport limitation by diffusion (Koch 1990). Extracellular bacterial enzymes diffuse spherically outward from the secreting cell but adsorb strongly to particles, such as their target substrate, which promotes local action that most benefits that cell (Vetter et al. 1998). Diffusion of hydrolyzate will be spherical from the point of hydrolysis, resulting in only partial return to the organism that secreted the enzyme (Fig. 1). The inefficient procurement of nutrition due to these sequential, spherical diffusion steps probably restricts net cost-effective foraging ambits to ca. 10 μm for individuals (Vetter et al. 1998). Dependence on diffusion, coupled to high partitioning to the solid phase (of enzymes, at least), implies slow transport for enzymes and hydrolyzates.

This inefficiency must lead to sharing of digestive effort and gain, by both individual cells and conspecific clonal populations (or genets, Andrews 1991), with other bacterial species in a mixed-species bacterial community. Mucus envelopes may enhance retention of both DA and digestive product for bacteria (Plante et al. 1990). Alternatively, monoclonal "swarming" of bacteria, for example, myxobacteria found in soil (but not marine) environments, can enhance net benefit to the genet from free extracellular enzymes (Reichenbach 1984). The lack of water in soils also allows bacteria to concentrate enzymes locally. The higher diffusivity of digestive agents and products in sediments, however, must reduce the marginal return on digestive agent secretion to sedimentary bacterial genets.

ANIMALS

Animals enhance transport of food particles, digestive enzymes, hydrolyzate, and consuming cells by advective means. Animals' gut enzymes are largely dissolved (Mayer et al. 1997), increasing

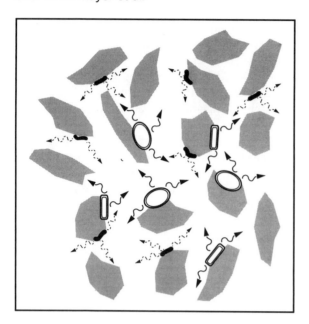

Fig. 1. Inoculation of food-containing sediment by two species of bacterial cells (rods and ovals), forming two genets. Each cell secretes cell-free enzymes (arrows with solid lines), which diffuse spherically, encountering food patches (black coatings on gray sediment grains) from which then diffuses hydrolyzate (arrows with dashed lines). Hydrolyzate is shared among the two genets. Figure adapted from Fig. 1 of Vetter et al. (1998).

effective foraging distance between food substrate and the secreting and absorbing cells well beyond the 10 μm viable for bacteria. Upon absorption at the gut wall, advection in the circulatory system takes over the transport of hydrolyzate to absorbing cells throughout the individual. This combination of partitioning into the fluid phase and fluid advection implies faster transport of enzymes and hydrolyzates than for bacteria, and routinely increases profitable foraging distances between digestion and consuming cells to centimeters in macrofauna—three orders of magnitude greater than bacteria.

The enclosed digestive geometry of animals provides an opportunity for greater retention of secreted digestive agents. Losses in a flow-through digestive system may be in dissolved or adsorbed form (i.e., adsorbed to transiting sediment). Loss via dissolved form is likely restricted through resorption, as evidenced by strongly lower enzyme and surfactant activities in hindguts relative to midguts (Mayer et al. 1997), or control of fluid transport. Gut fluid retention (Jumars 1993; Mayer et al. 1997) is currently being quantified in our lab. Adsorption to transiting sediment, which is poorly understood, was addressed here in two ways.

Experimental Methods and Results

In order to evaluate export of enzymes from macrofaunal digestive systems, we measured enzyme activities in ambient sediments and fecal material of *Arenicola marina* (Sheepscot estuary, Maine). Ten paired sediment samples from feeding pits and fecal mounds were collected and stored at -80°C until analysis. Activities of three enzymes—esterase, lipase, and protease—were measured by fluorescence assay (Mayer 1989; Mayer et al. 1997). Fecal ejecta had the same or lower enzyme activities as the ambient sediments (Fig. 2), indicating no significant export of gut enzyme activity by the animals. In the case of protease there was a significant decrease in activity in the feces.

In a second series of measurements, we examined changes in enzyme activity and surfactant concentration after gut fluid incubations with different size fractions of sediments. Subtidal estuarine sediments were collected from the Damariscotta estuary, on the mid-coast of Maine, and size-separated by wet sieving into < 10, 10–63, and 63–250 μm fractions. Digestive fluids of two deposit feeders, *Arenicola marina* (from Lubec, Maine) and *Parastichopus californicus* (from Puget Sound, Washington), were obtained as described in Mayer et al. (1997). Each species commonly inhabits sandy environments. The sediment size fractions were incubated with aliquots of the two gut

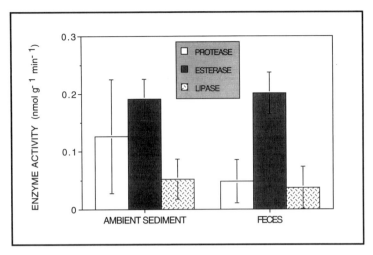

Fig. 2. Extracellular enzyme activities (nmol g^{-1} min^{-1}) in ambient sediments (feeding pits) and feces of *Arenicola marina*. Bars are standard errors.

fluids for 0.5 h at room temperature, and then assayed for enzyme activity (protease, esterase, lipase, using the fluorescent substrate methods of Mayer et al. 1997) and surfactancy (using contact angle titration assays of Mayer et al. 1997). Gut fluid with no added sediment served as a control.

Minor, though measurable, decreases in most enzyme activities were found in the gut fluid from *A. marina* (Fig. 3). These losses increased with the amount of sediment surface area added to the gut fluid. The role of surface area (or a correlate such as organic carbon concentration, Mayer 1994) was separated from other factors potentially varying among size fractions. This was achieved by adding different mass amounts of various size fractions of sediment to obtain the same surface area per milliliter of gut fluid. The greatest fractional loss of activity occurred with lipase. With *P. californicus* gut fluid, there was no loss of any enzyme activity except for lipase. The relatively high losses of lipase are consistent with the need for this enzyme to adsorb to lipid-water interfaces in order to function. Our enzyme assay results do not allow distinction between inactivation of enzymes and adsorption of enzymes onto the sediments.

Surfactancy was assessed with contact angles of gut fluid on Parafilm, and titrating with clean seawater to test for presence of micelles (Fig. 4). The *P. californicus* gut fluid showed evidence of micelles: in a two-phase titration plot the contact angle remained fairly constant with dilution below 100% gut fluid, until an inflection point at which the contact angle increased with further dilution. This inflection point is termed the critical micelle dilution factor (CMD), and represents the dilution at which all micelles initially present in the gut fluid disappear. Neither initial contact angles nor the shape of subsequent titration plots changed significantly as a result of the adsorption experiments. An exception was a small increase in the CMD in one of the triplicates using the < 10 μm fraction at the highest surface area loading (Fig. 4C). This increase would correspond with adsorption on the order of 15–20% of the surfactant. *A. marina* gut fluid showed little evidence for micelles in the raw gut fluid, which is atypical for individuals of this species. No changes in initial contact angles or titration plots were observed after sediment incubation, also consistent with little or no adsorption of surfactant. While the contact angle approach only crudely assesses surfactant concentration, these results do indicate a lack of strong adsorption of the compounds responsible for surfactancy.

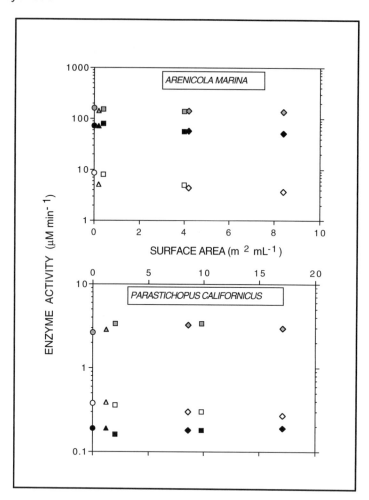

Fig. 3. Enzyme activities before and after incubation of midgut fluids of *Arenicola marina* (top) and *Parastichopus californicus* (bottom) with size fractions of a marine sediment. Gray symbols: esterase, white symbols: lipase, and black symbols: protease. Circles represent the control (no sediment), triangles are the 63–250 μm fraction, squares are the 10–63 μm fraction, and diamonds represent the < 10 μm fraction. Significant losses of activity after incubation were found for lipase in *Parastichopus* gut fluid (p = 0.0024) and protease in *Arenicola* gut fluid (p = 0.013).

With the finest size-fractions (i.e., 8–17 m^2 ml^{-1}), the amount of surface area incubated with these gut fluids likely exceeds the normal in vivo processing conditions of these animals by an order of magnitude. Furthermore, these experiments represent midgut conditions; hindgut concentrations of digestive agents are much lower (Mayer et al. 1997), which may allow desorption of material adsorbed in the midgut. Thus minimal loss of dissolved enzyme activity during gut passage probably occurs in vivo. These results are consistent with the fecal pellet enzyme activity assays and indicate that deposit feeders do not incur significant loss of enzyme proteins. As the same conclusion applies to the surfactancy, we conclude that little {DA} is lost by adsorption to transiting sediment in these species.

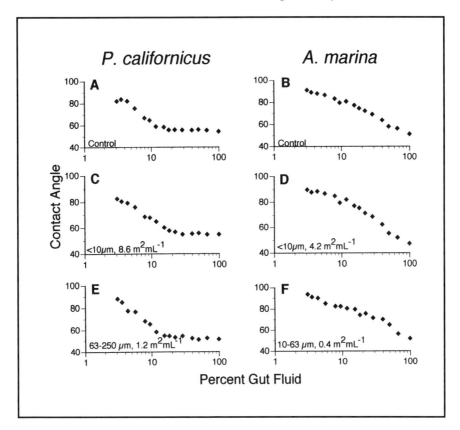

Fig. 4. Surfactant activity, as measured by contact-angle titrations, after incubation of midgut fluids of *A. marina* (C, E) and *P. californicus* (D, F) with size fractions of a marine sediment. Controls (no sediment) are given by A and B, respectively. The X-axis values represents dilution of original gut fluid; 100% is pure gut fluid and smaller values are after subsequent dilutions. C represents the most extreme case of apparent surfactant adsorption, with the change in slope (i.e., CMD) increasing from its control value of ca. 15% gut fluid to about 22% gut fluid. E represents the typical *A. marina* titration plot after sediment incubation, with no apparent change in CMD.

Animals have a similar advantage (relative to bacteria) in absorption efficiency of hydrolyzate: before releasing gut fluids to the environment the hydrolyzate is exposed to absorptive gut epithelia. By reducing dissolved hydrolyzate via gut wall absorption, re-equilibration between gut fluid and sediment will promote continued desorption of hydrolyzate that might have adsorbed onto sediment particles. Adsorptive loss of hydrolyzate onto transiting sediment should be especially reduced by lowered concentrations in the hindgut, again displacing sorption equilibria toward the dissolved phase, which facilitates further absorption.

A strong analogy exists between the digestive approaches of deposit-feeding animals and laundry technology (Fig. 5). Each seeks to dissolve a subset of the organic matter associated with particles. Each process has a wash cycle using enzymes and surfactants, that work together to dissolve, selectively, the desired organic compounds while leaving the rest of the particulate matrix

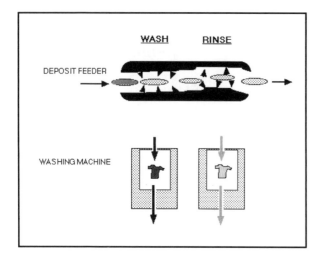

Fig. 5. "Wash-and-rinse" cycles in deposit feeders and washing machines. In deposit feeders the dirty material (pellets) is input at the mouth, with net secretion (arrowheads) of digestive agents in the foregut and midgut. Gut wall absorption occurs throughout the gut, but it is completed in the hindgut by rinse cycles induced either by bringing clean water from the posterior or by simple absorptive removal of dissolved materials that displaces sorption equilibria toward desorption from sediment surfaces. In a washing machine, digestive agents are physically introduced in the wash cycle and then physically advected away. Subsequent rinse cycles of advectively introduced clean water induce desorption of remaining adsorbed dirt and cleaning agents.

(inorganic and organic) intact. The selectivity of each process is less than perfect, resulting in faded clothes and bioavailable pollutants. Sorption equilibria will cause some desirable organic materials to remain stuck to transiting particles. Rinse cycles with clean water, which can be enhanced in invertebrate hindguts by anal inspiration, promote displacement of adsorbed material toward the aqueous phase. One adaptive result of rinse cycles in deposit feeders may be to remove chemical cues that would attract predators. Material in the aqueous phase is then absorbed in the animal system, which is in contrast to the laundry system where extracts are discarded. There are unexplored possibilities in this analogy. For example, commercial laundry formulations add chemicals to be adsorptively retained by clothes and stimulate responses from other biota after the laundry event; analogous deposits can be hypothesized for fecal ejecta from animals.

Retention of Digestive Agents Rewards Higher {DA}

Bacteria and animals apparently have developed fundamentally different extracellular enzymes, in terms of adsorbability, to improve the net rate of nutrition gained. Their different efficiencies at retention of digestive agents and products should result in different net returns of product per investment of DA (Fig. 6). Gross and net returns of hydrolyzate due to enzyme secretion should increase until hydrolysis sites are saturated (Fig. 6A, B). However, the net return should be greater for animals (Fig. 6A) than for bacteria (Fig.6B), because of greater retention of both secreted enzymes and hydrolyzate. Sedimentary bacteria likely operate at enzyme activities below saturation of substrate, as evidenced by enhanced hydrolyzate release upon addition of fungal proteases (Mayer et al. 1995).

Surfactants should exhibit different net yield curves. The purpose of surfactants found in animal guts is not well understood (Mayer et al. 1997), but their common occurrence at levels above the critical micelle concentration (CMC) suggests that the micelle form is critical. They probably act to partition food substrates, digestive agents, and hydrolyzates into the solution phase. Hence gain should increase markedly above a secretion rate necessary to maintain the CMC, leading to a sigmoid gain curve (Fig. 6C). For an open hydraulic system, such as dispersed bacteria, a high rate of

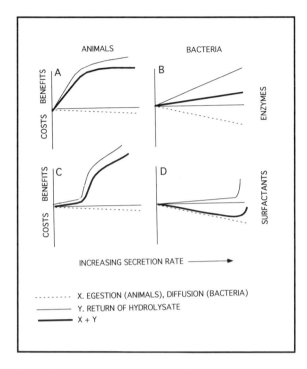

Fig. 6. Costs and benefits (mass time^{-1}) of secretion rate (mass time^{-1}) of digestive agents (enzymes or surfactants) to bacteria versus animals. Costs (dashed lines) result from spherical diffusion (bacteria) and fecal egestion (animals); animals suffer less cost because they recover digestive agents. Benefits include gross gain of hydrolyzate (thin lines) and net gain after loss of digestive agent (heavy lines). Special cases: (A) We assume saturation of hydrolytic sites at some high level of enzyme secretion; (B) Saturation of hydrolytic sites does not occur; (C) Gain increases only when critical micelle concentration (CMC, that concentration above which micelles form) is exceeded, due to increased transport of slightly soluble lipid digestates; (D) Achieving the CMC, and consequent gain in benefits, takes greater secretion rate for bacteria than for animals because of diffusive loss.

continuous surfactant secretion would be necessary to maintain these concentrations above the CMC. In a partially closed gut system, especially with fluid retention, the secretion rate necessary to maintain concentrations above the CMC should be lower. This advantage should be especially important for nutritional materials requiring micelles for solubilization or transport (e.g., lipids).

The guts of deposit-feeder taxa such as polychaetes, echiurans, and sipunculids have enzyme activities that are 10^2–10^3 times higher than the extracellular bacterial enzyme activities found in sediments, although deposit-feeding holothuroids exhibit activities similar to bacterial levels (Mayer et al. 1997; unpublished data). All deposit feeders examined so far, however, show evidence for surfactant micelles (Mayer et al. 1997), so that the combination of enzymes and surfactancy in holothuroids may cause a {DA} that is more intense than that exhibited by sedimentary bacteria.

Biomass and Time Constraints

The standing stock of sedimentary bacteria is usually on the order of 10^9 bacteria cc^{-1} (Schmidt et al. 1998). This concentration is so common that it suggests a cap on biomass, although any firm explanation for its ubiquity is not settled. Nevertheless, this standing stock sets an upper limit to bacterial food utilization per unit of time. Expressed as protein, the major cellular constituent, 10^9 bacteria cc^{-1} is equivalent to ca. 0.05 mg-bacterial-protein cc^{-1} (Mayer et al. 1995). If all members of the community are active and their maximum doubling time is on the order of 1 h with an anabolism:catabolism ratio of 0.25:0.75, then this community cannot eat more than 0.2 mg-protein cc^{-1} h^{-1}. This figure is almost certainly an overestimate because of the inactivity of most sedimentary bacteria.

Deposit feeders have gut residence times similar to bacterial doubling times. However, their advective transport (gathering food, distributing digestate) and storage (e.g., goblet cells) systems allow them to apply the capability of a much larger biomass, in terms of digestive action and consuming cells, to a parcel of sediment than can the bacteria resident in that parcel. Thus animals should have a kinetic advantage in digesting food at concentrations above that of the resident bacteria on short time scales. At least some benthic animals are apparently "tuned" to grow positively only above this threshold flux rate of 0.2 mg-protein cc^{-1} h^{-1} (Taghon & Greene 1990; Tsutsumi et al. 1990). There is, as yet, no physiological budgeting that can explain minimum food quality requirements for deposit feeders. There may be other explanations, but it is possible that this threshold flux simply reflects an inability of animals to compete with bacteria at concentrations below this value. The ubiquity of bacterial densities of ca. 10^9 bacteria cc^{-1} indicates their ability to maintain biomass at input fluxes of < 0.1 mg-protein cc^{-1} h^{-1}.

Relative Success, Particle Selection, and Bioturbation

Obviously both digestive strategies lead to sparagmos in sediments, and many factors besides digestion control the relative biomass and activity of animals versus bacteria (Andrews 1991). Metazoans in soft sediments usually appear to account for less than half of secondary production (Gerlach 1978; Probert 1986; Riddle et al. 1990; Piepenburg et al. 1995), except in areas with energetic flow regimes (Dye 1981). However, estimates generally rely on biomass measurements coupled to production:biomass (P:B) ratios, and have considerable uncertainty due to ignorance of the percent of active bacteria.

Food disappears more rapidly under animal than under bacterial attack. *P. californicus* gut fluid digests protein more quickly than native extracellular enzymes (Mayer et al. 1995). Half-lives of proteinaceous materials are on the order of a gut residence time under metazoan digestion (Carey & Mayer 1990) and on the order of weeks to months under microbial attack (e.g., Mayer & Rice 1992). The differences in these half-lives are reduced only slightly by taking into account assimilation of some fraction of the protein by the metazoans.

If animals are geared, digestively, for high rates of substrate dissolution, then deposit feeding should succeed if the animals can obtain nutrients from a parcel of sediment more quickly than can bacteria. Their relative advantage must proceed from more rapid kinetics of dissolution (due to higher {DA}) than is possible by the standing stock of bacteria generally observed. One obvious implication for feeding strategy is to select microzones where food concentration, of a quality amenable to hydrolysis on time scales of 10^1–10^3 min, is particularly high. Such selection (reviewed by Lopez & Levinton 1987; Jumars 1993) occurs at particle to patch scales, utilizing physical and chemical attributes of the particles as well as natural sorting mechanisms such as resuspension (Muschenheim 1987). Many animals are observed to focus on regions of high-quality particles, mining the steady flux to the sediment-water interface or deeper deposits in turbidites (Griggs et al. 1969). The food source for many conveyor-belt, subsurface deposit feeders is more problematic, although it may represent either fresh material hoed down burrows or accumulated as adsorbed remains from fundamentally inefficient (Vetter et al. 1998) microbial action.

Surficial deposit feeding provides a temporary opportunity for patch selection in the vertical, before removal of surficial food enrichments by either bacterial inoculation from below or downward mixing. Kanneworff and Christensen (1986) accordingly found rapid feeding response to spring bloom detritus by macrofauna, followed by burial and subsequent utilization by bacteria. Burial of detritus has the dual jeopardy to animals of dilution plus the acceleration of bacterial inoculation (mixing the food into the inoculum rather than vice versa). The relative advantage of surficial deposit-feeding should be greatest in areas with relatively little sediment dilution of the food-fall—either by resuspended sediment mixing with food in the nepheloid layer or physical mixing of the upper 0.1–1.0 cm in the sedimentary bed. Animal ingestion causes sediment mixing and hindguts promote bacterial growth (Plante et al. 1989) so that animal feeding may shift advantage toward the prokaryotes, although mixing can benefit some metazoans (e.g., subsurface deposit feeders). Initial dilution may increase bioturbation rates as bulk feeding increases in response to lowered concentration (Taghon & Greene 1990), while continued dilution may bring food levels to a threshold value below which prokaryotes dominate. Animal feeding has long been thought to stimulate microbial activity (cf. Yingst & Rhoads 1980), but here this stimulation is provided a different rationale. A coherent understanding of the relationship between food inputs and sediment mixing, important for paleooceanography, will benefit from a systematic examination of how dilution versus enrichment of food particles facilitate the partitioning of mass fluxes between bacteria and animals.

Biomass itself is concentrated labile organic matter, and its consumption (carnivory, herbivory, and bacterivory) is therefore best achieved by metazoans. Bacterivory as a principal source of nutrition, however, can only be achieved by smaller animals which can select particles at very small scales; at larger size scales the concentration of nutrition is too dilute (Cammen 1980). Carnivory, herbivory, and bacterivory are not known as bacterial feeding strategies in the sea, perhaps due in part to their low {DA} capabilities. Terrestrial myxobacteria, with higher {DA}, are bacterivorous.

Conclusions and Analogies

In the context of Eq. 1, animals are designed for relatively high {DA}, which relies on higher concentrations and mobility of digestive agents. They also appear to be adapted to higher $\Sigma^Q k_Q S_Q$, which can be due to enriched concentrations and/or greater lability of polymer substrate.

Applying a military analogy, dispersed bacterial cells resemble guerilla warfare while animals are structured "armies" of cells. Armies, as a form of warfare, are effective against concentrated targets, and developed with the rise of agriculture and towns. They do not do well in regions of dilute resources unless connected by supply lines to a concentrated resource base (hence scorched earth tactics in military defense).

Returning to the battery analogy, food resource utilization by organisms can be parameterized in terms of wattage (W; Peters 1983), which can be decomposed

$$W = EI = \text{potential} \times \text{current} \tag{3}$$

where the potential (E) is the difference between oxidant and reductant redox states and current (I) is the mass flow of metabolized food ($\Sigma^Q k_Q S_Q$). Oxygen-requiring animals are high-voltage consumers. Sedimentary animals seem to be tuned also for high current, requiring high food concentration and/or lability, so that animals are high-wattage consumers. In areas of low food concentration or lability (hence low I), or poor oxidants at depth in sediments (hence low E), the battery discharge is carried out primarily by low-wattage bacterial communities. The digestive approach of the metazoan "washing machine" is suited to its high wattage requirements. Thus metazoans emerge or arrive to discharge food enrichments, in space and in time, where oxygen is also available. It is intriguing to consider how metazoan evolution in the Proterozoic–Paleozoic might have been guided by current versus voltage.

Acknowledgments

We thank L. Schick for laboratory assistance, G. King and I. Kornfield for discussions, and the reviewers for their comments on the manuscript. This work was supported by the National Science Foundation–Biological Oceanography, and is contribution #330 from the Darling Marine Center.

Literature Cited

Andrews, J. H. 1991. Comparative Ecology of Microorganisms and Macroorganisms. Springer Verlag, New York.

Berner, R. A. 1980. Early Diagenesis: A Theoretical Approach. Princeton University Press, Princeton, New Jersey.

Cammen, L. M. 1980. The significance of microbial carbon in the nutrition of the deposit feeding polychaete *Nereis succinea*. *Marine Biology* 61:9–20.

Carey, D. A. & L .M. Mayer. 1990. Nutrient uptake by a deposit-feeding enteropneust: Nitrogenous sources. *Marine Ecology–Progress Series* 63:79–84.

Dye, A. H. 1981. A study of benthic oxygen consumption on exposed sandy beaches. *Estuarine, Coastal and Shelf Science* 13:671–680.

Gerlach, S. A. 1978. Food-chain relationships in subtidal silty sand marine sediments and the role of meiofauna in stimulating bacterial productivity. *Oecologia* 33:55–69.

Graf, G., W. Bengtsson, U. Diesner, R. Schulz, & H. Theede. 1982. Benthic response to sedimentation of a spring phytoplankton bloom: Process and budget. *Marine Biology* 67:201–208.

Griggs, G. B., A. G. Carey, Jr., & L. D. Kulm. 1969. Deep-sea sedimentation and sediment-fauna interaction in Cascadia Channel and on Cascadia Abyssal Plain. *Deep-Sea Research* 16:157–170.

Jumars, P. A. 1993. Gourmands of mud: Diet selection in marine deposit feeders, p. 124–156. *In* Hughes, R. N. (ed.), Mechanisms of Diet Choice. Blackwell, Oxford.

Jumars, P. A. & R. A. Wheatcroft. 1989. Responses of benthos to changing food quality and quantity, with a focus on deposit feeding and bioturbation, p. 235–253. *In* Berger, W. & G. Wefer (eds.), Productivity of the Ocean: Present and Past. Wiley, Chichester.

Kanneworff, E. & J. Christensen. 1986. Benthic community respiration in relation to sedimentation of phytoplankton in the Oresund. *Ophelia* 26:269–284.

Koch, A. L. 1990. Diffusion: The crucial process in many stages of the biology of bacteria. *Advances in Microbial Ecology* 11:37–70.

Lehtonen, K. K. & A. B. Andersin. 1998. Population dynamics, response to sedimentation and role in benthic metabolism of the amphipod *Monoporeia affinis* in an open-sea area of the northern Baltic Sea. *Marine Ecology–Progress Series* 168:71–85.

Lopez, G. R. & J. S. Levinton. 1987. Ecology of deposit-feeding animals in marine sediments. *The Quarterly Review of Biology* 62:235–260.

Mayer, L. M. 1989. Extracellular proteolytic activity in the sediments of an intertidal mudflat. *Limnology and Oceanography* 34:973–981.

Mayer, L. M. 1994. Surface area control of organic carbon accumulation in continental shelf sediments. *Geochimica et Cosmochimica Acta* 58:1,271–1,284.

Mayer, L. M. & D. L. Rice. 1992. Early diagenesis of protein: A seasonal study. *Limnology and Oceanography* 37:280–295.

Mayer, L. M., L. L. Schick, T. Sawyer, C. Plante, P. A. Jumars, & R. L. Self. 1995. Bioavailable amino acids in sediments: A biomimetic, kinetics-based approach. *Limnology and Oceanography* 40:511–520.

Mayer, L. M., L. L. Schick, R. F. L. Self, P. A. Jumars, R. H. Findlay, Z. Chen, & S. Sampson. 1997. Digestive environments of benthic macroinvertebrate guts: Enzymes, surfactants, and dissolved organic matter. *Journal of Marine Research* 55:785–812.

Muschenheim, D. K. 1987. The dynamics of near-bed seston flux and suspension-feeding benthos. *Journal of Marine Research* 45:473–496.

Paglia, C. 1990. Sexual Personae. Vintage Books, New York.

Peters, R. H. 1983. The Ecological Implications of Body Size. Cambridge Press, Cambridge.

Piepenburg, D., T. H. Blackburn, C. F. von Dorrien, J. Gutt, P. O. J. Hall, S. Hulth, M. A. Kendall, K. W. Opalinski, E. Rachor, & M. K. Schmid. 1995. Partitioning of benthic community respiration in the Arctic (northwestern Barents Sea). *Marine Ecology–Progress Series* 118:199–213.

Plante, C. J., P. A. Jumars, & J. A. Baross. 1989. Rapid bacterial growth in the hindgut of a marine deposit feeder. *Microbial Ecology* 18:29–44.

Plante, C. J., P. A. Jumars, & J. A. Baross. 1990. Digestive associations between marine detritivores and bacteria. *Annual Review of Ecology and Systematics* 21:93–127.

Probert, P. K. 1986. Energy transfer through the shelf benthos off the west coast of South Island, New Zealand. *New Zealand Journal of Marine and Freshwater Research* 20:407–417.

Reichenbach, H. 1984. Myxobacteria: A most peculiar group of social prokaryotes, p. 1–50. *In* Rosenburg, E. (ed.), Myxobacteria: Development and Cell Interactions. Springer-Verlag, New York.

Riddle, M. J., D. M. Alongi, P. K. Dayton, J. A. Hansen, & D. W. Klumpp. 1990. Detrital pathways in a coral reef lagoon. I. Macrofaunal biomass and estimates of production. *Marine Biology* 104:109–118.

Schmidt, J. L., J. W. Deming, P. A. Jumars, & R. G. Keil. 1998. Constancy of bacterial abundance in surficial marine sediments. *Limnology and Oceanography* 43:976–982.

Schwinghamer, P. 1983. Generating ecological hypotheses from biomass spectra using causal analysis: A benthic example. *Marine Ecology–Progress Series* 13:151–166.

Smith, C. R., I. D. Walsh, & R. A. Jahnke. 1992. Adding biology to one-dimensional models of sediment-carbon degradation: The multi-B approach, p. 395–400. *In* Rowe, G. R. & V. Pariente (eds.), Deep-Sea Food Chains and the Global Carbon Cycle. Kluwer, Dordrecht.

Taghon, G. L. & R. R. Greene. 1990. Effects of sediment-protein concentration on feeding and growth rates of *Abarenicola pacifica* Healy et Wells (Polychaeta: Arenicolidae). *Journal of Experimental Marine Biology and Ecology* 136:197–216.

Tsutsumi, J., S. Fukunaga, N. Fujita, & M. Sumida. 1990. Relationship between growth of *Capitella* sp. and organic enrichment of the sediment. *Marine Ecology–Progress Series* 63:157–162.

Ugolev, A. M. 1972. Membrane digestion and food assimilation processes in the animal kingdom. *Journal of Evolutionary Biochemistry and Physiology* 8:238–247.

Vetter, Y. A., J. W. Deming, P. A. Jumars, & B. B. Krieger-Brockett. 1998. A predictive model of bacterial foraging by means of freely released extracellular enzymes. *Microbial Ecology* 36:75–92.

Weissburg, M. J. & R. K. Zimmer-Faust. 1994. Odor plumes and how blue crabs use them in finding prey. *Journal of Experimental Marine Biology and Ecology* 197:349–375.

Yingst, J. Y. & D. C. Rhoads. 1980. The role of bioturbation in the enhancement of bacterial growth rates in marine sediments, p. 407–421. *In* Tenore, K. R. & B. C. Coull (eds.), Marine Benthic Dynamics, University of South Carolina Press, Columbia, SC.

In Vivo Characterization of the Gut Chemistry of Small Deposit-Feeding Polychaetes

Michael J. Ahrens* and Glenn R. Lopez

Abstract: *The digestive tracts of deposit-feeding organisms constitute diagenetic microenvironments whose biogeochemical conditions may differ considerably from those in the surrounding sediment. To quantify the chemical properties inside the intestines of small (< 2 cm) deposit-feeding polychaetes, we developed protocols for measuring pH, redox potential (Eh), and extracellular protease activity using noninvasive, microfluorimetric and colorimetric techniques. Juvenile* Nereis succinea *were used as test species. We report here preliminary observations on the gut pH, Eh, and protease activity of* N. succinea, *and on gut pH of three other deposit-feeding worms,* Streblospio benedicti, Polydora cornuta, *and* Pygospio *sp. A method to determine gut surfactancy, currently under development, is also discussed. Measurements were performed by feeding reactive tracers to animals and using video microscopy and image analysis to quantify changes in fluorescent or colorimetric properties inside their intestines. Gut pH was measured by the pH-indicator fluorescein. All worms analyzed exhibited moderately acidic to neutral mid guts (pH 5.5–7.5), with a gut pH consistently below the pH of seawater (pH 8). Gut redox of* N. succinea *was estimated by feeding the worms tetrazolium salts of decreasing reducibility and analyzing their feces for the presence of colored formazans. Large* N. succinea *were capable of reducing neotetrazolium (NT, Eh = -170 mV) and small individuals were only capable of consistently reducing iodonitrotetrazolium violet (INT, Eh = -90 mV). Gut protease activity of* N. succinea, *quantified by the hydrolysis of ingested BODIPY-FL labeled casein over time, was pronounced and appeared to encounter substrate limitation within a few minutes after ingestion, which seemed to elicit renewed ingestion activity. The in vivo methods described here may be expandable to a broader range of ecophysiological questions concerned with the digestive capabilities of animals and how these may be influenced by ontogenetic and environmental factors.*

Introduction

In this paper we focus on the characterization of the chemical conditions inside the guts of small deposit-feeding polychaetes. Three physicochemical parameters—pH, redox potential (Eh), and protease activity—known to affect digestive kinetics in larger deposit feeders were studied. We report preliminary observations on gut pH, Eh, and protease activity of *Nereis (Neanthes) succinea* (Leukart), and on gut pH of three other deposit-feeding worms, *Streblospio benedicti* (Webster), *Polydora cornuta* (Bosc), and *Pygospio* sp. (Claparède), species common to temperate salt marshes on the East Coast of the United States. The digestive parameters investigated were measured with newly developed, noninvasive, in vivo fluorimetric or colorimetric techniques, which are particularly suitable for examining small organisms.

The reason for our interest in guts is that differences between gut chemistry and ambient conditions are likely to influence a variety of biogeochemical processes in sediments, including mineral stability, molecular partitioning between particular and dissolved phases, and contaminant bioavailability. Prediction of these processes requires a better understanding of gut chemistry and its variability in deposit feeders. Body size is known to be an important source of variation for many physiological processes, including respiration, metabolism, and behavior, and the size-dependence of deposit-feeding strategy has been the subject of several recent studies (Forbes 1989; Forbes & Lopez 1990; Penry & Jumars 1990; Hentschel 1996, 1998a,b; Shimeta 1996). Notwithstanding, size-related differences in gut chemistry of deposit feeders are hardly recognized. So far, the published literature on the gut chemistry of deposit feeders has focused exclusively on cross-phyletic comparisons (of mostly pH and enzymes) among larger polychaete and holothuroid species, mainly because these animals were more practical to manipulate (Jeuniaux 1969; Longbottom 1970; Vonk & Western 1984; Plante & Jumars 1992; Mayer et al. 1997). Small deposit feeders greatly outnumber large deposit feeders in nature, and may be at least as important to sediment diagenesis as the large deposit feeders.

Ultimately, we are interested in how digestive kinetics scale with body size, that is, do small or juvenile deposit feeders have the same digestive strategy as large deposit feeders, or is there a shift in digestive physiology as organisms grow, perhaps from short, fast, and aggressive guts to longer, slower, and more moderate ones? Although this paper will not give a full answer to this intriguing question, it makes a first step toward this goal by presenting several newly developed methods for measuring pH, redox potential, and protease activity in the guts of deposit feeders too small to sample with the existing conventional methods.

PH

The majority of deposit-feeding animals analyzed so far possess neutral to weakly acidic guts (Table 1), with a pH that is 1–2 orders of magnitude lower than seawater. An acidic gut pH is likely to affect a large number of geochemical reactions. For instance, low pH increases the solubility of many minerals, and may facilitate particle dissolution of carbonate sediments (Emery et al. 1954; but see Hammond 1981 for contrary observations). Low pH could also dissolve mucus and other organic matrices and improve accessibility to enzymatic attack. Most importantly, the majority of the digestive enzymes of marine and other invertebrates studied so far show pH optima in the neutral to slightly acidic range very similar to the pH of their gut fluids (Vonk 1964; Longbottom 1970). Lastly, acidic gut pH is likely to increase the solubility of some particle-reactive metals, as suggested by Bryan and Hummerstone (1971) and Rhoads (1974). For example, acidic pH increased the solubility of Cu and Pb (Campbell & Stokes 1985), Am and Zn (Fisher & Teyssié), and Ag and Zn (Wang et al. 1999), thereby potentially enhancing the bioavailability of these elements. It has been hypothesized that, owing to their inherent "openness," deposit feeders are rather ineffective at regulating their gut pH, and that gut conditions are largely a function of what they ingest and of the microorganisms residing in the intestine (Plante et al. 1989; Plante & Jumars 1992). Previous measurements of gut pH were performed exclusively on large (> 2 cm) organisms, to permit the use of pH electrodes or the extraction of gut contents for external determination of gut pH. It is interesting to note in this regard, that most of the lower gut pH values (i.e., < pH 6) listed in Table 1 were measured by older, colorimetric procedures, whereas electrode measurements usually rendered neutral pH values. In any case, invasive methods, like inserting electrodes into an organism or dissecting its intestine, are prone to artifacts caused by the entrainment of fluids from outside the alimentary canal or due to disturbance of the organism. Furthermore, any kind of extraction procedure will blur small-scale spatial gradients that might exist within different regions of the gut.

The pH-sensitive microfluorimetric method described in this paper offers several advantages over electrode measurements or over alternative procedures that rely on the extraction of gut contents. The method is well suited for small organisms, it is nondestructive, it avoids the measurement artifacts characteristic of electrodes (e.g., liquid junction potentials and contamination; see Kolthoff 1941 and Perrin & Dempsey 1974) or resulting from dissection, and it gives detailed spatial information. The use of pH-sensitive microfluorimetry in marine research was first applied by Pond et al. (1995) to measure gut pH of planktonic copepods, using the fluophore BCECF. We present here for the first time gut pH data for small deposit-feeding polychaetes measured by a microfluorimetric technique.

REDOX POTENTIAL (EH)

That deposit-feeding organisms actively regulate their gut Eh seems improbable. It is much more likely that redox conditions are controlled to a large extent by the gut microflora. For this reason, guts may be expected to be less reducing in small animals than in larger ones, since enteric bacteria have less time to deplete dissolved oxygen or other electron acceptors. Similar to pH, Eh differences between gut and surrounding may have implications for mineral stability and desorption processes, since redox conditions affect the solubility of common matrix material (e.g., iron oxides and sulfides). Knowledge of gut redox conditions allows us to assess the likelihood of redox-sensitive diagenetic reactions that may alter the sediment matrix and thus lead to greater bioavailability of previously matrix-associated substances (e.g., trace metals, hydrophobic organic contaminants, and organic matter). What is the magnitude of redox changes that a particle undergoes while transiting through the gut of a deposit-feeding organism? Plante and Jumars (1992), using an Eh microelectrode, found Eh decreased by 100–200 mV along the gut of the holothuroid *Molpadia intermedia* and the polychaetes *Abarenicola pacifica* and *Eupolymnia heterobranchia*. They ascribed this reduction to bacterial metabolism. Griscom (pers. comm. 1998) found that gut contents of juvenile *Macoma balthica* possessed a smaller fraction of reduced Fe^{2+} than did those of adult *M. balthica*.

ENZYME ACTIVITY

Digestive enzymes in benthic animals have been studied qualitatively by numerous investigators (Kermack 1955; Jeuniaux 1969 [review]; Longbottom 1970; Kristensen 1972; Gelder 1984; Michel et al. 1984; Vonk & Western 1984 [review]). However, these early investigators commonly did not determine hydrolysis rates but only the presence-absence of selected enzymes (exceptions are Longbottom 1970; Michel 1977; Michel et al.1984) using homogenates of gut tissue, rather than the gut fluid itself (Feral 1989), and disregarding size differences. Only recently have the hydrolytic activities of proteases and lipases in gut fluid extracts of several larger polychaetes and holothuroids been quantified by Mayer et al. (1997). Gut fluids were pooled from several individuals; and juveniles were not investigated. Protease activity was found to be high in detritivorous polychaetes, whereas lipase activity was relatively low, confirming earlier studies by Kay (1974) and Michel et al. (1984). Compared to other digestive enzymes (e.g., lipases and carbohydrases), proteases are the most likely candidate to display differences in activity between juvenile and adults, since food protein content commonly constitutes the predominant nutritive limitation to deposit feeders (Tenore 1977). Our efforts to measure digestive enzyme activity in small deposit feeders therefore focused on proteases.

Table 1. Gut pH of deposit-feeding invertebrates, arranged by increasing pH (H = holothuroid, P = polychaete E= echinoid; full or empty indicates gut fullness; pH given as range or geometric mean ± 1 sd). Method of pH measurement indicated by C = colorometric, U = potenometric (electrode), F = Fluorometric, NA = not available.

Organism	Gut fullness	Gut pH	Method	Reference
Holothuria floridana [H]	empty	4.75–7.0	NA	Mayor (1924)
Stichopus moebii [H] (now *Isostichopus badionotus*)	full	4.8–5.5	C	Crozier (1918)
Holuthuria bivittata [H]	empty	5.0–6.0	C	Yamanouti (1939)
Stichopus moebii [H] (now *Isostichopus badionotus*)	empty	5.0–6.5	C	Crozier (1918)
Holothuria stellati [H]		5.1–5.6	C	Oomen (1926)
Holothuria tubulosa [H]		5.1–5.6	C	Oomen (1926)
Stichopus chloronotus [H]	empty	5.3–5.9	C	Yamanouti (1939)
Holothuria atra [H]	starved	5.3–7.4	NA	Trefz (1956) in Bakus (1973)
Arenicola marina [P]		5.4–7.0	C	Kermack (1955)
Holothuria bivittata [H]	full	5.5–6.0	C	Yamanouti (1939)
Stichopus japonicus [H]	empty	5.6–5.9	NA	Tanaka (1958)
Holothuria atra [H]	full	5.6–7.4	NA	Trefz (1956) in Bakus (1973)
Stichopus variegatus [H]	full	5.7–6.6	C	Yamanouti (1939)
Holothuria japonicus [H]	full	6.0–6.2	NA	Tanaka (1958)
Holuthuria vitiens [H]	full	5.9–6.5	C	Yamanouti (1939)
Holothuria scabra [H]	full	6.0	C	Yamanouti (1939)
Stichopus variegatus [H]	empty	6.0–6.2	C	Yamanouti (1939)
Holothuria atra [H]	full	6.0–6.6	C	Yamanouti (1939)
Stichopus chloronotus [H]	full	6.0–6.6	C	Yamanouti (1939)
Amphitrite johnstoni [P]	full	6.0–7.2	C	Dales (1955)
Arenicola marina [P]		6.2	U	Longbottom (1970)
Pygospio sp. [P]	full	6.2 ± 0.3	F	Ahrens & Lopez (this work)
Polydora cornuta [P]	full	6.3 ± 0.5	F	Ahrens & Lopez (this work)
Abarenicola pacifica [P]		6.3–8.1	U	Plante & Jumars (1992)
Travisia foetida [P]		6.3–8.2	U	Plante & Jumars (1992)

Table 1. Continued.

Organism	Gut fullness	Gut pH	Method	Reference
Thelepus crispus [P]		6.3–8.9	U	Plante & Jumars (1992)
Actinopyga agassizi [H]	starved	6.4 ± 0.3	U	Hammond (1981)
Holothuria flavo-maculata [H]	full	6.5	C	Yamanouti (1939)
Holothuria thomasi [H]	starved	6.5 ± 0.2	U	Hammond (1981)
Holothuria edulis [H]	full	6.5–6.7	C	Yamanouti (1939)
Holothuria spp. [H]		6.5–7.0	NA	Barth et al. (1968)
Pectinaria gouldii [P]	full	6.7	U	Ahrens et al. (In press)
Holothuria arenicola [H]	starved	6.7 ± 0.1	U	Hammond (1981)
Holothuria mexicana [H]	starved	6.7 ± 0.2	U	Hammond (1981)
Isostichopus badionotus [H]	starved	6.7 ± 0.2	U	Hammond (1981)
Nereis succinea [P]-juvenile	full	6.7 ± 0.4	F	Ahrens & Lopez (this work)
Nereis succinea [P]-adult	full	6.7 ± 0.4	U	Ahrens (unpubl. obs.)
Eupolymnia heterobranchia [P]		6.8–7.6	U	Plante & Jumars (1992)
Streblospio benedicti [P]	full	6.9 ± 0.3	F	Ahrens & Lopez (this work)
Holothuria floridana [H]	full	7.0	NA	Mayor (1924)
Amphitrite johnstoni [P]	empty	7.2–7.4	C	Dales (1955)
Plagiobrissus grandis [E]	starved	7.2 ± 0.1	U	Hammond (1981)
Molpadia intermedia [H]		7.2–7.8	U	Plante & Jumars (1992)
Thyone briareus [H] (now Sclerodactyla briareus)		7.2–8.2	C	van der Heyde (1922)
Brisaster latrifrons [E]		7.3–7.5	U	Plante & Jumars (1992)
Caudina chilensis [H] (now Paracaudina chilensis)		7.3–7.7	C	Sawano (1927)
Holothuria mexicana [H]	full	7.4	U	Hammond (1981)
Isostichopus badionotus [H]	full	7.4	U	Hammond (1981)
Meoma ventricosa [E]		7.5 ± 0.1	U	Hammond (1981)
Pectinaria koreni [P]		"neutral to weakly alkaline"	NA	Brasil (1904)
Arenicola marina [P]		"neutral to alkaline"	NA	Brasil (1903)

Materials and Methods

The surface deposit-feeding polychaetes *Nereis succinea, Polydora cornuta,* and *Streblospio benedicti* were collected from Flax Pond, Long Island, New York (USA). The spionid polychaete *Pygospio* sp., also a surface deposit feeder, was collected from Shinnecock Bay, Long Island, New York. Worms were isolated in 12-well culture plates (ϕ 2.2 cm, Falcon) and starved for a minimum of 1 d prior to experiments. Body length of worms used for pH and protease measurements ranged between approximately 3 and 20 mm. Worms larger than 25 mm commonly tended to be too opaque for fluorimetric measurements. For gut redox measurements, juvenile (< 2 cm) and large (ca. 4–5 cm) *N. succinea* were used.

PH

Measuring pH using fluorescein was suggested to us by D. Walt, Tufts University. The principle of pH measurement using fluorescein relies on fluorescein's emission intensity being strongly pH-sensitive when excited at 495 nm, and pH-independent at 450 nm excitation wavelengths (Haugland 1996). pH measurements were performed by obtaining sister images of the fluorescein-containing specimen at 495 nm and 450 nm excitation wavelengths and calculating the 495 nm:450 nm intensity ratio. This ratio is highly pH-dependent (high at alkaline pH and low at acidic pH), and can be calibrated accurately to standards of known pH. The advantage of using the ratio of two wavelengths is that several unwanted factors influencing fluorescence intensity cancel out because they affect both wavelengths similarly. Some of these factors are inhomogeneous dye concentration, photobleaching, quenching, uneven illumination intensity, and different tissue thickness (Bright et al. 1987). Thus, by using the ratio of 495 nm to 450 nm, one attains a concentration-independent and considerably more stable pH reading than when using the 495 nm wavelength alone.

Worms, unfed for 1–2 d, were fed ZYMOSAN A bioparticles fluorescein conjugate (Molecular Probes Inc.), a commercially available product of freeze-dried yeast cells (*Saccharomyces cerevisiae*), covalently labeled with fluorescein dye. Worms were allowed to ingest the yeast cells at their leisure, either pure or mixed with mud or small glass beads (ϕ 20–50 μm). The ratiometric procedure readily normalized for differences in brightness among worms ingesting different amounts of fluorescein-yeast conjugate. Only one to a few mouthfuls of fluorescein-yeast conjugate (approximately 0.1 mg) needed to be ingested to obtain a usable fluorescent signal from an animal's gut. After ingesting the dye, worms were immobilized in gelatin (0.1 g KNOX gelatin in 1 ml seawater) to reduce movement and gut peristalsis for the ensuing image capture. Additionally, immediately prior to image capture, worms were chilled in a freezer for approximately 2 min, to further reduce movement. Separate controls verified that this short temperature shock had no detectable effect on fluorescein's emission ratio, and thus on pH. Gut fluorescence of specimens was usually measured within 1–2 h after feeding. Since worms were not fed during the imaging procedure, they tended to retain yeast cells much longer (several hours) than the typical gut residence time for sediments (30–60 min) for worms of this size.

Images of specimens were acquired with a Zeiss epifluorescence microscope at 3–10X magnification, using either a Cohu 4915 or a Pulnix 745E CCD video camera connected to a computer (Power Macintosh 7300/200 MHz) via a Scion LG 3 frame grabber board. A Zeiss 50 W HBO mercury arc lamp was used as a light source. Images of fluorescing guts were captured in sequence at 495 nm and 450 nm excitation wavelengths and processed by the software application NIH

Image (developed at the US National Institutes of Health and available on the Internet at http://rsb.inf.nih.gov/nih-image/). To increase signal-to-noise, frame-averaging (commonly 6–8 frames), either on-chip (Cohu) or on-board (Pulnix) was performed. Excitation filters were custom manufactured (Chroma Technologies) to the following specifications: 450 ± 10 nm and 495 ± 5 nm. A wider bandwidth was chosen for the 450 nm filter to compensate for the overall weaker fluorescence intensity of fluorescein at the 450 nm wavelength. Emission intensity was measured at 530 ± 15 nm (filter by OMEGA Optical).

Average image background intensity was subtracted from each raw image in a pixel-by-pixel arithmetic operation. For *N. succinea*, minor worm autofluorescence, originating from the cuticle and not attributable to fluorescein, was observed in a few individuals. This effect, labeled "wormground" (in analogy to "background"), was eliminated by subtracting the average gray value of a fluorescein-absent portion of the worm (e.g., parapodia region) from each image. In any case, wormground intensity was commonly only a few gray values bright, and less than 25% of background brightness. After background and wormground correction, the 495 nm image was divided by the 450 nm image to render a 495 nm:450 nm ratio image.

The 495 nm:450 nm ratio image was converted to pseudocolor and calibrated against a series of buffered seawater standards of known pH, ranging from pH 4 to pH 8. Seawater buffers (100 mM) were prepared using organic buffers MES ($pK_a = 6.1$), MOPS ($pK_a = 7.2$) and HEPES ($pK_a = 7.5$), dissolved in glass-fiber-filtered Flax Pond seawater (salinity approximately 28‰). Dye stock was prepared by suspending 1 mg ZYMOSAN A bioparticles in 250 μl distilled water, and keeping the suspension in the dark and at 1°C to avoid chemical and bacterial degradation. For calibration, a 1:1 suspension of dye stock and buffered seawater standard were drawn into borosilicate glass capillaries, to serve as gut analogs. Borosilicate glass was chosen because of its minimal effect on pH, compared with other glass types. Images of capillaries were acquired, corrected, and ratioed in the same manner as the gut images. The average gray values of the ratio of the 495 nm to 450 nm capillary images were regressed onto pH using a third-degree polynomial fit. This calibration curve closely approximated the sigmoid curve typical of the fluorescein ratio (Fig. 1). Fluorescein's excitation ratio changed most prominently from about pH 4.5 to pH 7.5. Over this pH range, the 495 nm:450 nm ratio increased about 6–7 fold in a typical calibration (Fig. 1). Because the ratio approximated a constant at high acidic and high alkaline pH respectively, interpolation by a third-degree polynomial became inaccurate below pH 5 and above pH 7.5. Thus, gut pH measurements were considered reliable only if they fell within this range. Even when performing ratioing, some photobleaching was observed, a common phenomenon with fluorescein and most other fluorescent dyes, due to photochemical degradation of the fluophore. However, even under constant illumination, reproducible pH readings could be obtained from the same region of the specimen for at least 5 min. Light exposure was commonly kept to less than a few seconds by shuttering the illumination source except during frame capture. This also avoided unnecessary physiological stress to the animal.

REDOX POTENTIAL (EH)

Tetrazolium salts have been used in numerous studies as indicators of biological reducing systems (Altman 1976). Dissolved and practically colorless in unreduced form, they irreversibly precipitate under reducing conditions, forming insoluble, colored formazans. Jumars (1993) fed tetrazolium salts adsorbed to Sephadex beads to *Pseudopolydora kempi japonica* (Imajima &

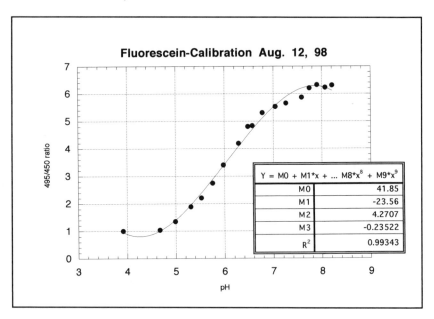

Fig. 1. The pH calibration curve of fluorescein. The 495 nm:450 nm emission ratio of fluorescein-labeled yeast cells (ZYMOSAN bioparticles) in buffered seawater standards.

Hartman), and was able to observe anteriorward transport of gut fluid by following reduction fronts. Although the ability of tetrazolium salts to accurately indicate Eh is controversial (Altman 1976), they may be used as indicators for estimating a lower boundary of redox conditions inside an animal's gut. By feeding the worms a suite of tetrazolium salts of decreasing reducibility, it is possible to bracket the redox conditions prevailing in their gut. It should be noted, however, that since the reduction to the colored formazan is an irreversible reaction, tetrazolium salts are more likely to reflect extreme redox history rather than current or average conditions.

Four tetrazolium salts of decreasing reducibility (nitroblue tetrazolium [NBT], Eh = -50 mV; iodonitrotetrazolium violet [INT], Eh = -90 mV; neotetrazolium [NT], Eh = -170 mV; tetrazolium violet [TT], Eh = -440 mV; Eh corresponding to 50% reduction at pH 7.2, as tabulated by Altman 1976) were loaded onto Chelex beads (Na-form, mesh 50–100 [ϕ = 150–300 μm], Biorad), coated with bovine serum albumin for greater palatability, and fed to adult and juvenile *Nereis succinea*. Chelex, being a weakly acidic cation exchanger, readily chelated the positively charged tetrazolium ion, and brought it into a particulate form suitable for feeding. The exact ratio of tetrazolium salt to Chelex was not determined; approximately 0.1 g of Chelex were soaked overnight in a solution of approximately 0.1 mg tetrazolium salt in 1 ml distilled water (or 70% ethanol), and subsequently rinsed in distilled water, to remove unbound tetrazolium dye. After feeding, worms were transferred to clean chambers, and fed on glass beads for depuration. Worms were kept under subdued light (to minimize photoreduction of the tetrazolium salts). Feces were analyzed qualitatively for colored precipitates (green, brown, or purple), indicative of reduction. Uningested tetrazolium-Chelex beads were kept in the same chamber as worms, to serve as controls.

PROTEASE ACTIVITY

Protease activity was estimated inside the guts of *Nereis succinea* by following the hydrolysis of a model substrate, BODIPY FL-labeled casein (Enz Chek Protease kit, Molecular Probes). BODIPY FL (excitation max: 505 nm, emission max: 513 nm) is molecularly quenched when attached to casein, but fluoresces bright green upon hydrolysis of the casein molecule. The observable fluorescence increase is proportional to the amount of casein hydrolyzed, and is linear over time when substrate is supplied in excess (Haugland 1996). BODIPY FL-casein was prepared for feeding by dissolving 50 µl BODIPY FL stock (200 µg BODIPY FL + 200 µl deionized water + 200 µl 50 mM phosphate buffer) in 100 µl 20% gelatin solution (200 mg KNOX gelatin + 1 ml seawater). Pieces of hardened gelatin were then fed to starved animals maintained inside transparent glass capillaries (VitroCom). Worms readily ingested the gelatin-substrate mixture, usually within less than 1 min of delivery. Worms were observed under a fluorescence microscope (1X objective, excitation filter: 495 nm, emission filter: 530 nm), and images captured every 20 s for approx. 30 min (employing on-chip integration at 100 frames image^{-1}, capture time approx. 3.5 s image^{-1}). We used the same imaging system as for pH measurements. Hydrolysis inside the worm's gut was measured as increase of green (530 nm) fluorescence intensity over time, integrated over the entire image, and normalized to total fluorescent gut area (pixel area). Calibrations of protease activity were performed by adding Protease K (Sigma) solutions of known activity to gelatin-casein cubes (size: approximately 1 mm^3) and calibrating the rate of increase of the (area-normalized) green signal with the apparent hydrolysis rates determined inside the guts of individual worms.

Results

PH

All worms analyzed had average gut pH values between 6.1 and 7.0 (Table 2), and differed significantly from ambient seawater (pH 7.8–8.2). *Nereis succinea* and *Streblospio benedicti* exhibited fairly uniform pH (no axial gradients along the mid gut), with a pH range of 6.0–7.3 for *Nereis succinea* and pH 6.5–7.0 for *Streblospio benedicti. Pygospio* sp. displayed a pH range of 5.7–6.9 in the midgut, but a pH close to seawater in the esophageal region (Fig. 2). *Polydora cornuta*'s gut pH ranged between pH 5.7 and 7.4. One individual we measured showed a noticeable pH increase, from pH 5.7 in the midgut to pH 6.3 in the hindgut, indicating possible dilution by seawater entering through its anus.

REDOX POTENTIAL (EH)

Small *N. succinea* (1–2 cm body length) were able to reduce the tetrazolium compound iodonitrotetrazolium violet (INT, Eh = -90 mV), but not neotetrazolium (NT, Eh = -170 mV), the salt tested with the next lower reducibility than INT. Chelex beads, used for delivery of INT, turned amber-colored during gut passage, while uningested beads remained colorless. This color change, which indicated the formation of the reduced formazan, was observed in both small (< 2 cm) and large (> 2 cm) *N. succinea* that ingested the dye. Although reduction was not confirmed in every worm tested, most *N. succinea* were capable of reducing INT. Some worms that defecated INT-labeled Chelex beads without a noticeable color change, produced fecal pellets with a sheath of purple INT-formazan several hours after defecating INT-labeled Chelex beads. The time for this occurrence

Table 2. In vivo gut pH (geometric mean ± 1 sd).

Species	Foregut	Midgut	Hindgut
Nereis succinea (< 2 cm)	6.9 ± 0.4 (n = 4)	6.7 ± 0.3 (n = 10)	6.6 ± 0.5 (n = 8)
Streblospio benedicti	6.7 (n = 1)	6.7 ± 0.1 (n = 2)	7.0 ± 0.4 (n = 2)
Pygospio sp.	6.3 ± 0.3 (n = 9)	6.2 ± 0.3 (n = 12)	6.1 ± 0.2 (n = 11)
Polydora cornuta	6.1 ± 0.4 (n = 4)	6.2 ± 0.7 (n = 5)	6.5 ± 0.4 (n = 6)

was considerably longer than the typical particle gut transit time of 30–60 min for small *N. succinea*. From this we concluded that some of the INT compound must have desorbed during gut passage, was eventually reduced inside the gut, precipitated, and finally defecated with later gut contents. While juvenile *N. succinea* (1–2 cm body length) were incapable of reducing neotetrazolium (NT) in their guts, adult worms readily reduced NT-labeled Chelex beads to a yellow-greenish color during gut transit. Uningested beads remained colorless.

Nereis succinea showed a low preference for the tetrazolium-labeled Chelex beads. This may be due to the characteristic bitter taste of tetrazolium salts and the somewhat large size of the Chelex beads used for delivery (ϕ approx. 150–300 μm). The only way to induce worms to feed on tetrazolium-labeled beads was to present them with a mix of labeled Chelex and regular glass beads, soaked in bovine serum albumin. This procedure usually overwhelmed a worm's aversion to the dye, and motivated it to ingest a few Chelex beads along with the glass beads. In some instances, however, even when using this "bovine-bait" approach, certain worms refused to ingest the Chelex beads and actually averted themselves from the food source.

PROTEASE ACTIVITY

N. succinea worms readily ingested the BODIPY-casein-gelatin substrate presented to them and showed an increase in green (530 nm) fluorescence intensity over time, due to hydrolysis of BODIPY-labeled casein in their guts, confirming the presence of proteases (Fig. 3). Due to sporadic movement of the worms during the 30-min observation period, hydrolysis data appear rather scattered (Fig. 4). Disregarding data points from frames during which the animal moved (usually indicated by a discontinuity in the slope) rendered an overall increasing trend in BODIPY fluorescence with time. Protease activity within the gut of *N. succinea*, calculated from the increase of (body-area normalized) green fluorescence over time, varied by approximately one order of magnitude among different individuals. The maximum observed rate was approximately similar to that of a 1 unit protease K solution (hydrolysis of 1μmol of casein min^{-1}; Fig. 4). The hydrolysis rate remained linear for approximately 5–10 min after ingestion of the BODIPY-casein substrate, then decreased. In one worm (N7, Fig. 5), a reduced hydrolysis rate was followed by renewed ingestion activity by the worm, which resulted in a rebound of the hydrolysis rate.

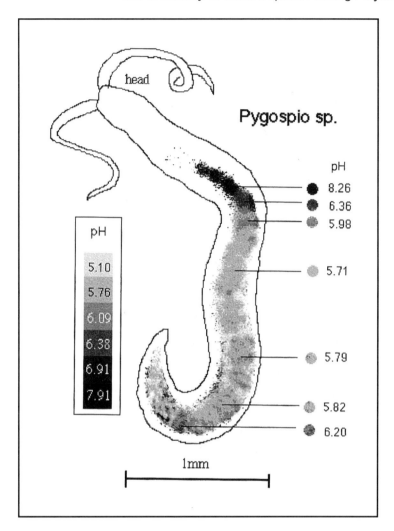

Fig. 2. Gut pH of *Pygospio* sp., after ingestion of fluorescein-labeled yeast cells. The 495 nm:450 nm excitation ratio was converted to gray scale and calibrated to buffers of known pH (note: pseudocolor images of the gut pH of this and other polychaetes may be viewed on the web at http://alpha1.msrc.sunysb.edu/~ahrens/).

Discussion

PH

We were able to produce detailed pseudocolor images of gut pH with high spatial resolution (color images may be viewed on the web at http://alpha1.msrc.sunysb.edu/~ahrens/). Although pixel-to-pixel variability was often considerable (due to high noise and imaging artifacts), averaging of several pixels rendered fairly homogenous, spatially reproducible gut pH values. In the proximity of

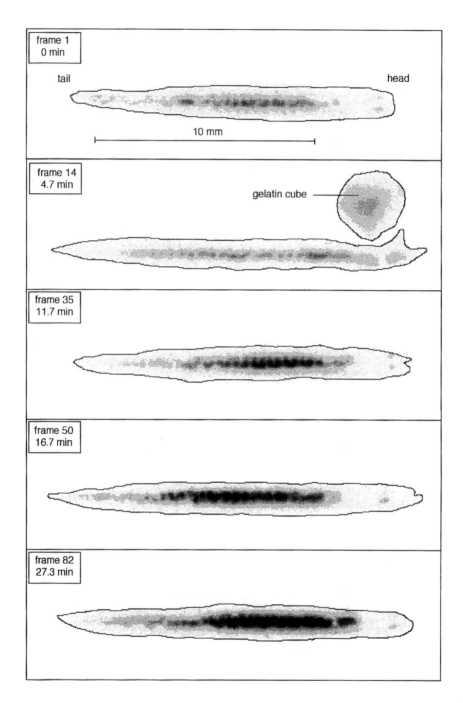

Fig. 3 Casein hydrolysis in the gut of *Nereis succinea*. BODIPY-FL fluorescence intensity (of fore, mid, and hindgut) imaged over 82 frames (= 27.3 min, capturing interval 20 s) displayed in gray scale (darker = greater intensity). BODIPY-casein, dissolved in gelatin, was added at T = 4.7 min (gelatin cube). Representative frames, taken at T = 11.7, 16.7, and 23.7 min, show a progressive increase in gut fluorescence. During feeding, the worm was maintained in a glass capillary tube, head oriented toward the right.

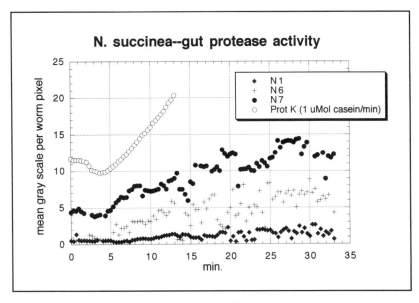

Fig. 4. BODIPY-casein hydrolysis of three specimens of *Nereis succinea*, displayed as integrated 530 nm fluorescence normalized to worm area (= mean pixel intensity). Frame capture interval 20 s. Worm s were free to move during frame capture. Hydrolysis rate of 1 μmoles casein min^{-1}. (O) is shown for comparison.

segment borders, blood vessels, or the body wall, pH values differed markedly from average pH values. Furthermore, slight movement by the worm (e.g., by gut peristalsis) could result in unusable data in certain areas. These edge and movement artifacts were ascribed to optical nonlinearities (e.g., light scattering and refraction) and imperfect registering of sister images. For this reason, pH data in the proximity of boundaries or in active organisms were excluded from further analysis. The inhibition of movement was an important prerequisite for achieving spatially accurate pH data using the ratiometric method in its current state. The possible artifacts resulting from chilling worms for 2 min in a freezer and immobilization in gelatin cannot be assessed at this point, but they certainly represent an additional source of variation to measures of their digestive performance. By decreasing the interval between capture of sequential images to a duration that is shorter than an organism's inherent activity pattern (e.g., by using a more light sensitive camera and/or an automatic filter changer), it may be possible to eliminate the immobilization and chilling requirement in the future. The gut pH imaging method was capable of measuring gut pH with an accuracy of approximately 0.5 pH units. This accuracy was much lower than that of a pH microelectrode; however, the microfluorimetric approach allowed estimation of the acidity inside guts of living animals too small to sample with an electrode. Contrary to Pond et al. (1995), who measured gut pH in copepods by injecting the pH-sensitive fluophore via the anus, our approach allowed animals to freely ingest the dye together with food, which reduced disturbance considerably. This method should be suitable for measuring gut pH in many other small animals. The only prerequisites are the successful ingestion of the dye-yeast conjugate and low opacity and minimal autofluorescence of the organism. Since animals need to ingest only minute amounts of dye-yeast conjugate for the ensuing measurement, unusually high stimulation of gut microbial activity due to the ingested yeast cells seems improbable. However, this does not rule out the possibility that gut microbes may exert a central role in

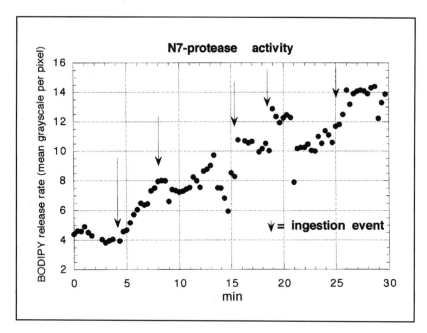

Fig. 5. Protease activity of one individual of *Nereis succinea* (N7, same individual as in Fig. 4). Arrows indicate ingestion events.

maintaining gut pH, and that there may be distinct differences in gut pH between starved and well-fed worms, similar to the observations on copepods by Pond et al. (1995). Despite 1–2 d of starvation prior to measurements, worms continued to have sediment in their guts, and thus considered "full" (in Table 1).

The pH sensitivity range of fluorescein conveniently bracketed the gut pH range encountered in the four worm species studied. All specimens analyzed exhibited average midgut and hindgut pH values of 6.1 to 7.0. We did not encounter any single gut pH value below pH 5.5. However, gut pH values consistently below pH 6 were observed in some benthic harpacticoids (*Coullana canadensis*) measured concurrently (M. Ahrens unpubl. obs.). Spurious pH values above pH 8.5 occurred near tissue boundaries and were discarded as artifacts (fluorescein's fluorescence saturates around pH 8). Average gut pH values in this study were less than one pH unit lower than the values of pH 6.5–7.8 measured by Plante and Jumars (1992), among the highest in the literature on deposit feeding. Assuming our measurements are accurate, gut pH in small deposit feeders does not appear to be considerably lower than in large animals. Nevertheless, comparing gut pH data in *N. succinea* over a wide size range (using a microelectrode), we have detected a weak trend of decreasing gut acidity with increasing body length (M. Ahrens in prep.)

Our results confirm the classic view that marine invertebrates, in general, possess only moderately acidic guts (Rhoads 1974), but, nonetheless, with a pH clearly different from seawater (i.e., pH 8). On the other hand, sediment pH generally tends to be lower than seawater pH. Typical sediment pH values range between pH 6.5–8.0 in limnic and coastal sediments (Gorham 1960; Fisher & Matisoff 1981) and pH 7.4–7.7 in deep-sea sediments (Cai et al. 1995). The slightly acidic to neutral condition in deposit feeder guts could, thus, reflect primarily ambient sediment conditions and consequently the metabolic action of sediment microbes rather than an active physiological effort by the animal. If average gut pH ranges between pH 6 and 7, sediment and associated organic matter, metals, and anthropogenic contaminants may undergo up to a 10–100 fold change in the hydrogen ion concentration when transiting the gut of these species. This change in acidity could alter various chemical equilibria that may be present in the sediment prior to ingestion. Model calculations performed by us suggest that a gut pH of 6 may measurably affect the saturation of $CaCO_3$ in a time span comparable to typical gut transit times of one to a few hours, and should lead to the gradual dissolution of carbonate minerals. However, evidence for this process has been controversial (see Hammond 1981). Low pH may also enhance desorption of particle-bound metals (Wang et al. 1999), either by increasing the solubility of the metal or by altering the composition of the sediment matrix. The moderately acidic environment of the gut may be favorable for maximizing enzyme activity; invertebrate's digestive enzymes generally have pH optima in the weakly acidic pH range (Vonk 1964).

All four polychaete species investigated concentrated the fluorescein-labeled yeast cells primarily in mid and posterior gut segments. Thus, pH data are restricted to these regions only, but we may speculate that pharyngeal and esophageal pH are probably only of negligible importance to food digestion, since food spends very little time there. Further, while observing *N. succinea* and *Pygospio* sp. we noticed that glass particles transited the gut faster than fluorescein-labeled yeast particles. The prolonged retention of yeast particles over inert glass particles without any food value is suggestive of some kind of intestinal particle selection mechanism.

EH

Feeding tetrazolium dyes bound to Chelex beads to deposit-feeding worms appears to be a feasible method to estimate their gut redox conditions. Preliminary results indicate that *N. succinea* has slightly reducing gut conditions. The observation that most individuals were capable of reducing iodonitrotetrazolium violet (INT) and in some adult worms even neotetrazolium (NT), suggests that Eh in these worms is on the order of -90 mV, based on the tabulation of half-way potentials by Altman (1976). INT-labeled Chelex beads, having transited through a gut once, could be reduced further by exposure to strong light. From this, we inferred that INT reduction did not proceed to completion during gut passage. This, and the finding that only larger specimens of *N. succinea* were able to reduce the tetrazolium salt with the next lower reducibility (NT), leads us to conclude that redox conditions in small *N. succinea* are only mildly reducing, and probably not much lower than -90 mV. However, this value is only a rough estimate since the degree of color change was not quantified. The more reducing conditions in larger worms may be due to longer gut transit time (GTT). GTT is approximately 30 min for a 1-cm juvenile, but it can be longer than 300 min (> 5 h) for a large adult (> 5 cm length) at 24°C (M. Ahrens pers. obs.). Plante and Jumars (1992) have measured gut Eh values between -144 and +295 mV in four adult deposit-feeding polychaetes.

It should again be noted that since tetrazolium salts change color irreversibly upon reduction, they are more likely to reflect extreme redox history rather than average or present conditions. In other words, finding colored beads in feces gives no indication whether reducing conditions prevail uniformly throughout the gut or only in localized microzones, similar to the reduced microzones in sediments. In any case, redox conditions inside the gut of *N. succinea* were distinctly more reducing than in the ambient chamber water, since uningested control beads remained unchanged. Bovine serum albumin (BSA), added to improve palatability, caused no enhanced reduction of control beads. However, it may be possible that uningested control beads experienced exceptionally mild microbial exposure since bacterial abundances in seawater are commonly considerably lower than in sediments or guts. Also, added BSA may have had a stronger stimulatory effect on microbes in the gut than in the chamber water, ultimately leading to greater redox differences between gut and controls. An improvement to future experiments might be to feed worms a mixture of tetrazolium beads and sediment, rather than just beads, and to draw controls from the uningested fraction of this mixture. Future studies might also attempt to quantify the relative contribution of gut microbes and the host organism to gut redox conditions. Perhaps this could be accomplished by feeding food that contains antibiotics and/or fungicides. If gut Eh were primarily controlled by gut microbes, then low gut Eh might be an indicator of high bacterial activity and thus poor efficiency of food transfer to the animal. Gut redox conditions would thus depend on the amount of bacterially labile food that is ingested. For a discussion of the potential interactions between deposit feeders and gut microbes see Plante et al. (1990).

Since tetrazolium salts can be toxic at higher concentrations, it is advisable to keep the amount that is added to an animal's feeding chamber as low as possible. Commonly, only a few beads need to be ingested to detect reduction reliably. If worms are left with a large number of tetrazolium-labeled beads over several hours, the whole organism will eventually be dyed purple and start demonstrating unusual behavior. We therefore removed worms promptly from their feeding chambers after successful ingestion and depurated them in separate chambers. In future experiments, it may be advisable to use smaller Chelex beads (e.g., 200–400 mesh, $\phi = 37–75$ μm), instead of the 50–100 mesh used ($\phi = 150–300$ μm), to approximate more closely the smaller size preference of worms (Brown 1986; Self & Jumars 1988). Finally, it should be noted that it is important to minimize the amount of light exposure in order to avoid artifacts due to photoreduction. We recommend feeding in the dark.

PROTEASE ACTIVITY

Nereis succinea displayed measurable protease activity throughout the gut, with an apparent maximum in the mid-gut region. This corroborates results by Michel (1977) and Mayer et al. (1997) who also found protease activity peaked in the anterior to mid intestine of the holothuroid *Parastichopus californicus* and the polychaetes *Thelepus crispus* and *Sabellaria alveolata.* Ingested BODIPY-casein was quickly hydrolyzed by *N. succinea,* leading to an increase of BODIPY fluorescence over time. The finding that the apparent hydrolysis rate began to diminish approximately 5–10 min after ingestion in *N. succinea,* suggests that gut proteases may be capable of hydrolyzing quite a large fraction of ingested protein within the typical gut residence time of 30–60 min. A decreased hydrolysis rate could entail a reduced absorption rate, and may be a cue to the worm to seek additional food. Figure 5 suggests how gut kinetics may directly affect feeding behavior: A worm, seeking to maximize its rate of hydrolytic and absorptive gain of a limiting nutrient (= protein), will ingest food until it brings the hydrolysis rate back to zero order, thereby literally achieving substrate saturation. This ability to observe directly the digestive process of living animals in real time is one of the great assets of our fluorimetric method, and provides a powerful tool for investigating temporal and spatial plasticity of digestive enzymatic processes. It is highly probable that the protein loading in BODIPY-casein-laced gelatin food was considerably higher than typical protein loading in natural sediment, thus saturating digestive proteases (which might normally not occur).

In future work, we plan to measure other digestive enzymes besides proteases. A protocol for measuring lipase activity is currently under development. The ratio of gut protease to lipase activity has been used by Mayer et al. (1997) to distinguish sediment feeders from live cell feeders. Digestive fluids from sediment feeders commonly exhibit high protease-to-lipase ratios, whereas the inverse holds for live cell feeders. Since Mayer et al. (1997) focused exclusively on larger individuals, one may only speculate about the enzymatic capabilities of juvenile deposit feeders. Extrapolating Mayer's findings to juveniles, one would predict the protease-to-lipase ratio to be diminished in juveniles that feed primarily on live cells and labile organic matter. On the other hand, it seems equally plausible that protease activity may be elevated in juveniles in order to compensate for shorter gut transit time and to meet their high protein requirements for growth. A comparison of gut protease-to-lipase activities between small and large individuals of the same species would be an interesting study to test for ontogenetic shifts in feeding strategy.

General Applicability

The results we present in this paper were obtained primarily from one species, *Nereis succinea*, and based on a small number of specimens. Thus, we are, at present, unable to state how representative our measurements are for the entire species, or a particular developmental stage. The in vivo methods described here give us a tool to probe the gut chemistry of animals too small to measure by the commonly employed invasive methods. Since all probes are introduced into the gut naturally, via ingestion, the potential artifacts of injection or dissection are eliminated and disturbance to organisms is greatly reduced. Nevertheless, our current methodology is not free of caveats. To achieve higher fluorescence intensity in guts, worms were prevented from ingesting natural sediment prior to experiments (in redox and protease measurements) and then fed primarily with protein-coated glass beads as a sediment mimic. Because the guts of deposit feeders commonly contain sediment, measurements of gut chemistry in sediment-devoid guts may not be representative of natural conditions. Thus, it seems advisable to let worms ingest at least some sediment prior to measurements, even if this may entail a certain degradation of the fluorescence signal. Also, the feeding media used for tracer delivery or for enhancing palatability (e.g., yeast cells, bovine serum, gelatin) should ideally constitute only a small fraction of the gut contents to avoid unusual stimulation or inhibition of the organism's physiology or its intestinal microflora. Furthermore, the immobilization of worms, in addition to the observation under high light intensity, may induce unusual stress on the animal and possibly alter its physiological state. Although the disturbances due to the measurement procedure cannot be removed entirely, using equipment with higher light sensitivity may reduce exposure time and light intensity and thus diminish some of the methodological artifacts.

We are currently also working on a protocol for quantifying gut surfactancy in small deposit feeders. Surfactants have been detected in the gut fluids of a number of large deposit-feeding species (Mayer et al. 1997) and other benthic invertebrates (Vonk 1969; Tugwell & Branch 1992) and appear to play a central role in the uptake of sediment-bound hydrophobic compounds. Work in progress on *N. succinea*, employing glass beads coated with the fluorescent dye Nile Red (Nile Blue A Oxazone), has confirmed the presence of surfactant micelles in the gut, although calibration remains a problem.

Conclusions

Gut pH in *Nereis succinea*, *Streblospio benedicti*, *Pygospio* sp., and *Polydora cornuta* deviate measurably from conditions typical at the sediment surface. Worm guts are mildly acidic, and in *N. succinea* (the only species for which we also measured Eh, enzyme activity and surfactancy), mildly reducing, displaying protease activity and presence of surfactants. These conditions are likely to be conducive to a number of early diagenetic processes, like desorption and degradation of organic matter, the alteration of certain mineral phases, and possibly desorption of sediment-sorbed trace elements or organic compounds. Preliminary observations on *N. succinea* suggest that gut chemistry may be quite variable among individuals of the same species, which could be the result of size differences or different feeding history. Our current data indicate that small worms may have slightly lower gut pH, but that their gut chemistry is not altogether considerably different from larger worms.

The usefulness of the in vivo techniques described extends beyond the study of small deposit feeders' gut chemistry. The methods are likely to be applicable to other aquatic invertebrates. They provide a means for studying how digestive chemistry varies over space and time inside a living animal and enable us to examine in greater detail how food quality and availability may influence digestive physiology. Ultimately this knowledge may help us to better understand digestive strategies of organisms and how these may affect biogeochemical cycles.

Acknowledgments

This work was supported grant OCE 9711793 from the National Science Foundation. We thank B. Hentschel and C. Plante for improvements to the manuscript. This is publication no. 1209 by the Marine Sciences Research Center.

Literature Cited

Ahrens, M. J., J. Hertz, E. M. Lamoureux, G. R. Lopez, A. E. McElroy, & B. J. Brownawell. In press. The effect on body size of digestive chemistry and Absorption effencies of food and sediment-bound organic contaminants in *Nereis* (*Neanthes*) *succinea*. *Marine Ecology Progress Series*

Altman, F. P. 1976. Tetrazolium salts and formazans. *Progress in Histochemistry and Cytochemistry* 9(3):1–57.

Bakus, G. J. (1973). The biology and ecology of tropical holothurians, Ch. 10, p. 324–367. *In* Jones, O.A. & R. Endean (eds.), Biology and Geology of Coral Reefs, Series 1 (Biology). Academic Press, New York.

Barth, R., L. Bonel-Ribas, & M. L. Osorio E Castro. 1968. Observacoes no conteudo intestinal de holoturias. *Publicacao do Instituto de Pesquisas da Marinha* 9:1–15.

Brasil, L. 1903. Origine et role de la secretion des coecums oesophagiens de l'arenicole. *Archives de Zoologie Experimentale et Generale* (notes et revue) 4(1):vi–xiii.

Brasil, L. 1904. Contribution a la connaissance de l'appareil digestif des Annelides Polychetes. *Archives de Zoologie Experimentale et Generale* 4(2):91–255.

Bright, G. R., G. W. Fisher, J. Rogowska, & D. L. Taylor. 1987. Fluorescence ratio imaging microscopy: Temporal and spatial measurements of cytoplasmic pH. *Journal of Cell Biology* 104:1,019–1,033.

Brown, S. L. 1986. Feces of intertidal benthic invertebrates: Influences of particle selection in feeding on trace element concentration. *Marine Ecology Progress Series* 28:219–231.

Bryan, G. W. & L. G. Hummerstone. 1971. Adaptation of the polychaete *Nereis diversicolor* to estuarine sediments containing high concentrations of heavy metals. I. General observations and adaptation to copper. *Journal of the Marine Biological Association of the United Kingdom* 51:845–863.

Cai, W.-J., C. E. Reimers, & T. Shaw. 1995. Microelectrode studies of organic carbon degradation and calcite dissolution at the California Continental rise site. *Geochimica et Cosmochimica Acta* 59(3):497–511.

Campbell, P. G. C. & P. M. Stokes. 1985. Acidification and toxicity of metals to aquatic biota. *Canadian Journal of Fisheries and Aquatic Sciences* 42:2034–2049.

Crozier, W. J. 1918. The amount of bottom material ingested by holothurians (Stichopus). *Journal of Experimental Zoology* 26:379–389.

Dales, R. P. 1955. Feeding and digestion in terebellid polychaetes. *Journal of the Marine Biological Association of the United Kingdom* 34:55–79.

Emery, K. O., J. I. Tracey, Jr., & H. S. Ladd. 1954. Geology of Bikini and Nearby Atolls. *United States Geological Survey Professional Papers* 260-A.

Feral, J.-P. 1989. Activity of the principal digestive enzymes in the detritivorous apodous holothuroid *Leptosynapta galliennei* and two other shallow-water holothuroids. *Marine Biology* 101:367–379.

Fisher, J. B. & G. Matisoff. 1981. High resolution vertical profiles of pH in recent sediments. *Hydrobiologia* 79:277–284.

Fisher, N. S. & J. L. Teyssié. 1986. Influence of food composition on the biokinetics and tissue distribution of zinc and americum in mussels. *Marine Ecology Progress Series* 28:197–207.

Forbes, T. L. 1989. The importance of size-dependent physiological processes in the ecology of the deposit-feeding polychaete *Capitella* species 1. Ph.D. dissertation, Marine Sciences Research Center, State University of New York, Stony Brook, New York.

Forbes, T. L. & G. R. Lopez. 1990. Ontogenetic changes in individual growth and egestion rates in the deposit-feeding polychaete *Capitella* sp. 1. *Journal of Experimental Marine Biology and Ecology* 143(3):209–220.

Gelder, S. R. 1984. Diet and histophysiology of the alimentary canal of *Lumbricillus lineatus* (Oligochaeta, Enchytracidae). Proceedings of the Second International Symposium on Aquatic Oligochaete Biology. G. Bonomi & C. Erseus (eds.). *Hydrobiologia* 115:71–81.

Gorham, E. 1960. The pH of fresh soils and soil solutions. *Ecology* 41:563.

Hammond, L. S. 1981. An analysis of grain size modification in biogenic carbonate sediments by deposit-feeding holothurians and echinoids (Echinodermata). *Limnology and Oceanography* 25(5):898–906.

Haugland, R. P. (1996). Handbook of Fluorescent Probes and Research Chemicals (Molecular Probes Inc. Catalog), Spence, M. T. Z (ed.). 6th edition, Eugene, Oregon.

Hentschel, B. T. 1996. Ontogenetic changes in particle-size selection by deposit-feeding spionid polychaetes: The influence of palp size on particle contact. *Journal of Experimental Marine Biology and Ecology* 206(1–2):1–24.

Hentschel, B. T. 1998a. Intraspecific variations in delta ^{13}C indicate ontogenetic diet changes in deposit-feeding polychaetes. *Ecology* 79(4):1357–1370.

Hentschel, B. T. 1998b. Spectrofluorometric quantification of neutral and polar lipids suggests a food-related recruitment bottleneck for juveniles of a deposit-feeding polychaete population. *Limnology and Oceanography* 43(3):543–549.

Jeuniaux, C. 1969. Nutrition and digestion, p. 69–91. *In* Florkin, M. & B. R. Scheer (eds.), Chemical Ecology 4, Academic Press, New York.

Jumars, P. A. 1993. Gourmands of mud: Diet selection in marine deposit feeders, p. 124–156. *In* Hughes, R. N. (ed.), Mechanisms of Diet Choice. Blackwell Scientific Publishers, Oxford, England.

Kay, D. G. 1974. The distribution of the digestive enzymes in the gut of the polychaete *Neanthes virens*. *Comparative Biochemistry and Physiology* 47A:573–582.

Kermack, D. M. 1955. The anatomy and physiology of the gut of the polychaete *Arenicola marina* (L.) *Proceedings of the Zoological Society of London* 125:347–381.

Kolthoff, I. M. 1941. pH and Electro Titrations; The Colorimetric and Potentiometric Determination of pH. Potentiometry, Conductometry, and Voltammetry (Polarography). Outline of Electrometric Titrations. 2nd edition. Wiley & Sons, New York.

Kristensen, J. H. 1972. Carbohydrases of some marine invertebrates with notes on their food and on the natural occurrence of the carbohydrates studied. *Marine Biology* 14(2):130–142.

Longbottom, M. R. 1970. Distribution of the digestive enzymes in the gut of *Arenicola marina*. *Journal of the Marine Biological Association of the United Kingdom* 50:121–128.

Mayer L.M., L. L. Schick, R. F. L. Self, P. A. Jumars, R. H. Findlay, Z. Chen, & S. Sampson. 1997. Digestive environments of benthic macroinvertebrate guts: Enzymes, surfactants and dissolved organic matter. *Journal of Marine Research* 55:785–812.

Mayor, A. G. 1924. Causes which produce stable conditions in the depth of the floors of pacific fringing reef-flats. *Carnegie Institution Washington Publication (Serial)* 340:27–36.

Michel, C. 1977. Tissular localization of the digestive proteases in the sedentary polychaetous annelid *Sabellaria alveolata*. *Marine Biology* 44(3):265–273.

Michel, C., M. Bhaud, P. Boumati, & S. Halpern. 1984. Physiology of the digestive tract of the sedentary polychaete *Terebellides stroemi*. *Marine Biology* 83:17–31.

Oomen, H. A. P. C. 1926. Verdauungsphysiologische Studien an Holothurien. *Pubblicazioni della Stazione Zoologica di Napoli* 7:215–297.

Penry D. L. & P. A. Jumars. 1990. Gut architecture, digestive constraints and feeding ecology of deposit-feeding and carnivorous polychaetes. *Oecologia* 82:1–11.

Perrin D. D. & B. Dempsey. 1974. Buffers for pH and metal ion control. Chapman and Hall, London.

Plante, C. J. & P. A. Jumars. 1992. The microbial environment of marine deposit-feeder guts characterized via microelectrodes. *Microbial Ecology* 23:257–277.

Plante C. J., P. A. Jumars, & J. A. Baross. 1989. Rapid bacterial growth in the hindgut of a marine deposit feeder. *Microbial Ecology* 18:29–44.

Plante C. J., P. A. Jumars, & J. A. Baross. 1990. Digestive associations between marine detritivores and bacteria. *Annual Review of Ecology and Systematics* 21:93–127.

Pond , D. W., R. P. Harris, & C. Brownlee. 1995. A microinjection technique using a pH-sensitive dye to determine the gut pH of *Calanus helgolandicus*. *Marine Biology* 123:75–79.

Rhoads, D. C. 1974. Organism-sediment relations on the muddy sea floor. *Oceanography and Marine Biology Annual Review* 12:263–300.

Sawano, E. 1927. On the digestive enzyme of *Caudina chilensis*. *Science Reports of the Tohoku University*, Series 4 (Biology) 3(1):205–218.

Self, R. F. L. & P. A. Jumars. 1988. Cross-phyletic patterns of particle selection by deposit feeders. *Journal of Marine Research* 46:119–143.

Shimeta, J. 1996. Particle-size selection by *Pseudopolydora paucibranchiata* (Polychaeta: Spionidae) in suspension feeding and in deposit feeding: Influences of ontogeny and flow speed. *Marine Biology* 126(3):479–488.

Tanaka, Y. 1958. Feeding and digestive processes of *Stichopus japonicus*. *Bulletin of the Faculty of Fisheries*, Hokkaido University 9:14–26.

Tenore, K. R. 1977. Growth of *Capitella capitata* cultured on various levels of detritus derived from different sources. *Limnology and Oceanography* 22:936–941.

Trefz, S. M. 1956. Office of Naval Research Project NR 165-264, p. 1 (as cited in Bakus 1973)

Tugwell, S. & G. M. Branch. 1992. Effects of herbivore gut surfactants on kelp polyphenol defenses. *Ecology* 73(1):205–215.

van der Heyde, H. C. 1922. On the physiology of digestion, respiration and excretion in echinoderms. Academic thesis, University of Amsterdam, C. De Boer, Den Helder.

Vonk, H. J. 1964. Comparative biochemistry of digestive mechanisms, Ch. 7, p. 347–402. *In* Florkin, M. & H. S. Mason (eds.), Comparative Biochemistry Vol. VI, Cells and Organisms. Academic Press, Orlando, Florida.

Vonk, H. J. 1969. The properties of some emulsifiers in the digestive fluids of invertebrates. *Comparative Biochemistry and Physiology* 29:361–371.

Vonk, H. J. & J. R. H. Western 1984. Comparative Biochemistry and Physiology of Enzymatic Digestion. Academic Press, Orlando, Florida.

Wang W.-X., I. Stupakoff, & N. S. Fisher 1999. Bioavailability of dissolved and sediment-bound metals to a marine deposit-feeding polychaete. *Marine Ecology Progress Series* 178:281–293.

Yamanouti, T. 1939. Ecological and physiological studies on holothurians in the coral reef of Palao Islands. *Palao Tropical Biological Station Studies* 1:603–636.

Unpublished Materials

Griscom, S. 1998. Personal communication. Marine Sciences Research Center, State University of New York, Stony Book, New York 11794–5000 USA.

The Role of Fecal Pellet Deposition by Leaf-Eating Sesarmid Crabs on Litter Decomposition in a Mangrove Sediment (Phuket, Thailand)

Erik Kristensen* and Randi Pilgaard

Abstract: *The decay of uneaten green (GL) and 24-h water-leached leaves (WL) of Rhizophora apiculata was compared with anoxic decomposition of fecal pellets (FP) from the sesarmid crab Neoepisesarma versicolor feeding on a diet Rhizophora apiculata leaves. The leaf and fecal pellet materials were mixed into anoxic mangrove sediment and contained in gas-tight glass vials ("jars"). Subsets of the jars were sacrificed at regular intervals over 20–30 d to determine temporal changes in porewater solutes (DIC (CO_2), NH_4^+, SO_4^{2-}, and DOC). The concentration change over time of all solutes exhibited one linear phase in unamended jars (sediment only, control)and one or two linear phases in amended jars. Sulfate reduction was the dominant respiration process as determined from ΔDIC: ΔSO_4^{2-} ratios of about 2. In jars amended with green leaves (GL), instant initial leaching of DOC was responsible for high carbon mineralization due to sulfate reduction during the linear phase 1. During the second linear phase, carbon mineralization ceased in these jars (GL-amended) due to sulfate depletion. No such leaching occurred in FP-amended and WL-amended jars. In the former, crabs assimilated all leachable carbon from ingested leaf material before defecation. In the latter, an amount of DOC equivalent to the initial DOC recovered from GL-amended jars was lost during the 24-h leaching period . The lack of DOC leching in FP- and WL-amended jars indicated that fermentative DOC production controlled carbon mineralization at a rate 50% lower than during linear phase 1 in GL-amended jars. The low production and in some cases rapid initial disappearance of NH_4^+ in amended jars indicated nitrogen-limited bacterial growth. Decay of the particulate fraction (POC) of fecal pellets was 3–10 times higher than the background (unamended jars) and up to 2 times higher than the decay of leaf-POC (excluding the contribution of leached DOC). Sesarmid crabs are important for the carbon and nutrient cycling in mangrove forests. By eating green leaves in the tree canopy, crabs retain leaf POC and assimilate DOC (and probably DON). The rate of leaf POC decay in sediments is enhanced after maceration and passage through crab guts compared with that of shredded fresh leaves.*

Introduction

Mangrove leaf litter is considered the basis of food webs in mangrove forests (Giddins et al. 1986; Robertson 1986; Twilley et al. 1997). However, rates and pathways of litter turnover may vary among ecological types of mangrove swamps (Twilley et al. 1986) and continental regions (McIvor

369

& Smith 1995). Variation in litter dynamics among mangrove areas within regions may be explained by differences in geophysical processes, such as tides and river flow (Twilley et al. 1997). Regional differences are considered of biological origin. The detritus food web in forests of the Caribbean region is driven primarily by microbial processes (Odum & Heald 1972; McIvor & Smith 1995), whereas herbivorous crabs are important litter processors in the Indo-West Pacific region (Robertson 1986; McIvor & Smith 1995; Lee 1997).

Litter consumption and burial by crabs may have a profound impact on the organic matter flow and nutrient dynamics within mangrove forests and affect exchange with the adjacent coastal zone (Robertson 1986; Twilley et al. 1997). A number of studies from Australia and Southeast Asia have documented that sesarmid crabs (Grapsidae) may consume 28–79% of the annual litterfall (Kofoed et al. 1985; Robertson 1986; Robertson & Daniel 1989). More recently, it has been shown that these crabs may remove 20–40% of the litterfall in East African mangrove forests (Slim et al. 1997). The remaining litter is either exported to adjacent waterways (Boto & Bunt 1981; Chansang & Poovachiranon 1990), decomposed aerobically at the sediment surface, or decomposed anaerobically within the sediment (Alongi 1989; Robertson et al. 1992). The proportion of litter entering each pathway is largely determined by mangrove geomorphology, tidal flushing, and type of litter (Robertson et al. 1992). Much of the litter handled by crabs will eventually enter the microbial food chain, either in the form of fecal material or as uneaten remains buried in the sediment (Giddins et al. 1986; Robertson 1986; Lee 1997). Intact leaves are, in contrast to fecal pellets, readily exported by tides before sinking and thus lost from the mangrove ecosystem.

Leaf litter decomposition in mangrove environments is well documented (e.g., Robertson et al. 1992), but only a little is known about the fate of leaf detritus after passage through crabs (Neilson & Richards 1989; Lee 1997). Since fecal pellets of sesarmid crabs are deposited on or in the sediment, the rate and pathway of anoxic fecal pellet decomposition may be important for the carbon and nutrient status of mangrove environments. In the present study, anoxic decomposition of fecal pellets from the leaf-eating sesarmid crab *Neoepisesarma versicolor* feeding on a diet of *Rhizophora apiculata* leaves is compared with the decomposition of uneaten green leaves and leaves leached for 24 h in water. Rates of decomposition of fecal material and litter material buried in anoxic sediment are quantified as DIC (CO_2) and NH_4^+ production, SO_4^{2-} consumption, and DOC dynamics over time. The results are evaluated and discussed in relation to mangrove litter dynamics.

Materials and Methods

STUDY SITE

Sediment cores, *Rhizophora apiculata* leaves, and sesarmid crabs were collected in the Bang Rong mangrove forest (8°03' N, 98°25' E) on the northeast coast of the island of Phuket, Thailand (Fig. 1). The forest is a 1-km wide, tide-dominated, fringing mangrove lagoon and stretches about 2 km inland. There are no major freshwater discharges to the forest (except during heavy rains), and tidal water exchange with the ocean occurs through one main channel fed by smaller creeks along the length axis of the forest. Tidal range in the area is about 2.5 m. The annual average temperature and salinity are 28°C and 35‰, respectively. The vegetation within the forest is dominated by *Rhizophora apiculata* and *R. mucronata,* with a canopy height of 5–10 m. The most abundant benthic fauna in the forest are crabs belonging to the genera *Uca* (fiddler crabs) and *Neoepisesarma* (sesarmid crabs). All samplings for the experiments in this study were made in the central part of the mangrove forest, with the origin at a mudflat in the main channel (station MC, Fig. 1).

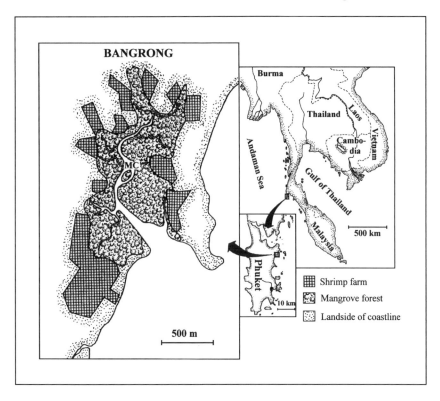

Fig. 1. Map of the Bang Rong mangrove forest with station MC indicated.

CRAB HANDLING AND FECAL PELLET COLLECTION

Specimens of the leaf-eating crab *Neoepisesarma versicolor*, were caught during the night within the mangrove forest adjacent to station MC. Due to their habit of eating leaves in the tree canopy at night, sesarmid crabs were picked off the trees by hand. No attempt was made to estimate *N. versicolor* density in the Bang Rong mangrove forest; however, in a nearby forest, Ao Nam Bor, the density has been estimated to be 2–5 crabs m^{-2} (Kofoed et al. 1985). Thirty crabs were brought to the laboratory and kept two by two in 40-l aquaria containing seawater to a depth of a few millimeters. Green *Rhizophora apiculata* leaves without grazing marks, which had been picked from the trees during crab catching, were rinsed briefly in seawater and stored frozen until used as crab feed.

The crabs were starved for 4 d to empty their guts. Fecal pellets were removed and the water was replaced once every day. All fecal pellets collected during the starvation period were discarded. Subsequently, the crabs in each aquarium were fed two thawed green *Rhizophora apiculata* leaves. Fecal pellets were collected frequently and stored frozen for later use in the "jar" experiments. The water and leaves were replaced once a day. The fecal pellet harvesting procedure continued for 1.5 mo.

"Jar" Experiments

Three sets of anaerobic incubations using 20-ml "jars"(gas-tight glass vials) were made to compare the impact of fecal pellet and green leaf depositions on anaerobic sediment processes. In JAR1, the vertical distribution of anaerobic carbon and nitrogen transformations was determined in the upper 20 cm (0–1, 2–3, 6–8, 14–16 cm) of unamended sediments at station MC over 17 d. This experiment was conducted both during the dry season (January) and the rainy season (August). In JAR2, the impact of added fecal pellets and green leaves on sediment carbon and nitrogen transformations were examined over 32 d in the 0–2 and 5–7 cm depth intervals of sediment from station MC. JAR3 included the same treatments as JAR2 plus an additional series with water-leached leaves. JAR3 was only examined over 21 d.

Sediment from station MC for the jar experiments was sampled by hand at low tide using 8 cm id. and 25 cm long core liners. Subsequently, a number of sediment cores (4–8) were sectioned into the chosen depth intervals in air. The slices from each depth interval were pooled and mixed thoroughly. All sediment handling was done rapidly to minimize oxygen exposure. Cores sampled for JAR1 were sectioned immediately after sampling, whereas the cores used in JAR2 and JAR3 were stored for 10 d in a darkened seawater bath before sectioning in order to reduce the impact of indigenous labile organic matter. In JAR2, the sediment mixtures were split into three batches; the first received 2.1% wet weight fecal pellets (FP, about 1% dry weight), the second received 1.1% wet weight chopped green leaves (see later) (GL, about 1% dry weight), and the third was maintained as an unamended control. In JAR3, the sediment mixtures was split into four batches; the first received 1.0% wet weight fecal pellets (FP, about 0.5% dry weight), the second received 0.6% wet weight chopped green leaves (GL, about 0.5% dry weight), the third received 0.7% wet weight water-leached and chopped green leaves (see later) (WL, about 0.5% dry weight), and the fourth was maintained as an unamended control.

Chopped leaves added to the sediment mixtures were treated as follows: Green leaves (GL) - fresh leaves were rinsed briefly in seawater, cut into small pieces (less than 3 x 3 mm), and rinsed briefly in seawater. Water-leached leaves (WL) - fresh leaves were cut into small pieces (less than 3 x 3 mm), soaked in seawater (22.5 ml per g wet wt leaf material) for 24 h, and rinsed in seawater. The water used in the leaching procedure was stored frozen for later DOC analysis. All the pretreated leaf materials were stored frozen until use in the jar experiments. The leaves were chopped to facilitate the comparison between microbial attack on leaf material and fecal pellets because in the latter case, leaf material was macerated during crab feeding.

The homogeneous sediment mixture from each batch was transferred into a number of 20-ml glass scintillation vials (8, 30, and 20 vials in JAR1, JAR2, and JAR3, respectively), which were capped with no head space, taped to prevent oxygen intrusion and incubated in the dark at 28°C. Two vials (denoted "jars") from each depth were sacrificed at 3–7 d intervals for determination of porewater DIC (TCO_2), SO_4^{2-}, DOC, and NH_4^+ concentrations. Porewater was extracted from the jars by centrifugation at 3,000 rpm for 15 min. After GF/F filtering, DIC was analyzed immediately by the flow injection-diffusion cell technique of Hall and Aller (1992), and SO_4^{2-}, DOC, and NH_4^+ subsamples were stored frozen for subsequent analysis. SO_4^{2-} was measured by ion chromatography with a Dionex auto-suppressed anion system (IonPac AS4A-SC column and ASRS suppressor). DOC was quantified by a Shimadzu TOC-5000 Total Organic Carbon Analyzer after acidification with 2 M HCl (pH < 2) to remove dissolved inorganic carbon. NH_4^+ was analyzed manually by the salicylate-hypochlorite method (Bower & Holm-Hansen 1980).

The rates of production or consumption of a particular solute was calculated based on the slopes of regressions between solute concentration and time. Only periods of linear concentration change with time were used subsequently.

Solid Phase Characteristics

Cores taken at station MC with 5 cm id. and 25 cm long core liners were used for determination of sediment characteristics. The cores were cut into the same depth intervals as in JAR1. Water content of the sediment sections and of samples of leaf and fecal materials were determined as the weight loss after drying at 105°C for 12 h. Sediment porosity was calculated from wet density and water content. Subsamples of the 105°C dried materials were analyzed for particulate organic carbon (POC) and nitrogen (PON) according to Kristensen and Andersen (1987) using a Carlo Erba CHNS analyzer. Protein content of the added plant materials and fecal pellets was analyzed spectro-photometrically by the Coomassie blue-technique after sodium hydroxide digestion (Bradford 1976; Mayer et al. 1986).

Results

Characteristics of Sediment, Leaf, and Fecal Materials

The sediment at station MC was composed of grey-brown silt (50% of particles < 63 μm) with a median particle size of about 70 μm down to at least 20 cm depth (E. Kristensen unpublished data). Porosity varied from 0.8 at the surface to a constant level of 0.7 below 3 cm depth. The anoxic subsurface sediment was never black and sulfidic. The organic carbon (POC) content varied between 2.3 and 2.8 mmol (g dw)$^{-1}$ with no specific depth pattern (Table 1). The organic nitrogen (PON) content, on the other hand, decreased by about 40% (p < 0.01) from the surface to 14–16 cm depth, causing the sedimentary C:N ratio to increase from 19 to 30.

Fecal pellets contained more water (80 %) than did green leaves (64 %) and sediment (49–58%), whereas the leaching procedure increased the water content of leaves to about 72% (Table 1). Fecal pellets contained only about 80% on dry weight basis and 44% on wet weight basis (p < 0.01),

Table 1. Water, particulate organic carbon (POC), and nitrogen (PON) contents of sediment, leaf, and fecal materials used in the three jar experiments. Sediment values are given as mean ± sd (n = 3–4). Leaf and fecal values are given as mean ± range of two determinations.

	Water (%)	POC (mmol (g dw)$^{-1}$)	PON (mmol (g dw)$^{-1}$)	Protein (mg (g dw)$^{-1}$)	POC:PON (mol)
Sediment (0–2 cm)	57.9 ±1.1	2.78 ± 0.19	0.143 ± 0.018	-	19.4
Sediment (5–7 cm)	49.0 ± 0.9	2.30 ± 0.13	0.097 ± 0.010	-	23.7
Sediment (14–16 cm)	49.2 ± 2.5	2.51 ± 0.14	0.083 ± 0.009	-	30.2
Green leaves	63.5 ± 1.5	38.92 ± 0.25	1.079 ± 0.064	60.6± 0.1	36.1
Fecal pellets	80.3 ± 0.3	30.83 ± 0.08	0.871 ± 0.050	31.9 ± 2.2	35.4
Leached leaves	72.3	37.25	1.257	67.9 ± 2.1	29.6

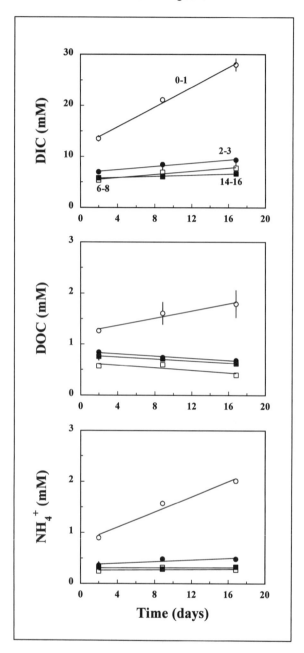

Fig. 2. Examples of changes in concentration of porewater DIC (upper), DOC (middle), and NH_4^+ (lower) with time in JAR1 (January incubations). Each data point represent the mean ± range of duplicate jars. Numbers in upper panel indicate sediment depth intervals in centimeters. Lines are drawn according to least squares linear regression.

respectively, of the POC and PON content of green leaves, which were the food source for the crabs. However, the C:N ratio of both materials remained almost the same, 36.1 and 35.4, respectively. The almost two times higher protein content of green leaves than of fecal pellets, suggested that a large fraction of the nitrogen in fecal pellets was of nonprotein origin. The water-leaching procedure seemed to reduce the POC content of dried leaves by about 4%, but increased PON and protein content by about 17% and 12%, respectively, thus reducing the C:N ratio to about 30. However, when the data from leached leaves were extrapolated to wet weight, the former contained only 73% of the carbon, 88% of the nitrogen, and 85% of the protein in green leaves. Accordingly, they lost 3.9 mmol C and 46 µmol N per g wet wt. (C:N = 85) during the 24-h leaching period. About 82% of the lost nitrogen was protein-N (assuming that 16% of the protein weight is nitrogen). The recovered DOC in the water after leaching accounted for 2.65 mmol per g wet wt., equivalent to 19% of the initial leaf carbon.

REACTION RATES IN UNAMENDED SEDIMENT (JAR1)

The concentration change appeared to be resonably linear for all solutes in JAR1 from day 2 to day 17 (Fig. 2). The highest rates of carbon and nitrogen reactions at station MC always occurred in the surface layer of the sediment. Since there was no significant seasonal difference, the data from the two sampling periods were pooled and are presented in Fig. 3. Net DIC production in the 0–1 cm depth interval was more than 6 times higher (p < 0.01) than the rates in the deeper sediment layers (Fig. 3). For DOC the reaction in both seasons changed direction

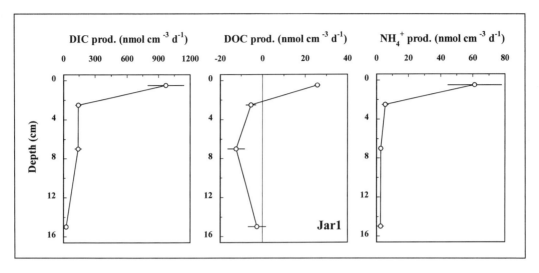

Fig. 3. Rates of DIC (left), DOC (middle), and NH_4^+ (right) production with depth in the sediment at station MC. Data are given as mean values (± range) of JAR1 results from the dry (January) and the wet (August) season.

(p < 0.01), from a production in the 0–1 cm interval to an uptake with a rate of about one third below this depth. The decrease in NH_4^+ production from the 0–1 cm interval to the sediment below was more dramatic (a factor of at least 11) than for DIC (Fig. 3). The C:N ratios of dissolved mineralization products increased from 16 to 53 with depth from 0–1 cm to 6–8 cm.Below this depth C:N decreased to about 10.

REACTION RATES IN AMENDED SEDIMENT (JAR2, JAR3)

The concentration changes were linear with time for all solutes in the unamended control jars (C) in JAR2 and JAR3 (30-d and 20-d periods, respectively), indicating constant reaction rates when no organic matter is added (Figs. 4 & 5). The reaction rates of DIC and NH_4^+ in control jars (C) were twice as fast (p < 0.01) in JAR3 than JAR2 (Table 2); the former with rates similar to JAR1 in the upper layer and the latter with rates similar to JAR1 in the deep layer. Furthermore, the production rates of DIC and NH_4^+ were generally 2–3 times higher in the 0–2 cm than the 5–7 cm layer (p < 0.01). DOC was produced at low rates in all control jars, less than 20% of the DIC production rates. Sulfate reduction (Figs. 4 & 5) appeared to be the predominant source for DIC in all control jars as indicated by C:S ratios between DIC production and SO_4^{2-} removal of 1.64–2.00.

In the amended jars the concentration changes were more dramatic and in some cases exhibited two different linear phases. Sulfate reduction was the primary source of DIC; the temporal patterns of SO_4^{2-} (Figs. 4 & 5) were almost complete mirror images of DIC, with a mean C:S ratio of 2.3 ± 0.9. The concentration change of carbon solutes (DIC and DOC) was rapid and linear during the initial 12–16 d (phase 1) in both the leaf-amended (GL) and fecal-amended (FP) JAR2 incubations (Fig. 4). Subsequently, the rates ceased (DIC, p < 0.001)) or changed direction (DOC, p < 0.05) and exhibited a new linear pattern to the end (phase 2). Only the leaf-amended (GL) treatment in the JAR3 experiment showed a significant (p < 0.001) two-phase pattern (Fig. 5). Phase 1 for DIC and DOC lasted only 7 d in the 0–2 cm layer, whereas this phase appeared to last for 13 d for DIC in the 5–7 cm layer. Unfortunately, no DOC results are available for the 5–7 cm layer in JAR3 due to analytical problems.

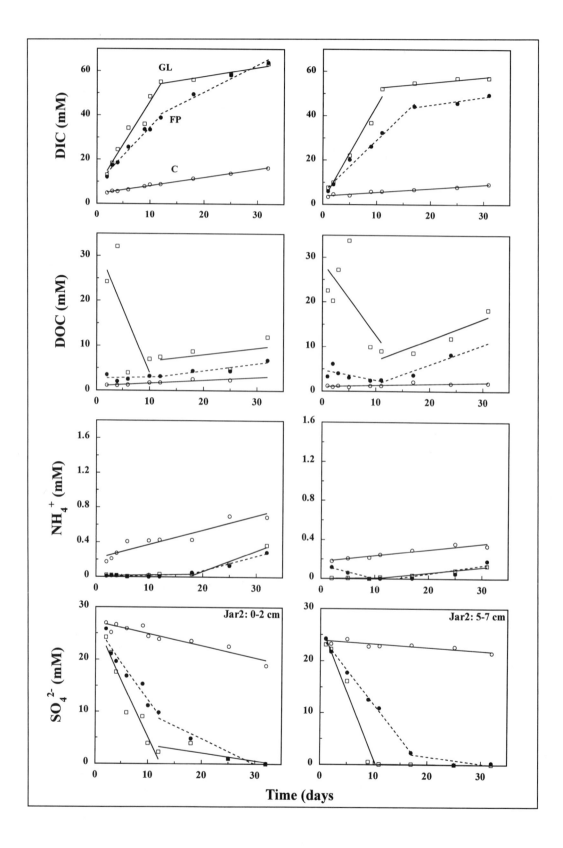

Fig. 4. Temporal changes in concentration of porewater DIC, DOC, NH_4^+, and SO_4^{2-} in JAR2. The left panel is based on sediment from the upper 0–2 cm and the right panel represent sediment from the 5–7 cm depth interval. Treatments are C – control, GL – green leaf addition, and FP – fecal pellet addition. Data points represent the mean of duplicate jars. The range (not shown) was generally within 10% of the mean. Lines are drawn according to least squares linear regression.

Rates of DIC production during phase 1 of GL treatments in JAR2 and JAR3 was 8 to 24 times higher ($p < 0.001$) than the controls (C) (Table 2). During phase 2 the production of DIC was reduced to a level similar to (JAR2, $p > 0.10$) or below (JAR3, $p < 0.001$) the controls (C). The low carbon mineralization during this phase was substantiated by the low and almost constant concentration of SO_4^{2-} (< 5 mM in JAR2 and 5–10 mM in JAR3). The DIC production during phase 1 of FP treatments in JAR2 was only 66% (0–2 cm) and 55% (5–7 cm) of the rates in the GL treatment ($p < 0.01$; Table 2). The DIC production during phase 2 was reduced 48% and 15%, respectively, of those in phase 1 ($p < 0.001$). The low carbon mineralization rates during this phase in FP-amended JAR2 was associated with diminished uptake and partial depletion of SO_4^{2-}. Jars amended with fecal pellets (FP) and leached leaves (WL) in JAR3 exhibited similar and constant DIC production throughout the experimental period (Table 2) but at rates of only about 30% (0–2 cm, $p < 0.001$) and 57% (5–7 cm, $p < 0.01$) of those in phase 1 of GL jars.

The net consumption of DOC during phase 1 of GL treatments was high (from high initial concentrations of 25–30 mM, Figs. 4 & 5) in both JAR2 and JAR3 incubations (Table 2). The DOC consumption was not significantly difference from the DIC production in the 0–2 cm layer. In the 5–7 cm layer, however, the loss of DOC during phase 1 (JAR2) only accounted for one third of the DIC evolution ($p < 0.01$). During phase 2 of the GL treatments the production of DOC was inversely correlated with the DIC production. DOC was produced or consumed at low rates in the FP and WL treatments, with rates generally not significantly different from the controls (C). The relatively high production rates during phase 2 of FP treatments in JAR2 ($p < 0.05$) exhibited a similar inverse relationship between DOC and DIC production as found for the GL treatment.

The concentration changes of NH_4^+ in amended jars were less dramatic, but more erratic, than for the carbon species (Figs. 4 & 5). NH_4^+ was practically zero during the first 11–18 d in phase 1 of the amended treatments in JAR2, except for a consumption from 0.1 mM to 0 in FP-amended jars at 5–7 cm (Table 2). Subsequently, NH_4^+ increased linearly (phase 2) to the end at rates similar to those in control jars (C). Samples for NH_4^+ analysis from JAR3 were lost after d 14, which hampered the long-term evaluation of this compound. Concentrations were generally high in the 0–2 cm layer, but there was a great deal of scatter in the data and no particular trend (i.e., rates not significantly different from zero). However, in the GL and FP treatments of the 5–7 cm layer in JAR3, NH_4^+ showed a rapid consumption from 0.1–0.2 mM to 0 within the first 2–3 d, with no subsequent changes, whereas a more gradual decline occurred in the WL treatment, reaching 0 at d 14 (Fig. 5 & Table 2).

Discussion

The present study compares biogeochemical responses and reaction kinetics after addition to mangrove sediment of two different food sources and fecal pellets of sesarmid crabs. The experimental set-up (jar) was designed to remove the natural variability caused by factors such as macrofauna bioturbation and roots of vascular plants and is not an attempt to simulate in situ conditions. Mixing and homogenization may temporarily destroy the natural chemical, physical, and biological structure of sediments, but it has been shown (Sun et al. 1991) that effects of such manipulations are short term (days) relative to an incubation time of weeks. This is also substantiated in the present study by the linearity of

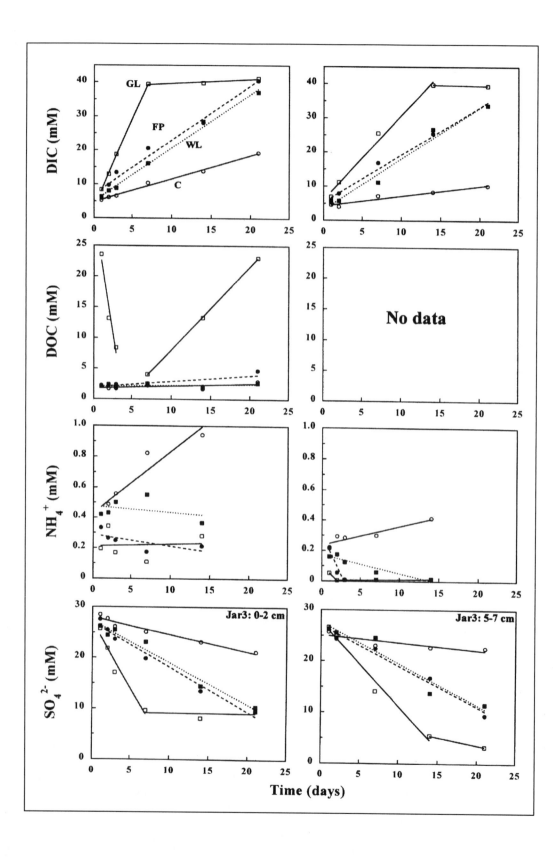

Fig. 5. Temporal changes in concentration of porewater DIC, DOC, NH_4^+, and SO_4^{2-} in JAR3. The left panel is based on sediment from the upper 0–2 cm and the right panel represent sediment from the 5–7 cm depth interval. Treatments are C – control, GL – green leaf addition, FP – fecal pellet addition, and WL – water-leached leaf addition. Data points represent the mean of duplicate jars. The range (not shown) was generally within 10% of the mean. Lines are drawn according to least squares linear regression.

concentration changes with time in all unamended jars (Figs. 2, 4, & 5). Others have shown that the jar approach provides a measure of reaction rates that is comparable with sediment-water fluxes and sulfate reduction assays in a variety of marine sediments, ranging from deep oceanic silts (Canfield et al. 1993; Kristensen et al. 1999) to shallow coastal sands (Kristensen & Hansen 1995) and muds (Holmer & Kristensen 1994).

At first glance, the depth-integrated areal rate of DIC production in the JAR1 experiment (29 mmol m^{-2} d^{-1}) is low compared with measured sediment-water fluxes (53.0 mmol m^{-2} d^{-1}, Kristensen et al. 2000). However, when the large influx of DOC (-20 mmol m^{-2} d^{-1}, Kristensen et al. 2000) across the sediment-water interface is included in the carbon budget, the two measures compare well. Accordingly, carbon transformations are very dynamic in the surface layer of this mangrove sediment; as is also evident from the rapidly decreasing vertical reaction profiles (Fig. 3). Depth-integrated sulfate reduction (measured by the $^{35}SO_4^{2-}$ injection technique, Jørgensen 1978), converted to DIC by multiplying with 2, only accounted for 15–25% (Kristensen et al. 2000) of the depth-integrated jar rates. The apparently dominant role of the oxidized surface layer in total carbon mineralization in this mangrove sediment is justified by the fact that the volume-specific sulfate reduction rates are similar when measured by the $^{35}SO_4^{2-}$ injection and the jar technique in layers deeper than 3 cm. All mineralization processes in the closed anaerobic jars, even in the otherwise oxidized surface layer, are due to sulfate reduction; but anoxic jar mineralization of fresh organic matter is likely to be as fast as oxic and suboxic mineralization (Kristensen et al. 1995). No comparison can be made between jar rates and fluxes of NH_4^+ because nitrification affects the budget in the latter, but not the former, approach.

The high concentration of DOC observed initially in jars amended with green leaves (GL) was caused by rapid leaching. It should be noted, however, that the leaching process may be enhanced by the leaf chopping procedure. By extrapolating the porewater DOC of GL treatments to day 0 using the linear regression during phase 1, carbon initially leached as DOC was estimated to account for 12.5% (JAR2) and 21.2% (JAR3) of the added green leaf carbon. For comparison, the water-leached (WL) leaves lost 19–27% of their carbon content as DOC during the 24-h leaching period but apparently nothing further in jars when it was mixed into the sediment. DOC loss during the chopping and rinsing procedure may be responsible for the lower values estimated from GL jars. Other studies have shown that leaves of mangrove trees generally lose 30–40% of their weight by leaching after being submerged in seawater (Benner et al. 1986; Camilleri & Ribi 1986). Living mangrove leaves contain about 50% lignocellulose, a highly refractory structural complex consisting of the aromatic heteropolymer lignin, in close physical and covalent association with the polysaccharides, cellulose, and hemicellulose (Benner & Hodson 1985). The nonlignocellulose components of the leaves are mostly soluble carbohydrates that rapidly leach from the leaves after submersion in water (Neilson & Richards 1989). However, tannins and other phenolic compounds with microbial inhibitory potential may account for a significant fraction (18%) of the dissolved organic matter in mangrove leachate (Benner et al. 1986).

The DOC concentration in jars treated with fecal pellets (FP) were similar to or only slightly higher than in control jars (C), indicating that the maceration during ingestion and passage through crab intestines must have removed the leachable fraction. Accordingly, assimilation efficiencies reported for sesarmid

Table 2. Dissolved carbon (DC = DIC and DOC) and NH_4^+ evolution in anaerobic mangrove sediment jars enriched with green leaves of *Rhizophora apiculata* (GL), fecal pellets of the sesarmid crab *Neoepisesarma versicolor* (FP), and water-leached leaves of *Rhizophora apiculata* (WL). C indicates unamended control jars. Results are given from JAR2 and JAR3, with representative data from JAR1. For GL and FP (only JAR2) the incubation periods are separated in two linear phases (subscript 1 and 2) according to the dramatic temporal changes in concentration pattern. Rates ± se (nmol cm^{-3} d^{-1}) are obtained by least squares linear regression.

		0–2 cm			5–7 cm		
		JAR1	JAR2	JAR3	JAR1	JAR2	JAR3
DIC	C	556 ± 100	294 ± 7	533 ± 23	141 ± 32	137 ± 12	231 ± 32
	GL₁	–	3,071 ± 302	4,063 ± 62	–	3,352 ± 269	1,940 ± 154
	GL₂	–	319 ± 68	90 ± 31	–	191 ± 46	-37 ± ?
	FP₁	–	2,030 ± 184	1,251 ± 92	–	1,844 ± 115	1,092 ± 69
	FP₂	–	969 ± 112	–	–	282 ± 93	–
	WL	–	–	1,233 ± 42	–	–	1,137 ± 102
DOC	C	10 ± 2	51 ± 9	29 ± 18	-12 ± 4	24 ± 7	–
	GL₁	–	-2,217 ± 1,528	-5,902 ± 1,268	–	-1,272 ± 691	–
	GL₂	–	114 ± 170	1,057 ± 13	–	365 ± 116	–
	FP₁	–	24 ± 61	75 ± 31	–	-176 ± 100	–
	FP₂	–	127 ± 38	–	–	346 ± 104	–
	WL	–	–	13 ± 15	–	–	–
NH₄⁺	C	33.3 ± 9.4	14.8 ± 2.1	36.2 ± 6.5	2.7 ± 1.5	5.4 ± 0.7	11.2 ± 2.6
	GL₁	–	1.1 ± 0.5	1.3 ± 8.9	–	0.6 ± 0.8	0.0 (-43.1)[1]
	GL₂	–	21.3 ± 4.2	–	–	5.2 ± 0.6	–
	FP₁	–	1.7 ± 0.9	-6.8 ± 4.3	–	-11.9 ± 2.1	0.0 (-94.4)[2]
	FP₂	–	15.2 ± 2.8	–	–	7.6 ± 2.7	–
	WL	–	–	-3.7 ± 6.8	–	–	10.8 ± 2.0

[1] Number in bracket indicate the rate obtained during the initial 2 d of incubation.
[2] Number in bracket indicate the rate obtained during the initial 3 d of incubation.

crabs eating mangrove leaves (35–43%, Kofoed et al. 1985; 35%, Giddins et al. 1986; 50% , Lee 1997), are less than twice the loss found by water-leaching. Accordingly, Neilson and Richards (1989) reported a significant loss (7–10%) of carbohydrates from the insoluble fraction of *Ceriops tagal* leaves by cellulase activity during passage through the gut of the sesarmid crab *Neosarmatium smithi*. *N. versicolor* may also digest a fraction of the structural components, but, as shown here, they primarily assimilate the leachable fraction. It should be noted that DOC may be lost by leaching from the intact leaves offered to the crabs in the experimental aquaria prior to ingestion and lost from the mandibles by "sloppy feeding" (Camilleri 1989). The nutritive value of mangrove leachates is substantiated by the results of Camilleri and Ribi (1986), who found that particulate matter formed from the leachates is readily utilized as food by copepods, amphipods, isopods, crabs, and shrimps.

The distinct two-phase concentration pattern with time in jars amended with green leaves (GL) and with fecal pellets (FP) in JAR2 indicates a dramatic change in the chemical and biological conditions (Figs. 4 & 5). The simultaneous rapid DOC loss (from 30 mM to less than 10 mM within 10 d) and high DIC production during phase 1 in GL jars was probably caused by rapid mineralization of leached DOC in the anaerobic sediment. A similar rapid degradation of mangrove leachate under aerobic conditions, with a high conversion efficiency into microbial biomass of 64–94%, was found by Benner et al. (1986). The similarity of DOC consumption and DIC production in the 0–2 cm interval in phase 1 of GL jars suggested that all DOC was mineralized. The calculated DOC consumption may, however, be underestimated by simultaneous, but unaccounted for fermentative DOC production from the particulate pool. Anyway, Linley and Newell (1984) and Benner et al. (1988) have reported conversion efficiencies as low as ~10% and ~30%, respectively, for microbial communities degrading *Spartina* detritus in marine sediments. That microbial growth actually occurred is evident from the rapid initial consumption and generally low concentrations of NH_4^+ in all jars (Figs. 4 & 5). The low availability of NH_4^+ and the high C:N ratio of the organic pools suggests, in accordance with Boto and Wellington (1983), that nitrogen probably was a limiting factor for microbial growth in the present mangrove sediment.

It is rather surprising that sulfate reducers were able to mineralize the leached DOC instantly. Other studies have indicated that sulfate reducers need a lag or adaptation time of about 14 d before DOC leached from added plant materials is consumed (e.g., Kristensen & Hansen 1995). One explanation for this discrepancy could be conditioning of the microbial community. When detritus-associated microorganisms have been conditioned to DOC of a specific composition, mineralization will occur instantly, whereas new microbial strains or enzyme composition must develop to a certain level before a new substrate will be mineralized. The microbial community in the sediment used here has most likely been exposed to DOC originating from leached leaves and fecal pellets in the recent past and the microorganisms should then be ready to mineralize this substrate instantly. Benner et al. (1986) suggested that conditioning also allowed microorganisms from mangrove sediments to utilize leachates with high concentrations of potentially inhibitory phenolic compounds.

Although no net leaching was observed during phase 1 in the sediment amended with fecal pellets and leached leaves, the DIC production was 50–70% of that in the GL sediment, indicating a close coupling between fermentative DOC production and mineralization to DIC in these sediments (Kristensen & Hansen 1995). During phase 2 in the GL and partly in the FP (JAR2) sediment, the fermentative production of DOC and NH_4^+ was significant because microbial respiration (i.e., sulfate reduction) and growth ceased, probably due to sulfate limitation. In JAR2 the DIC + DOC accumulation in phase 2 was 1.2–2.0 times higher in FP than GL jars, suggesting that the POC of the FP substrate after 10-d burial in the sediment was fermented faster to DOC than in the GL substrate (Table 2). In JAR3 (0–2 cm), where sulfate limitation (phase 2) only occurred in the GL sediment, the reactivity to fermenters (DOC + DIC production) of the FP substrate was only about 10% higher than GL and WL. It should be noted here that the initial chopping of GL and WL materials does not fully simulate the maceration performed by crabs during ingestion. The handling and grinding by crab mandibles increases the relative surface area of the leaf material in the gut and deposited fecal pellets and thus the area available for microbial attack.

The low net production of dissolved NH_4^+ during phase 2 in GL and FP treatments of JAR2 as depicted from (DIC + DOC):NH_4^+ ratios of 20.3–106.9 and 72.1–245.6 suggests that microbial assimilation of NH_4^+ still remained active. The FP sediment appeared to be more devoid of reactive nitrogen initially than the GL sediment, which probably is a consequence of efficient absorption of available leaf protein during passage through crab intestines. Preferential nitrogen retention of detritus has previously been observed for a variety of marine invertebrates (Newell 1965; Johannes & Satomi 1966; Kofoed 1975; Anderson 1994). Lee (1997), on the other hand, found that the crab *Sesarma messa* assimilates carbon in preference for nitrogen when feeding on *Rhizophora stylosa* leaves. In any case, consumption of green leaf material by crabs would not supply the C:N ratio of 17 in the diet necessary to maintain animal tissue (Russel-Hunter 1970) because the leaves offered had a C:N ratio of about 36 (Table 1). Under natural conditions the nitrogen required by crabs is likely to come from microorganisms that they remove from the mud surface by picking detrital material (Robertson 1986). Alternatively, crabs may obtain nitrogen by selectively feeding on decaying leaves (at the sediment surface) with a C:N ratio < 25 (Lee 1989).

Based on the recovery of carbon in the form of DIC, ingestion and fecal pellet deposition of mangrove leaves by sesarmid crabs generally reduced the initial mineralization of litter carbon by 25–50%. However by subtracting the contribution of initial DOC leaching, the loss of particulate carbon from fecal material in FP jars was more than twice that from green leaves in GL jars during both phase 1 and 2. After digestion in the crab intestine, the fecal material has a C:N stoichiometry similar to leached leaves but a slightly higher decomposition potential. Fecal pellets have long been known to be important as sites for microbial activity. Hargrave (1976) reported that egested fecal material of marine invertebrates is rapidly colonized by microorganisms, with maximum degradation occurring within the first 5 d. This generalization includes primarily species from shallow coastal areas in the temperate zone feeding on more labile plant materials (e.g., algae) and not sesarmid crabs living on a mangrove leaf diet. Anyway, the deposition of fecal pellets of sesarmid crabs to the mangrove sediment increases particulate carbon decomposition by a factor of 3–10 above the background activity and up to 2 times higher than leached green leaves deposited in the sediment. Maceration during ingestion, passage through the gut, and contact with digestive enzymes therefore appears to facilitate microbial attack on the insoluble carbohydrates.

In conclusion, sesarmid crabs are important for the nutrient retainment in mangrove environments by their leaf-eating habit and subsequent fecal pellet deposition to the sediment. By eating green leaves in the tree canopy, crabs retain leaf POC and assimilate DOC and probably DON, which would otherwise have been lost to the ocean via tidal currents. Crabs appear to assimilate primarily the leachable fraction of leaves. The decomposition pattern of fecal pellets and leached leaves is remarkably similar, whereas the leaching of DOC from newly detached leaves enhances the initial decomposition dramatically. The rate of leaf POC loss in the sediment is enhanced after maceration and passage through crab guts compared with chopped intact leaves.

Acknowledgments

We are grateful to the staff af PMBC for providing facilities and invaluable assistance during this study. We thank H. Brandt for technical assistance. This work was supported by DANIDA (SCP-project) and SNF (grant #9601423).

Literature Cited

Alongi, D. M. 1989. The role of soft-bottom benthic communities in tropical mangrove and coral reef ecosystems. *Reviews in Aquatic Sciences* 1:243–280.

Anderson, T. R. 1994. Relating C:N ratios in zooplankton food and faecal pellets using a biochemical model. *Journal of Experimental Marine Biology and Ecology* 184:183–199.

Benner, R. & R. E. Hodson. 1985. Microbial degradation of the leachable and lignocellulosic components of leaves and wood from *Rhizophora mangle* in a tropical mangrove swamp. *Marine Ecology Progress Series* 23:221–230.

Benner, R., J. Lay, E. K'ness, & R.E. Hodson. 1988. Carbon conversion efficiency for bacterial growth on lignocellulose: Implications for detritus-based food webs. *Limnology and Oceanography* 33:1514–1526.

Benner, R., E. R. Peele, & R. E. Hodson. 1986. Microbial utilization of dissolved organic matter from leaves of the red mangrove, *Rhizophora mangle*, in the Fresh Creek estuary, Bahamas. *Estuarine, Coastal and Shelf Science* 23:607–619.

Boto, K. G. & J. S. Bunt. 1981. Tidal export of particulate organic matter from a northern Australian mangrove system. *Estuarine, Coastal and Shelf Science* 13:247–255.

Boto, K. G. & J. T. Wellington. 1983. Phosphorus and nitrogen nutritional status of a northern Australian mangrove forest. *Marine Ecology Progress Series* 11:63–69.

Bower, C. E. & T. Holm-Hansen. 1980. A salicylate-hypochlorite method for determining ammonia in seawater. *Canadian Journal of Fisheries and Aquatic Science* 37:794–798.

Bradford, M. M. 1976. A rapid and sensitive method for the quantification of microgram quantities of protein utilizing the principle of protein-dye binding. *Analytical Biochemistry* 72:248–254.

Canfield, D. E., B. Thamdrup, & J. W. Hansen. 1993. The anaerobic degradation of organic matter in Danish coastal sediments: Fe reduction, Mn reduction and sulfate reduction. *Geochimica et Cosmochimica Acta* 57:3867–3883.

Chansang, H. & S. Poovachiranon. 1990. The fate of mangrove litter in a mangrove forest on Ko Yao Yai, southern Thailand. *Phuket Marine Biological Research Bulletin* 54:33–46.

Camilleri, J. 1989. Leaf choice by crustaceans in a mangrove forest in Queensland. *Marine Biology* 102:453–459.

Camilleri, J. C. & G. Ribi. 1986. Leaching of dissolved organic carbon (DOC) from deal leaves, formation of flakes from DOC, and feeding on flakes by crustaceans in mangroves. *Marine Biology* 91:337–344.

Giddins, R. L., J. S. Lucas, M. J. Neilson, & G. N. Richards. 1986. Feeding ecology of the mangrove crab *Neosarmatium smithi* (Crustacea: Decapoda: Sesarmidae). *Marine Ecology Progress Series* 33:147–155.

Hall, P. O. J. & R. C. Aller. 1992. Rapid,small-volume flow injection analysis for ΣCO_2 and NH_4^+ in marine and fresh waters. *Limnology and Oceanography* 37:1113–1118.

Hargrave, B. T. 1976. The central role of invertebrate faeces in sediment decomposition, p. 301–321. *In* Anderson, J. M. & A. Macfadyen (eds.), The Role of Terrestrial and Aquatic Organisms in Decomposition Processes. Blackwell Scientific Publications, Oxford.

Holmer, M. & E. Kristensen 1994. Coexistence of sulfate reduction and methane production in an organic-rich sediment. *Marine Ecology Progress Series* 107:177–184.

Johannes, R. E. & M. Satomi. 1966. Composition and nutritive value of fecal pellets of a marine crustacean. *Limnology and Oceanography* 11:191–197.

Jørgensen, B. B. 1978. A comparison of methods for the quantification of bacterial sulfate reduction in coastal marine sediments. *Geomicrobiology Journal* 1:11–27.

Kofoed, L. H. 1975. The feeding biology of *Hydrobia ventrosa* (Montagu). I. The assimilation of different components of the food. *Journal of Experimental Marine Biology and Ecology* 19:233–241.

Kofoed, L. H., S. Madsen, K. Olsen, J. Dalsgaard, & M. B. Jørgensen. 1985. The role of sesarmid crabs in the breakdown of mangal leaves. Report of the experimental work of the tropical marine biology study group, Odense University, Odense.

Kristensen, E. & F. Ø. Andersen. 1987. Determination of organic carbon in marine sediments: Comparison of two CHN-analyzer methods. *Journal of Experimental Marine Biology and Ecology* 109:15–23.

Kristensen, E., F. Ø. Andersen, M. Holmboe, M. Homer, & N. Thongtham. 2000. Carbon and nitrogen mineralization in sediments of the Bangrong mangrove area, Phuket, Thailand. *Aquatic Microbial Ecology* 22: 199–213.

Kristensen, E., S. I. Ahmed, & A. H. Devol. 1995. Aerobic and anaerobic decomposition of organic matter in marine sediment: Which is fastest? *Limnology and Oceanography* 40:1430–1437.

Kristensen,E., A. H. Devol, & H. E. Hartnett. 1999. Carbon and nitrogen diagenesis in sediments on the continental slope of the eastern tropical and temperate North Pacific. *Continental Shelf Research* 19: 1331–1351.

Kristensen, E. & K. Hansen. 1995. Decay of plant detritus in organic-poor marine sediment: Production rates and stoichiometry of dissolved C and N compounds. *Journal of Marine Research* 53:675–702.

Lee, S. Y. 1989. The importance of sesarminae crabs *Chiromanthes* spp. and inundation frequency on mangrove (*Kandelia candel* (L.) Druce) leaf litter turnover in a Hong Kong tidal shrimp pond. *Journal of Experimental Marine Biology and Ecology* 131:23–43.

Lee, S.Y. 1997. Potential trophic importance of the faecal material of the mangrove sesarmid crab *Sesarma messa*. *Marine Ecology Progress Series* 159:275–284.

Linley, E. A. S. & R. C. Newell. 1984. Estimates of bacterial growth yields based on plant detritus. *Bulletin of Marine Science* 35:409–425.

Mayer, L. M., L. L. Schick, & F. W. Setchell. 1986. Measurement of protein in nearshore marine sediments. *Marine Ecology Progress Series* 30:159–165.

McIvor, C. C. & T. J. Smith. 1995. Differences in the crab fauna of mangrove areas at a southwest Florida and a northeast Australian location: Implications for leaf litter processing. *Estuaries* 18:591–597.

Neilson, M. J. & G. N. Richards. 1989. Chemical composition of degrading mangrove leaf litter and changes produced after consumption by mangrove crab *Neosarmation smithi* (Crustacea: Decapoda: Sesarmidae). *Journal of Chemical Ecology* 15:1267–1283.

Newell, R. 1965. The role of detritus in the nutrition of two marine deposit feeders, the prosobranch *Hydrobia ulvae* and the bivalve *Macoma balthica*. *Proceedings of the Zoological Society of London* 144:25–45.

Odum, W. E. & E. J. Heald. 1972. Trophic analysis of an estuarine mangrove community. *Bulletin of Marine Science* 22:671–737.

Robertson, A. I. 1986. Leaf-burying crabs: Their influence on energy flow and export from mixed mangrove forests (*Rhizophora* spp.) in northeastern Australia. *Journal of Experimental Marine Biology and Ecology* 102:237–248.

Robertson, A. I. & P. A. Daniel. 1989. The influence of crabs on litter processing in high intertidal mangrove forests in tropical Australia. *Oecologia* 78:191–198.

Robertson, A. I., D. M. Alongi, & K. G. Boto. 1992. Food chains and carbon fluxes, p. 293–326. *In* Robertson, A. I. & D. M. Alongi (eds.), Tropical Mangrove Ecosystems. American Geophysical Union, Washington, D.C.

Russel-Hunter, W. D. 1970. Aquatic Productivity: An Introduction to Some Basic Aspects of Biological Oceanography and Limnology. Collier-MacMillan, London.

Slim, F. J., M. A. Hemminga, C. Ochieng, N. T. Jannink, E. Cocheret de la Moriniere, & G. van der Velde. 1997. Leaf litter removal by the snail *Terebralia palustris* (Linnaeus) and sesarmid crabs in an east African mangrove forest (Gazi Bay, Kenya). *Journal of Experimental Marine Biology and Ecology* 215:35–48.

Sun, M., R. C. Aller, & C. Lee. 1991. Early diagenesis of chlorophyll-a in Long Island Sound sediments: A measure of carbon flux and particle reworking. *Journal of Marine Research* 49:379–401.

Twilley, R. R., A. E. Lugo, & C. Patterson-Zucca. 1986. Production, standing crop, and decomposition of litter in basin mangrove forests in southwest Florida. *Ecology* 67:670–683.

Twilley, R. R., M. Pozo, V. H. Garcia, V. H. Rivera-Monroy, R. Zambrano, & A. Bodero. 1997. Litter dynamics in riverine mangrove forests in the Guayas River estuary, Ecuador. *Oecologia* 111:109–122.

Feeding Processes of Bivalves:
Connecting the Gut to the Ecosystem

Jeffrey S. Levinton*, J. Evan Ward, Sandra E. Shumway,
and Shirley M. Baker

Abstract: *Bivalves exist in dense populations and may strongly affect the seston of coastal and estuarine water columns. Conversely, the seston content of such systems may affect the feeding behavior, particle processing, and digestive strategies of suspension feeders, both as individuals and within suspension-feeder communities such as oyster and mussel beds. It is an important objective to develop conceptual and mathematical models to describe both of these sets of interactions, but it is equally important to connect individual bivalve feeding selectivity and water column processes. Many models that consider feeding rates exist, often in the aquaculture literature, but the incorporation of selectivity needs to be addressed. In order to produce such models, we must know much more about particle selectivity and rates of particle processing in response to availability in the water column. We describe some methods and results demonstrating that (1) rate-limiting steps within bivalve feeding compartments might affect particle processing from the water column; (2) particle selectivity might strongly affect the composition of the seston in estuaries; and (3) resuspension of pseudofeces influences the role of bivalves in affecting the seston.*

Introduction

Our purpose in this essay is to argue the necessity for connecting the factors that determine particle transfer functions within the various pallial cavity compartments of a suspension-feeding bivalve (Levinton et al. 1996) with the factors that regulate the abundance and particle composition of the water column seston. Bivalve molluscs are efficient suspension feeders and can strongly modulate the seston of estuaries, in those with sluggishly circulating coastal waters (Dame 1993a,b) as well as vigorously circulated coastal waters that are nevertheless trapped in circulating cells close to shore by confinements such as offshore bars (McLachlan 1980). For example, very dense coastal diatom blooms off the coasts of Oregon, USA, and South Africa are set off by a special set of circumstances that include restricted circulation and bivalve excretion (Lewin et al. 1975; McLachlan 1980).

The intimate relationship between particle processing by suspension-feeding individuals and the changing composition of seston suggests a number of questions:

(1) Does systematic rejection-acceptance of particles (i.e., selectivity among particle types), combined with resuspension processes, strongly affect the composition of particle types in the water column?

(2) Within bivalves, does particle transfer between organs in the pallial cavity (gills, palps, digestive tract) create rate-limiting steps that result in feedback to the seston in the water column?

Some of these questions have already been asked of the relationships between zooplankton grazers and phytoplankton dynamics (e.g., Frost 1972, 1987; Steele 1974). Also, there is a rich literature attempting to connect bivalve feeding to total phytoplankton concentration, especially in the context of mariculture. While we discuss some of the literature next, our main focus in this paper is to present new empirical approaches to bivalve feeding selectivity and adjustments of feeding rate that can eventually be integrated into more standard models of bivalve total feeding effects on estuaries and long residence time coastal waters. Are bivalves restricted to just feeding at a certain rate? Do they respond to changes in the seston and does this in turn affect the water column? These are questions that can be answered with studies of bivalve functional ecology and modeling of estuarine systems.

Past Work and Models on Bivalve Feeding Effects on Ecosystems

The objective of bivalve mariculture is to grow as many bivalves in a confined space as fast as possible, while maintaining a convenient means of recovering animals for harvest. Models predicting carrying capacity contrast the filtration rates of bivalve populations with factors that control phytoplankton supply rate (Grant et al. 1993; Heral 1993; Prins et al. 1998; Smaal et al. 1998). This objective is concordant with the objective of understanding how a "natural bivalve population" might influence the seston of an estuary or how the seston might affect the bivalve population dynamics and individual growth (Cloern 1982; Officer et al. 1982; Herman 1993).

To assess the interaction of a bivalve population with an estuary one must develop an ecosystem-based budget that relates phytoplankton production, water exchange, and bivalve feeding rate (e.g., Grant et al. 1993; Herman 1993; Prins et al. 1998). The simplest model extrapolates individual feeding rate, pseudofeces production rate, and resuspension to ecosystem and population levels. When compared with phytoplankton growth rate, a steady-state bivalve population size can be calculated if we know individual feeding rate and conversion efficiencies, which will enable calculation of scope for growth as a measure of individual energy balance. The natural ecosystem analogue to this was pioneered by Officer et al. (1982). Equations were developed that predicted phytoplankton change as a function of bivalve population density, and bivalve population density as a function of mortality and feeding rate. While it is doubtful that this model is very useful for bivalve population growth rate, it was still an excellent formulation to understand the potential for bivalve effects on phytoplankton dynamics and for the potential of bivalve self-limitation by overfiltration. A number of mariculture-based models have been developed to predict the so-called carrying capacity of a localized area. Some models use relatively few variables to model bivalves and water-column organisms (Grant et al. 1993), while others are elaborate with as many as 60 parameters that require extensive measurement (Campbell & Newell 1998).

Even with such models, many complexities complicate a simple extrapolation of individual feeding rates to the ecosystem. For example bivalve beds produce depletion boundary layers (Wildish & Kristmanson 1984; Frechétte et al. 1989) and variations in water column stratification therefore require strong adjustments to such a simple model. Phytoplankton abundance varies seasonally in most locales, as does temperature. The concordance of a phytoplankton bloom with a seasonal larval release may strongly influence recruitment success.

It would be extremely useful if we could develop generalizations concerning the water exchange and phytoplankton dynamics that determine the importance of bivalve populations to the water column and vice versa. Dame and Prins (1998) employed Herman's (1993) simple model to characterize the carrying capacity of nearshore systems. They used three parameters to compare properties of estuaries (our notation differs somewhat):

Water residence time (T) is the time that it takes for the volume or mass of water, W, in a basin to be replaced by exchange at rate dW/dt with another body of water, such as open coastal waters.

$$T = \frac{W}{dW/dt} \qquad \text{(units: hours = g/g h}^{-1}\text{)}$$

Primary production time was defined (Dame & Prins 1998) as the ratio of yearly-averaged phytoplankton biomass (B) to primary production (P).

$$PPT = \frac{B}{P} \qquad \text{(units: hours = g/g h}^{-1} = \text{h)}$$

Finally, **bivalve population clearance time** CT (Smaal & Prins 1993) is the time needed to filter out one volume of the basin in question.

Assuming that phytoplankton production is the same inside and outside the water body in question, the carrying capacity clearance time, CT_K, can be related to the first two parameters:

$$CT_K \propto \frac{1}{PPT \times T}$$

As water residence time increases, phytoplankton influx from the outside will diminish, giving bivalves a greater opportunity to clear the water column. PPT and T are not independent of each other because as T decreases, the relative influence of PPT within the estuary diminishes. As PPT increases, the possibility for clearance will similarly increase. If open coastal water is relatively depleted in phytoplankton, then there will a complex interplay between water exchange from outside the estuary and phytoplankton growth within.

As mentioned, Clearance Time is probably a much more complex function of PPT, T, and other variables, as there are likely to be nonlinearities when relating clearance rates between large and small basins. Large basins (e.g., Chesapeake Bay) may have restricted circulation within the estuary and vertical stratification, making much of the phytoplankton inaccessible to bivalves on the bottom. Also, large basins probably have proportionally less water exchange with the open sea than smaller basins. Basins with relatively low primary production may find that bivalves reduce feeding rate as phytoplankton biomass falls below a threshold.

There are also, unfortunately, other parameters. For example, suitable habitat area should increase with basin size, most likely with area. If the semi-enclosed basin has a characteristic linear size scale, L, then the carrying capacity clearance time should be

$$CT_K \propto \frac{1}{L^2}$$

It is likely that larger basins have longer replenishment times: $T \propto L^2$, and therefore

$$CT_K \propto \frac{1}{T}$$

which might make *PPT* the only relevant parameter, because the effect of increasing bivalve space would be counteracted by the decreased replenishment of phytoplankton from the outside of the semi-enclosed basin.

Dame and Prins (1998) examined available data from a number of estuaries. While it is rather difficult to make effective comparisons, as localized effects (e.g., the very small oyster population in Chesapeake Bay) prevent the use of simple parameterization, they found that T is positively correlated with *PPT*. Large bivalve populations apparently thrive under conditions where $T < 2,400$ h and $PPT < 240$ h. My calculations suggest that there is a positive correlation between T and basin area, although Chesapeake Bay is off scale, perhaps owing to its large size and relatively restricted opening.

Two Connected Models

Modeling total feeding and the effects on total phytoplankton is important, but food quality is equally important to bivalves. Bivalve food quality must be described in terms of the composition of the seston, which includes both living and nonliving components. To connect bivalve functional morphology to processes affecting seston dynamics, we have constructed conceptual models to describe the dynamics of particle processing within a bivalve and within the water column. The challenge is then to connect the two systems, by theory and empirical study, to ask whether reciprocal effects are possible.

BIVALVE COMPARTMENT MODEL

Levinton et al. (1996) developed the beginnings of a compartment model for bivalve molluscs (Fig. 1a), which can be likened, with some modifications, to a box model in oceanography. In the bivalve compartment model, the input comes from the water column, with which an eventual connecting model must be made. Figure 1b shows examples of specific bivalve anatomies and their fit into the box model.

If the compartment model is a chain of reactions in equilibrium, then there should be a conservation of mass transfer from one compartment to the next. The rate of transfer from one box to the next (e.g., gill to palp) is

$$\frac{d \sum_i N_i}{dt} = \frac{\sum_i N_i}{T}$$

where N_i = the mass of particles of type i and T is the residence time of all particles in a compartment. We assume that all particles move in a single tract at the same velocity, and that the gill compartment therefore has a single residence time. This would apply to transport on the gills of *Mytilus* spp. (Ward et al. 1998a): particles are collected on the gill and transported ventrally. Alternatively, there may be some sorting within the gill, as occurs in oysters of the genus *Crassostrea* (Ward et al. 1998b) and in the zebra mussel, *Dreissena polymorpha* (Baker et al. 1998, 2000), and therefore different particle types may have differing residence times.

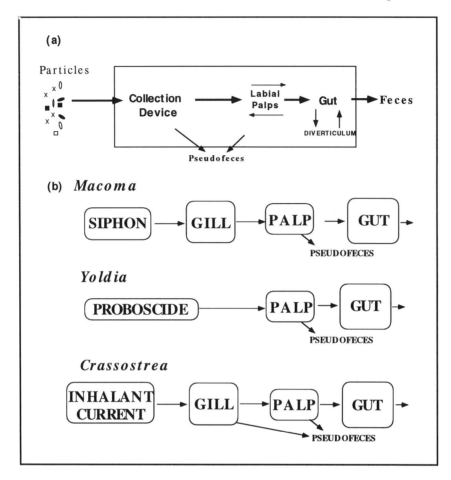

Fig. 1. (a) A bivalve compartment model, showing the important components and routes of transfer. (b) Three examples of the application of the compartment model to bivalves with differing anatomies and feeding styles (after Levinton et al. 1996).

This set of relationships would be in balance for the compartment model if the two degrees of freedom in a compartment–residence time and transfer rate to the next compartment–could be freely adjusted, or if a "downstream compartment" that is being overfilled or underfilled could exert feedback on "upstream" compartments. The simplest case would involve equality of residence times and transfer rates between compartments. Levinton et al. (1996), for example, demonstrated that residence times on gills and palps of *Macoma secta* are approximately equal, which implies that transfer rates are also equal. However, residence time in the gut is 60–240 times longer. To achieve balance with no other means implies that the mass of gut contents at steady state must be at least 60–240 times greater (or even more) than on the gills or palps (see Hylleberg & Gallucci 1975; Decho & Luoma 1991; Levinton et al. 1996). On the other hand, the equality of transfer from gills to palp and from palp to mouth suggests no feedbacks that cause cessation of particle transfer from gills to palps.

The balance model must accommodate at least two major sources of variability:

(1) Particle concentration entering the siphon may vary considerably, and individual species of suspension feeders appear to be adapted to variable particle concentrations (Iglesias et al. 1992).

(2) Food quality of the particles may vary, ranging from pure and readily digestible algae to virtually indigestible particles (Widdows et al. 1979; Berg & Newell 1986; Ward et al. 1998a,b), especially in nearshore habitats and in marsh creeks where relatively indigestible detrital particles and even toxic cells are mixed with far more digestible microorganisms.

Particle mixtures decline in food quality as the proportion of inert particles increases. Ingestion rate of the mussel *Mytilus edulis* increases with overall particle concentration, but then attains a plateau, suggesting a compensation of ingestion rate in response to high amounts of nonnutritious particles (Bayne et al. 1989). Such a plateau in ingestion may be compensated by the hard clam, *Mercenaria mercenarya,* by rejection of particles as pseudofeces or reduction of filtration rates (Bricelj & Malouf 1984). While cockles change filtration rate in response to changing particle concentrations of algae, they use production of pseudofeces to compensate for high concentrations of indigestible particles (Iglesias et al. 1992). While components of response have been studied in different species, it is likely that all suspension-feeding bivalves living in nearshore temperate environments have similar responses, with only the degree of response varying (Hawkins et al. 1998). Many nearshore bivalves can feed in the face of high particle concentrations, but the cockle *Cerastoderma edule* is particularly good at this, owing to its continual exposure to near-bottom resuspension of sediment (Navarro et al. 1998). At present it is not clear how various common bivalves can be ordered in ability to adjust to high particle loads or to select against nonnutritious particles.

The entire bivalve compartment system may be able to respond in unison to changes in seston concentration. For example, an optimal ingestion model suggests that ingestion rate should increase with increased food quality (Taghon 1981). As food quality increases (nutrients available per particle), gut throughput should increase because more food will be exposed to digestive enzymes and will therefore be absorbed. On the other hand, the cost of digestion and the maximum rate of digestion might impose limits on gut throughput (Willows 1992). If there is a unified compartment system response, then increased food quality should result in shorter gut residence times and increased transport rates on the gill and palp compartments up to the point of diminishing returns owing to the cost of digestion.

Taghon (1981) demonstrated an increased gut throughput in a benthic deposit-feeder with increased particle quality. Similarly, we have found an increased gill dorsal tract particle transport rate with increased food quality in the oyster *Crassostrea gigas* (Levinton & Ward unpublished data), and Bayne et al. (1987, 1989) found increased gut throughput with increased food concentration, which might be a surrogate for increased food quality. Cockles slow gut throughput when there is a large fraction of indigestible particles and gut residence time is minimized when nutritious food particles are available (Ibarrola et al. 1998). It may be that in these cases digestive efficiency is not limiting.

Clearance rate may also vary with food quality, although particles may often enter the mantle cavity in direct proportion to their occurrence in the water column (Baker et al. 1998; Ward et al. 1998a,b). The intake of water into the mantle cavity in bivalves does not seem to be adaptable to particle rejection. Therefore, pumping rate is the only mechanism of adjusting for total particle intake. Subsequent to entry into the mantle cavity, retention of particles depends upon mechanisms of sorting and rejection of particles.

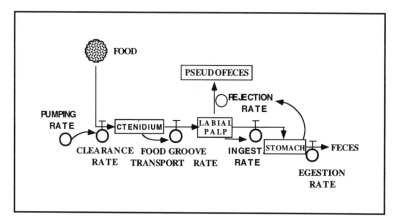

Fig. 2. The construction of a simple model of interactions in the bivalve compartment model, showing feedback loops that might affect the in between-compartment interactions.

On the other hand, changes in quantity and quality of the food supply might create imbalances among the compartments. For example, it may be relatively easy to move particles through the siphon, but handling toxic cells on the gill surfaces and ciliary tracts may be far more difficult, creating a rate-limiting step for particle processing. Similarly, retention of indigestible particles might slow digestive processes, even though these particles can be rapidly delivered to the mouth just as easily as digestible cells. Thus there is reason to believe that the variation in food quality and quantity seen in coastal waters, which minimally involves seasonal changes and strong spatial patchiness, might require adaptive responses beyond mere adjustments up and down of the entire compartmental system. This alternative model would argue that imbalances might be established that require feedback adjustments among the compartments. Thus pseudofeces rejection might work to allow a suitable throughput to the gut, but it is also possible that gut fullness or a maximum of digestive capacity provides a signal that causes the gill-palp system to reject particles as pseudofeces to differing degrees.

The transfer system inherent in bivalve compartments (Fig. 2) might be likened to a set of biochemical pathways, where threshold concentrations of a substrate may cause feedback and adjustments of the entire system (Levinton et al. 1996). Thus as the gut is overloaded, a feedback signal might be transferred to the gills and palps to reduce transfer rates. This might be accomplished by (a) shutting down incurrent or siphon activities; (b) reducing transport rates of particles collected on the gills; (c) rejecting more particles by gills and palps, so that they will not be ingested and saturate the gut. Bivalve pumping rate may change, but to a small degree, with changing particle concentration; a complete shutdown of siphon activities may occur, however, when there is no food or too high a particle concentration seems unlikely. That reductions in transport rates, in isolation from reduced pumping or increased rejection, are an option at high particle concentrations, if only because particles would accumulate on the gills and overload them. The reduced transport would only work if particles were rejected from the gills as pseudofeces. This mechanism could be employed without changing transport rates very much.

A final issue that must be taken into account is the large variation in anatomy of the compartments among bivalve species with differing gill, palp, and gut architectures. For example, gills range from organs that appear only to cleanse the mantle cavity (e.g., *Yoldia limatula*, Levinton et al. 1996) to those that have elaborate variations in gill form, ciliary transport direction, and even direction of particle transport (e.g., oysters, Atkins 1937; Ward et al. 1994). Oysters have complex plicate gills, capable of collecting particles and transporting them in different directions. Particles may be transported in plical troughs to a basal ciliated tract where particles are then moved anteriorly in a slurry to the palp (Ward et al. 1994), or they may be transported ventrally along the plical crests, where they are moved anteriorly in a mucus-string, also to the palp. Ventrally transported particles may be rejected as pseudofeces or processed further and ingested. By contrast, *Mytilus* gills do not appear to have a sufficiently complex gill structure to have such flexibility in ctenidial particle transport, although strong selectivity is still possible on the palps (Ward et al. 1998b).

Methods

Video endoscopy was employed to make qualitative observations of particle collection and transport. For oysters and mussels, methods followed Ward et al. (1998a) and for zebra mussels methods followed Baker et al. (2000). Discrimination among particles was accomplished by means of flow cytometry (see Shumway et al. 1990). The video endoscope was employed to permit, by means of a micromanipulator, the placement of a micropipet that could withdraw particles from different locations in the mantle cavity. Methods are described in detail in Ward et al. (1998a). When detrital particles were prepared they were sieved to place them within the size range of phytoplankton cells that were fed to bivalves concurrently. We also performed depletion experiments at different particle concentrations by sequentially sampling the chamber over time (either with bivalves or control) and running the withdrawn sample through a flow cytometer or Coulter counter according to methods reported in Ward et al. (1998a,b). Using the flow cytometer, it was possible to make direct comparisons of selectivity of identical size classes, even when the range of detrital particles was found to be greater than phytoplankton cells.

Results and Discussion

AN INTERNAL CONSTRAINT WITHIN THE BIVALVE COMPARTMENT SYSTEM

As mentioned above, bivalves are exposed to changing concentrations of suspended particles, and to varying combinations of particles of differing quality. The question arises whether bivalves respond to changing particle concentrations by making adjustments within the feeding compartment system. Figure 3 demonstrates the response of the oyster *Crassostrea virginica* to prolonged feeding on *Rhodomonas lens* at concentrations of 10^4 and 10^5 particles per ml. At 10^4 particles ml^{-1} there is a slight increase in pseudofeces production over a period of 9 h. This suggests that, as a result of increasing gut fullness, the gut is signaling the gill-palp system to increase rejection of otherwise nutritious particles as pseudofeces. The response at 10^5 cells ml^{-1} demonstrates a far steeper slope with time of food exposure, suggesting that the gut is sending a signal for higher degrees of rejection.

While we mainly wish to point out the gut fullness phenomenon as an internal constraint, the bivalve response to high concentrations of cells can also have an impact on the ecosystem. If, when particle concentrations are high, more cells are rejected as pseudofeces, two ecosystem effects may occur. If bottom currents are sluggish, then rejected cells will be added to the sediment, thus enhancing biodeposition of nutrient-rich material to the benthic deposit-feeding community. Biodeposition is known to strongly affect benthic processes in fresh and salt water (Izvekova Lvova-Katchanova 1972; Tenore et al. 1982; Wisniewski 1990; Reusch et al. 1994). Alternatively, bottom currents may be sufficient to resuspend biodeposits and restore them to the water column above (Rhoads & Young 1970). Zebra mussels, for example, efficiently remove all particles from the water column of the Hudson River nonselectively, but resuspension of biodeposits returns rejected and defecated material back to the water column above (Roditi et al. 1996; Baker et al. 1998).

AN ECOSYSTEM IMPACT UPON THE BIVALVE FEEDING-COMPARTMENT SYSTEM

Estuaries and back-bar lagoons of the eastern and southeastern United States are commonly bordered by marshes dominated by the cordgrass *Spartina alterniflora* (Redfield 1972), which develop especially well under conditions of rising sea level (Ranwell 1972). But little of the total production of *Spartina* is consumed by herbivores and most material enters detrital portions of the food web (Burkholder 1956; Odum & Smalley 1959). Salt marsh sediments often are composed of substantial amounts of detritus derived from *Spartina* (Levinton & Bianchi 1981; Levinton 1985; Lopez & Levinton 1987), suggesting that the combined processes of leaf decomposition, transport as suspended particles, and deposition dominate salt marsh systems. It follows, therefore, that both suspension and deposit feeders are exposed continually to a complex mixture of clays, living phytoplankon cells, and *Spartina* detrital particles. *Spartina* detritus is a particular challenge, because it is highly refractory for both deposit feeders and suspension feeders, as opposed to bacteria and many types of microalgae (Newell 1965; Lopez et al. 1977). While decomposing, *Spartina* detritus loses nitrogen much more rapidly than carbon, as opposed to phytoplankton detritus (Buchsbaum et al. 1991). This material, dominated by cellulose, is nutritionally very poor. The oyster *Crassostrea virginica* can absorb only 3% of the carbon in this food; an estuary like the Choptank River (part of the Chesapeake Bay system) probably provides not more than 1% of total carbon through this route (Crosby et al. 1989). Even the ribbed mussel, *Geukensia demissa*, which lives among salt marsh blades, can only absorb 9% of the carbon (Newell & Langdon 1990). It is not clear what proportion of particles in such an environment consist of phytoplankton cells or nonnutritive particles. This is further complicated by particles that derive from fresh and relatively digestible seaweed detritus (Levinton & McCartney 1991).

Thus, the *Spartina* estuarine and lagoonal ecosystems provide a food resource challenge to oysters, which are confronted with a large, but as of yet uncharacterized, proportion of refractory particles. This raises the question of whether oysters, which are often common in such estuarine environments, can actively select between particles that are nutritionally rich and poor. If they cannot, then all particles must enter the gut, posing a significantly different challenge. Pre-ingestive rejection involves perhaps additional investments of

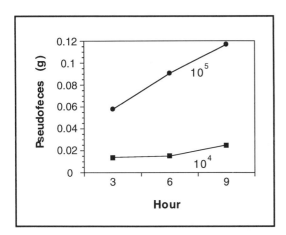

Fig. 3. Production of pseudofeces (g dry mass) by the oyster *Crassostrea virginica* after prolonged feeding at concentrations of 10^4 and 10^5 cells ml^{-1}. Note the increased slope with time at the higher food concen-tration. Number of replicate oysters for each treatment = 5. Data were analyzed with a repeated measures design. Pseudofecal production at 3 h was significantly different from production at 9 h ($p < 0.05$).

ciliary energy to separate and reject unsuitable *Spartina* particles, or losses of time in processing and sorting among particle types. Post-ingestive responses, however, might involve adjustments of gut transport rates, secretion of digestive enzymes, and degrees of sorting and diversion of particles into a digestive diverticulum. Previous work (Newell & Jordan 1983) demonstrates conclusively that pre-ingestive sorting can reduce the nutritive quality of the pseudofeces, in comparison with the food supply. But where does the sorting occur? Newell and Jordan suggested that the palps are the locus of sorting, which means that the gill could transport enormous quantities of material that would have to be handled by the next compartment.

To evaluate the nature of these responses, we examined the response of the oysters *Crassostrea virginica* from New York waters and *C. gigas*, cultured on rafts on San Juan Island, Washington. Both species have a plicated gill, with principal filaments that transport particles dorsally to a basal ciliated tract and ordinary filaments that transport particles either toward the basal (= dorsal) tract or ventrally toward a mucus-laden ventral tract (Fig. 4). We fed equal mixtures of aged *Spartina alterniflora* particles and cells of the red-colored flagellate *Rhodomonas lens* (Ward et al. 1998a,b). We sampled the water, basal tract, and ventral tract, and calculated an electivity index (Bayne et al. 1977), which represents the selection of *Rhodomonas* relative to *Spartina* (Fig. 4). *Rhodomonas* particles are sorted preferentially basally and are transported eventually to the mouth. *Spartina* particles by contrast are preferentially moved ventrally and are eventually moved toward the palp and are rejected as pseudofeces. The degree of selectivity was strong and suggests that sorting occurs efficiently on the gill. The palp did not participate in further sorting. These results suggest that adjustments are made before ingestion, which increases the quality of ingestive particles and increases the potential for digestive efficiency. Under other circumstances the palp may also play a role in selection. In this case, however, the gill did all of the work of selecting among particle types.

A BIVALVE FEEDING COMPARTMENT SYSTEM IMPACT UPON THE ECOSYSTEM

Presumably bivalves adjust filtration and ingestion rate, and select among different particle types to maximize nutrient uptake immediately and eventually to maximize fitness. Such behavioral adjustments, however, may affect the seston, given the intimate interaction between bivalves and shallow-water ecosystems (Dame 1993a,b). Oysters are dominant bivalves in estuaries and may have had an important effect on the seston of estuaries as large as Chesapeake Bay (Newell 1988). Unfortunately, most of our evidence concerning the controlling effect of benthic suspension-feeding bivalve grazing on nearshore ecosystems involve plausibility arguments, based upon reasonable models of grazing balanced against phytoplankton growth and mixing (Officer et al. 1982).

Combining such models with experimental introductions or removals of suspension-feeder populations would be more desirable, but this is obviously not practical, especially in the desirable form of replicate experiments in replicate estuaries. A surrogate might be so-called natural experiments, where an event has occurred in an ecosystem for which we have previously recorded survey data. For example, the introduction of the Asian clam, *Potamocorbula amurensis* (Carlton et al. 1990), caused major changes in the phytoplankton levels of San Francisco Bay. This invasion, however, was facilitated by a large reduction in salinity, which means that more than one factor might have been at work in affecting the phytoplankton. This problem often plagues such natural experiments.

The zebra mussel, *Dreissena polymorpha,* invaded North American fresh waters in the mid-1980s and formed dense benthic suspension-feeding populations with a high filtration capacity

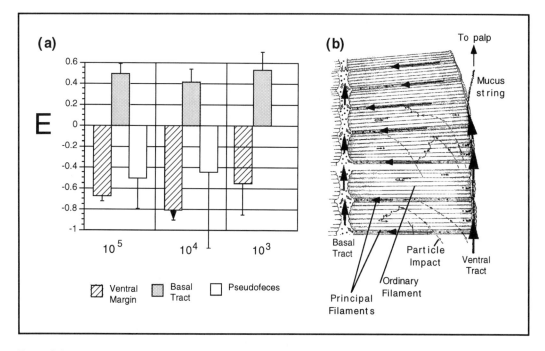

Fig. 4. Selectivity on the gills of the oyster *Crassostrea gigas*. (a) Electivity index, E, for the red-colored flagellate *Rhodomonas lens* at different cell concentrations ml⁻¹, relative to aged detritus of *Spartina alterniflora*, on the basal tract, ventral tract (= margin), and in pseudofeces, relative to food, in which the two particle types were presented in approximately equal numbers and concentrations. A positive value means that *R. lens* was enriched in the sample, relative to the food supplied (after Ward et al. 1998b). (b) Schematic of a gill lamella of *Crassostrea* with location of basal (= dorsal) and ventral marginal tract. Particles move basally mainly via slurries in the principal filaments. Ciliary movement on the ordinary filaments may be basal or ventral (after Ward et al. 1994).

(Reeders et al. 1989, 1993). Following their arrival, many water bodies experienced strong reductions in phytoplankton abundance (MacIsaac et al. 1992; Holland 1993; Leach 1993; Fahnenstiel et al. 1995).

The invasion of the Hudson River (Strayer & Smith 1993) is an interesting case, because work has been done both on the composition and abundance of the phytoplankton before and after the zebra mussel invasion in 1988 (Caraco et al. 1997). Despite the massive reduction in phytoplankton biomass in the Hudson River, water transparency increased by only 12%, owing to the persistence of nonliving particles, which dominated the seston (Caraco et al. 1997). Furthermore, dominance by cyanobacteria before the invasion has shifted to dominance by diatoms since the invasion. Thus there are a number of changes that might be related to bivalve behavior and to properties of the Hudson River ecosystem.

Baker et al. (1998, 2000) used endoscopy to examine the response of zebra mussels to exposure to mixtures of cyanobacteria (*Microcystis* sp.) formerly dominant in the Hudson River and nonliving vascular plant detritus, formed from grinding the cat-tail, *Typha,* and sieved to match the cyanobacterial cells in size. Particles were apparently trapped on the gills and transported ventrally to the gill margin. Near the margin, however, an obvious sorting occurred: *Microcystis* particles were carried to the marginal groove, transported anteriorly to the palp and ingested, whereas *Typha* detritus was transported somewhat above the groove to the palp and rejected as pseudofeces. This would explain why nonliving vascular detrital material might not be removed from the system, especially since the vigorous bottom currents that characterize Hudson River zebra mussel habitats would resuspend the delicate pseudofeces back into the water column.

Baker et al. (1998) used flow cytometry to examine the potential for zebra mussels to select among a variety of particle types. Clearance rates were higher in the presence of *Microcystis*, which was also preferentially ingested relative to other particle types, including several diatoms and green algal species. Bastviken et al. (1998), using more traditional counting of cells, found a similar preference for *Microcystis* cells. Baker et al. (1998) found that diatoms were rejected as diffuse pseudofeces, which were readily resuspended even in still water. A combination of selective feeding and resuspension therefore explained the shift in phytoplankton composition with no accompanying change in turbidity. The relatively slower doubling time of *Microcystis* may also have contributed to their decline relative to the faster reproducing diatoms.

Is selectivity independent of the particular combinations of particle types to which the bivalves are exposed? If so, we might expect a bivalve to select a particular favored phytoplankton species relative to others, and always selectively eliminate it from the spectrum of particle types. On the other hand, selectivity might vary with the spectrum of particle type. For example, zebra mussels preferentially ingested the diatom *Cyclotella* species when paired with *Thalassiosira pseudonana* in combination with three different particle types, but there was no preferential ingestion when *Cyclotella* was paired with *Thalassiosira* only (Baker et al. 1998). This suggests that, depending upon the antecedent phytoplankton assemblage, the zebra mussel invasion might bring about different trajectories of phytoplankton community composition.

Conclusion

It is well known that dense populations of suspension feeders can affect the particle concentration of the water column, but somewhat less appreciated is that other properties, such as species composition of the phytoplankton, and routes of supply of organic matter to benthos or water column may also be affected by suspension-feeding processes. Of course suspension feeders remove particles from the water column and dense populations should draw down the seston if the water column is shallow and well mixed (e.g., Cloern et al. 1982). But responses such as rejection of particles of poor nutritive quality and the feedback effect of gut fullness on ingestion rate also apparently can have strong effects on the seston. These effects depend greatly on the character of particle rejection and the rate of resuspension of biodeposits to the water column.

The above considerations suggest that a model is required that connects individual limitations and feedbacks, as depicted by the bivalve feeding-compartment model, and their relationships to water column stability, phytoplankton reproduction, resuspension, and water retention time in the estuary. Only then will we be able to completely understand the cycling of seston in nearshore waters dominated by benthic suspension-feeding activities.

Acknowledgments

We benefitted from the comments of two anonymous reviewers. We are grateful for the technical assistance of Terry Cucci, Michelle Gaston, Josepha Kurdziel, Lisa Milke, and Eun Joo Yi. We also thank Gary Wikfors for supply of phytoplankton cultures, and the Cornell Cooperative Extension Hatchery (Long Island,

New York) and Westcott Bay Oyster Farms (San Juan Island, Washington) for supply of oysters. This work was supported by the Hudson River Foundation and by the National Science Foundation. We thank Dennis Willows for providing facilities at the Friday Harbor Laboratories, the Bigelow Laboratory for Ocean Science for provision of flow cytometry facilities and the Becton Dickinson Corporation for provision of a flow cytometer for some of our work. We also would like to give our thanks to Don Rhoads for his many contributions to our understanding of benthic processes controlling the seston and the physiological condition of benthic bivalves.

Literature Cited

Atkins, D. 1937. On the ciliary mechanisms and interrelationships of lamellibranchs. I. New observations of sorting mechanisms. *Quarterly Journal of Microscopical Science* 79:181–308.

Baker, S. M., J. S. Levinton, J. P. Kurdziel, & S. E. Shumway. 1998. Selective feeding and biodeposition by zebra mussels and their relation to changes in phytoplankton composition and seston load. *Journal of Shellfish Research* 17:1,207–1,213.

Baker, S. M., J. S. Levinton, & J. E. Ward. 2000. Particle transport in the zebra mussel, *Dreissena polymorpha*. *Biological Bulletin* 199:116–125.

Bastviken, D. T. E., N. F. Caraco, & J. J. Cole. 1998. Experimental measurements of zebra mussel (*Dreissena polymorpha*) impacts on phytoplankton community composition. *Freshwater Biology* 39:375–386.

Bayne, B. L., A. J. S. Hawkins, & E. Navarro. 1987. Feeding and digestion by the mussel *Mytilus edulis* L. (Bivalvia: Mollusca) in mixtures of silt and algal cells at low concentrations. *Journal of Experimental Marine Biology and Ecology* 111:111–122.

Bayne, B. L., A. J. S. Hawkins, E. Navarro, & I. P. Iglesias. 1989. Effects of seston concentration on feeding digestion and growth in the mussel *Mytilus edulis*. *Marine Ecology Progress Series* 55:47–54.

Bayne, B. L., J. Widdows, & R. I. E. Newell. 1977. Physiological measurements on estuarine bivalve molluscs in the field, p. 57–68. *In* Keegan, B. F., P. Ó'Céidigh, & P. J. S. Boaden (eds.), Biology of Benthic Organisms. Pergamon Press, Oxford.

Berg, J. A. & R. I. E. Newell. 1986. Temporal and spatial variations in the composition of seston available to the suspension feeder *Crassostrea virginica*. *Estuarine, Coastal and Shelf Science* 23:375–386.

Bricelj, V. M. & R. E. Malouf. 1984. Influence of algal and suspended sediment concentrations on the feeding physiology of the hard clam, *Mercenaria mercenaria*. *Marine Biology* 84:155–165.

Buchsbaum, R., I. Valiela, T. Swain, M. Dzierzeski, & S. Allen. 1991. Available and refractory nitrogen in detritus of coastal vascular plants and macroalgae. *Marine Ecology Progress Series* 72:131–143.

Burkholder, P. 1956. Studies on the nutritive value of *Spartina* grass growing in the marsh areas of central Georgia. *Bulletin of the Torrey Canyon Botanical Club* 83:327–334.

Campbell, D. E. & C. R. Newell. 1998. MUSMOD, a production model for bottom culture of the blue mussel, *Mytilus edulis* L. *Journal of Experimental Marine Biology and Ecology* 219:171–203.

Caraco, N. F., J. J. Cole, P. A. Raymond, D. L. Strayer, M. L. Pace, S. E. G. Findlay, & D. T. Fischer. 1997. Zebra mussel invasion in a large, turbid, river: Phytoplankton response to increased grazing. *Ecology* 78:588–602.

Carlton, J. T., J. K. Thompson, L. E. Schemel, & F. H. Nichols. 1990. Remarkable invasion of San Francisco Bay (California, USA) by the Asian clam, *Potamocorbula amunensis*. I. Introduction and dispersal. *Marine Ecology – Progress Series* 66:81–94.

Cloern, J. E. 1982. Does the benthos control phytoplankton biomass in south San Francisco Bay? *Marine Ecolog –Progress Series* 9:191–202.

Crosby, M. P., C. J. Langdon, & R. I. E. Newell. 1989. Importance of refractory plant material to the carbon budget of the oyster *Crassostrea virginica*. *Marine Biology* 100:343–352.

Dame, R. F. (ed.) 1993a. Bivalve filter feeders and estuarine and coastal ecosystem processes: Conclusions, p. 567–569. *In* Dame, R. F. (ed.), Bivalve Filter Feeders in Estuarine and Coastal Ecosystems. Springer-Verlag, Berlin.

Dame, R. F. 1993b. Bivalve Filter Feeders in Estuarine and Coastal Ecosystem Processes. Springer-Verlag, Berlin.

Dame, R. F. & T. C. Prins. 1998. Bivalve carrying capacity in coastal ecosystems. *Aquatic Ecology* 31:409–421.

Decho, A. W. & S. N. Luoma. 1991. Time-courses in the retention of food material in the bivalves *Potamocorbula amurensis* and *Macoma balthica*: Significance to the absorption of carbon and chromium. *Marine Ecology – Progress Series* 78:303–314.

Fahnenstiel, G. L., G. A. Lang, T. F. Nalepa, & T. H. Johengen. 1995. Effects of the zebra mussel (*Dreissena polymorpha*) colonization on water quality parameters in Saginaw Bay, Lake Huron. *Journal of Great Lakes Research* 21:435–448.

Fréchette, M., C. A. Butman, & W. R. Geyer. 1989. The importance of boundary-layer flows in supplying phytoplankton to the benthic suspension feeder *Mytilus edulis* L. *Limnology and Oceanography* 34:19–36.

Frost, B. W. 1972. Effects of size and concentration of food particles on the feeding behavior of the marine planktonic copepod *Calanus pacificus*. *Limnology and Oceanography* 17:805–815.

Frost, B. W. 1987. Grazing control of phytoplankton stock in the open subArctic Pacific Ocean: A model assessing the role of mesozooplankton, particularly the large calanoid copepods *Neocalanus* spp. *Marine Ecology – Progress Series* 39:49–68.

Grant, J., M. Dowd, & K. Thompson. 1993. Perspectives on field studies and related biological models of bivalve growth and carrying capacity, p. 371–420. *In* Dame, R. F. (ed.), Bivalve Filter Feeders in Estuarine and Coastal Ecosystem Processes. Springer-Verlag, Berlin.

Hawkins, A. J. S., B. L. Bayne, S. Bougrier, M. Héral, J. I. P. Iglesias, E. Navarro, R. F. M. Smith, & M. B. Urrutia. 1998. Some general relationships in comparing the feeding physiology of suspension-feeding bivalve molluscs. *Journal of Experimental Marine Biology and Ecology* 219:87–103.

Héral, M. 1993. Why carrying capacity models are useful tools for management of bivalve molluscs culture, p. 455–475. *In* Dame, R. F. (ed.), Bivalve Filter Feeders in Estuarine and Coastal Ecosystem Processes. Springer-Verlag, Berlin.

Herman, P. M. J. 1993. A set of models to investigate the role of benthic suspension feeders in estuarine ecosystems, p. 421–454. *In* R. F. Dame (ed.), Bivalve Filter Feeders in Estuarine and Coastal Ecosystem Processes. Springer-Verlag, Berlin.

Holland, H. E. 1993. Changes in planktonic diatoms and water transparency in Hatchery Bay, Bass Island area, western Lake Erie since the establishment of the zebra mussel. *Journal of Great Lakes Research* 19:617–624.

Hylleberg, J. & V. F. Gallucci. 1975. Selectivity in feeding by the deposit-feeding bivalve *Macoma nasuta*. *Marine Biology* 33:167–178.

Ibarrola, I., E. Navarro, & J. I. P. Iglesias. 1998. Short-term adaptation of digestive processes in the cockle *Cerastoderma edule* exposed to different food quantity and quality. *Journal of Comparative Physiology B* 168:32–40.

Iglesias, J. I. P., E. Navarro, P. A. Jorna, & I. Armentia. 1992. Feeding, particle selection, and absorption in cockles, *Cerastoderma edule* (L.), exposed to variable conditions of food concentration and quality. *Journal of Experimental Marine Biology and Ecology* 162:177–198.

Izvekova, E. I. & A. A. Lvova-Katchanova. 1972. Sedimentation of suspended matter by *Dreissena polymorpha* Pallas and its subsequent utilization by chironomid larvae. *Polskie Archiwum Hydrobiologii* 19:203–210.

Leach, J. H. 1993. Impacts of the zebra mussel (*Dreissena polymorpha*) on water quality and fish spawning reefs in western Lake Erie, p. 381–397. *In* Nalepa, T. F. & D. W. Schloesser (eds.), Zebra Mussels: Biology, Impacts & Control. Lewis Publishers, Boca Raton, Florida.

Levinton, J. S. 1985. Complex interactions of a deposit feeder with its resources: Roles of density, a competitor & detrital addition in the growth and survival of the mud snail *Hydrobia totteni*. *Marine Ecology – Progress Series* 22:31–40.

Levinton, J. S. & T. S. Bianchi. 1981. Nutrition and food limitation of deposit-feeders. I. The role of microbes in the growth of mud snails. *Journal of Marine Research* 39:531–545.

Levinton, J. S. & M. M. McCartney. 1991. The use of photosynthetic pigments in sediments as a tracer for sources and fates of macrophyte organic matter. *Marine Ecology – Progress Series* 78:87–96.

Levinton, J. S., J. E. Ward, & S. E. Shumway. 1996. Biodynamics of particle processing in bivalve molluscs: Models, data and future directions. *Invertebrate Biology* 115:232–242.

Lewin, J. C., T. Hruby, & D. Mackas. 1975. Blooms of surf-zone diatoms along the coast of the Olympic Peninsula, Washington. V. Environmental conditions associated with blooms (1971 and 1972). *Estuarine and Coastal Marine Science* 3:229–242.

Lopez, G. R. & J. S. Levinton. 1987. Ecology of deposit-feeding animals in marine sediments. *Quarterly Review of Biology* 62:235–260.

Lopez, G. R., J. S. Levinton & L. B. Slobodkin. 1977. The effect of grazing by the detritivore *Orchestia grillus* on *Spartina* litter and its associated microbial community. *Oecologia* 30:111–127.

MacIsaac, H. J., W. G. Sprules, O. E. Johannsson, & J. H. Leach. 1992. Filtering impacts of larval and sessile zebra mussels (*Dreissena polymorpha*) in western Lake Erie. *Oecologia* 92:30–39.

McLachlan, A. 1980. Exposed sandy beaches as semi-enclosed ecosystems. *Marine Environmental Research* 4:59–63.

Navarro, E., M. B. Urrutia, J. I. P. Iglesias, & I. Ibarrola. 1998. Tidal variations in feeding, absorption and scope for growth of cockles (*Cerastoderma edule*) in the bay of Marennes-Oleron (France). *Vie et Milieu* 48:331–340.

Newell, R. C. 1965. The role of detritus in the nutrition of two marine deposit feeders, the prosobranch *Hydrobia ulvae* and the bivalve *Macoma balthica. Proceedings Zoolological Society of London* 144:25–45.

Newell, R. I. E. 1988. Ecological changes in Chesapeake Bay: Are they the result of overharvesting the American oyster, *Crassostrea virginica*?, p. 536–546. *In* Understanding the Estuary: Advances in Chesapeake Bay Research. Proceedings of a Conference. Chesapeake Bay Research Consortium, Baltimore, Maryland.

Newell, R. I. E. & S. J. Jordan. 1983. Preferential ingestion of organic material by the American oyster, *Crassostrea virginica. Marine Ecology Progress Series* 13:47–53.

Newell, R. I. E. & C. J. Langdon. 1990. Utilization of detritus and bacteria as food sources by two bivalve suspension-feeders, the oyster *Crassostrea virginica* and the mussel *Geukensia demissa. Marine Ecology – Progress Series* 58:299–310.

Odum, E. P. & A. E. Smalley. 1959. Comparison of population energy flow of a herbivorus and deposit-feeding invertebrate in a salt marsh ecosystem. *Proceedings of the National Academy of Sciences of the United States of America* 45:617–622.

Officer, C. B., T. J. Smayda, & R. Mann. 1982. Benthic filter feeding: A natural eutrophication control. *Marine Ecology – Progress Series* 9:203–210.

Prins, T. C., A. C. Small, & R. F. Dame. 1998. A review of the feedbacks between bivalve grazing and ecosystem processes. *Aquatic Ecology* 31:349–359.

Ranwell, D. S. 1972. Ecology of Salt Marshes and Sand Dunes. Chapman and Hall, London.

Redfield, A. C. 1972. Development of a New England saltmarsh. *Ecological Monographs* 42:201–237.

Reeders, H. H., A. B. d. Vaate, & R. Noordhuis. 1993. Potential of the zebra mussel (*Dreissena polymorpha*) for water management, p. 439–451. *In* Nalepa, T. F. & D. W. Schloesser (eds.), Zebra Mussels: Biology, Impacts, and Control. Lewis Publishers, Boca Raton, Florida.

Reeders, H. H., A. b. d. Vaate, & F. J. Slim. 1989. The filtration rate of *Dreissena polymorpha* (Bivalvia) in three Dutch lakes with reference to water quality management. *Freshwater Biology* 22:133–141.

Reusch, T. B. H., A. R. O. Chapman, & J. P. Groger. 1994. Blue mussels, *Mytilus edulis,* do not interfere with eelgrass, *Zostera marina,* but fertilize shoot growth through biodeposition. *Marine Ecology – Progress Series* 108:265–282.

Rhoads, D. C. & D. K. Young. 1970. The influence of deposit-feeding organisms on sediment stability and community trophic structure. *Journal of Marine Research* 28:150–178.

Roditi, H., N. F. Caraco, J. J. Cole, & D. L. Strayer. 1996. Filtration of Hudson River water by the zebra mussel (*Dreissena polymorpha*). *Estuaries* 19:824–832.

Shumway, S. E. 1990. A review of the effects of algal blooms on shellfish and aquaculture. *Journal of the World Aquaculture Society* 21:65–104.

Smaal, A. C. & T. C. Prins. 1993. The uptake of organic matter and the release of inorganic nutrients by bivalve suspension feeder beds, p. 272–298. *In* R. F. Dame (ed.), Bivalve Filter Feeders in Estuarine and Coastal Ecosystem Processes. Springer-Verlag, Berlin.

Smaal, A. C., T. C. Prins, N. Dankers, & B. Ball. 1998. Minimum requirements for modeling bivalve carrying capacity. *Aquatic Ecology* 31:423–428.

Steele, J. H. 1974. The Structure of Marine Ecosystems. Harvard University Press, Cambridge, Massachusetts.

Strayer, D. L. & L. C. Smith. 1993. Distribution of the zebra mussel (*Dreissena polymorpha*) in estuaries and brackish waters, p. 715–727. *In* Nalepa, T. F. & D. W. Schloesser (eds.), Zebra Mussels: Biology, Impacts, and Control. Lewis, Boca Raton, Florida.

Taghon, G. L. 1981. Beyond selection: Optimal ingestion rate as a function of food value. *American Naturalist* 118:202–214.

Tenore, K. R., L. F. Boyer, R. M. Cal, J. Corral, C. Garcia-Fernandez, N. Gonzalez, E. Gonzalez-Gurriaran, R. B. Hanson, J. Iglesias, M. Krom, E. Lopez-Jamar, J. McClain, M. M. Pamatmat, A. Perez, D. C. Rhoads, G. deSantiago, J. Tietjen, J. Westrich, & H. L. Windom. 1982. Coastal upwelling in the Rias Bajas, NW Spain: Contrasting the benthic regimes of the Rias de Arosa and de Muros. *Journal of Marine Research* 40:701–772.

Ward, J. E., J. S. Levinton, S. E. Shumway, & T. Cucci. 1998a. Particle sorting in bivalves: In vivo determination of the pallial organs of selection. *Marine Biology* 131:283–292.

Ward, J. E., J. S. Levinton, S. E. Shumway, & T. L. Cucci. 1998b. Site of particle selection in a bivalve mollusc. *Nature* 390:131–132.

Ward, J. E., R. I. E. Newell, R. J. Thompson, & B. A. MacDonald. 1994. In vivo studies of suspension-feeding processes in the eastern oyster, *Crassostrea virginica* (Gmelin). *Biological Bulletin* 186:221–240.

Widdows, J., P. Fieth, & C. M. Worrall. 1979. Relationships between seston, available food and feeding activity in the common mussel, *Mytilus edulis*. *Marine Biology* 50:195–207.

Wildish, D. J. & D. D. Kristmanson. 1984. Importance to mussels of the benthic boundary layer. *Canadian Journal of Fisheries and Aquatic Science* 41:1,618–1,625.

Willows, R. I. 1992. Optimal digestive investment: A model for filter feeders experiencing variable diets. *Limnology and Oceanography* 37:829–847.

Wisniewski, R. 1990. Shoals of *Dreissena polymorpha* as bio-processor of seston. *Hydrobiologia* 200/201:451–458.

Index

The page number refers to the first page of the chapter in which the index word or phrase occurs.